Mechanical Desktop® 4: Applying Designer and Assembly Modules

Mechanical Desktop® 4: Applying Designer and Assembly Modules

Autodesk Press

Thomson Learning™

Africa • Australia • Canada • Denmark • Japan • Mexico • New Zealand • Philippines
Puerto Rico • Singapore • Spain • United Kingdom • United States

NOTICE TO THE READER

Publisher does not warrant or guarantee any of the products described herein or perform any independent analysis in connection with any of the product information contained herein. Publisher does not assume, and expressly disclaims, any obligation to obtain and include information other than that provided to it by the manufacturer.

The reader is expressly warned to consider and adopt all safety precautions that might be indicated by the activities herein and to avoid all potential hazards. By following the instructions contained herein, the reader willingly assumes all risks in connection with such instructions.

The publisher makes no representation or warranties of any kind, including but not limited to, the warranties of fitness for particular purpose or merchantability, nor are any such representations implied with respect to the material set forth herein, and the publisher takes no responsibility with respect to such material. The publisher shall not be liable for any special, consequential, or exemplary damages resulting, in whole or part, from the readers' use of, or reliance upon, this material. Autodesk does not guarantee the performance of the software and Autodesk assumes no responsibility or liability for the performance of the software or for errors in this manual.

Trademarks

Autodesk, AutoCAD and the AutoCAD logo are registered trademarks of Autodesk, Inc. Thomson Learning uses "Autodesk Press" with permission from Autodesk, Inc. for certain purposes. Windows is a trademark of the Microsoft Corporation. All other product names are acknowledged as trademarks of their respective owners.

Autodesk Press Staff
Executive Director: Alar Elken
Executive Editor: Sandy Clark
Development: John Fisher
Executive Marketing Manager: Maura Theriault
Executive Production Manager: Mary Ellen Black
Production Coordinator: Jennifer Gaines
Art and Design Coordinator: Mary Beth Vought
Marketing Coordinator: Paula Collins
Technology Project Manager: Tom Smith

Cover design by Scott Keidong's Image Enterprises. AutoCAD images reprinted with permission from and under copyright of Autodesk.

Printed in Canada
6 7 8 9 10 XXX 04 03

For more information, contact
Autodesk Press
3 Columbia Circle, Box 15-015
Albany, New York USA 12212-15015;
or find us on the World Wide Web at http://www.autodeskpress.com

Library of Congress Cataloging-in-Publication Data

Banach, Daniel T.
Mechanical Desktop 4.0: applying designer and assembly modules / Daniel T. Banach.
p. cm.
ISBN 0-7668-1946-9
1. Engineering graphics. 2. Mechanical desktop. 3. Engineering design—Data Processing. 4. AutoCAD. I. Title.
T353 .B183 2000
620_.0042_02855369—dc21 99-053881
CIP

CONTENTS

INTRODUCTION

*Items are new or changed commands or procedures in Mechanical Desktop® 4

CHAPTER 3—SKETCH PLANES, BOOLEAN OPERATIONS, FILLETS, CHAMFERS, HOLES AND ARRAYS

CHAPTER 4—DRAWING VIEWS AND ANNOTATIONS

CHAPTER 8—ASSEMBLIES

INTRODUCTION

Welcome! If you are new to Mechanical Desktop or 3D design, you have just joined over 160,000 people already using Mechanical Desktop. If you are a current Mechanical Desktop user, you will find major enhancements in the software over the previous release. Look for an asterisk in front of the new and enhanced commands in the table of contents.

The chapters in this book follow the order in which you will create your own models and drawings. Each chapter introduces a set of topics and then takes you through a basic, step-by-step example. Each chapter builds on the material learned in the previous chapter(s). At the end of most chapters you will find practice exercises for you to complete on your own. They are based on real world parts used in different disciplines of design.

PRODUCT BACKGROUND

Mechanical Desktop 4.0 was written by Autodesk and runs inside of AutoCAD 2000. Mechanical Desktop is a 3D feature-based parametric solid modeler that allows you to create complex 3D parametric models and to generate 2D views from those models.

Mechanical Desktop consists of AutoCAD 2000 and four modules:

Designer: Feature-based parametric modeler. Part Modeling.

AutoSurf: Non Uniform Rational B-Splines (NURBS) surfaces. Surface Modeling.

Assembly: Manage and constrain assembled parts. Assembly Modeling.

Drawing Manager: 2D view layout and dimensioning for outputting engineering drawings.

REQUIREMENTS

This book assumes that you are running Mechanical Desktop 4.0 and that you are proficient with AutoCAD commands such as lines, arcs, circles, polylines, move, erase, grips etc. If you are not proficient in those areas, you may want to refer to the AutoCAD online help as needed.

BASICS OF 3D MODELING

If you are new to creating 3D models, you need to take time to evaluate what you are going to model and how you are going to approach it. When I evaluate a part, I look for the main basic shape. Is it flat or cylindrical in shape? Depending on the shape, I will take a different approach to the model. I try to start with a flat face if possible; it is easier to add other features to a flat face. If the model is cylindrical in shape, I look for the main profile or shape of the part and revolve or extrude that profile. After the main body is created, work on the other features and look to see how this shape will connect to the first part. Think of 3D modeling as working with building blocks: each block sits on another block, but remember that material can also be removed from the original solid.

TERMS AND PHRASES

To help you to better understand Mechanical Desktop, a few of the terms and phrases that will be used in the book are explained below.

Parametric Modeling: Parametric modeling is the ability to drive the size of the geometry by dimensions. For example, if you want to increase the length of a plate from 5" to 6", change the 5" dimension to 6" and the geometry will update. Think of it as the geometry along for a ride, driven by the dimensions. This is opposite to AutoCAD 2D dimensioning, known as associative dimensioning: as lines, arcs and circles are drawn, they are created to the exact length or size; when they are dimensioned, the dimension reflects the exact value of the geometry. If you want to change the size of the geometry, you stretch the geometry and the dimension automatically gets updated. Think of this as the dimension along for a ride, driven by the geometry.

Feature-based parametric modeling: Feature-based means that as you create your model, each hole, fillet, chamfer, extrusion etc. is an independent feature that can be edited or deleted.

Bi-directional Associativity: The model and the drawing views are linked. If the model changes, the drawing views will automatically update. And if the dimensions in a drawing view change, the model is updated and the drawing views are updated based on the updated part.

OVERVIEW OF PART CREATION

To get a better idea of the process that you will go through to create a part and its 2D views, refer to the steps outlined below. This is intended as an overview only; not all of the steps are required for every feature that is created.

1. Sketch the geometry using lines, arcs, polylines or splines.

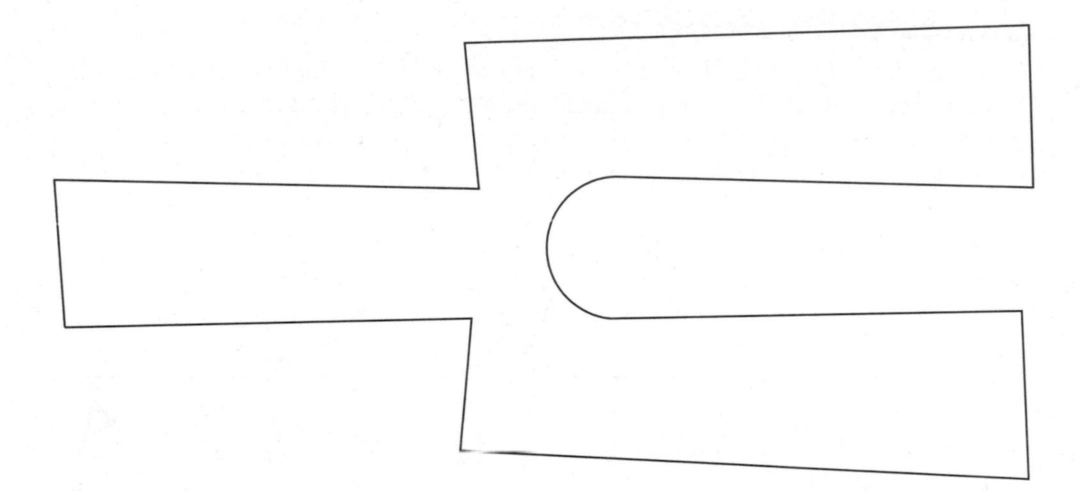

Figure 1

2. Profile the sketch. This analyzes the profile, and tells you how many dimensions or constraints are required to fully constrain the profile.

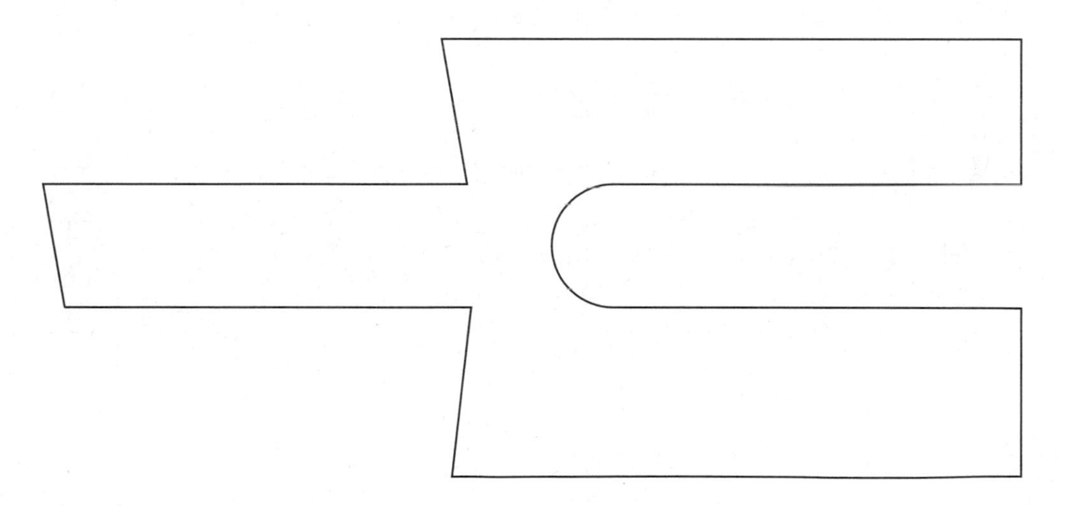

Figure 2

3. Add/remove constraints to control the behavior of the profile.

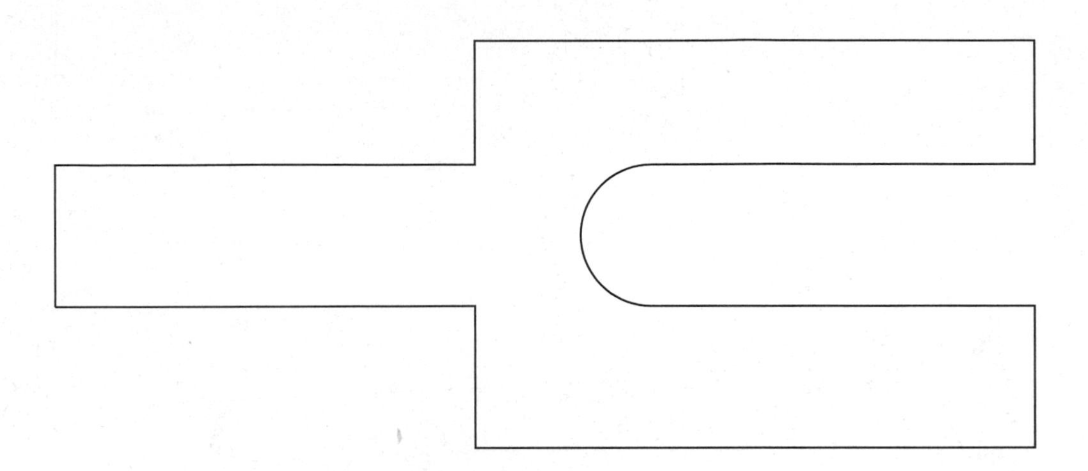

Figure 3

4. Add dimensions to control the size of the profile.

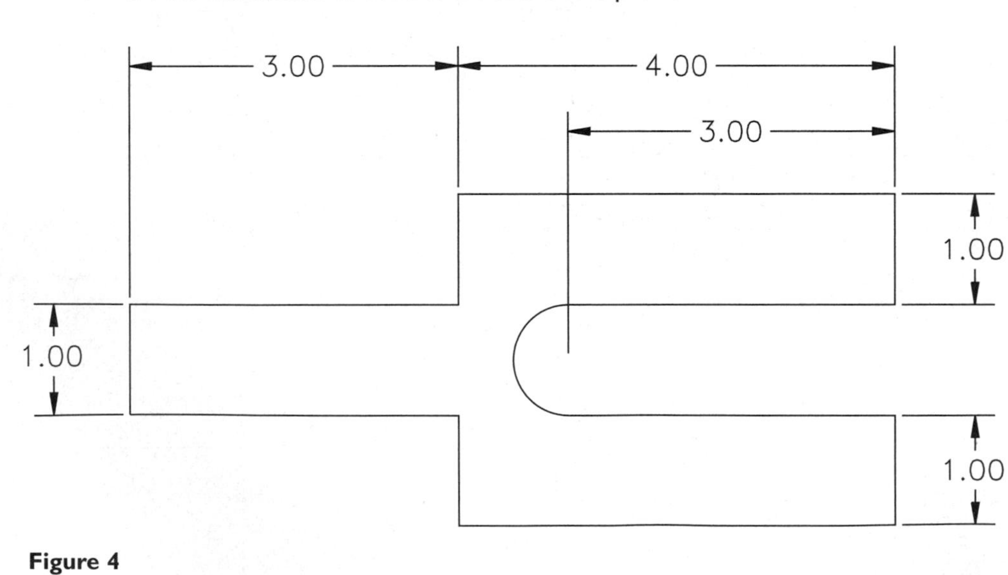

Figure 4

5. Extrude, revolve or sweep the profile into a solid. In Figure I.05 the profile was extruded.

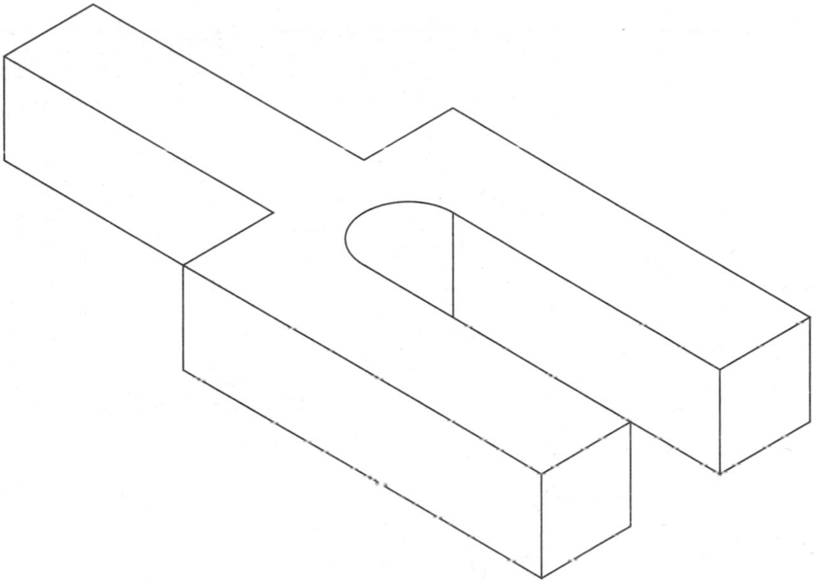

Figure 5

6. Add sketched features and placed features such as extrusions, holes, fillets and chamfers.

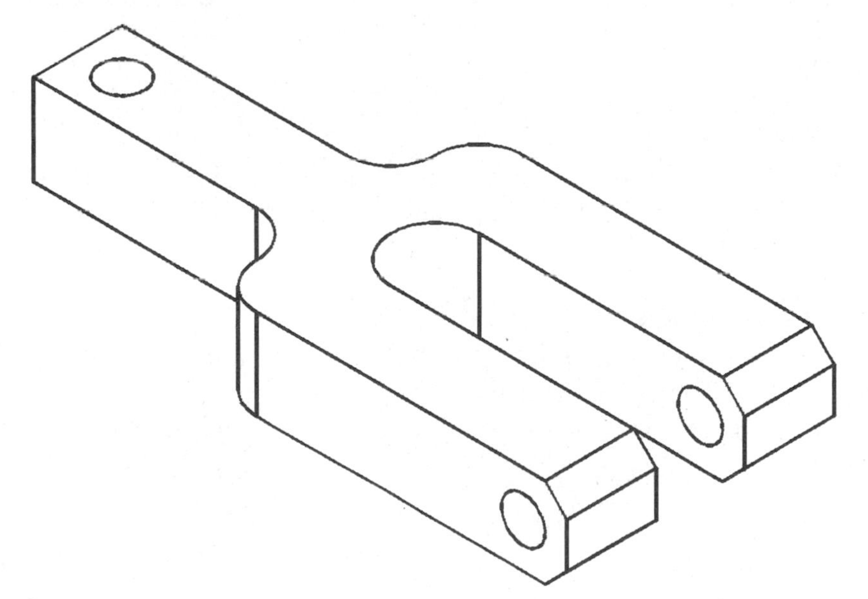

Figure 6

7. Create the 2D views and annotate the drawing.

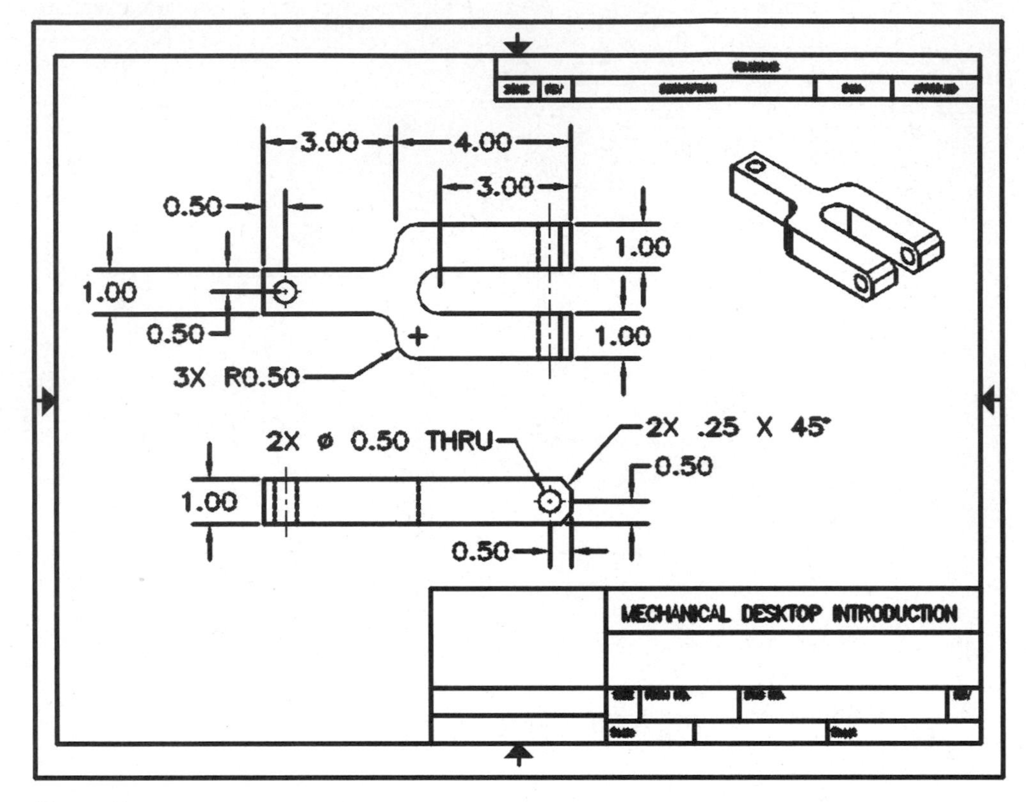

Figure 7

TOOLBARS

In this book, the toolbars will be shown in the default vertical orientation, but remember that this is a personal choice, and the toolbars can be placed and oriented anywhere you want.

BOOK'S INTENT

The intent of this book is to focus on the Designer and the Assembly modules in an applied, hands-on environment. This book guides you through the process of generating 3D parametric models, assembling parts and producing 2D views from them.

Each chapter will be broken into specific subjects, which are introduced and then followed by a short tutorial. When a new command is introduced, there is a figure showing the exact location of the icon in the specific toolbar as well as how to access the command through right-click menus. Read through each subject and then complete its tutorial while at the computer. At the end of each chapter, there are exercises you can complete on your own and review questions to reinforce the topics covered in that chapter.

SPECIAL SECTIONS

You will find sections marked:

Notes: Here you will find information that points out specific areas that will help you learn Mechanical Desktop.

Tips: Here you will find information that will assist you in generating better models.

BOOK NOTATIONS

- ENTER refers to Enter on the keyboard or right mouse click.
- Numbers in quotation marks are numbers that need to be typed in.
- Part and model both refer to a Mechanical Desktop part.
- Select refers to a pick by the left mouse button.
- Choose refers to a selection from a menu.
- Desktop browser and browser both refer to the Mechanical Desktop browser, where the history of the file is shown.
- Sketch refers to lines, arcs, circles and polylines drawn to define an outline shape of a feature.
- Profile refers to a sketch that has been profiled (analyzed) by Mechanical Desktop.

THE TUTORIALS AND EXERCISES

The best way to learn Mechanical Desktop is to practice the tutorials and exercises on the computer. Each book ships with a CD with drawings directories for each chapter.

Tip: Make a copy of your Mechanical Desktop 4.0 startup icon and change the properties of the "Start in:" directory to "c:\MD4Book\chapter??" (where "c" represents the disk drive where the files are located on your system and the "??" represents the chapter you are working on). If you update this startup location as you work through the book, you will reduce the number of picks required to open and save files.

As you go through the book you will find some of the exercises have already been started for you; they can be opened as noted in each exercise. As you go through each exercise, save the files as noted, because they may be required in a future exercise.

For clarity, some Figures will be shown with lines hidden and the text larger.

With AutoCAD, there are many ways to complete a part, and that is also true with Mechanical Desktop. After completing the exercises as noted, feel free to experiment with different methods.

FILES INCLUDED ON THE CD

You will find all the required drawings on the CD included with this book. Copy the files to your hard drive and remove the read-only property from all the files. It is suggested that you copy the files to C:\MD4Book.

ACKNOWLEDGMENTS

I would like to thank Ed O'Halloran of Autodesk, Inc., Novi, MI for performing the technical edit on the manuscript and for his help researching material for this book.

A special acknowledgment is due the following instructors, who reviewed the chapters in detail:

Jeffery R.Gibbs
Muskingum Area Technical College
Zanesville, OH

Steven Keith
De Anza
Cupertino, CA

Gary Masciadrelli
Springfield Technical Community College
Springfield, MA

David A. Probst
Pennsylvania College of Technology
Williamsport, PA

David W. Smith
Cincinnati State Technical & Community College
Cincinnati, OH

Thomas White
Shenendehowa Central School
Clifton Park, NY

DEDICATION

I dedicate this book to my wife, Cammi, and my children, Allison and Jonathan. Their love helped make this book possible.

CHAPTER 1

User Interface, Command Entry, Part Preferences, Sketching, Profiling, Constraining, and Dimensioning

To create a parametric solid, you always start with a 2-D profile. After a discussion about the user interface, command entry, and Mechanical Desktop options, this chapter will take you through the four steps in generating 2-D parametric profiles: drawing a rough 2-D sketch of the geometry, profiling (analyzing) the geometry, applying geometric constraints, adding parametric dimensions. The last part of the chapter will cover how to convert a 2D AutoCAD drawing into a Mechanical Desktop part.

AFTER COMPLETING THIS CHAPTER, YOU WILL BE ABLE TO:

- Understand Mechanical Desktop user interface.
- Describe Mechanical Desktop parts and desktop options.
- Describe how part settings (options) affect the creation of a part.
- Sketch a profile (outline) of a part.
- Profile a sketch.
- Understand what constraints are.
- Constrain a sketch.
- Dimension a sketch.
- Change a dimensions values in a sketch.
- Convert a 2D AutoCAD drawing into a Mechanical Desktop profile.

USER INTERFACE AND COMMAND ENTRY

In Mechanical Desktop 4.0, the user interface allows you to work intuitively by giving you onscreen graphics and messages that allow you to maintain focus on the screen. The user interface of Mechanical Desktop is Windows compliant and utilizes: toolbars, tool tips, a Mechanical Desktop browser (similar to Windows Explorer), context sensitive right mouse click menus and takes advantage of an IntelliMouse. Figure 1-1 shows the default Mechanical Desktop screen. The Mechanical Desktop browser (known in this text book as either the Mechanical Desktop browser or just the browser) is on the left side of the screen with the Mechanical Desktop toolbars located on the right side of the browser. The browser and toolbars can be moved or closed as needed. This text will show you the location of each command on the toolbar as well as telling you how to execute the command through right-click menus. Most of the commands can also be accessed through the browser by either double clicking or right clicking on the feature name. How commands are accessed is a personal preference.

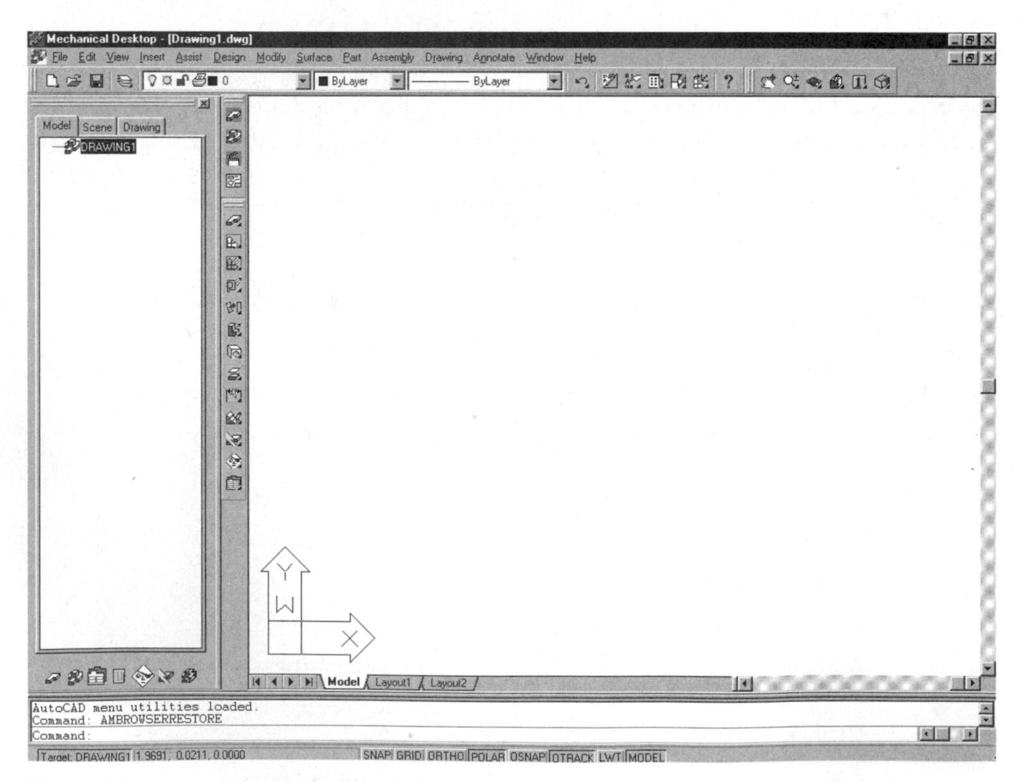

Figure 1.1

TOOLBARS

The Mechanical Desktop commands can be found in pull-down menus, in four toolbars, through a Desktop browser and by context sensitive right click menus. This book will focus on the toolbars and the right click menus. The default menu that Mechanical Desktop (amdt.mnu) uses, express toolbars; through the Desktop Main Toolbar, you can change the active toolbar by selecting the operation you want to work with—the mode will change to correspond to the toolbar. Express toolbars require fewer toolbars on the screen. Select from the four icons on the Desktop Application Toolbar to change the current Desktop toolbar. The following four figures show the express toolbars with their respective application toolbars. In this book, the toolbars will be shown in the default vertical orientation, but remember that this is a personal choice, and the toolbars can be placed and oriented anywhere you want.

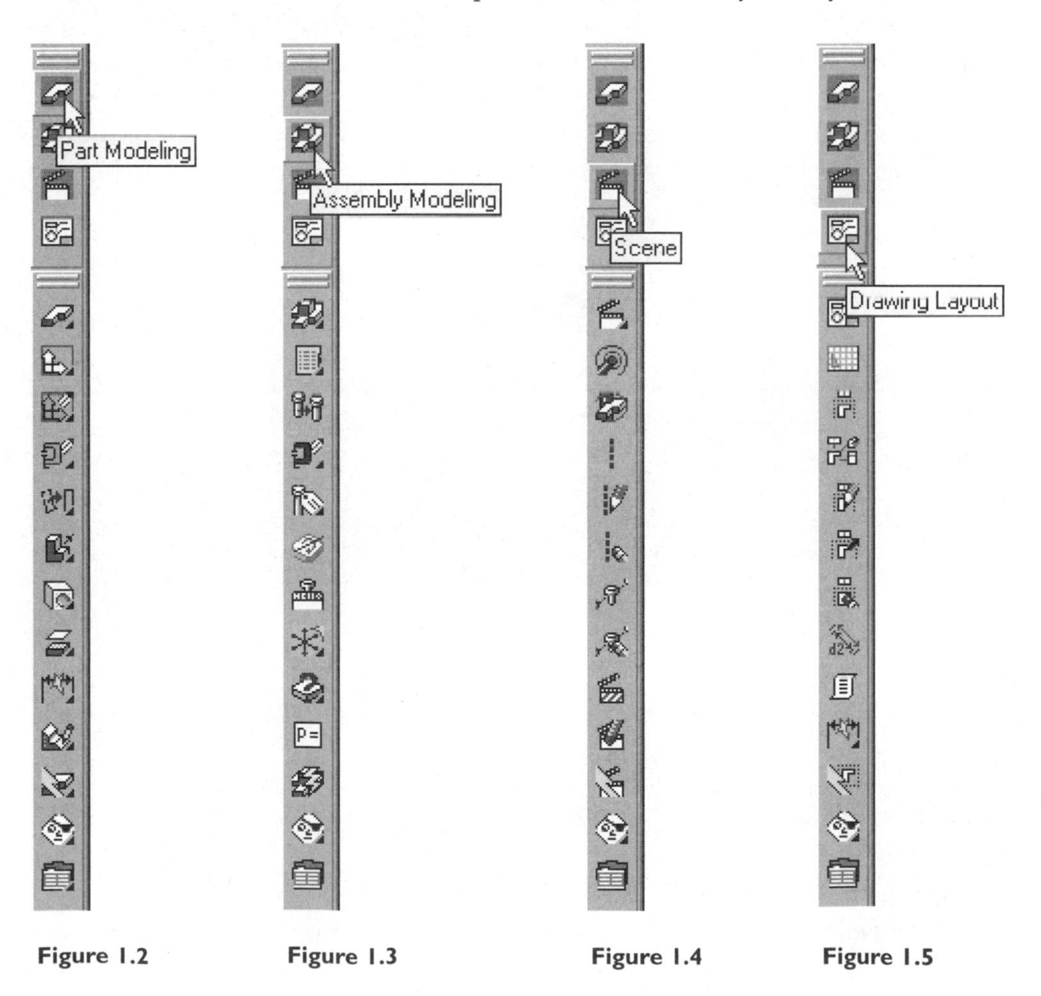

Figure 1.2 **Figure 1.3** **Figure 1.4** **Figure 1.5**

TOOL TIPS

When you are working with commands in Mechanical Desktop 4.0 and the cross hairs go over a part or feature, a tool tip will appear to identify it. Issue a Mechanical Desktop command and move the mouse over a part or feature and a tool tip will appear with the name of the profile, part name or feature name. After the correct feature is identified, select it with the mouse button and proceed as usual.

CONTEXT SENSITIVE RIGHT MOUSE CLICK MENUS

To display the options for a command, right mouse click in the drawing, browser or command line area and a pop-up menu with a set of the most likely commands will appear. Depending upon the mode that you are in you will get different menus. Figure 1.6 shows the menu that appears when you click in the drawing area while in model mode.

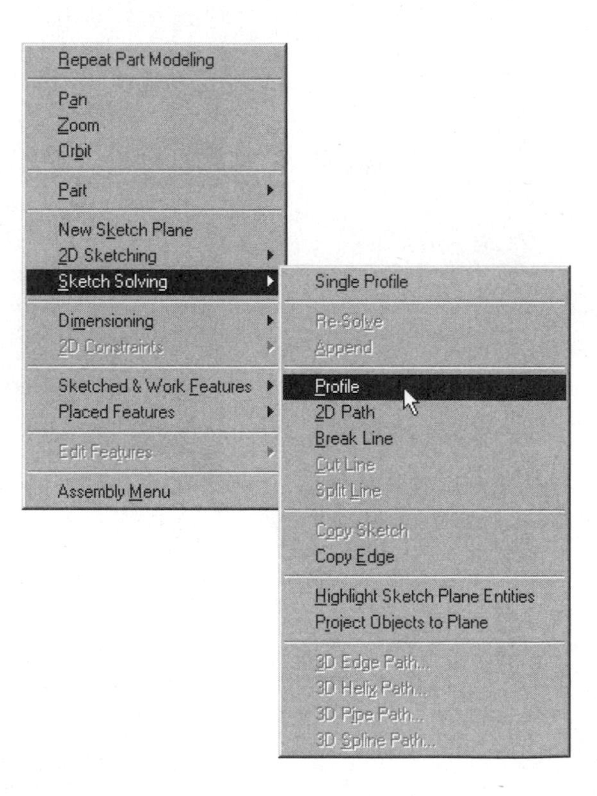

Figure 1.6

Note: Throughout the book when a command is introduced, you will be told how to access the command through right click menus and also be shown the icon's location on its toolbar.

INTELLIMOUSE

Mechanical Desktop 4.0 takes full advantage of an IntelliMouse. If you are using an IntelliMouse you can zoom in and out by scrolling the wheel. To pan, hold the wheel and pan in the direction you want. To zoom to the extents of the drawing double click on the wheel. You can still use a regular two or three button mouse; however, you will not get the functionality of an IntelliMouse.

MECHANICAL DESKTOP BROWSER

The Mechanical Desktop browser, by default, is located on the left side of the screen and can be used to create new parts, to copy existing parts, or edit and delete parts and features. The Desktop browser also shows the history of the file's creation, including features as well as other parts. The Mechanical Desktop browser can be moved and closed as you would any toolbar. To open the Mechanical Desktop browser issue the **AMBROWSER** command. The capabilities of the browser will be introduced in chapter 2.

Figure 1.7

MECHANICAL DESKTOP OPTIONS

The following sections give a description for the Mechanical Desktop Options, Desktop and Part tabs. As with AutoCAD, certain settings that you specify affect how the system operates and what you see on screen. In Mechanical Desktop, the command for specifying these settings is Edit Options **(AMOPTIONS)**. The Edit Options command can be issued by selecting the bottom icon from any of the Mechanical Desktop toolbars, pull downs menus, or the third from the left icon on the bottom of the Mechanical Desktop browser. Figure 1.8 shows the Edit Options AMOPTIONS icon on the bottom of the Mechanical Desktop browser. After selecting the icon, the Desktop Options dialog box appears, as shown in Figure 1.9. There are seven tabs—Desktop, Part, Assembly, Scene, Surface, Drawing, and Annotation—allowing changes for each specific area within the Mechanical Desktop. Depending on which toolbar the command was issued from, the corresponding tab will be active. The other tabs (Assembly, Scene, Drawing, and Annotation) will be covered in the chapter in which that is introduced. The surface tab subject matter is not covered in this book.

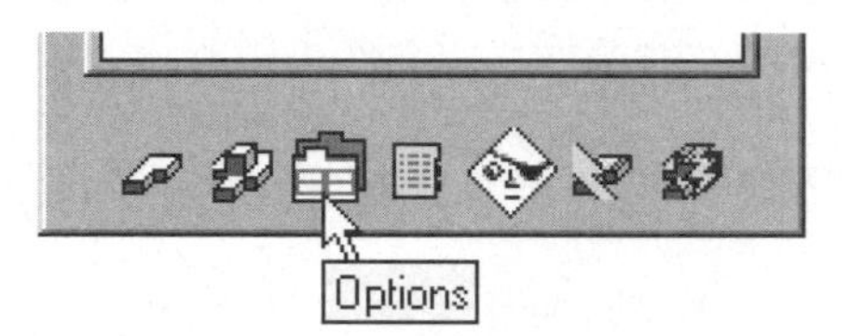

Figure 1.8

Figure 1.9

DESKTOP TAB

Synchronize Browser with Modes:

When checked, the browser will automatically update to the correct tab and mode that is selected in the corresponding express toolbar.

Synchronize Toolbars with Modes:

When checked, the toolbars will automatically update to the correct express toolbar that is selected in the corresponding tab in the browser.

Browser Font:

This option allows you to change the characteristics of the font used in the Desktop browser.

Dialog Controls:

You will set the values in this area that you will cycle through when giving values through dialog boxes. There are three areas to adjust Preset Values, Recent Values and Spinner Control Increments.

Preset Values:

You can predefine values for Distance, Angle, and Draft Angle from the drop down list. For given commands that require values for distance, angle or draft angle you can cycle through your preset values by clicking on the up or down arrows.

Recent Values:

Set the number of the recent values that you entered. These values will be used in the dialog boxes where they can be selected for input.

Spinner Control Increments:

Allows you to adjust the incremental values for the up and down arrows in specific dialog boxes. These values are used to adjust the angle and length in various commands. For example, when creating a helix with the 3D Path (Helix) command **AM3DPATH**, you can adjust the diameter by typing in a value or selecting the up and down arrow.

Drawing Units:

Set how the linear and angle dimensions will appear in the drawing views.

Show Messages:

Allows you to control whether or not you will see dialog boxes when doing specific tasks for external file editing and file saving/maintenance.

Desktop Audit Control:

There are three options for audit control.

Always:

This continuously looks at the parts whenever a file is opened, attached or inserted. Corrections are made when applicable so that any file corruption, if detected, will not be spread to other files. However, this auditing of parts can slow the performance of the system.

Inserted and Attached:

This will increase system performance but not audit any parts when a file is opened.

Never:

This will optimize the system but give you no part auditing at all.

Depending on your comfort level in using Mechanical Desktop, you may want to experiment with these options and see which one is best for you.

Note: All files that are being migrated from a previous release will be audited regardless of this setting.

AutoSAVE edited external files:

When checked, all edited external files will be autosaved at the increment set through the AutoCAD SAVETME command.

Object Selection:

Allows you to turn on and off the dynamic highlighting of faces and tool tips when editing. When Dynamic is checked and you issue the Create Sketch Plan command AMSKPLN and move the cursor across a given face, it will highlight. Or, when you are placing a feature such as a fillet, the edge will highlight. Face highlighting is covered in depth later in this chapter in the Face Highlighting section. When tool tips are checked and a command is issued, a tool tip will appear, identifying the part or feature where the cursor is located.

Desktop Symbol Size:

Allows you to maintain the size for Mechanical Desktop symbols, such as degrees of freedom and center of gravity, to remain a percentage of the viewport that it lies within. If changes are made to the size of these symbols, a regen will update its size on the screen.

Edit Fade Intensity:

When a part is edited in place, the other parts will fade as specified here. The higher the number, dimmer the non-editable part(s) will appear.

AutoUCS per View/Viewport

When checked, you can have a different active sketch plane set in different views.

PART TAB

Apply Constraint Rules:

If this check box is selected, Mechanical Desktop will apply geometric constraints to the sketch when the 2-D sketch is profiled. This will help reduce the number of dimensions and constraints required to fully constrain the sketch. The most common constraints are horizontal, vertical, and tangent. These, along with other constraints, will be covered later in this chapter.

Assume Rough Sketch:

If this box is not selected, the sketch will not change after being profiled. When the geometry has been drawn to the exact size or when profiling an existing 2D AutoCAD

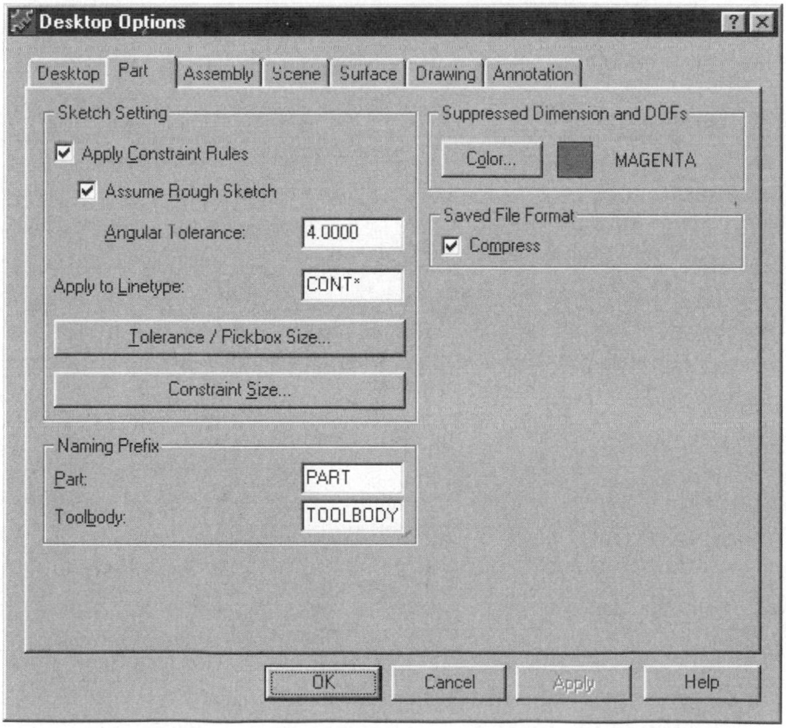

Figure 1.10

drawing, keep this box unchecked. If this check box is selected, the sketch profile will be analyzed on the basis of the angular tolerance and pickbox size. If the sketch's geometry falls within the angular tolerance, the geometry will snap to horizontal or vertical. For example, if two lines are within 4° of parallel, they will be changed and labeled parallel; or if a line is 3° from vertical, it will be changed to vertical.

Angular Tolerance:

The value specified in this text box is used when Mechanical Desktop is analyzing the sketch (the default value is 4). For example, any line that is sketched within 4° of horizontal or vertical will be made, respectively, horizontal or vertical. You can change this value to better reflect your geometry.

Apply to Linetype:

The linetype you select here is the linetype that will be turned into a solid; other linetypes will be used for construction purposes only. Construction lines are discussed further in Chapter 5.

Tolerance/Pickbox Size... :

This button allows you to adjust the size of your pickbox, used in closing a profile. If a sketch has a gap or overlap that is smaller than the pickbox size, it will close the gap. Remember that the pickbox size does not change as you zoom in or out. The further you are zoomed out, the larger the gap/overlap can be.

Constraint Size... :

This button allows you to change the height of the constraint symbols while displaying, editing, or deleting them.

Suppressed Dimension and DOFs:

You can change the color of dimensions that are referencing suppressed features here. Otherwise, the dimension(s) will take on the properties of the current layer. This setting will also control the color of the DOF symbol. DOF stands for degrees of freedom and is covered in chapter 8.

Saved File Format:

If the **Compress** check box is selected, the model will be saved in a compressed file. The knowledge of how the part was created will be tightly stored inside the file. The first time a compressed model is opened, it will be uncompressed and will rebuild itself.

Naming Prefixes:

The **Parts** text box in this area lists the default name that each part in a drawing will be given and then sequenced with a number. For example, PART1, PART2. Even if you use this default-naming scheme, you can type in a new name when a part is created.

The **Toolbody** text box identifies the name that a toolbody will be given when the **AMCOMBINE** command is used.

Any changes you make in the desktop and part tab are saved in the current drawing and affect geometry created *after* the changes have been made. If you want these settings to be used for other drawings, set them inside a template file. If, as you work with Mechanical Desktop, you are not getting the results that you anticipated, referring back to this section will help you to see if a setting needs to be adjusted.

Note: Make any changes in the Desktop Options dialog box *before* you profile your sketch.

SKETCH THE OUTLINE OF THE PART: STEP 1

All parametric 3-D models must first start with a 2-D sketch of the outline shape of the solid. A sketch can created with lines, arcs, circles, polylines or splines. When deciding what outline to start with, analyze what the finished shape will look like. Look for the shape that describes the part best. When looking for this outline, try to look for a flat face. It is usually easier to work on a flat face than on a curved edge, which can be difficult for new users. However, as you gain 3-D modeling experience, reflect back on how the model was created and think about other ways that the model could have been built. Just as with AutoCAD, there is usually more than one way to generate a given part.

When you are working in 2-D, you draw the geometry to the exact size. When sketching, you simply draw the geometry so it looks *close* to the desired shape and size; you do not need to be concerned about exact dimensional values.

Here are some guidelines that will help you generate good sketches.

Tips for Sketching: Select an outline that represents the part best. It is usually easier to work from a flat face.

Draw the geometry close to the finished size. For example, if you want a 2" square, do not draw a 200" square.

Create the sketch proportionate to the finished shape. For instance, to help maintain correct proportions for geometry that is about 10 x 5 in size, you could draw a 10 x 5 rectangle, sketch the geometry inside the rectangle, and then erase the rectangle. Another method would be to draw the first line segment using the AutoCAD technique called direct distance.

Draw the geometry so that it does not overlap. The geometry should start and end at the same point. See Figure 1-11.

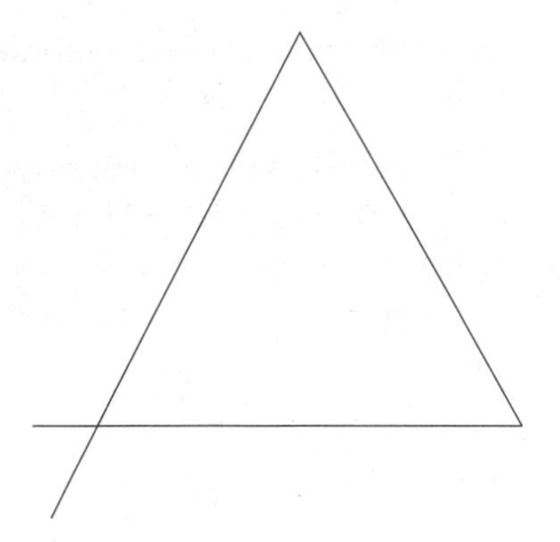

Figure 1.11

Do not allow the geometry to have a gap larger than the pickbox size. If the gap is smaller than the pickbox size, it will automatically be closed. See Figure 1-12.

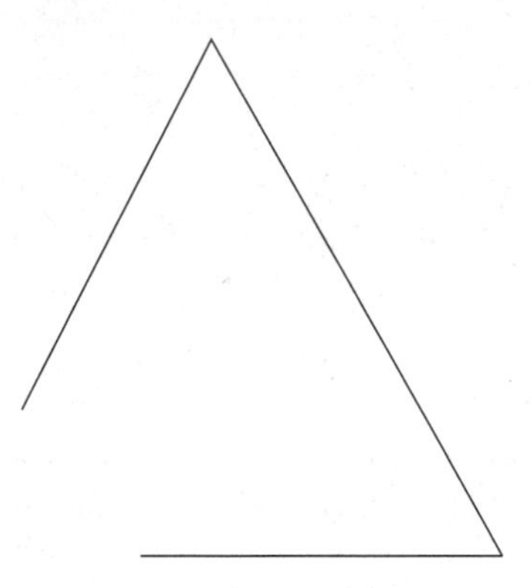

Figure 1.12

Keep the sketches simple. Leave out fillets and chamfers. They can easily be placed after the sketch is turned into a solid. The simpler the sketch, the fewer the number of constraints and dimensions that will be required to constrain the model.

Initially sketch thin areas larger. Once the sketch is fully constrained, you can go back and change the dimensions to smaller values.

PROFILE THE SKETCH: STEP 2

After the geometry is sketched, you will have Mechanical Desktop analyze it. This is referred to as *profiling*. Remember, the settings that you specified in the Desktop Options dialog box for angular tolerance, applying constraint rules, and assuming a rough sketch will be applied to the geometry in this step. Notice in the browser that a part name will appear after the sketch has been profiled. The sketch is now a Mechanical Desktop part.

Select the Profile a Sketch command **(AMPROFILE)** from the Part Modeling toolbar (see Figure 1.13), select the sketch using any AutoCAD selection technique, and then press ENTER.

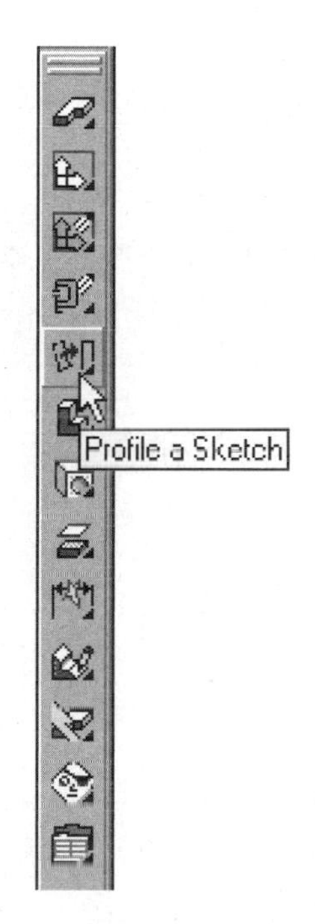

Figure 1.13

After the geometry is profiled, the number of dimensions or constraints required to solve the sketch will be displayed on the command line. If you receive a message:

```
Select edge to close profile: The sketch does not
form a closed profile
```

you can either select an existing edge on the part to close the sketch or press ENTER to exit the command, use grips to close the sketch, and then profile the sketch again. The method for closing an open profile by selecting an existing edge is covered in Chapter 6. An under-constrained sketch is a sketch that has objects that have not been given exact lengths or have constraints or relationships tied to it. A solved sketch or fully-constrained sketch has a dimension or constraint for all the objects.

The first point that you drew will be considered the fixed point. This *fixed point* is the point to which all geometry will shrink or grow from, depending on the dimensions that are added. You can delete, change or add more fixed points by issuing the Fix Point option of the **AMADDCON** command. This technique will be covered in the constraints section of this chapter.

Note: When selecting the sketch to be profiled, you can use any AutoCAD selection method: select individual entities, window, crossing, fence, etc.

PROFILING THE LAST OBJECT DRAWN

The Single Profile command **MNU_IPROFILE** will profile the last object that was drawn. Draw your sketch and then issue the Single Profile command from the Part Modeling toolbar, as shown in Figure 1.14, or right-click in the drawing area and from the pop-up menu, select Single Profile from the Sketch Solving menu.

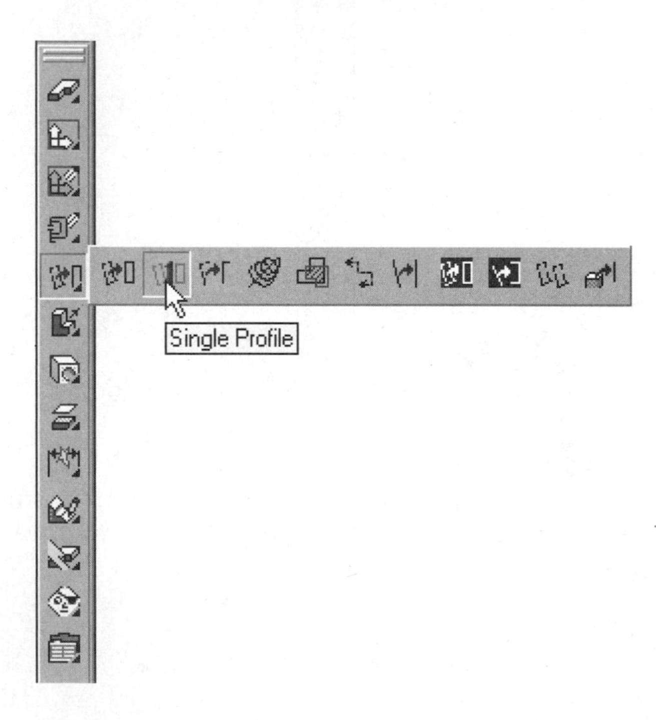

Figure 1.14

TUTORIAL 1-1—SKETCHING AND PROFILING (BOTTOM CLAMP)

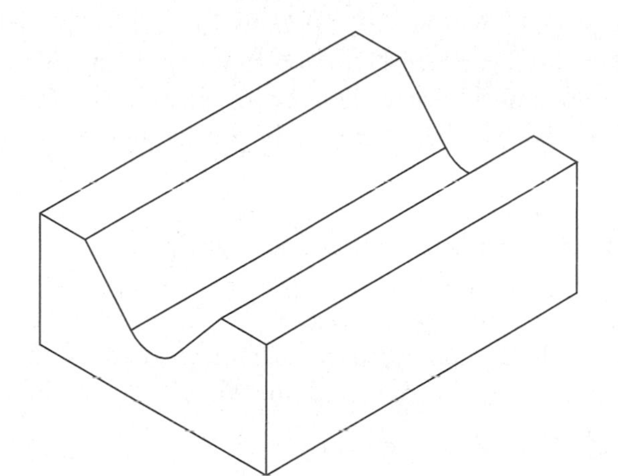

Figure 1.15

1. Start Mechanical Desktop.
2. Start a new drawing from scratch.
3. Sketch the geometry as shown in Figure 1.16; approximate size is 5 x 2.
4. Profile the sketch by issuing the Profile a Sketch command **AMPROFILE** or the Single Profile command and press ENTER.
5. If the geometry was sketched within 4° of horizontal and vertical, it should require seven dimensions or constraints. If your sketch requires more, you will add them in the next section.
6. Save the file with the following name: \Md4book\Chapter1\TU1-1.dwg.

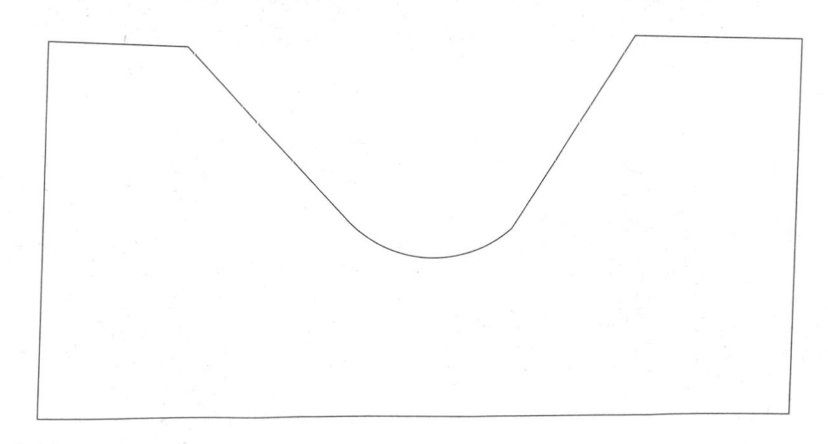

Figure 1.16

CONSTRAIN THE SKETCH: STEP 3

Once a sketch is profiled, Mechanical Desktop automatically analyzes it. If Apply Constraint Rules is checked in the Desktop Options dialog box (see Figure 1.10), Mechanical Desktop applies geometrical constraints. Constraints are used to apply behavior to specific objects or between objects. For example, you could apply a vertical constraint to a line and that line will always be vertical. You could apply a parallel constraint between two lines: those two lines will be parallel. As one line's angle would change so would the other. A line and arc could have a tangent constraint applied to them. When adding constraints, first select the segment of the sketch to be reoriented or resized then select the segment that the first segment will be linked to. If a segment of the sketch is already controlled by a parametric dimension and is selected first when applying a constraint, the command line will display the following message:

```
This constraint cannot be added. Existing
dimensions, constraints, or a fix constraint
prevent the constraint from being applied.
```

Mechanical Desktop 4.0 can apply 15 types of constraints. Figure 1.17 shows these constraint types and the letter symbols used to represent them.

Constraint	Letter	Example Figure
Tangent	T	Figure 1.18
Concentric	N	Figure 1.19
Collinear	C	Figure 1.20
Parallel	P	Figure 1.21
Perpendicular	L	Figure 1.22
Horizontal	H	Figure 1.23
Vertical	V	Figure 1.24
Project	J	Figures 1.25, 1.26
Join	None	Figure 1.27
Xvalue	X	Figure 1.28
Yvalue	Y	Figure 1.29
Radius	R	Figure 1.30
Equal Length	E	Figure 1.31
Mirror	M	Figure 1.32
Fix	F	Figure 1.33

Figure 1.17

Examples for each constraint follow, with the figures on the left showing the sketch before constraints and the figures on the right showing the sketch after the constraints have been applied. For clarity, all the constraint symbols in each example have been removed except those that are discussed in the example.

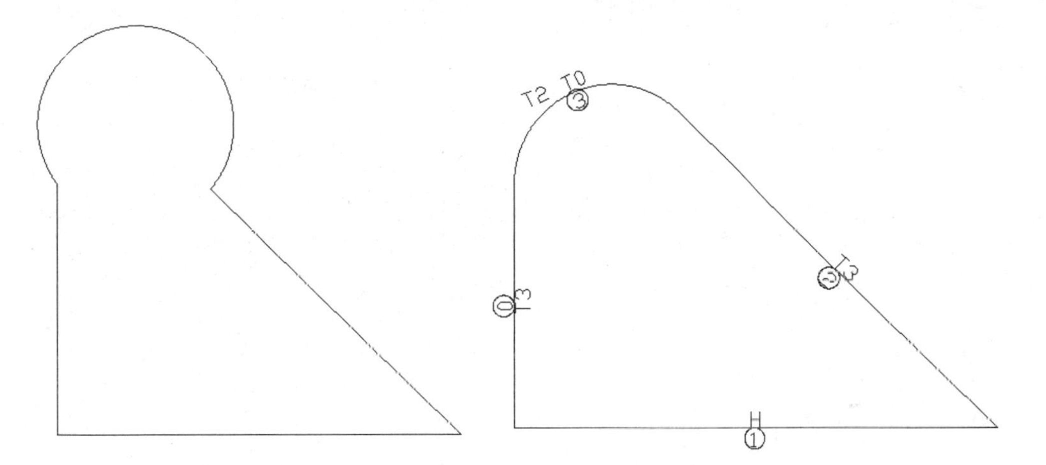

Figure 1.18 *T = Tangent constraint: An arc and a line will become tangent.*

Figure 1.19 *N = Concentric constraint: Arcs and/or circles will share the same center point.*

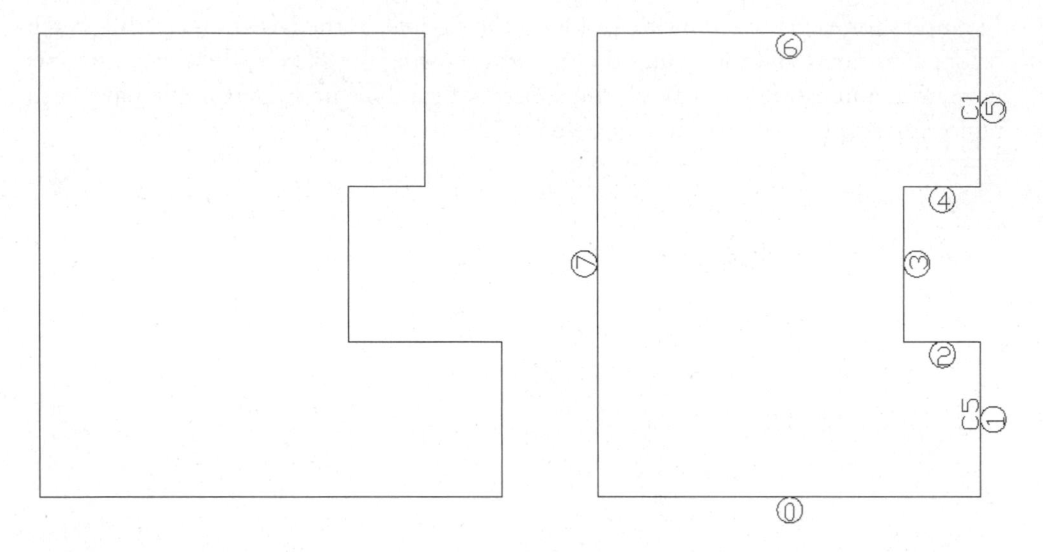

Figure 1.20 *C = Collinear constraint: Both lines will line up along a single line; if the first line moves, so will the second.*

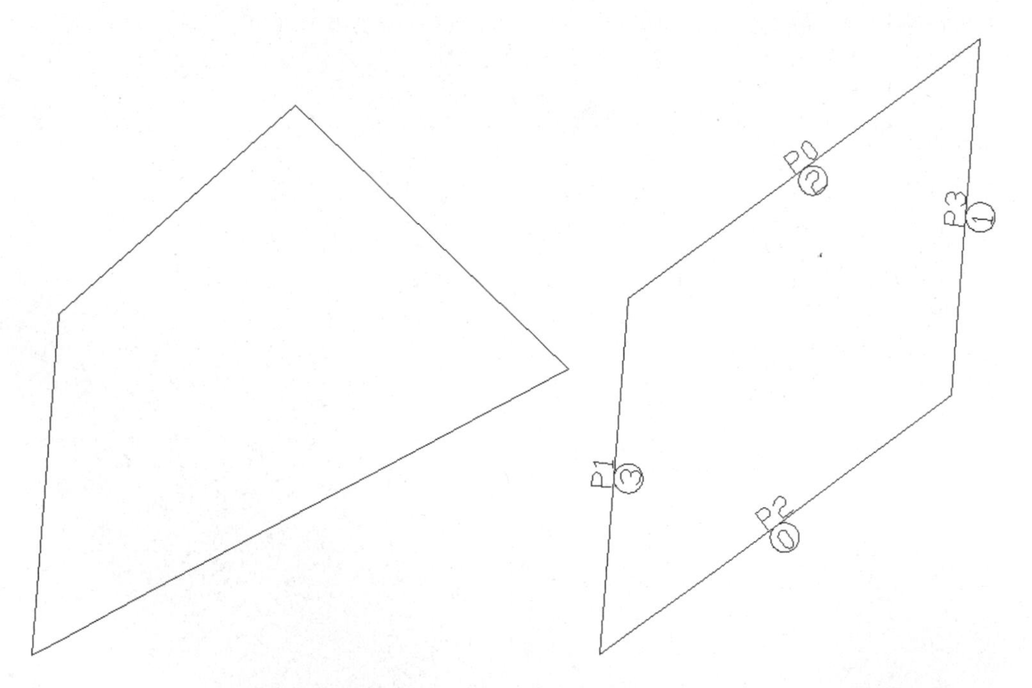

Figure 1.21 *P = Parallel constraint: Lines will be sketched exactly parallel to one another; that is, the first line you select will stay and the second will move to become parallel to the first.*

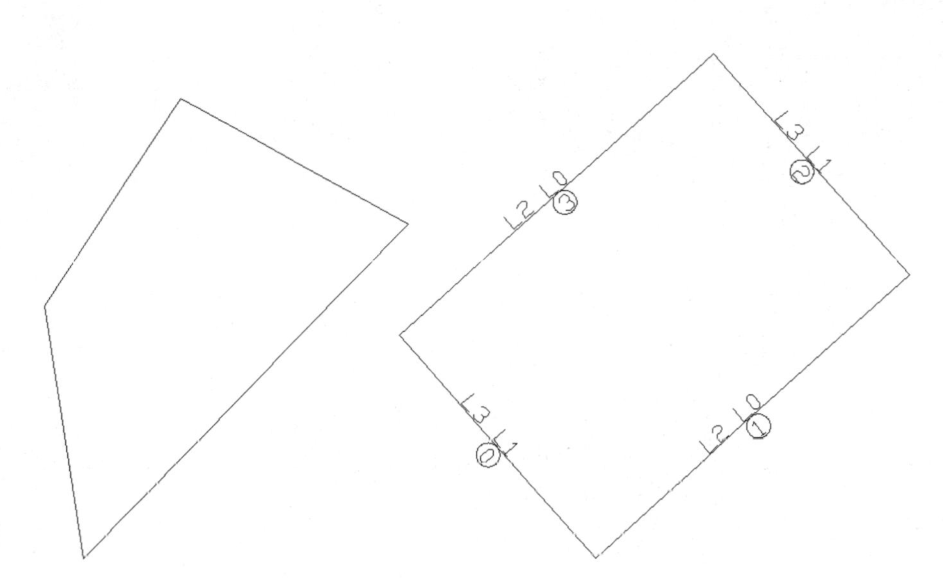

Figure 1.22 *L = Perpendicular constraint: Lines will be sketched at 90° to one another; that is, the first line you select will stay and the second will rotate until the angle between them is exactly 90°.*

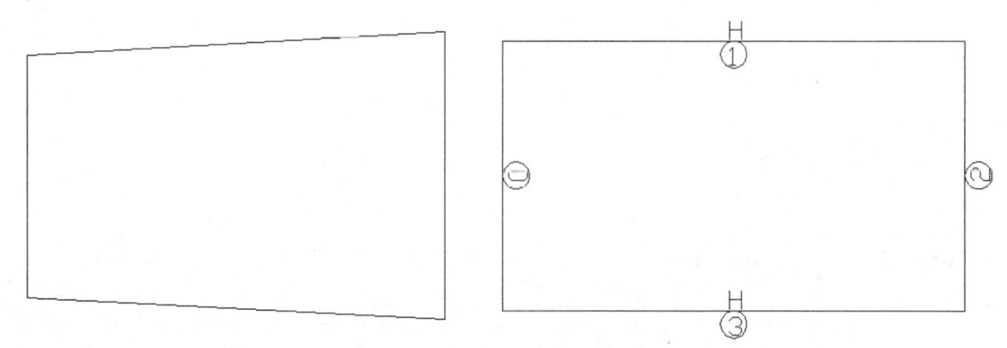

Figure 1.23 *H = Horizontal constraint: Lines are drawn parallel to the X axis.*

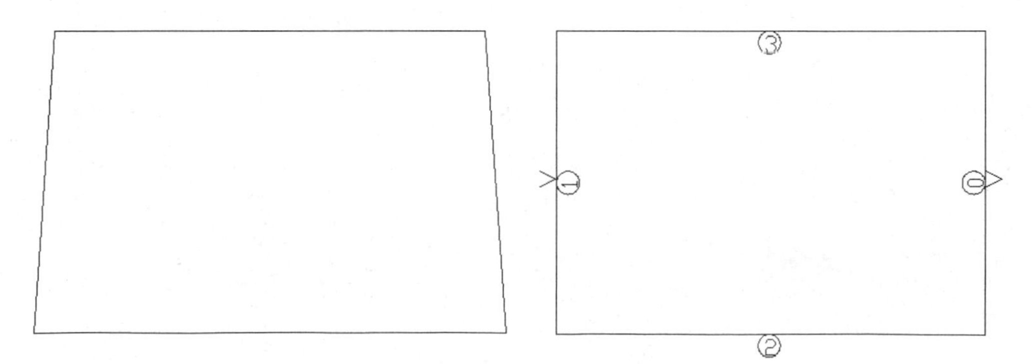

Figure 1.24 *V = Vertical constraint: Lines are drawn parallel to the Y axis.*

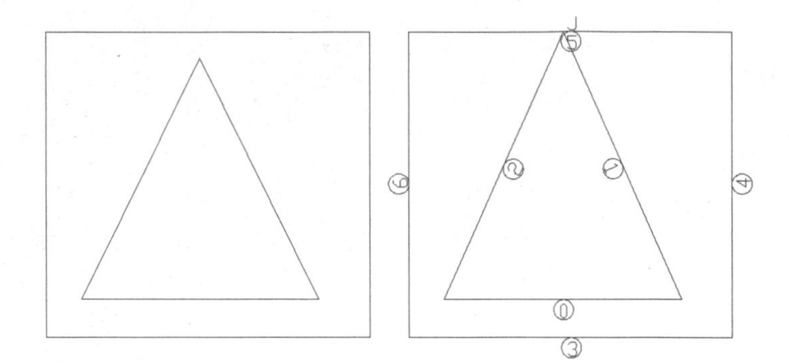

Figure 1.25

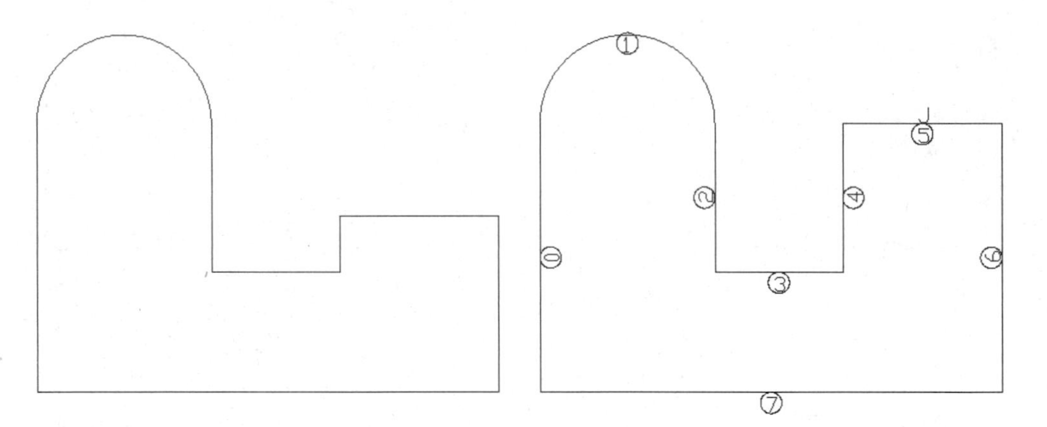

Figure 1.26 *J = Project constraint: Using an object snap, you select a point (endpoint, center point). Again using an object snap, you then select a line, arc, or a circle that this should touch. Depending on where you select the points, the geometry may move along the X or the Y axis. If the profile moves in an upredictable way, place a few dimensions first and then apply the project constraint. Figure 1.25 shows the top of the triangle projected onto the top of the square. Figure 1.26 shows the top of the right horizontal line projected to the top of the inside vertical line.*

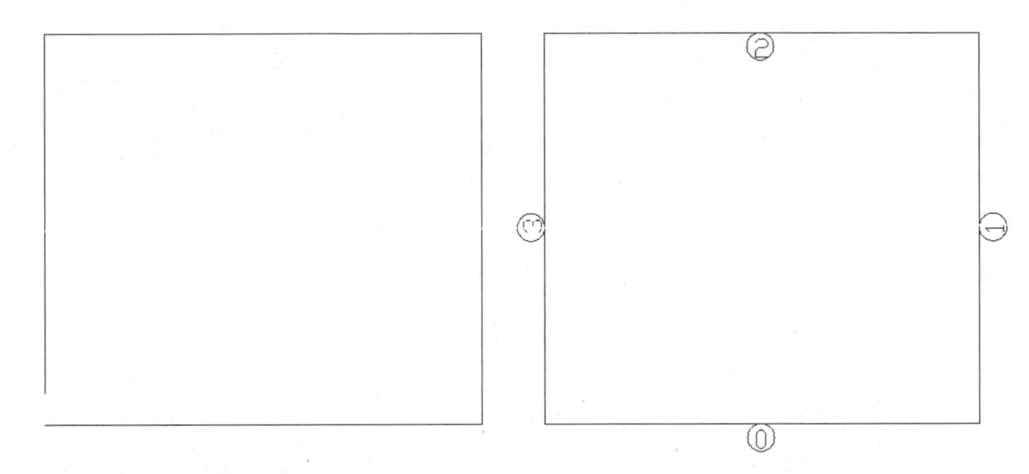

Figure 1.27 *Join constraint: The gap between two endpoints of arcs and/or lines will be closed. The Join constraint is applied automatically when the sketch is profiled and the gap is smaller than the AutoCAD pickbox size. There is no symbol for this constraint.*

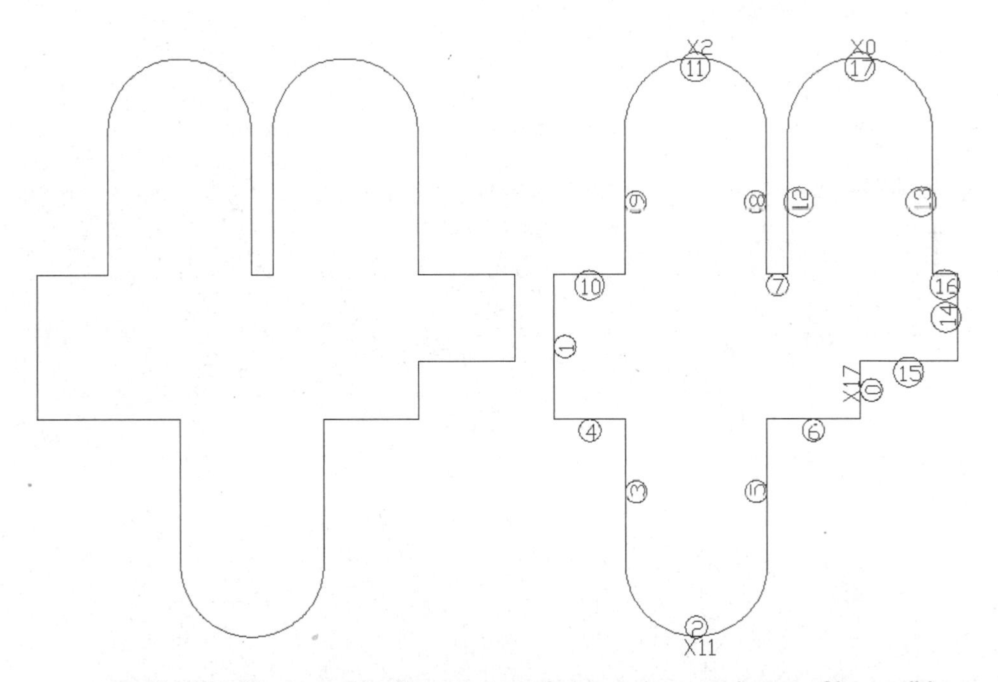

Figure 1.28 *X = X value constraint: Center points of arcs, circles or endpoints of lines will have the same X coordinate. Two center points, two end points or a center point and an end point can have the same X coordinate. Figure 1.28 shows the center points of the top and bottom arc on the left side aligned along the X coordinate. The center point of the top right arc and the endpoint of the inside vertical line have the same X coordinate.*

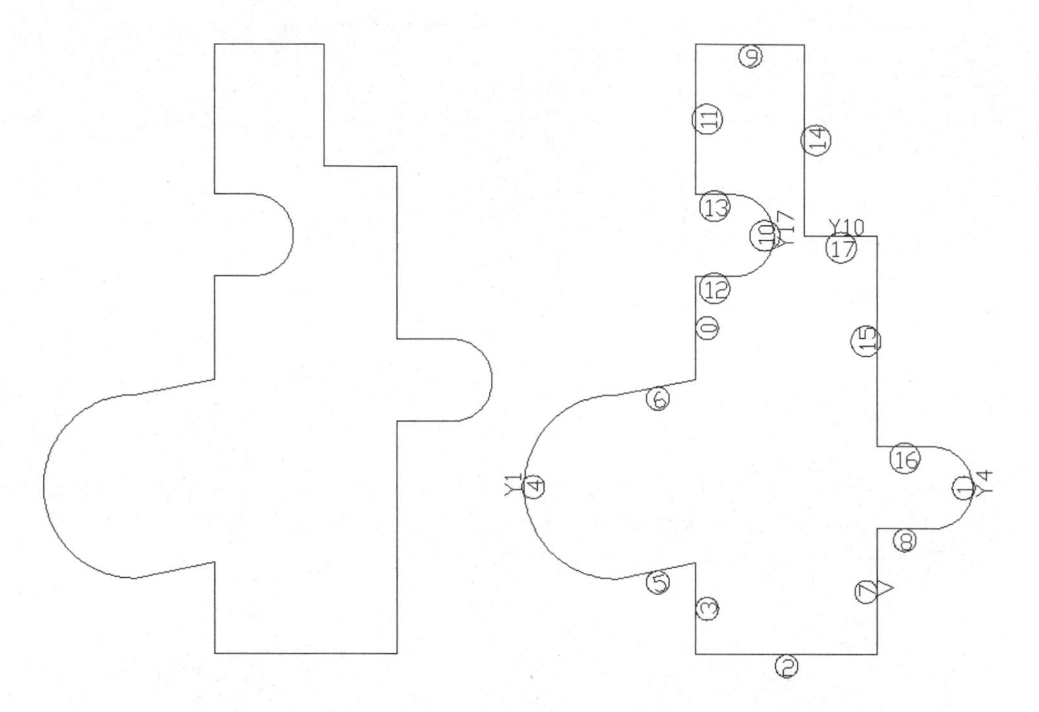

Figure 1.29 *Y = Y value constraint: Center points of arcs, circles or endpoints of lines will have the same Y coordinate. Two center points, two end points or a center point and an end point can have the same Y coordinate. Figure 1.29 shows the center points of the bottom two arcs aligned along the Y coordinate. The center point of the top left arc and the endpoint of the top inside horizontal line have the same Y coordinate.*

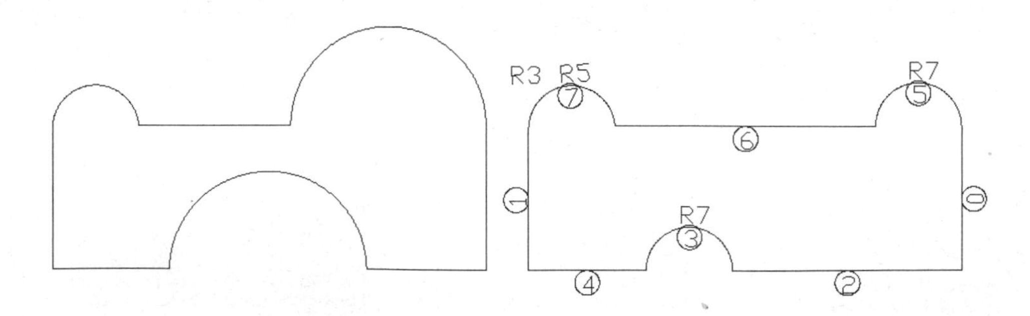

Figure 1.30 *R = Radius constraint: Circles and/or arcs will have the same radius. If this constraint is applied after one of the arcs is dimensioned, the second arc will take on the first arc's radius. If neither arc is dimensioned, both arcs will share the same value. The arc value shared is usually that of the second arc selected. For this reason it is better to dimension one arc and place a radial constraint to that arc and then to the other arc.*

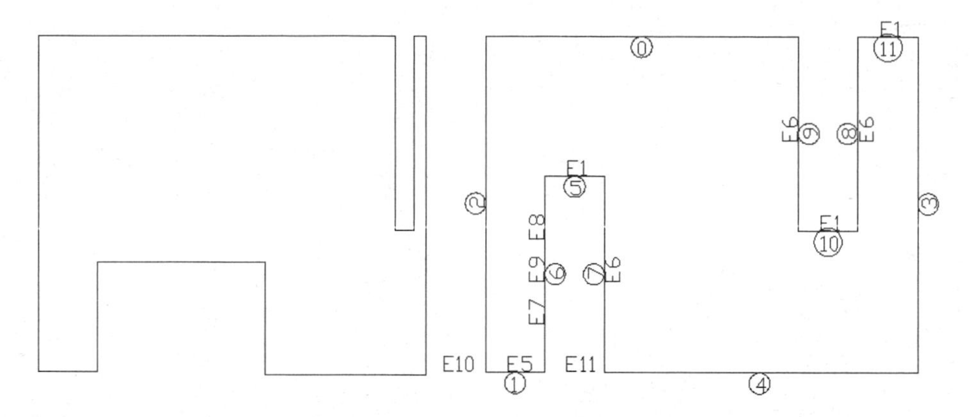

Figure 1.31 *E = Equal constraint: Line segments will be the same length. If one of the lengths changes so will the other segments that the equal constraint has been applied to.*

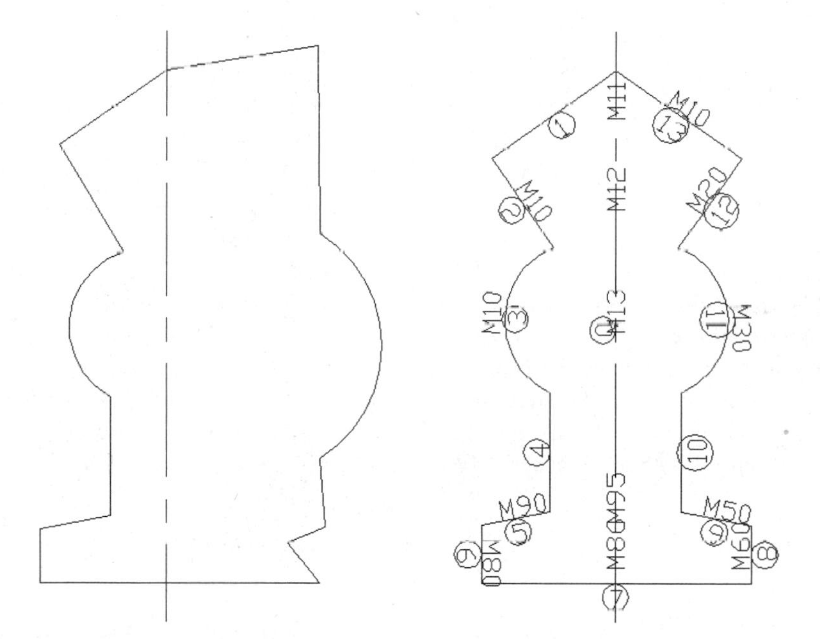

Figure 1.32 *M = Mirror constraint: Lines, arcs, circles, splines and ellipses will be made symentrical about an axis. When applying a mirror constraint you are prompted to select an axis to mirror about (this would be the center line through the center of the profile) it could be a line, spline, construction line or a part edge. After selecting the axis, select the object that you want to change then select the object that will be reflected. After the mirror constraint has been applied to two objects, as one changes so will the other. If other constraints exists on the profile they will take precedence. It may be necessary to delete a constraint before applying the mirror constraint. While in the command you do not need to reselect the axis to mirror about as Mechanical Desktop assumes you are using the same axis. Figure 1.32 shows the left side of the profile that was reflected to the right side.*

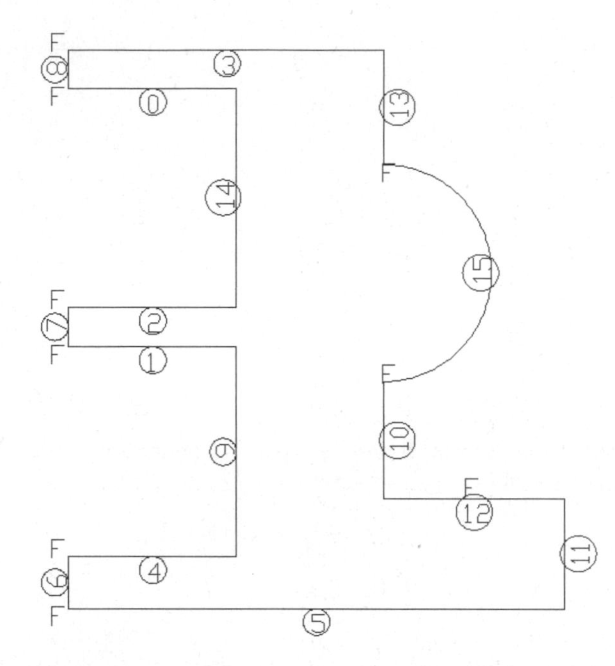

Figure 1.33 *F = Fix constraint: A fixed point(s) will prevent the selected point(s) from moving. By default, the first point drawn in a sketch will have a fix point constraint applied to it. The fixed point overrides any other constraint. Any endpoint, segment or center point of a line, arc, circle, spline segment or ellipse can be fixed. If you select near the endpoint of an object, the endpoint will be locked from moving. If you select near the midpoint of a segment the entire segment will be locked from moving. Multiple points in a profile can be fixed. If applying constaints and the profile is moving in directions that are undesirable you can apply fix constraints to hold points of the objects in place. You can always remove a fix constraint as needed. Deleting constraints will be covered later in this chapter. Figure 1.33 shows a narrow profile that has many fix point constraints applied; both the endpoints of lines, an arc and a line sgment.*

Note: The Project, *X* value, and *Y* value constraints are not applied when a sketch is profiled; they need to be added manually.

Tip: X and Y value constraints can usually be used instead of "0" length dimensions. When applying constraints between two objects, first select the object that you want to change, then select the object that you want the first to resemble.

SHOW CONSTRAINTS

After the sketch is profiled, the number of constraints and dimensions that are required to constrain the sketch fully will be displayed at the command line. To see the constraints that were applied, issue the Show Constraints command **AMSHOWCON** from the Part Modeling toolbar; see Figure 1.34 or right-click in the drawing area and from the pop-up menu, select Show Constraints from the 2D Constraints menu. After selecting the command you will see all the constraints that are applied to the profile. You will see numbers enclosed in circles going around the parts. These show the order in which the geometry was created, and they also function as labels for objects. The numbers then, serve to show relationships between different objects. For example, object 3 is parallel to object 8, or objects 3 and 4 are tangent. Don't worry if the numbers do not go around the part in sequence. The numbers only reflect the order in which the geometry was drawn. You will also see letters near the entities in the sketch. The letters tell the types of constraint that were applied (see Figure 1.17).

When you are finished viewing the constraints, press ENTER to return to the command line.

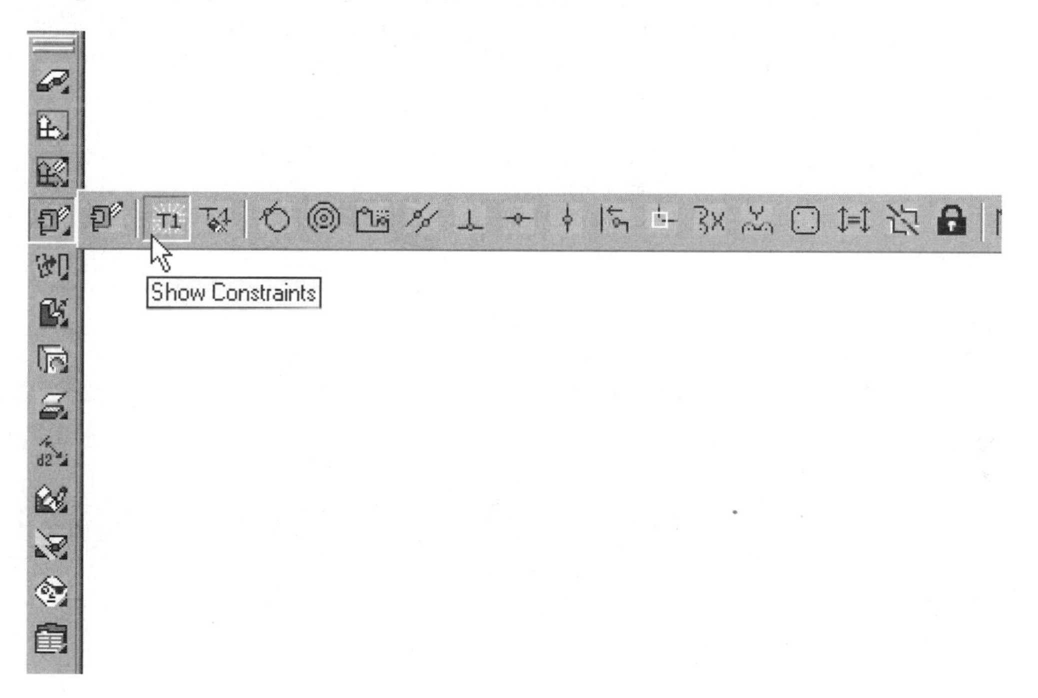

Figure 1.34

ADD CONSTRAINTS

If you need to add a constraint, select the particular constraint option from the Part Modeling toolbar, as shown in Figure 1.35 or right-click in the drawing area, and from the pop-up menu, select the constraint to add from the 2D Constraints menu. Constraints set behavior or relationships to the profile. As the constraints are added the number of dimensions or constraints that are required to fully constrain the sketch will be decreased. Mechanical Desktop will not allow you to over constrain the sketch or add duplicate constraints. If you add a constraint that would conflict with another you will be warned that this constraint will over-constrain the sketch. An example would be to add a vertical constraint to a line that already has a horizontal constraint. If you try to add a constraint to an object that already exists you will be alerted at the command line that this constraint already exists.

Note: If you click on the first icon, the toolbar will float on the screen until closed. After selecting the constraint option, you will be returned to the drawing, where you can select the geometry to which to apply the constraint. You will stay in this option until you press ENTER, and then you will be given the option at the command line either to select a different constraint to apply or to press ENTER to exit the command. As the constraints are applied, the number of constraints and dimensions required to solve the sketch will decrease.

Tip: When applying mirror constraints it may be easier to first delete all fixed constraints. When applying an equal, radius or mirror constraint, select the object that will be changed first and then select the object whose properties will be copied.

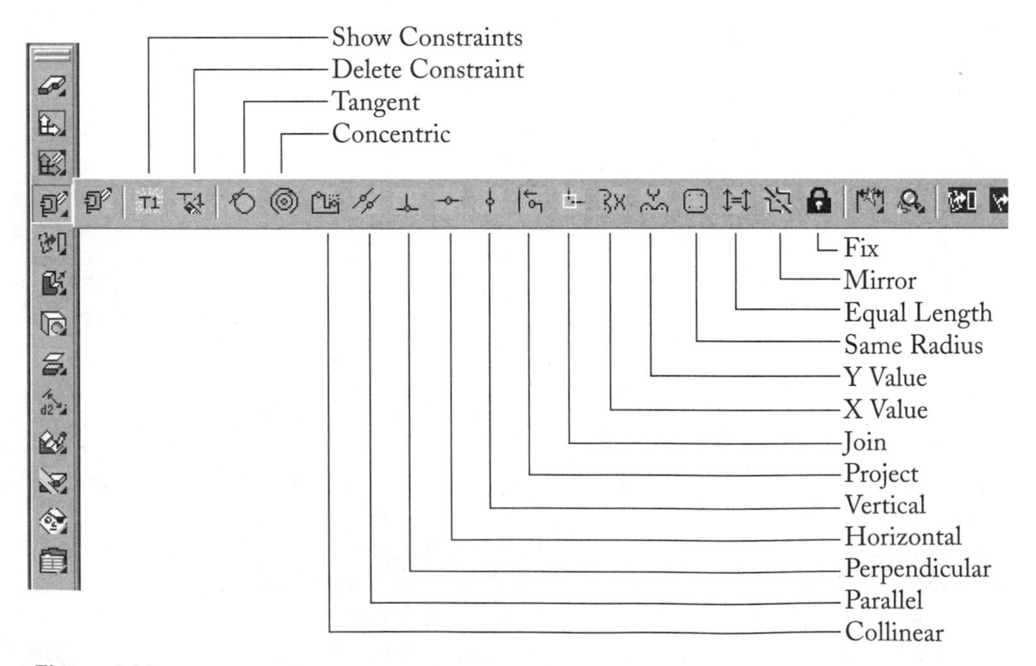

Figure 1.35

DELETE CONSTRAINTS

While working you may want to delete or change the type of a constraint on an object. Before adding the new constraint you must first delete the existing constraint. To delete a constraint on a sketch, issue the Delete Constraints command **AMDELCON** command from the Part Modeling toolbar, as shown in Figure 1.35 or right-click in the drawing area and from the pop-up menu, select **Delete Constraints** from the 2D Constraints menu. After you have issued the command, the constraint information will appear if there is only one sketch. Otherwise, you will be prompted to select a sketch. If the constraint symbols appear too small or too large, press S and press ENTER to adjust the size of the constraints. To delete a constraint, simply select the constraint symbol located on the given line, arc, or circle. After you select the constraint symbol, the number of constraints and dimensions required to constrain the profile will be increased. To delete another constraint, select that constraint. To exit the command, press ENTER.

RE-SOLVING PROFILES

When you are working on a profile and you are not sure how many more dimensions or constraints are required to fully constrain the profile, you can issue the Re-Solve Sketch command **MNU_RE_SOLVE** from the Part Modeling toolbar, as shown in Figure 1.36 or right-click in the drawing area and from the pop-up menu, select **Re-Solve** from the **Sketch Solving** menu. The command line will read:

```
Solved underconstrained sketch requiring X
dimensions or constraints.
```

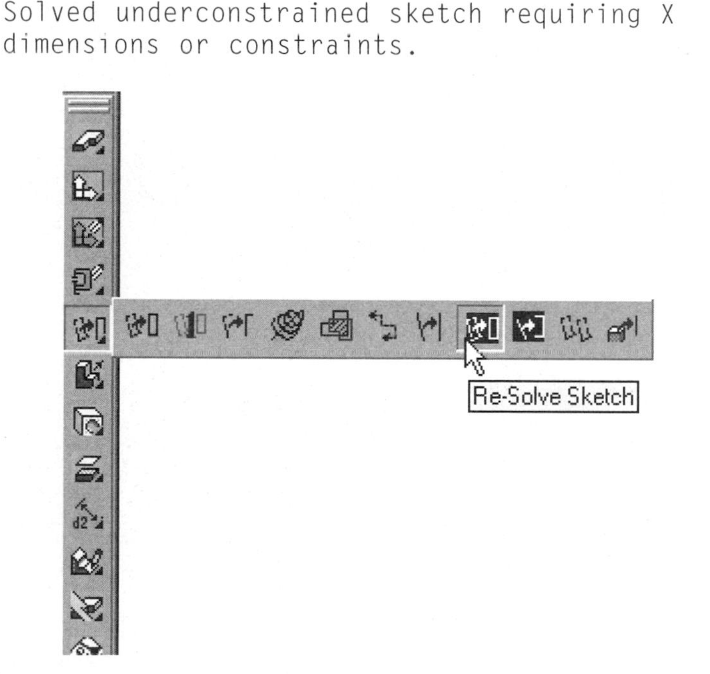

Figure 1.36

Mechanical Desktop will highlight the objects in the profile that are not yet constrained.

GRIP EDITING

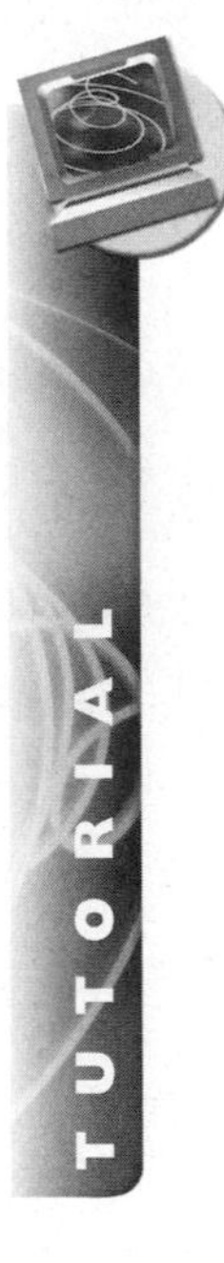

Another method of helping figure out what needs to be constrained on a sketch is to grip edit the sketch. After selecting an object, highlight a point and then try to move it to another location. If the point stretches it is under-constrained. While grip editing, constraints take precedence. For example if you drew a square, profiled it and two horizontal and vertical constraints were applied, it then grip edited one of the corners. You could change the size of the square but the lines would maintain their horizontal and vertical constraints. When grip editing an object that has a mirror or parallel constraint applied to it the other object will move at the same time.

TUTORIAL 1.1 (CONTINUED)—CONSTRAINING (BOTTOM CLAMP)

1. If the current file is not \Md4book\TU1-1.dwg, then open it now.
2. Issue the **Re-Solve Sketch** command to verify the number of constraints and dimensions that are required to constrain the sketch.
3. Use the Show Constraints command **AMSHOWCON** to show the constraints that were automatically applied. Issue the command and then press A to display all constraints. Your drawing should look like Figure 1.37. If you are missing a constraint, you can add it in step 3. When you are done looking at the constraints, press ENTER to exit the command.
4. Add missing constraints as needed, issuing the necessary command(s) from the 2-D Constraints toolbar or the right click menu.

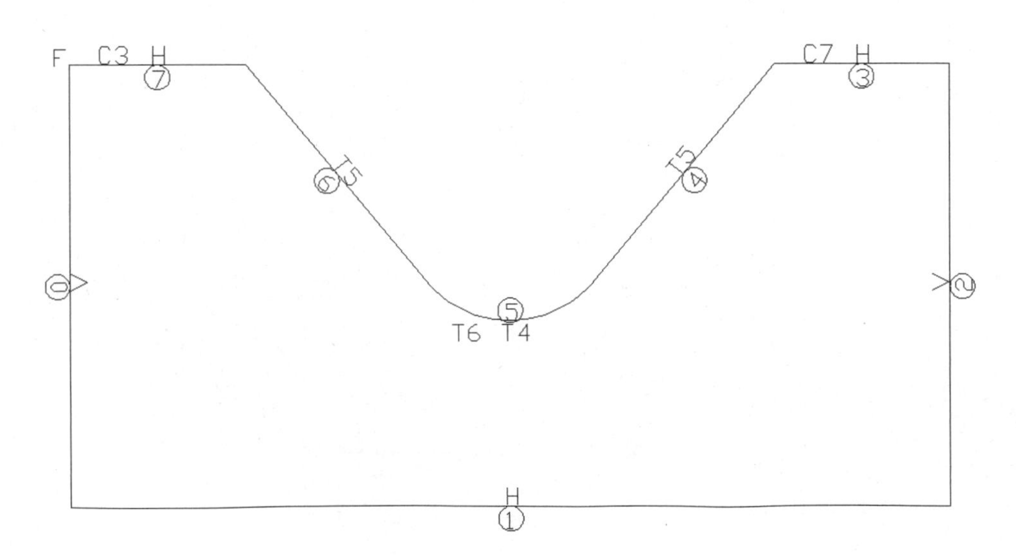

Figure 1.37

5. Make sure that there is only one fix constraint: it should be located in the upper left corner. If there is another fix constraint delete it.
6. For the next four steps practice grip editing. Grip edit the bottom left corner of the profile. Stretch the point up, down, in and out. The bottom horizontal line will only move up and down as you grip edit the corner because the fix point is located in the upper left corner and a vertical constraint on that line. Both of those constraints are preventing movement along the X axis.
7. Try Grip editing the upper left corner of the profile. You will be prevented from grip editing the corner because of the fix point.
8. Delete the fix point constraint that is on the upper left corner of the profile.
9. Now grip edit the upper left corner. As you move the point along the X axis the left side of the profile will move in and out. As you move the point up and down both of the top lines will move together because there is a collinear constraint between the two lines.
10. Save the file.

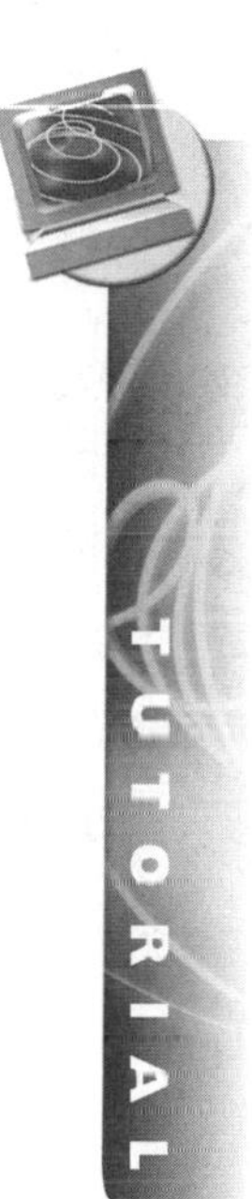

TUTORIAL 1.2—CONSTRAINING

1. Start a new drawing from scratch.
7. Sketch the geometry as shown in Figure 1.38; approximate size is 6 x 6.
8. Profile the sketch by issuing the Profile a Sketch command **AMPROFILE** or the Single Profile command and press ENTER.

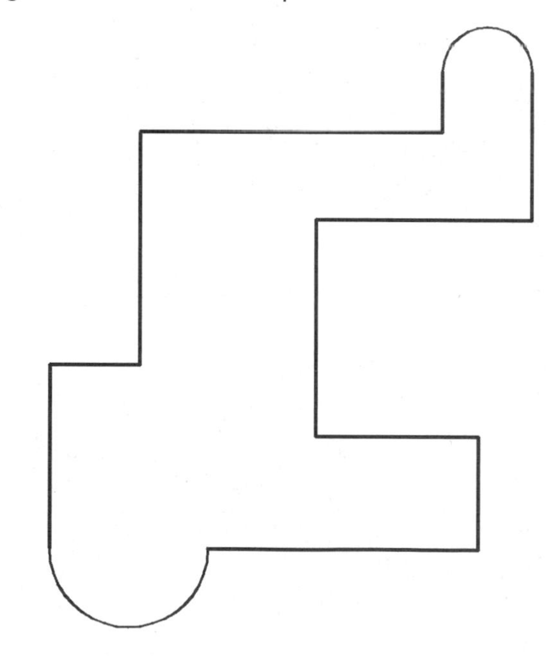

Figure 1.38

2. If the geometry was sketched within 4° of horizontal and vertical and the lines and arcs were nearly tangent, it should require twelve dimensions or constraints. If your sketch requires more, you will add them next. If it required less you may have had some collinear constraints added automatically.
3. Show the constraints using the Show Constraints command **AMSHOWCON**.

Note: Using the "S" option under AMSHOWCON will only allow you to select a specific segment of the sketch to show the constraints. The "S" option only works for **AMDELCON**.

4. If there is a collinear constraint "C" delete it.
5. Place a fix constraint in the lower left corner, if there is another fix constraint, delete it.
6. Add a collinear constraint; first select the top inside vertical line and then the left inside vertical line.
7. Apply a collinear constraint to the two bottom horizontal lines, selecting the line away from the fixed point first.
8. When complete, your screen should resemble Figure 1.39 with the constraints shown. If you are missing constraints that are shown, add them now.

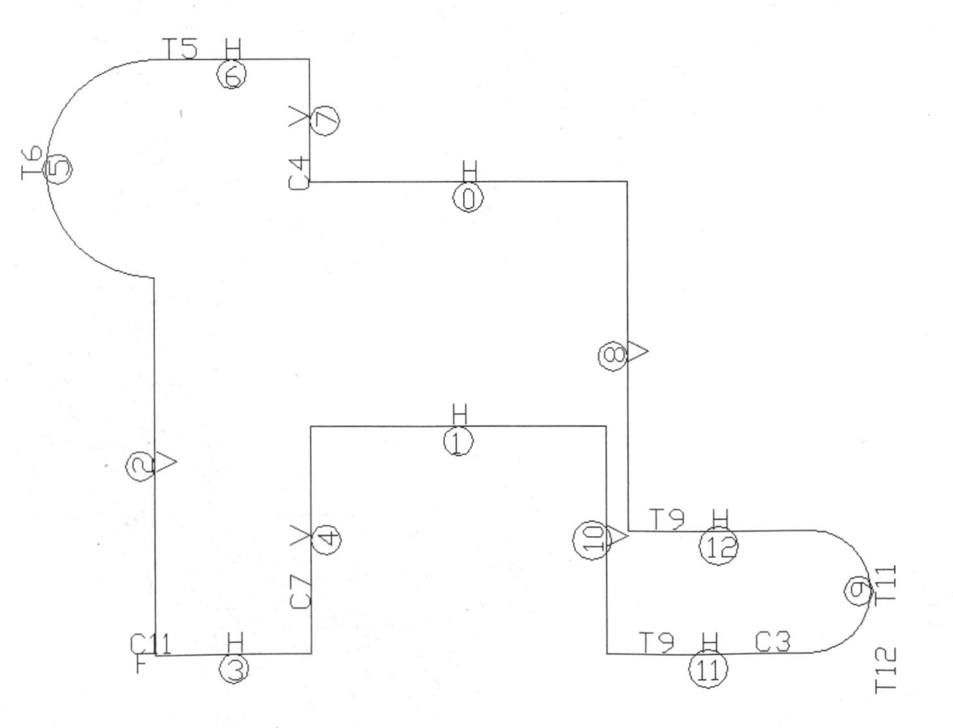

Figure 1.39

9. Apply a Y value constraint to the lower right arc and the second from the bottom horizontal line. Select the arc first, as it is the object that we want reoriented.
10. Apply a Y value constraint to the upper left arc and the second from the top horizontal line. Select the arc first, as it is the object that we want reoriented.
11. Apply an equal length constraint to one of the lower inside vertical lines and the lower left horizontal line. Select the vertical line first, as it is the object that we want reoriented.
12. Apply an X value constraint to the upper left arc and the left most vertical line. Select the arc first, as it is the object that we want reoriented.
13. Apply an equal length constraint to the second from the bottom horizontal line and the lower left horizontal line. Select the second from the bottom horizontal line first, as it is the object that we want reoriented. When complete your screen should resemble Figure 1.40 with the constraints shown.

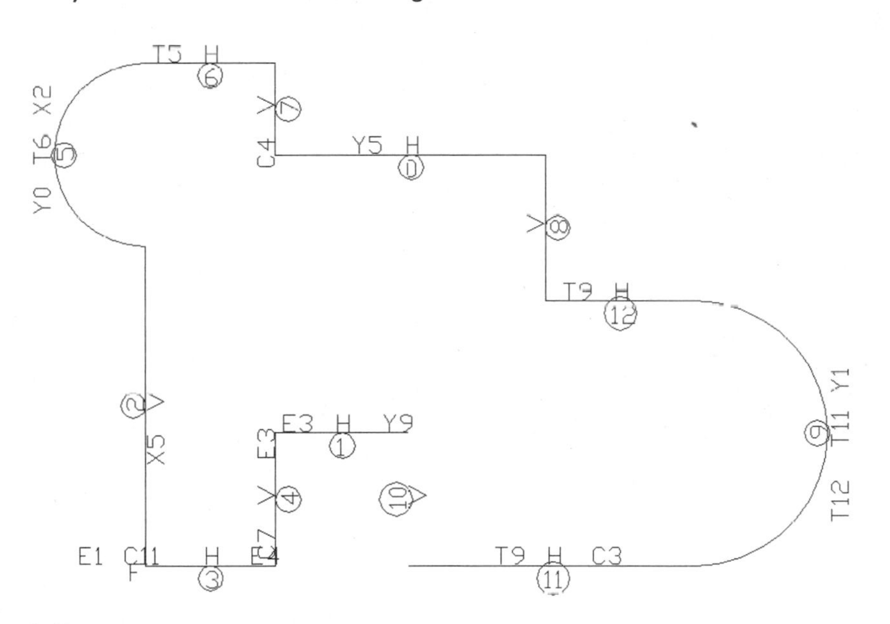

Figure 1.40

14. Save the file as \Md4book\Chapter1\TU1-2.dwg.

TUTORIAL 1.3 MIRROR CONSTRAINT

In this tutorial, the objects have already been sketched and profiled. This sketch contains a construction line (the centerline). Construction lines will be covered in Chapter 6.

1. Open the file \Md4book\Chapter1\TU1-3.dwg.
2. Issue the mirror constraint option of the **AMADDCON** command. After issuing the command, select the centerline, then the top right-angled line and then the top left angled line.
3. Continue applying the mirror constraint to the objects on the right side to the corresponding objects on the left side. Work your way down the sketch applying the mirror constraint to all five objects.
4. Apply a vertical constraint to the lower left angled line. The lower right-angled line will also now be vertical. When complete, your screen should resemble Figure 1.41 shown with the constraints visible.
5. Turn off all object snaps.
6. Practice grip editing. Grip edit the arcs and then a few of the lines.
7. Save the file.

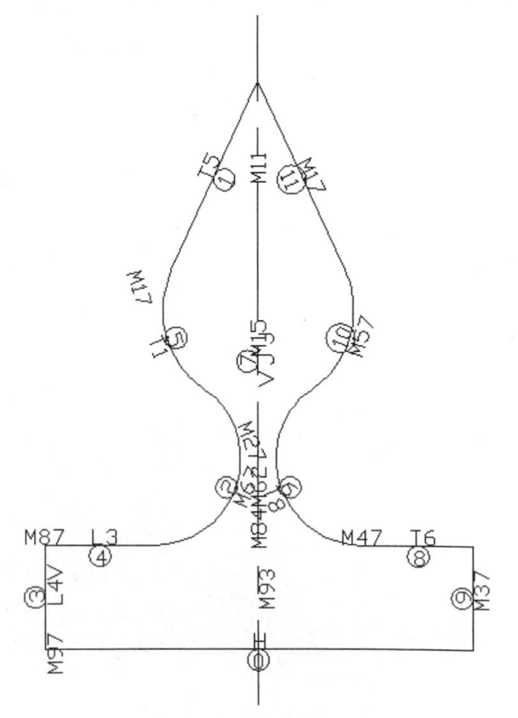

Figure 1.41

DIMENSION THE PROFILED SKETCH: STEP 4

The last step is to dimension the profiled sketch. The dimensions you place will control the size of the model and will also appear in the drawing views when they are generated. Don't be overly concerned about where the dimensions are placed, because usually they are not exactly where they need to be when the drawing views are finally laid out. They can be relocated easily after the drawing views have been created or this can be automated by setting "Automatically Arrange Dimensions" through the Edit Options **(AMOPTIONS)** command under the annotation option. This setting will be covered in chapter 4. Try to avoid having the extension lines go through the part as this will require more clean-up. With experience you will place the dimensions in better locations, requiring less cleanup when the drawing views are created. When selecting lines to place dimensions, pick near the side from which you anticipate the dimensions originating in the drawing views.

All parametric dimensions are created with a single command, New Dimension **AMPARDIM**, from the Part Modeling toolbar (see Figure 1.42). Or you can right-click in the drawing area and from the pop-up menu, select **New Dimension** from the **Dimensioning** menu. There is no need to use object snaps when placing these dimensions. When you select a line, it will snap to the nearest endpoint; when an arc or circle is selected, it will snap to its center point.

There are two techniques for dimensioning a line: You can select near two endpoints and then select a location for the dimension. Or if you want to dimension a single line, select the line anywhere and then select a location for that dimension.

To create an angular dimension, select near the midpoint of the two lines and then select a point for the dimension location.

To dimension an arc or a circle, select on its circumference and then select a location outside it. By default when an arc is dimensioned, the result is a radius dimension; when a circle is dimensioned, the result is a diameter dimension.

If the wrong dimension type appears after you select a placement point, you can change it at the command prompt by typing in the capital letter of the constraint type you want. The following string shows what the command prompt will look like:

```
[Undo/Hor/Ver/Align/Par/aNgle/Ord/Diameter/pLace]
<2.7536>:
```

For example, if a vertical dimension appears and you want an aligned dimension, press A and then press ENTER to change its type. Once the correct dimension type is on the screen, you can either press ENTER to accept the value in the command line or type in a new value and press ENTER.

When inputting values, type in the exact value; do ***not*** round up or down. The accuracy shown in the dimension is from the current dimension style. Mechanical Desktop models are accurate to six decimal places; for example, 1.0625 is more accurate than 1.06. When placing dimensions, it is recommended that you place the smallest dimensions first; this will prevent the geometry from flipping in the wrong direction.

As the sketch is constrained and dimensioned, the number of constraints and dimensions required will decrease until the sketch profile is fully constrained. With Mechanical Desktop it is not required that a sketch be fully solved. However it is recommended to fully constrain a sketch. A fully constrained sketch will allow you to anticipate how the sketch may change, while changing an under constrained sketch allows for ambiguous changes.

Figure 1.43, together with the following explanations, shows how to create linear, angular, and radial dimensions.

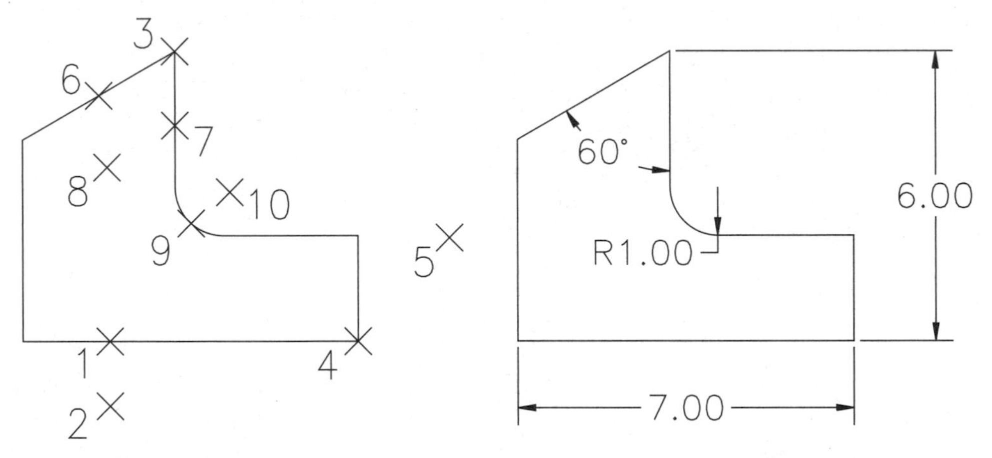

Figure 1.43

LINEAR DIMENSIONS

Select a line and then a placement point (points 1 and 2). Or select near two different endpoints and then select a placement point (points 3, 4, and 5).

ANGULAR DIMENSIONS

Select near the midpoints of the two lines you want dimensioned and then select a placement point (points 6, 7, and 8). Or if a linear dimension appears, press N at the command line to get an angle dimension.

RADIAL DIMENSIONS

Select the arc or circle to dimension and then select a placement point (points 9 and 10).

ORDINATE DIMENSIONS

To create an ordinate dimension, first select near the point on the sketch that you want to be the zero point and then select a location point. A linear dimension will appear. Then press O and then press ENTER. The dimension will change to an ordinate dimension with a value of zero. Any dimension placed after that will be an ordinate dimension related to the zero dimension. If placing an ordinate dimension based on a different zero dimension than that which the last ordinate dimension was based, select its extension line for the first selection and then pick the point to dimension. You will need to place a separate zero ordinate dimension in both the *X* and the *Y* directions.

CHANGING DIMENSIONS

To change a dimension's value, issue the Change Dimension command **AMMODDIM** from the Part Modeling toolbar (see Figure 1.44) or right-click in the drawing area and from the pop-up menu, select **Edit Dimension** from the **Dimensioning** menu. Then select the dimension you want to change. Its current value will be displayed on the command line. Type in a new value and press ENTER. To change another dimension's value, select the dimension, or to exit the command, press ENTER. After the command is exited, the sketch will automatically be updated to reflect this new value.

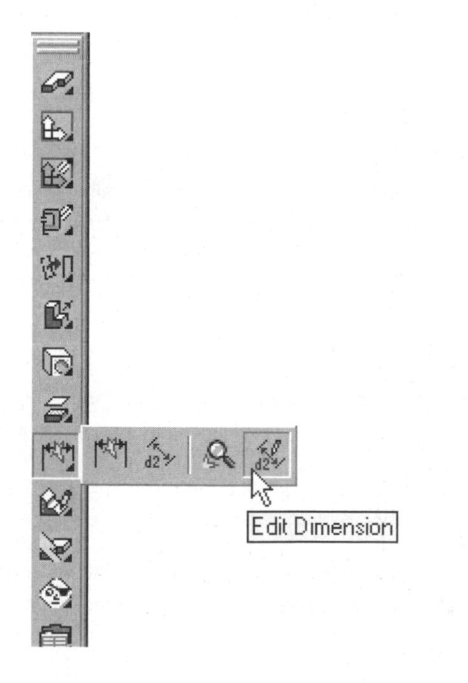

Figure 1.44

Another method for changing a dimension's value is to double click on the dimension that you want to change. A dialog box will appear like the one shown in Figure 1.45. Type in a new value in the expression area and then select OK and the profile will be updated. The information in the rest of the dialog box will be covered in chapter 6 under the power edit section.

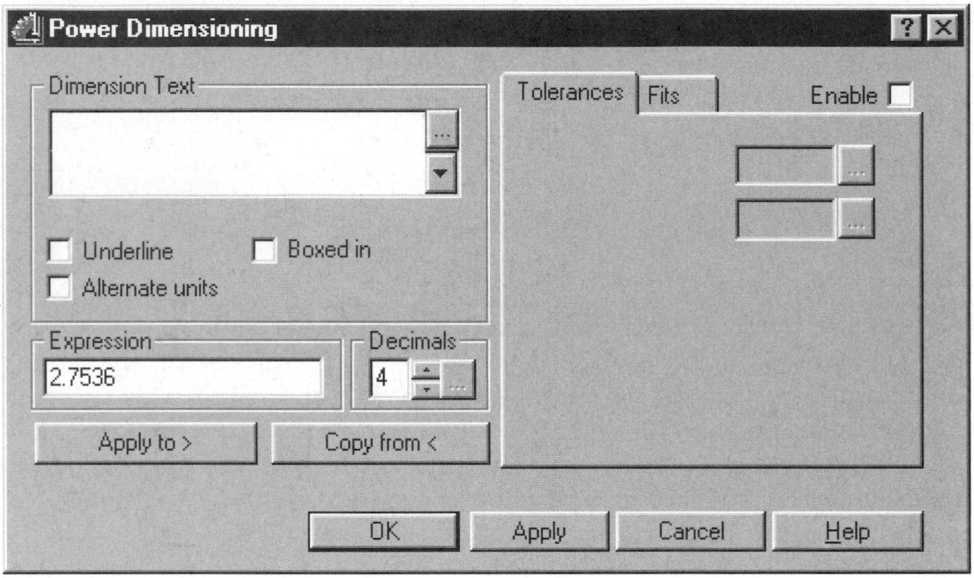

Figure 1.45

Note: When you input dimensional values, it is recommended that you use decimal values, not fractions. Fractions are interpreted literally. For example, 2-1/2 is interpreted as 2 minus 1/2. If you must use fractions, use a + sign, for example, 2+1/2.

Dimensions are created with the current dimension style (number of decimal places, text height, etc.).

You do not have to fully solve (dimension or constrain) your sketch. Mechanical Desktop permits the use of under-constrained profiles. As you go through this book, you will learn how to return to a sketch profile or profile and add or delete dimensions and constraints.

Tip: When selecting lines to place dimensions, pick near the side from which you anticipate the dimensions originating in the drawing views.

When typing in values, type in the exact values; do *not* round up or down. The accuracy shown in the dimension is from the current dimension style. Mechanical Desktop models are accurate to six decimal places; for example, 1.0625 is more accurate than 1.06.

Place the smallest dimensions first. This will prevent the geometry from flipping in the wrong direction.

When creating profiles with thin areas, it may help to draw them larger. Once the sketch is fully constrained, you can go back and change the dimensions to smaller values. Or fix these points and after the profile is early constrained delete the fixed points and then dimension them.

TUTORIAL

TUTORIAL 1.1 (CONTINUED)—DIMENSIONING (BOTTOM CLAMP)

1. Open the file \Md4book\Chapter1\TU1-1.dwg.
2. To add dimensions, issue the New Dimension command **(AMPARDIM)**. Dimension the profile as shown in Figure 1.46. Your dimension's style may differ, depending on the settings in your current dimension style.

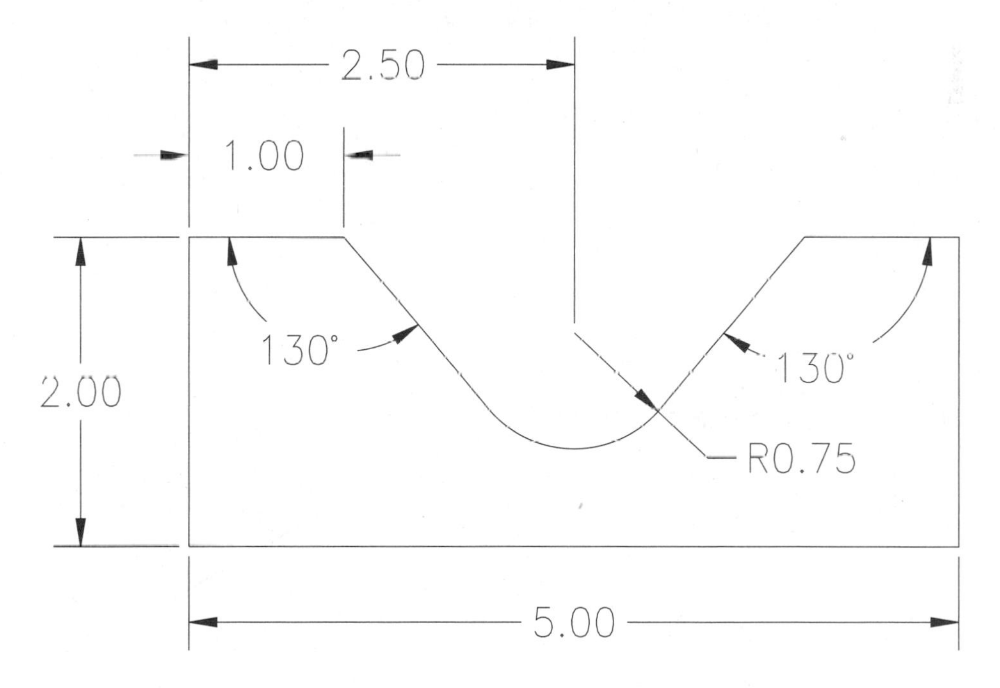

Figure 1.46

3. To change a dimension, issue the Change Dimension command **(AMMODDIM)**. Select the 2.00 dimension and change it to 2.5, and change the .75 radius to .625. Press ENTER to exit the command. The completed profile should look like Figure 1.47.

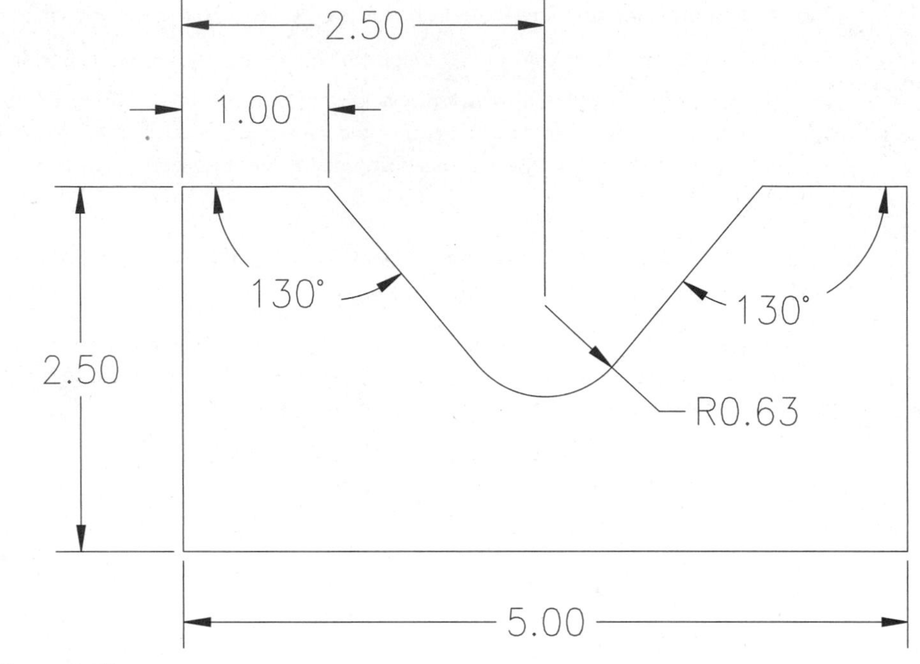

Figure 1.47

4. Save the file.

TUTORIAL 1.4—SKETCHING, PROFILING, CONSTRAINING, AND DIMENSIONING (TOP CLAMP)

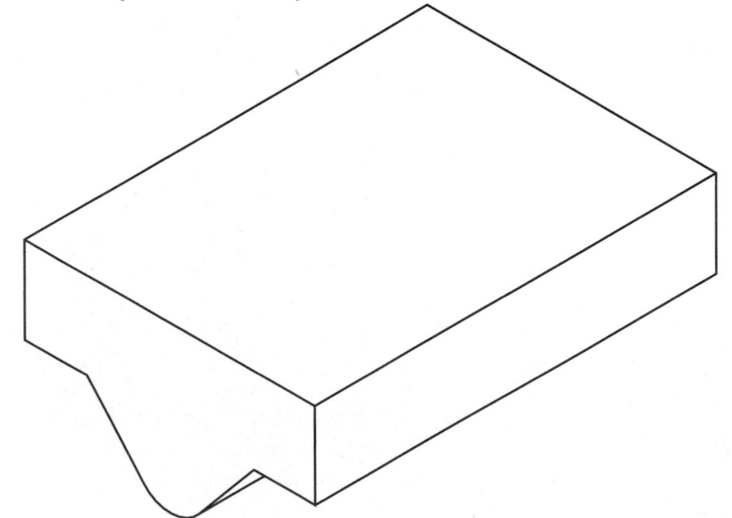

Figure 1.48

1. Start a new drawing.
2. Sketch the geometry as shown in Figure 1.49. The approximate size is 5 x 4.

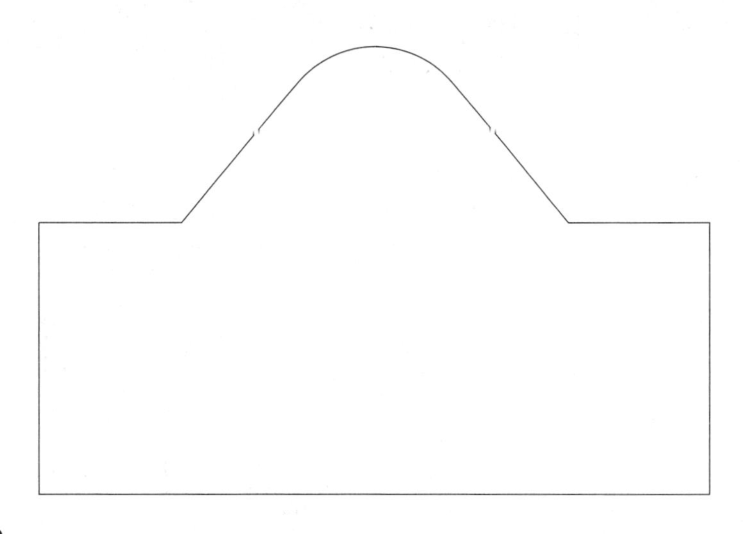

Figure 1.49

3. Profile the geometry. The sketch should require at least seven dimensions or constraints.
4. Show the constraints. Your sketch should resemble Figure 1.50.

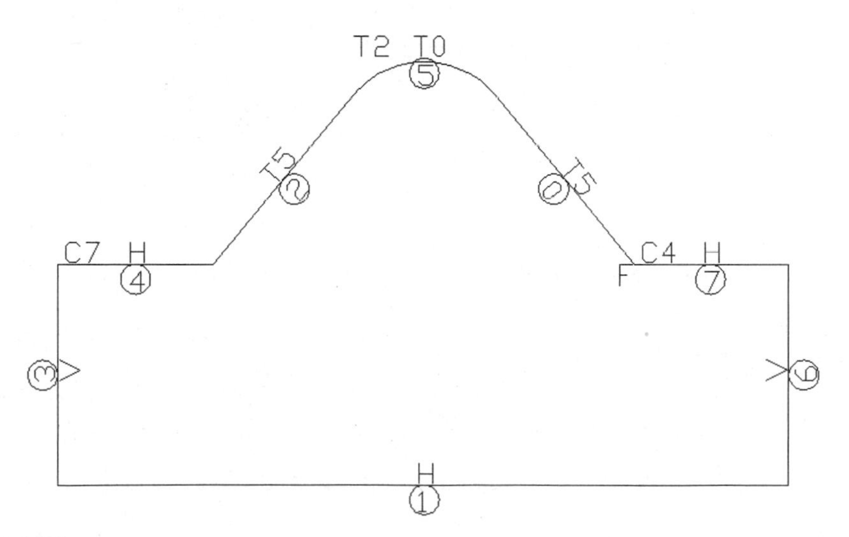

Figure 1.50

5. Add constraints if required.
6. Add the dimensions shown in Figure 1.51.

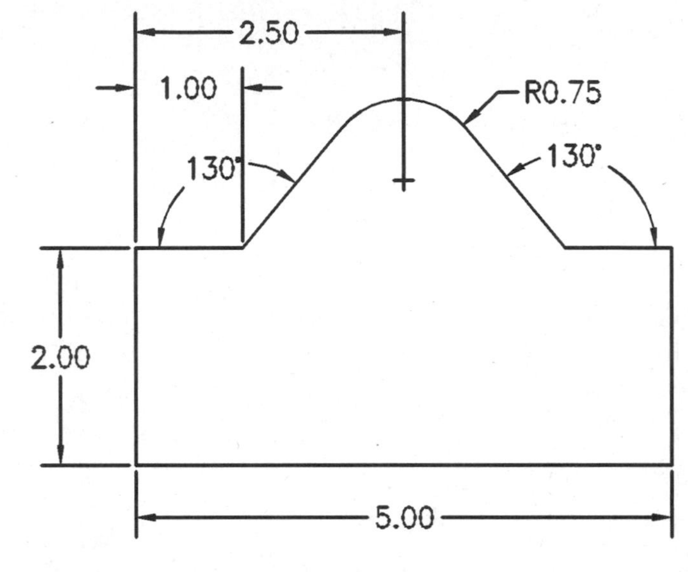

Figure 1.51

7. Change the 2.00 dimension to 1.50, the .75 radius to .63, and the 1.00 dimension to 1.06. The completed profile should resemble Figure 1.52.

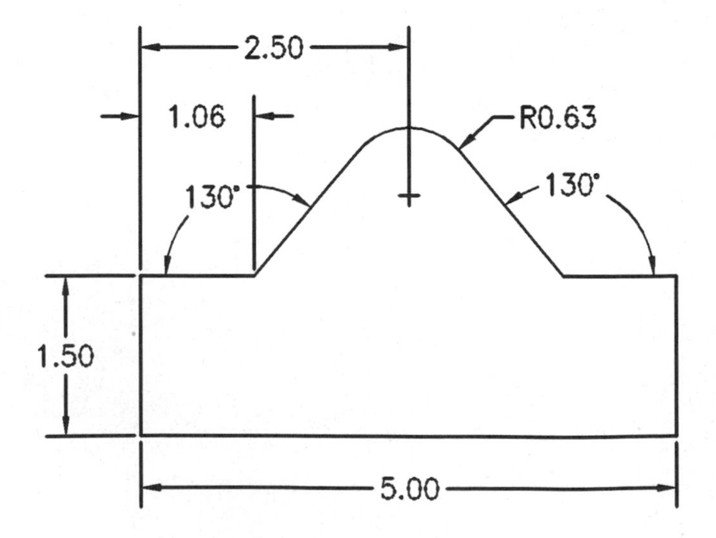

Figure 1.52

CONVERTING EXISTING 2D DRAWINGS TO PROFILES

In this section you will learn how to convert an existing 2D drawing to Mechanical Desktop parts. Even though you may not always be able to automatically convert an entire drawing, you can take the main shape with its dimensions, profile it and turn it into a part. To create a new 3D file from a 2D AutoCAD drawing you will first copy the information from the 2D drawing and place it into a new 3D file. First open the 2D drawing and then either drag and drop the 2D objects that best define the shape with its dimensions into a new file. Another method would be to WBLOCK or copy to the clipboard then insert or paste the geometry into a new drawing. After the geometry is inserted into the new drawing explode the block. The copied geometry needs to follow the same rules as sketches: no large gaps or overlaps, and the dimensions need to have been placed to the endpoints and center of the geometry. Then profile the geometry, along with all the dimensions. If you get an error message:

```
Highlighted dimension could not be attached
```

this means that the highlighted dimensions have not been created correctly or would form an over-constrained sketch. Erase the highlighted dimension and re-profile the sketch and dimensions. The dimensions that were profiled will automatically become parametric dimensions. Next, add any missing dimensions or constraints and then create a part. Drawings that have islands in them will be covered in Chapter 6.

TUTORIAL 1.5—CONVERTING AN AUTOCAD 2D DRAWING TO A 3D PART

1. Open the file \Md4book \Chapter1\TU1-5.dwg.
2. Start a new drawing from scratch.
3. Drag, copy to the clipboard (CTRL+C) or WBLOCK the profile of the top view (drawn as polyline) along with all the dimensions.
4. Drop or paste the geometry and dimensions from the clipboard, using ctrl+v or insert them from the file you WBLOCKed out. When complete, the new drawing should resemble Figure 1.53.
5. If the objects were inserted, explode the block with the EXPLODE command.
6. Profile the geometry and the dimensions. The sketch should be fully constrained and the dimensions should be parametric dimensions.
7. Change a few dimensions with the Change Dimension command **(AMMODDIM)** to prove that they are now parametric dimensions.
8. Change a few of the dimensions with the change dimension command. When done, change the dimensions back to their original values.
9. Save the file as \Md4book \Chapter1\TU1-5-3D.dwg.

TUTORIAL

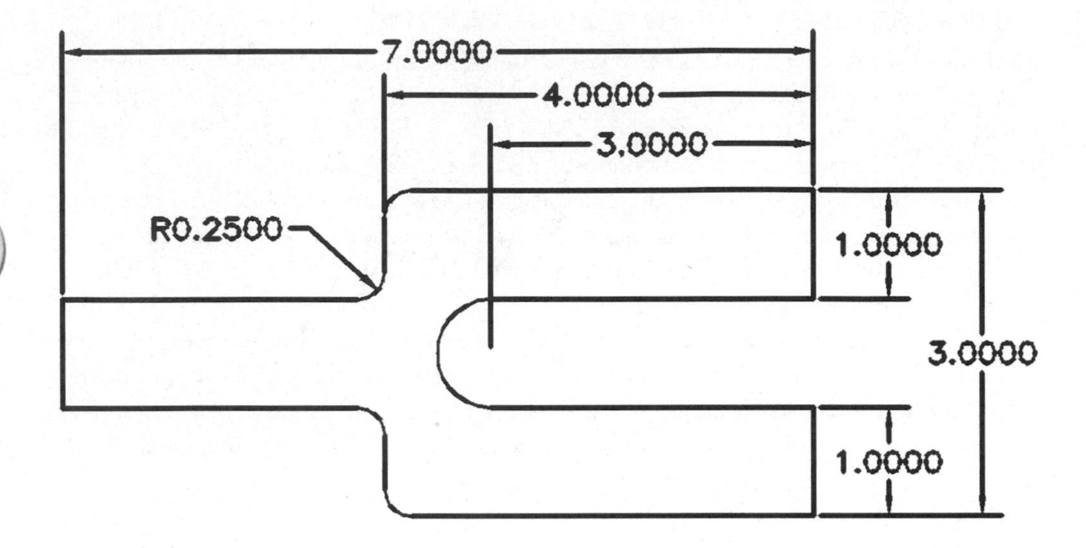

Figure 1.53

EXERCISES

The following exercises are intended to challenge you by providing problems that are open-ended. As with the kinds of the problems you'll encounter in work situations, there are multiple ways to arrive at the intended solution. In the end, your solution should match the drawing shown. For each of the following exercises, start a new drawing. When you have finished the drawing, save the file using the name supplied in the exercise. You will use the finished profiles in later chapters.

Exercise 1.1—Bracket

Save the finished drawing (Figure 1.54) as \Md4book\Chapter1\Bracket.dwg.

Exercise 1.2—Guide

Save the finished drawing (Figure 1.55) as Md4book\Chapter1\Guide.dwg. (Hint: Use Radial, X Value, and Y Value constraints).

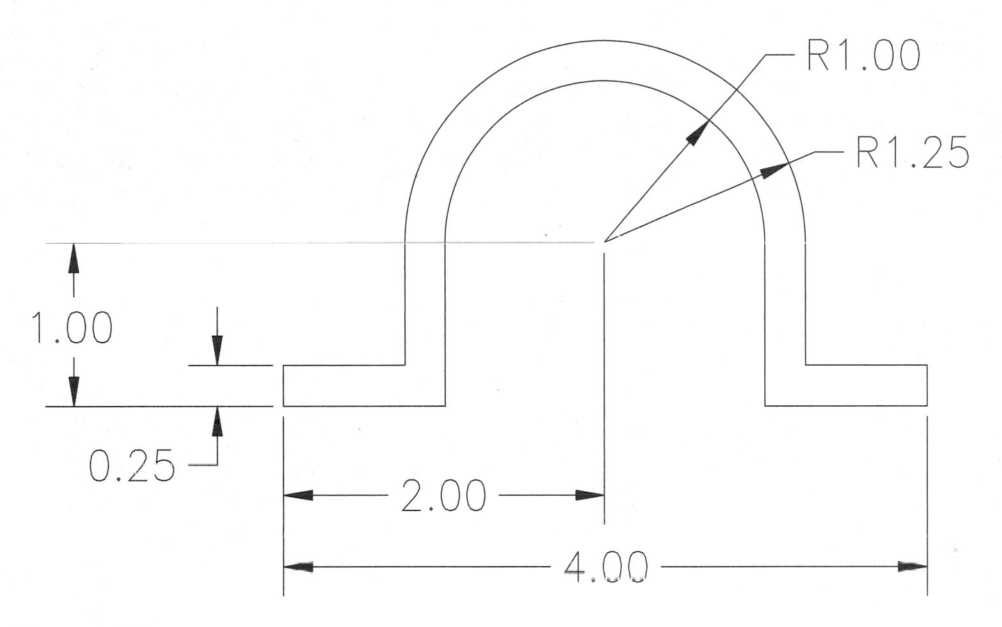

Figure 1.54

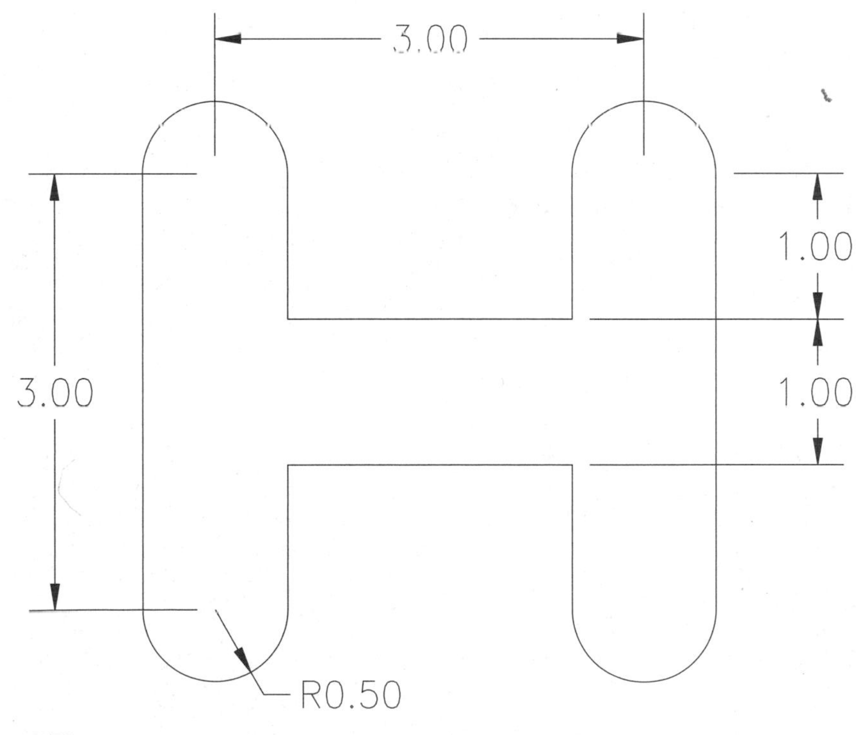

Figure 1.55

EXERCISE

EXERCISE

Exercise 1.3—Foot

Save the finished drawing (Figure 1.56) as \Md4book\Chapter1\Foot.dwg. Hint: Use the X value and Y value constraint to center the arc to the left vertical and top horizontal lines. Depending upon the order in which you sketched the objects, you may need to delete and replace the Fixed point.

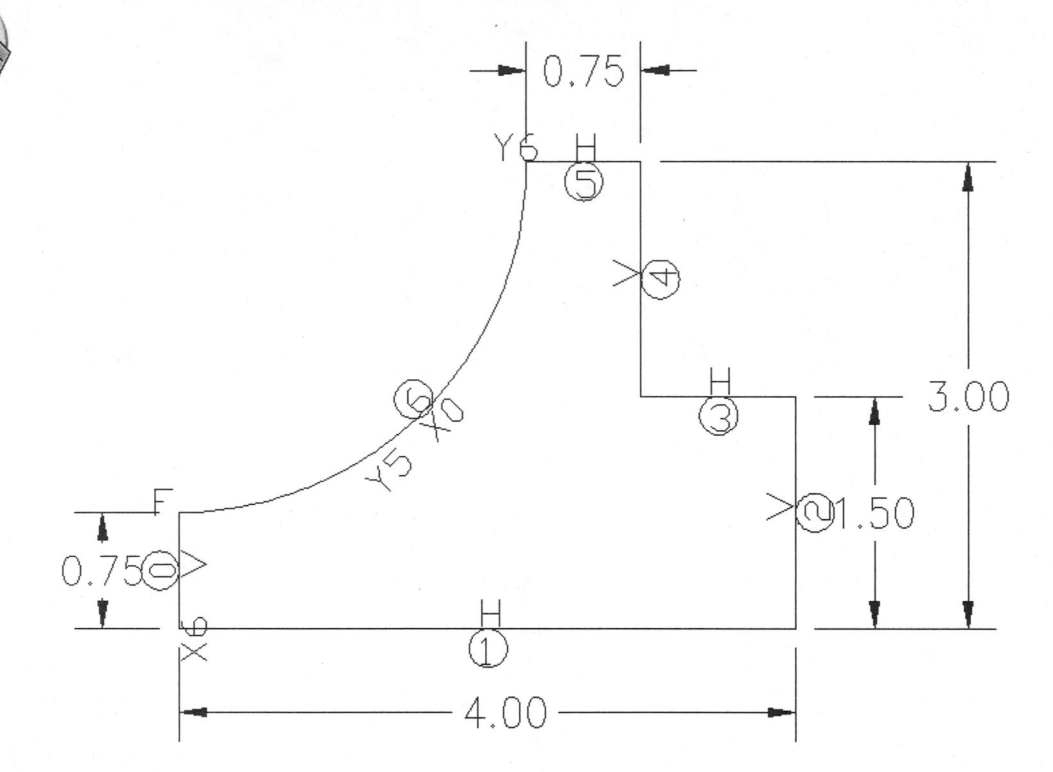

Figure 1.56

Exercise 1.4—Lamp (quarter profile of a lamp)

Save the finished drawing (Figure 1.57) as \Md4book\Chapter1\Lamp.dwg.

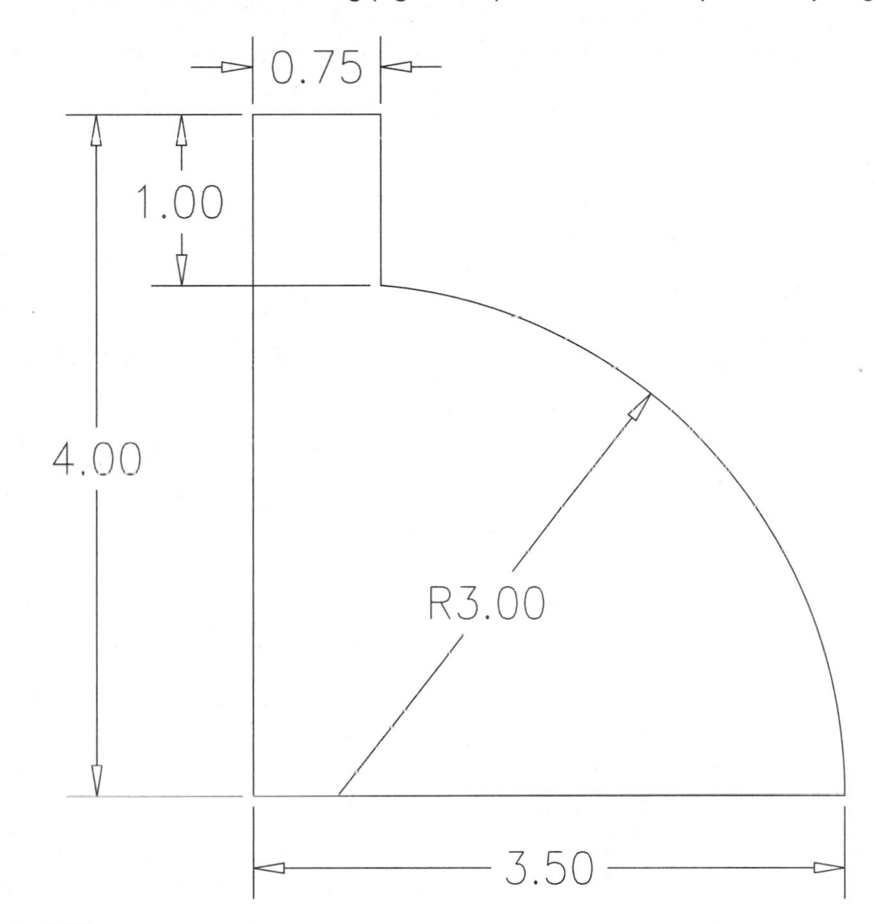

Figure 1.57

REVIEW QUESTIONS

1. List four items that can be set in the Desktop Options under the Desktop or Part tab.
2. When drawing a sketch there can be a gap in the sketch. How big can this gap be?
3. A profile does not need to be fully constrained. T or F?
4. If a sketch is drawn to exact size, does it need to be profiled?
5. There can be multiple fix constraints on a sketch. T or F?
6. Can a constraint be removed? If so, how?
7. When an angled dimension is required, a horizontal dimension appears. Can this be changed to an angle dimension? If so, how?
8. When an arc is dimensioned, is the default a radius or a diameter dimension?
9. To how many decimal places is a Mechanical Desktop part accurate?
10. After a profile is fully constrained, a dimension's value cannot be changed. T or F?

CHAPTER 2

Viewing, Extruding, Revolving and Editing Parts

After you have drawn, profiled, constrained, and dimensioned the sketch, your next step is to turn that sketch into a 3-D part. This chapter takes you through the options for viewing a part from different viewpoints, as well as the options for creating, extruding, revolving, and editing parts base features.

AFTER COMPLETING THIS CHAPTER, YOU WILL BE ABLE TO:

- View a part from different viewpoints.
- Understand what a feature is.
- Use the Mechanical Desktop Browser to edit and create parts.
- Extrude a profile into a part.
- Revolve a profile into a part.
- Edit features of a part.
- Edit the sketch of a base feature.
- Append objects to the sketch.

VIEWING A MODEL FROM DIFFERENT VIEWPOINTS

Up to now you have worked with 2-D sketches in the *XY* plane (plan view), as you have done in 2-D AutoCAD. The next step is to give the 2-D sketch depth in the *Z* plane. To see 3-D geometry, you must be able to view the model from different viewpoints. The next section will guide you through the most common methods of viewing objects from different viewpoints using Mechanical Desktop.

OPTIONS FOR VIEWING THE MODEL

AutoCAD has many options for viewing the model from different perspectives. One option is to use predefined model views. With Mechanical Desktop, you have two other options for viewing the model: number keys and dynamic rotation.

Model Views

To use predefined views, select 3D Views from the View pull-down menu and select the desired option from the cascading menu.

Number Keys

When you choose the number keys option for viewing the part, you can use the number keys on the keyboard to change the screen layout and viewpoint.

The **1**, **2**, **3**, and **4** keys split the screen into corresponding numbers of viewports, each with different viewpoints:

Key:	Viewport(s):
1	One (taken from the last active viewport)
2	Two (top and isometric views)
3	Three (top, front, and isometric views)
4	Four (top, front, side, and isometric views)

The following keys provide the following views:

Key:	View:
5	Top
6	Front
7	Side
8	Southeast isometric
88	Southwest isometric
9	Plan view of the current sketch plane

To use number keys, choose the desired number at the command prompt and press ENTER.

Dynamic Rotation

The second option for viewing the model in Mechanical Desktop is to rotate the model dynamically. To rotate a model dynamically first select the part or parts to rotate and select one of the rotation commands described below. Then select a point with the left mouse button. Hold the button down while moving the mouse. The model rotates in the mouse's direction. When you release the mouse button, the model stops rotating. To return to the original position press ENTER and select Reset View from the pop-up menu. To accept the current position, press ESC or ENTER and select Exit from the pop-up menu. The Mechanical View toolbar shown with the rotation option flown out in Figure 2.1 has six icons and five of them have fly-out menus. Below each icon and its fly-out menu is described. Another method for accessing the rotation commands and options is to issue the 3D Orbit command and then right click and select from the options from the pop-up menu.

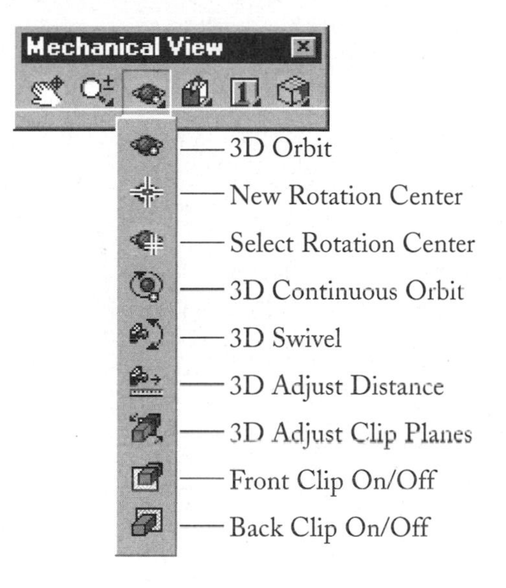

Figure 2.1

The first icon allows you to pan the part. The second icon and its fly-out icons contain zooming commands. The third icon on the Mechanical View toolbar allow you to:

(1st icon 3D Orbit) Rotate the model(s) about its geo-center.

(2nd icon New Rotation Center) This icon will center the geometry on screen before orbit. (The orbit command works off the center of the screen, not the part(s). You will only notice a change if the geometry is off center.

(3rd icon Select Rotation Center) Rotate the model around a selected point.

(4th icon 3D Continuous Orbit) Rotate the model continuously in the direction and speed the mouse moved before letting go of the left mouse button. To stop the rotation select another point with the mouse button.

(5th icon 3D Swivel) Gives the appearance that a camera is revolving around the part.

(6th icon 3D Adjust Distance) Adjusts the distance that the part is to the camera.

(7th–9th icon) Allows you to adjust and turn on or off clipping planes. Clipping planes can be used to dynamically see inside a shaded part or parts.

TUTORIAL

The fourth icon has preset views.

The fifth icon allows you to save your screen configuration and replay it. After getting the screen in the orientation that you want to save, select from the fly-out the red number 1, 2 or 3. To change the screen back to this configuration, select the black corresponding number from the fly-out. The last four icons on the fly-out will split the screen into 1, 2, 3 or 4 viewports respectively.

The sixth icon and its fly-out icons on the Mechanical View toolbar let you view the model in shaded, transparent, wireframe, hidden mode or shaded with edges visible. Selecting the top icon will toggle the part(s) from wireframe to shaded appearance.

TUTORIAL 2.1—VIEWING A MODEL

1. Open the file \Md4book\Chapter2\TU2-1.dwg.
2. Use the number keys to change view configurations.
3. When you feel comfortable changing views, return to one viewport by pressing the **1** key, then pressing ENTER.
4. To return to plan view, press the 9 key, and then press ENTER.
5. Dynamically rotate the parts by using the 3D Orbit icon from the Mechanical View toolbar.
6. Dynamically rotate the parts by using the New Rotation Center icon from the Mechanical View toolbar.
7. Dynamically rotate the parts by using the Select Rotation Center icon from the Mechanical View toolbar.
8. Alternate between shaded, transparent, wireframe, and hidden modes, and rotate the model dynamically in each. To change between shaded modes while in the 3D orbit command, right click and from the pop up menu select Shaded Modes and select the option you want.
9. Press ESC to exit the current command.
10. Select the magenta sphere, issue the 3D Orbit command and then rotate the sphere.
11. While still rotating the sphere right click and from the More pop up menu select Continuous Orbit. Hold down the left mouse button and move the mouse in any direction and then depress the mouse button while the mouse is still moving.
12. Practice putting the part into continuous orbit by moving the mouse faster, slower and with different mouse movements.
13. When you feel comfortable rotating the part save the file.

Tips: You can work with and edit parts in shaded, transparent, wireframe, or hidden mode.

Use your IntelliMouse wheel to zoom and pan wireframe or shaded mode.

To rotate a select number of parts, select them while not in a command then issue a rotation command.

To change between the different shaded modes while in the 3D orbit command right click and from the pop up menu select Shaded Modes and then select the option you want.

To quickly change from a wireframe to shaded mode or from shaded to a wireframe mode select the Toggle Shading / Wireframe icon on the Mechanical View toolbar.

UNDERSTANDING WHAT A FEATURE IS

In the first chapter, you created a sketch and profiled it. The next step is to turn the profile into a 3D part. The first profile of a part that is turned into a solid is referred to as a base feature or base part. Base features will be introduced in the next section. There are also sketched features, where you draw a sketch on a planar face or work plane and either add or subtract material from an existing part. Extruding, revolving, sweeping or lofting can create these sketched features on a part. You can also create placed features such as fillets, chamfers, and holes. Placed features require a base part to exist; they are added to the part. Sketched and placed features will be covered in Chapter 3. Work planes, work axis and work points are features as well and will be covered in Chapter 5. Features are the building blocks that create a part. For example, a plate with a hole in it would have a base feature representing the plate and a hole feature representing the hole. As the features are added to the part, they will appear in the browser, showing the history of the part (the order in which the features were created). Features can be edited, deleted, or reordered from the part as required. The next few chapters will go through many methods of adding and deleting features.

USING THE BROWSER FOR CREATING AND EDITING

By default the Mechanical Desktop browser is docked along the left side of the screen and displays the history of the file. In the browser you can help create, edit, rename, copy, delete and reorder features and parts. You can expand or collapse the history of the part(s) (the order in which the features and parts were created) by selecting the + and - on the left side of the part name in the Desktop browser. As parts grow in complexity, so will the information found in the browser. Dependent features will be indented to show that they are related to the top level. This is referred to as a parent-child relationship. The child cannot exist without the parent and is dependent on the parent. For example, if a hole is created in an extruded rectangle and the extrusion is then deleted, the hole will also be deleted. Each feature is given a default name. For example, a blind extrusion would be named ExtrusionBlind1, and the number will sequence as you add like features. The browser can also help you locate parts and features in the drawing area. Select the feature or part name in the browser with your left mouse button and it will be highlighted in the drawing area. The browser itself acts like a toolbar, except that it can be resized while docked. If the browser is not visible on the screen, you can display it by issuing the Desktop Browser command **(AMBROWSER)** or by selecting Desktop Browser from the View pull-down menu under the Display menu. To close the browser select the "**X**" in the upper right corner of the browser or by selecting Desktop Browser from the View pull-down menu under the Display menu which will toggle the check mark. There are three possible tabs along the top of the browser; Assembly, Scene and Drawing. By default when you issue the AutoCAD NEW command to create a new drawing, you will be started in the assembly tab. In this mode you can create one or more parts. The histories of the part or parts are placed in the Assembly tab. When a part that is created in the assembly tab is attached into another file it will come in as a sub-assembly. If you want to have only a single part in a file, you can start a new file with the New Part File option from the File pull-down menu. Then the assembly tab will be replaced with a part tab and the Scene tab will not be present. In this mode you can only have a single part and when this file is attached to another file it will be recognized as a part.

The specific functionality of the browser will be covered throughout the book in the sections were it pertains. However a basic rule is to either right-click or double click on the feature's name to edit or perform a function on the feature.

Note: When you start a new file with the NEW command, you can have one or more parts in the same file.

When you start a new file with the New Part File option from the File pull-down menu, you can only have a single part in that file.

To find a feature in a part, select the feature's name in the browser and it will be highlighted in the drawing.

USING THE XY ORIENTATION IN 3-D

Now that you have sketched, profiled, and constrained the model, and viewed and rotated it from different perspectives, you are ready to create a 3-D part. When you work in 3-D, it is important that the **UCSICON** (icon with horizontal and vertical arrows in the lower left corner of the screen) be turned on. Many 2-D AutoCAD users turn this icon off because the *XY* orientation does not change. When working in 3-D, you will continually be changing the location of the *XY* plane and will use the **UCSICON** as a visual aid. If this icon is off, use the **On** option of the **UCSICON** command.

The **W** just below the **Y**, which refers to the World Coordinate System, is the default setting. If the W is not below the Y, you are in a user-defined coordinate system (UCS), which means the UCS has been moved. The UCS and how it relates to Mechanical Desktop are covered in more detail in Chapter 3.

EXTRUDING THE PROFILE

The most common method for creating a part is to extrude the profile, which gives the profile depth along the *Z* axis. Before extruding, it is helpful to be in an isometric view, because Mechanical Desktop uses an arrow (negative or positive) to show you the direction of the extrusion. If you are viewing the profile from directly above, and the arrow is coming toward you, you will see a circle with a dot in its center; if it is going away from you, you will see a circle with a cross in it.

To extrude a profile, select the Mechanical Desktop Extrude command **(AMEXTRUDE)** on the Part Modeling toolbar (see Figure 2.2) or right-click in the drawing area and from the pop-up menu, select Extrude from the Sketched & Work Features menu. After you issue the command, the Extrusion dialog box appears (see Figure 2.3).

The Extrusion dialog box has five sections: Operation, Termination, Distance, Draft angle, and Flip. When changes are made in the dialog box the graphics on the right side of the dialog box will change showing an example of what the result would look like. You could also click on the graphics and you will cycle through all the options.

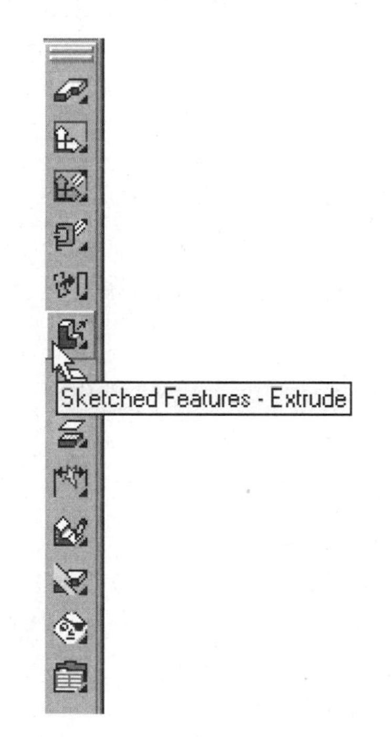

Figure 2.2

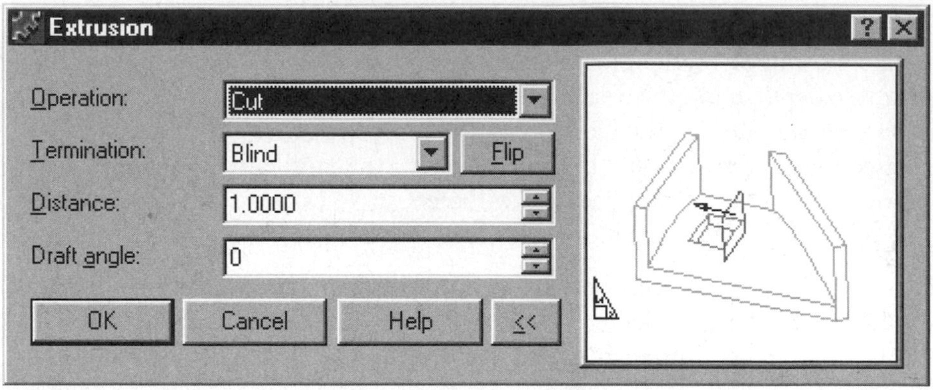

Figure 2.3

Operation

If this is the first profile, it is referred to as a base feature and the operation section will be grayed out. The operation will default to Base. If this is a profile on an existing part you will specify if this profile will add or remove material to or from an existing part. Once the base feature has been established you will have four options to choose from: Cut, Join, Intersect, and Split. These operations will be covered in Chapter 3.

Option:	Function:
Base	Serves as the default setting for the first extrusion. Creates a base part to or from which other operations will add or remove material.

Termination

The Termination section of the Extrusion dialog box, determines the extent of the extrusion. There are six options to choose from. Like the Operation section, this section has options that are grayed out until a base part exists. These grayed-out items are covered in Chapter 3.

Option	Function
Blind	Extrudes a specific distance in the positive or negative Z direction.
Through	Go all the way through the part in one direction.
Mid Plane	Goes equal distances in the negative and positive directions. For example, if the extrusion distance is 2", the extrusion goes 1" in both the negative and positive Z directions.
Mid Plane Through	Goes through the part in both the negative and positive directions.
To Face/Plane	Continues until the profile reaches a specific face that is contoured or plane that is flat.
From To	Starts at one plane/face and stops at another.

Distance

In the Distance section of the Extrusion dialog box, you specify the extrusion distance for only the blind and mid plane termination types.

Draft Angle

Type the draft angle. A 0 draft angle extrudes the profile straight outward, a negative number tapers the profile inward and a positive number expands the profile outward (in the manufacturing industry this is known as a reverse or negative draft). A draft angle can be applied to all termination types.

Flip

After executing the Mechanical Desktop Extrude command, an arrow will appear on the part showing the direction of the extrusion. Selecting the Flip button will change this direction. The flip option applies only to the base, blind or through extrusions. The length of the arrow will change to represent the value of the distance area in the dialog box.

Note: When you convert a profile to a part, all dimensions disappear. For more information on editing parts or features, see "Editing 3-D Parts" later in this chapter.

To extend the draft angle out from the part, give the draft angle a positive number. This is also known as reverse draft.

In the next two tutorials you will use the base operation and the blind and mid plane termination types. Tutorials covering the other operation and termination types will be covered in Chapter 3 after learning how to make different faces the active sketch plane.

TUTORIAL 2.2—EXTRUDING THE PROFILE

1. Start a new drawing from scratch.
2. Draw a circle.
3. Profile the circle with the **AMPROFILE** command.
4. Add a 2" diameter dimension to the circle.
5. Change to an isometric view by pressing **8**. When complete, your screen should resemble Figure 2.4.

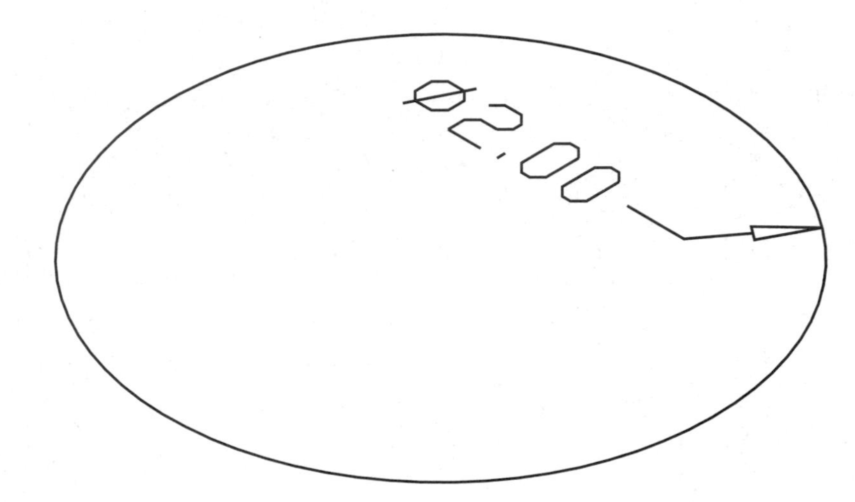

Figure 2.4

6. Issue the Extrude command **(AMEXTRUDE)**. The Extrusion dialog box appears.
7. Input the following data:
   Operation = Base (default setting and can not be changed)
   Termination = Blind (default setting)
   Distance = 4
   Draft Angle = -5
8. Select the Flip button to change the arrow's direction.
9. Again select the Flip button to change the arrow's direction.
10. Change the Distance to 2. Watch the length of the arrow change.

11. Select OK to execute the extrusion. When complete, your screen should resemble Figure 2-5.

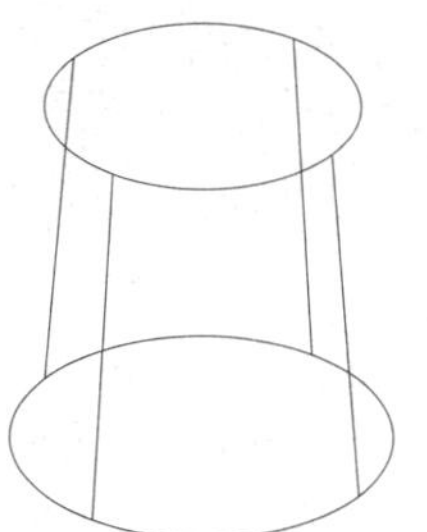

Figure 2.5

12. Save your file as \Md4book\Chapter2\TU2-2.

TUTORIAL 2.3—EXTRUDING THE PROFILE MID-PLANE

1. Open the file \Md4book\Chapter2\TU2-3.dwg.
2. Issue the Extrude command **(AMEXTRUDE)**. The Extrusion dialog box appears. You do not need to profile or add dimensions, because those steps have already been taken.
3. Input the following data:
   Operation = Base (default setting)
   Termination = Mid Plane
   Distance = 2
   Draft Angle = -5
4. Select OK. Your drawing should resemble Figure 2.6. As you can see, the profile extrudes equally in the positive and negative directions. As a result, you are not prompted to change the extrusion direction.

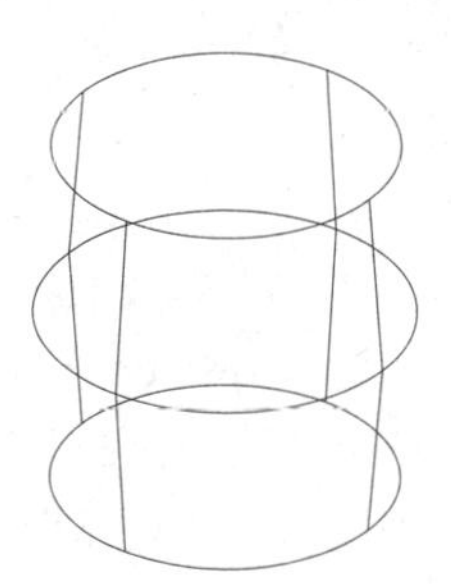

Figure 2.6

5. Save the file.

TUTORIAL

REVOLVING THE PROFILE

The second method for creating a part is to revolve a profile around a straight edge or axis. Revolve can be used to create cylindrical parts or features. To revolve a profile you will follow the same steps you did to extrude the profile (sketch, profile, constrain, and dimension). Then select the Revolve command **(AMREVOLVE)** from the Part Modeling toolbar (see Figure 2.7) or right-click in the drawing area, and from the pop-up menu select, Revolve from the Sketched & Work Features menu. After issuing the command the Revolution dialog box appears (see Figure 2.8). The Revolution dialog box has four sections: Operation, Termination, Angle, and Flip.

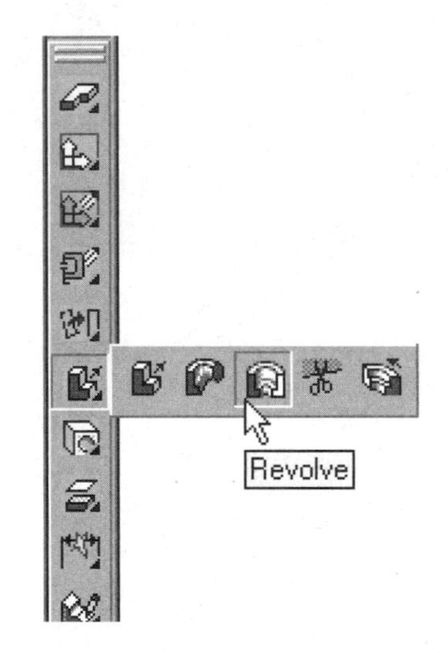

Figure 2.7

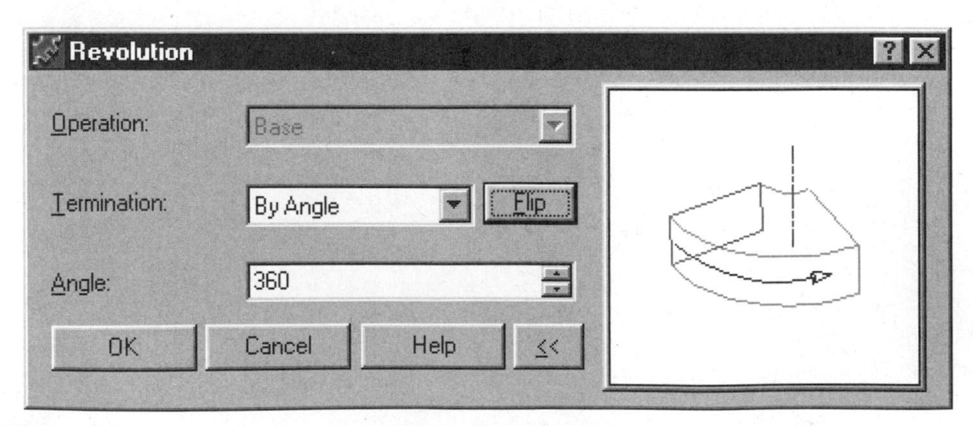

Figure 2.8

Operation

If this is the first profile in the part it is referred to a base feature and the operation section will be grayed out and operation will say Base. If this is a profile on an existing part you will specify if this profile will add or remove material to or from an existing part. Once the base feature has been established you will have four options to choose from: Cut, Join, Intersect, and Split, these operations will be covered in Chapter 3.

Option:	Function:
Base	Serves as the default setting for the first revolution. Creates a base part to or from which other operations will add or remove material.

Termination

The Termination section of the Revolution dialog box determines the extent of the revolution. There are four termination options. Like the Operation section, this section has options that are grayed out until a base part exists. These grayed-out items will be covered in Chapter 3.

Option	Function
By Angle	Revolve a specific angle in the positive or negative direction. An arrow directs you to choose the direction in which to revolve.
Mid Plane	Goes equal distances in the negative and positive directions. For example, if the revolution angle is 180°, the revolution will go 90° in both directions.
To Face/Plane	Continues until the profile reaches a specific face that is contoured or plane that is flat.
From To	Starts at one plane/face and stops at another plane/face.

Flip

After executing the Mechanical Desktop Revolve command an arrow will appear on the part showing the direction of the revolve. Selecting the Flip button will change this direction. The flip option applies only to By Angle and To Face/Plane termination type for revolve. The length of the arrow will change to represent the value of the angle area in the dialog box.

Angle

In this section, you denote the number of degrees the revolve will travel. The default is 360°.

Note: There is not an option for draft angle with revolve.

After inputting all data in the Revolution dialog box, select **OK**. Mechanical Desktop prompts you to select a revolution axis around which to revolve the profile (Chapter 6 discusses constructions geometry, which result from revolving a profile around an axis that will not become part of the part.) Select the edge and press ENTER to revolve the profile and exit the command.

TUTORIAL

TUTORIAL 2.4—REVOLVE BY ANGLE

1. Start a new drawing from scratch.
2. Draw a triangle, profile it and dimension it as shown in Figure 2.9.

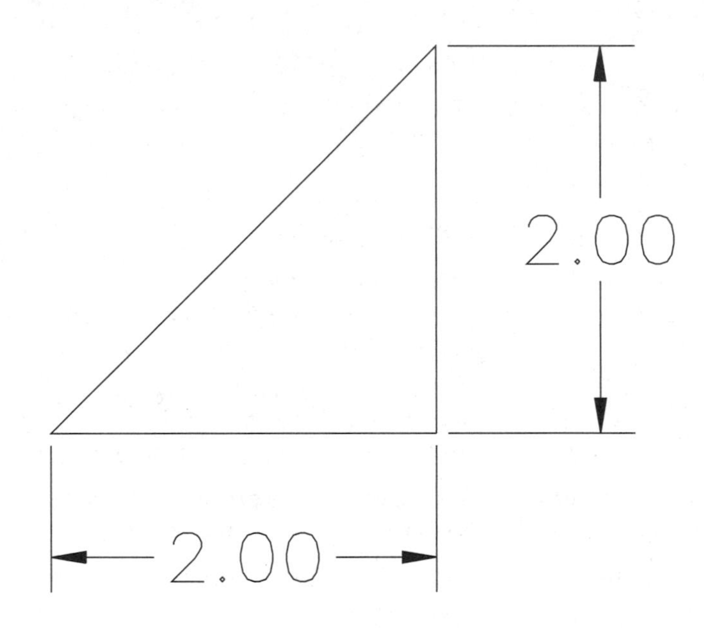

Figure 2.9

3. Change to an isometric view using **8**.
4. Issue the Revolve command **(AMREVOLVE)**.
5. Select the vertical line as the revolution axis.

   The Revolution dialog box appears; input the following data.
   Operation = Base (default setting)
   Termination = By Angle (default setting)
   Angle = 90
6. Select the Flip button to change the direction of the arrow.

7. Select the Flip button to change the direction of the arrow back.
8. Change the Angle to 180 and watch the length of the arrow change.
9. Select OK. When complete your screen should resemble figure 2.10

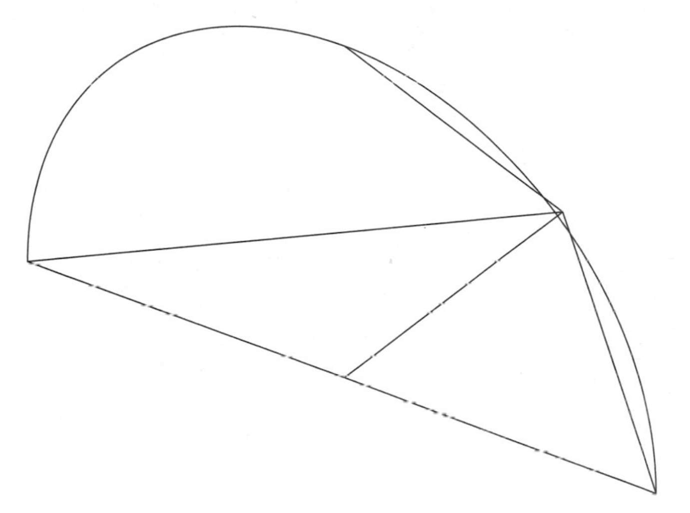

Figure 2.10

10. Save your drawing as \Md4book\Chapter2\TU2-4.

TUTORIAL 2.5—REVOLVING THE PROFILE 360°

1. Open the file \Md4book\Chapter2\TU2-5.
2. Issue the Revolve command **(AMREVOLVE)**. You do not need to profile or add dimensions because these steps have already been taken.
3. Select the vertical line as the revolution axis.

   The Revolution dialog box appears.

   Input the following data:
   Operation = Base (default setting)
   Termination = By Angle (default setting)
   Angle = 360
4. Select OK. When complete, your screen should resemble Figure 2.11.
5. Save the file.

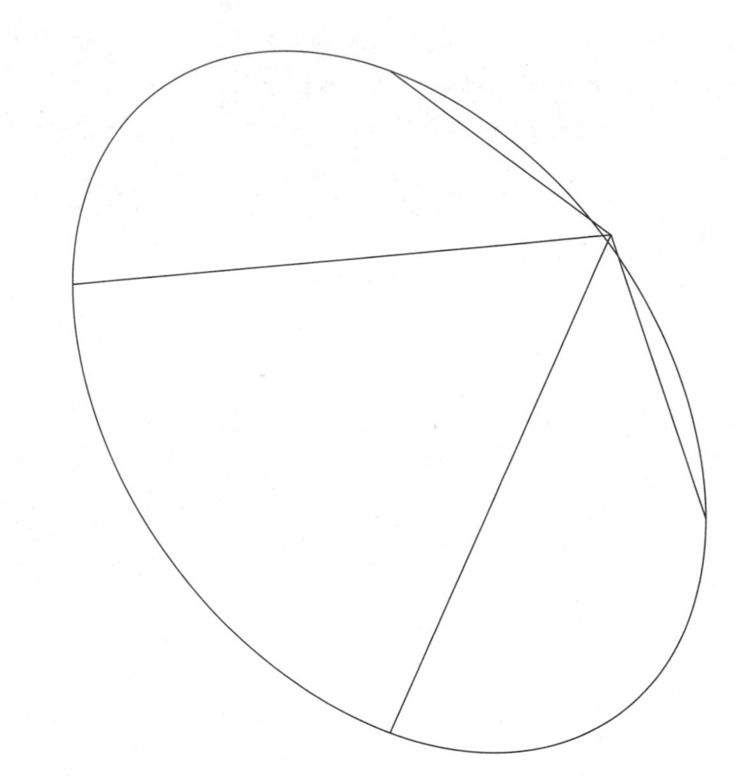

Figure 2.11

TUTORIAL

TUTORIAL 2.6—REVOLVING THE PROFILE MID PLANE

1. Open the file \Md4book\Chapter2\TU2-6.
2. Issue the Revolve command **(AMREVOLVE)**. You do not need to profile or add dimensions because these steps have already been taken.
3. Select the vertical line as the revolution line.

   The Revolution dialog box appears.

   Input the following data:
   Operation = Base (default setting)
   Termination = Mid Plane
   Angle 90
4. Select OK. When complete, your screen should resemble Figure 2.12, which shows the part with lines hidden.

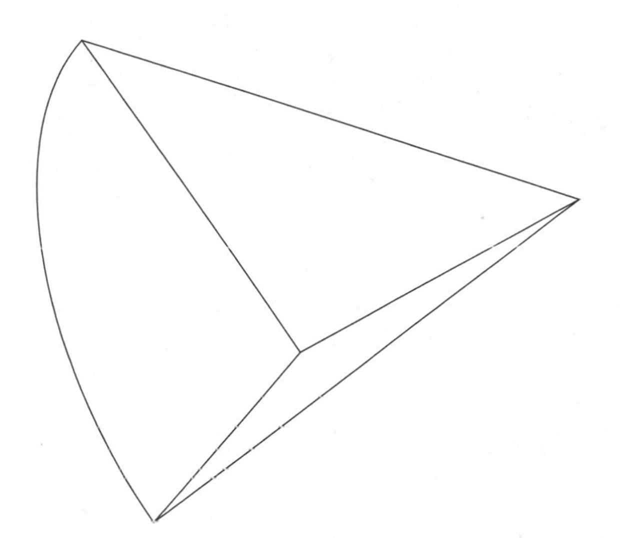

Figure 2.12

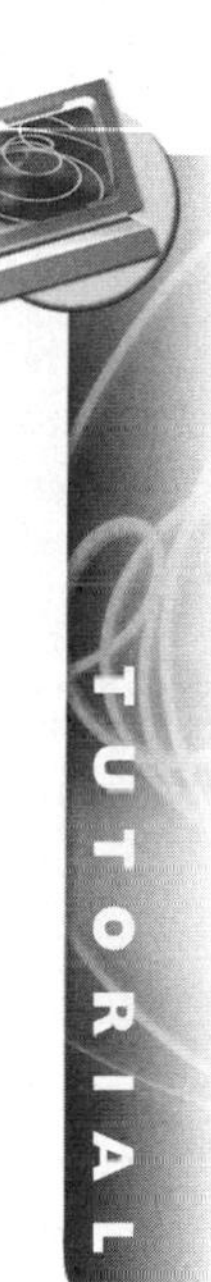

5. Save the file.

EDITING FEATURES IN PARTS (FEATURE EDITING)

After you generate a part, all the dimensions that were in the profile disappear. This part is called a base *feature*. If you need to edit the distance, angle, draft angle, operation or termination type of a base feature you will use the **AMEDITFEAT** command. The are four methods accessing the command in addition to typing it.

1. In the browser, double-click on the feature's name.
2. In the browser right-click on the feature's name and choose Edit from the pop-up menu that appears.
3. From the Part Modeling toolbar select the Edit Feature command as shown in Figure 2.13 and select the part.
4. Right-click in the drawing area and from the pop-up menu, select Edit from the Edit Features menu.

If you double clicked on the feature's name in the browser the dialog box that was used to create the feature, that feature will appear as well as the dimensions that were on the profile. Otherwise, select the feature to edit and the dialog box that was used to create the feature will appear as well as the dimensions that were on the profile. In the dialog box change the data that you want. Everything can be changed except the base operation. When complete, select the OK button. Then, to edit a dimension that was placed on a profile, select the dimension to change and type a new value. To edit another dimension, select it or press ENTER to exit the command. The dimensions that were

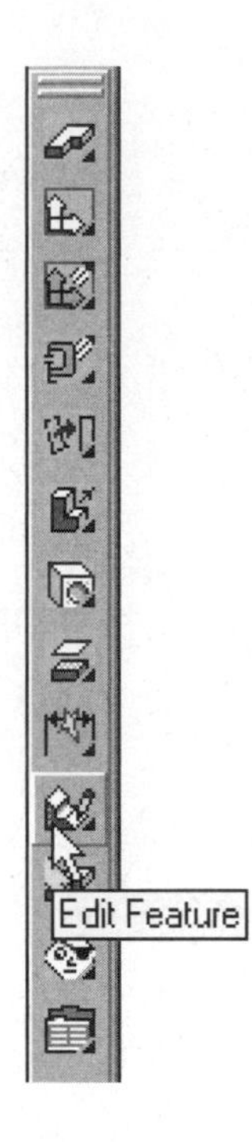

Figure 2.13

placed on the profile will be the color of the current layer or dimension color that was set using dimension styles. The value of the dimensions that were placed in the dialog box will have the color that the AutoCAD variable GRIPCOLOR is set to. If you double clicked on the features name in the browser the part will automatically be updated to reflect the changes. Otherwise the part is in an edited state and needs to be updated using the update command **AMUPDATE**. After issuing the update command you can enter to update the active part or type L and enter to update all the parts in the file. While a feature is in an edited state it will have a yellow background in the browser. The **AMUPDATE** icon resembles a lightning bolt and can be found on the Part Modeling toolbar or on the bottom of the browser as shown in Figure 2.14. Always update your part before adding or removing other features as this will help reduce the chance of part corruption. If you try to save your part in an edited state you will be warned that the part needs updating and the SAVE command will be cancelled.

There are four other options for the **AMEDITFEAT** command: Independent array instance, Sketch, Surfcut and Toolbody. Independent array instance will be covered in Chapter 3, Sketch will be covered in the next section, Surfcut will not be covered in this book and Toolbody will be covered in Chapter 7.

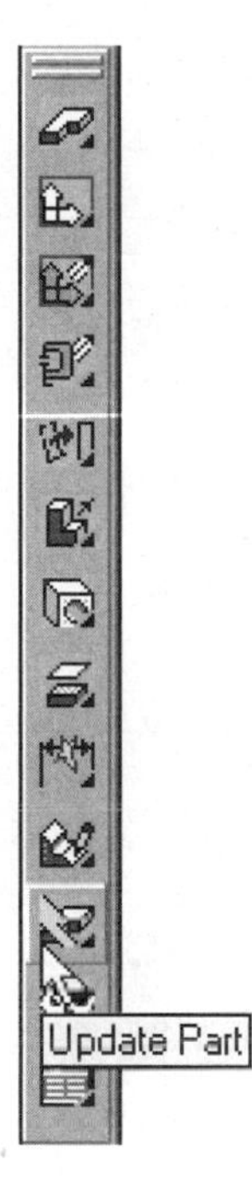

Figure 2.14

TUTORIAL

Editing Features in Different Shaded Modes

Parts can be edited in shaded, wireframe or hidden modes. While in shaded mode the dimension(s) will appear on top or through the part depending upon their placement. If you are in hidden mode you will temporarily be placed in wireframe mode and when the editing is complete you will be returned to hidden mode.

Notes: The Update Part command **(AMUPDATE)** can also be issued from the bottom of the browser.

While a feature is in an edited state it will have a yellow background in the browser.

If you double click on a part's name in the browser you do not select the part to edit and the part will automatically update after editing.

TUTORIAL 2.7—EDITING BASE FEATURES

1. Start a new drawing from scratch.
2. Draw a rectangle.
3. Profile the rectangle.
4. Add dimensions to the rectangle as shown in Figure 2.15.

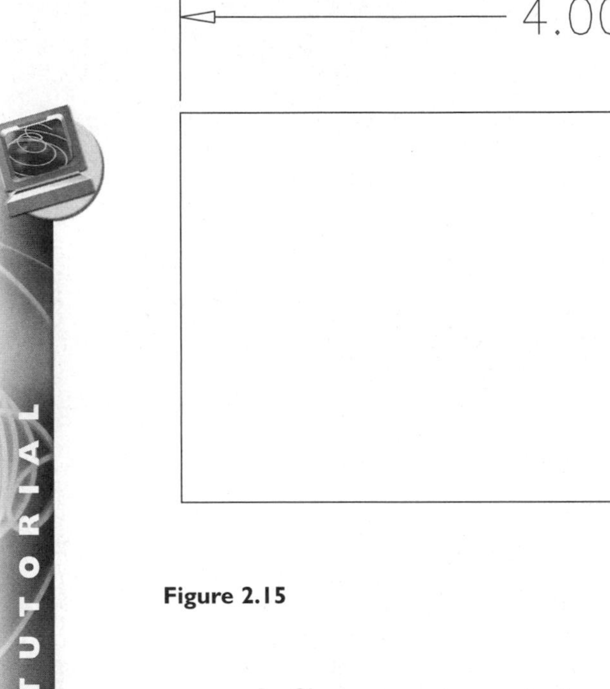

2.00

Figure 2.15

TUTORIAL

5. Change to an isometric view using **8**.
6. Issue the Extrude command **(AMEXTRUDE)**.
7. Extrude the rectangle **1** unit with **0** degrees of draft and in the default direction and select OK.
8. Issue the Edit Feature command **(AMEDITFEAT)** from the Part Modeling toolbar or right-click in the drawing area and from the pop-up menu, select Edit from the Edit Features menu.
9. Select the part.
10. In the Extrusion dialog box select the Flip button to change the extrusion direction (you may need to move the dialog box to see the arrow's new direction), change the distance to 1.5 and change the Draft angle to –5 and select the OK button to close the dialog box.
11. Select the **4** inch length dimension, type "3", and press ENTER.
12. Select the **2** inch width dimension, type "1.25", and press ENTER.
13. At the Select Object prompt, press ENTER to exit the command.
14. Issue the Update Part command **(AMUPDATE)** and ENTER. When complete, your screen should resemble Figure 2.16.

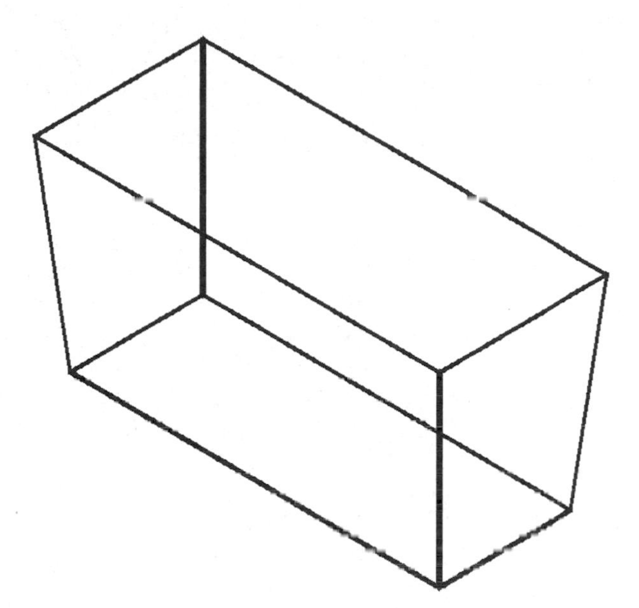

Figure 2.16

15. In the browser double click on the feature name "ExtrusionBlind1".
16. In the Extrusion dialog box select the Flip button to change the extrusion direction (you may need to move the dialog box to see the arrow's new direction) and select the OK button to close the dialog box.
17. Select the **3** inch length dimension, type "5", and press ENTER.
18. Select the **1.25** inch width dimension, type "2", and press ENTER.
19. Select the **1.5** inch extrusion dimension, type "2", and press ENTER.
20. Select the **-5** inch draft angle dimension, type "10", and press ENTER.
21. Press ENTER to complete the command and the part will automatically update and your screen should resemble Figure 2.17.
22. Save your file as \Md4book\Chapter2\TU2-7.dwg.

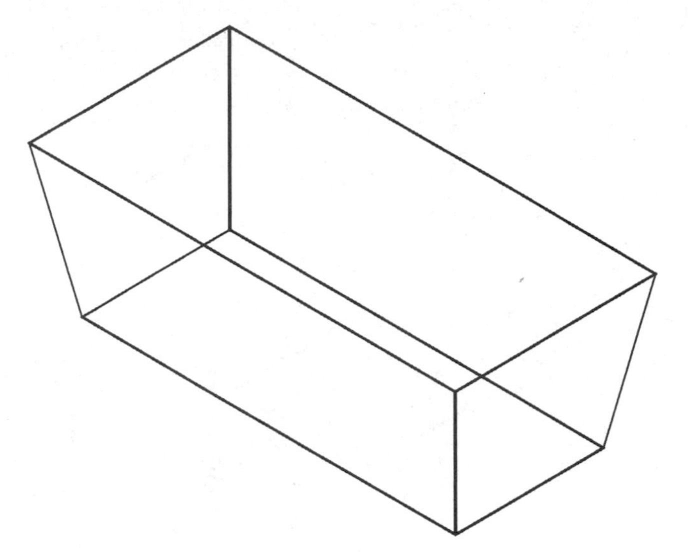

Figure 2.17

EDIT A FEATURE'S SKETCH

While working on a part you may need to add or delete a dimension or constraint to the original 2D sketch. However, while the part is in 3D you can only edit dimension values. To edit the original 2-D sketch of a selected feature there are three ways to edit a sketch in addition to typing in the command **AMEDITFEAT**.

1. Issue the Edit Feature command **(AMEDITFEAT)** from the Part Modeling toolbar, then press S and ENTER, and then select the part to edit.
2. In the browser right click on the features name that you want to edit and select Edit Sketch from the pop-up menu.
3. In the browser, expand the feature tree that you want to edit and either double click on the profile name or browser right click on the profile name and select Edit Sketch from the pop-up menu.

After issuing the **AMEDITFEAT** with the sketch option, the part disappears and the 2-D profile appears with the dimensions it had before it became a part. At this point you can add or delete constraints and dimensions, just as you would if it were a profile. After making the changes, issue the Update Part command **(AMUPDATE)**. The part regenerates, reflecting the changes to the sketch.

ADDING AND REMOVING OBJECTS FROM A FEATURE'S SKETCH

Also while in the sketch mode you can both add and remove objects from the sketch. The ability to append to a sketch also works on sketches that have not been extruded, revolved, swept or lofted. While in the sketch mode of the **AMEDITFEAT** command or while a sketch has yet to be turned into a solid you can add and delete objects (lines, polylines, arcs, circles and splines) to the sketch. Use any AutoCAD technique to modify the sketch (erase, trim, extend, fillet, etc.). The same rules apply to appending a sketch as to creating a new sketch. The gap and overlap needs to be smaller than the pickbox size. As you erase objects from the sketch and if there are dimensions associated to it and the dimensions are no longer valid for the sketch they too will be erased. The entire sketch can also be erased and replaced with an entirely new sketch. When replacing an entire sketch, other features that would be consumed by the new objects should be deleted first and recreated. Once the sketch has been modified correctly issue the Append to Sketch command **(AMRSOLVESK)** from the Part Modeling toolbar, as shown in Figure 2.5 or right-click in the drawing area and from the pop-up menu select Append Sketch or right-click on the profile name that is being edited in the browser and select Append Profile from the pop-up menu. Select the new objects and then add dimensions or constraints as needed. If the profile was extruded, revolved, swept or lofted you will need to update the part for the change to take effect.

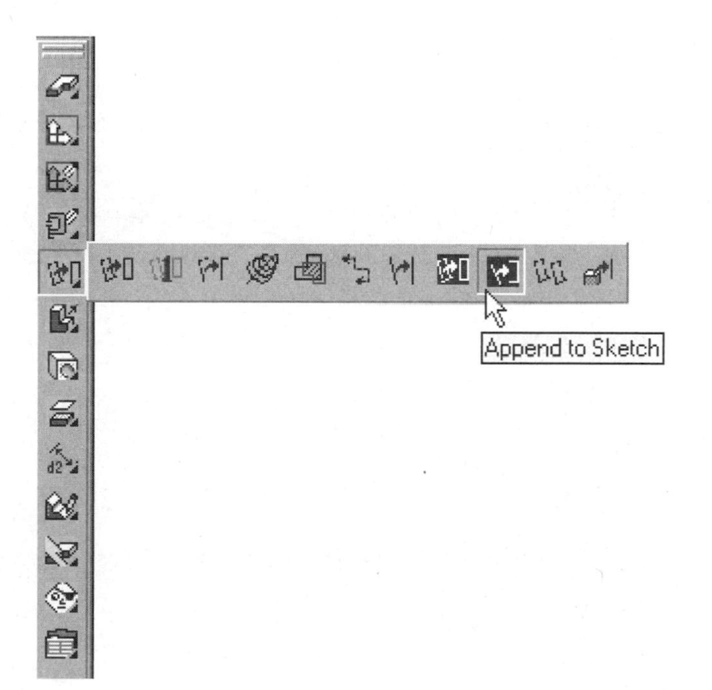

Figure 2.18

Notes: While editing a sketch, you can add and delete constraints and dimensions. The rules for working with 2-D profiles apply in this edit mode.

To remove a dimension from a profile, use the Erase command.

TUTORIAL 2.8—EDITING A FEATURE'S SKETCH AND ADDING AND REMOVING OBJECTS FROM A SKETCH OBJECT

1. Open the file \Md4book\Chapter2\TU2-8.dwg.
2. Issue the Edit Feature command **(AMEDITFEAT)** from the Part Modeling toolbar then type S and ENTER and select the extrusion.
3. Issue the Delete Constraints command **(AMDELCON)**. The constraints will appear on the sketch.
4. Delete both vertical constraints by selecting both **V**s.
5. Press ENTER to exit the command. The constraints will disappear.
6. Add two 70° dimensions as shown in Figure 2.19.

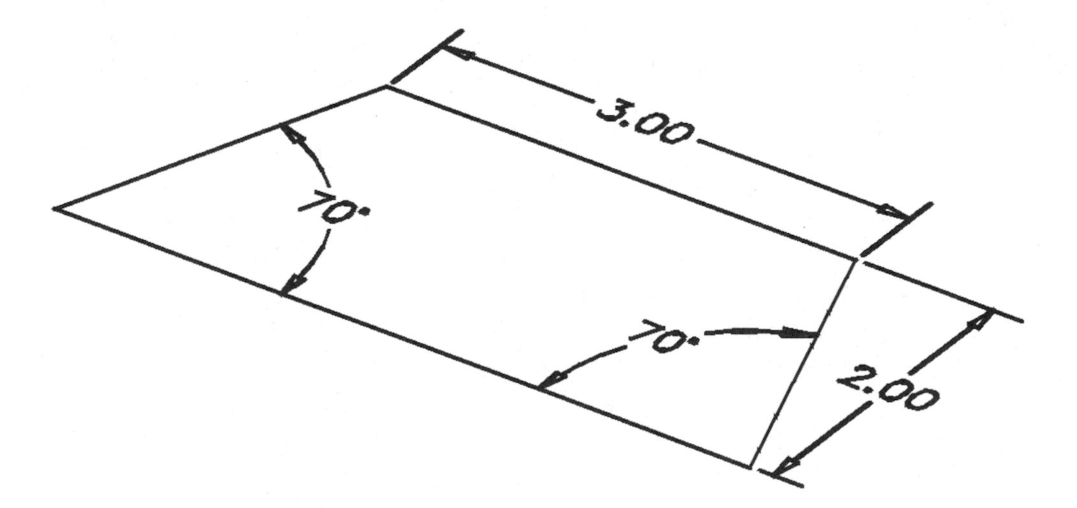

Figure 2.19

7. Update the part. When complete, your screen should resemble Figure 2.20.

TUTORIAL

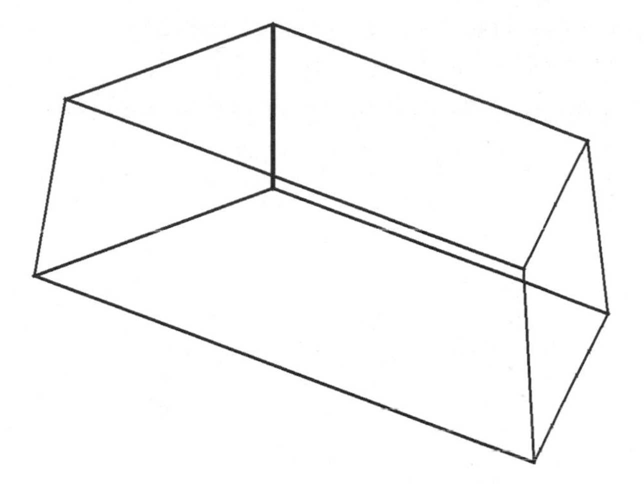

Figure 2.20

8. In the browser, expand the feature tree. Right click on the feature name "ExtrusionBlind1" or "Profile1" and select Edit Sketch from the pop-up menu.
9. Erase the angled line and 70° dimension on the left side of the sketch.
10. Create a fillet on the left side of the sketch. Hint: use a radius of "0" to fillet between two parallel lines and select near the left side of the top line first then the select the bottom line. When complete your screen should resemble Figure 2.21.

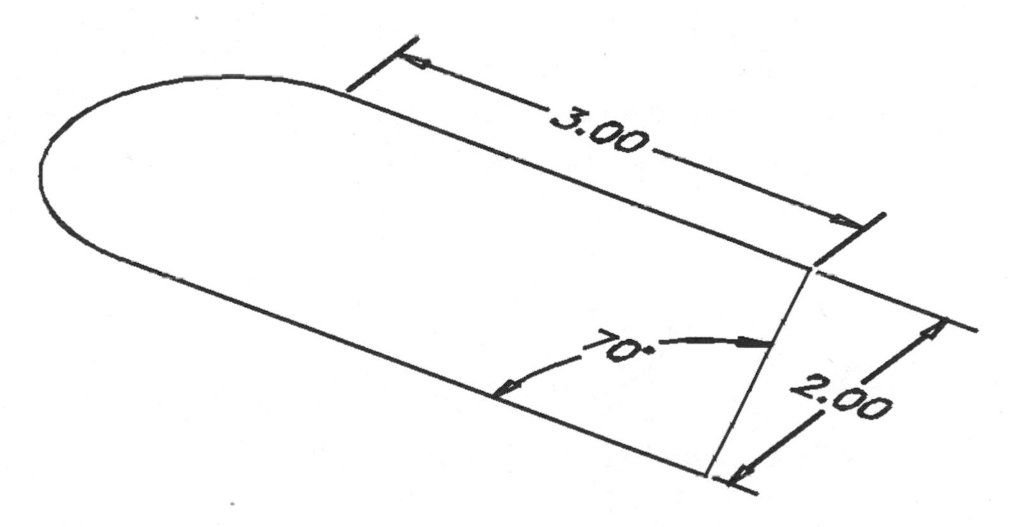

Figure 2.21

TUTORIAL

11. Issue the Append to Sketch command **(AMRSOLVESK)** and select the arc you just drew.
12. Update the part. When complete, your screen should resemble Figure 2.22.

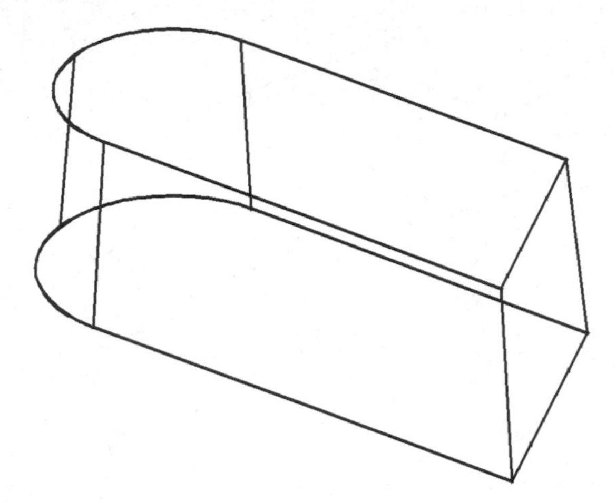

Figure 2.22

13. In the browser, double click on the name "Profile1".
14. Erase all the objects in the sketch including the dimensions.
15. Draw a circle where the sketch was.
16. Issue the Append to Sketch command **(AMRSOLVESK)** and select the circle you just drew.
17. Add a 3" diameter dimension to the circle.
18. Update the part. When complete, your screen should resemble Figure 2.23.

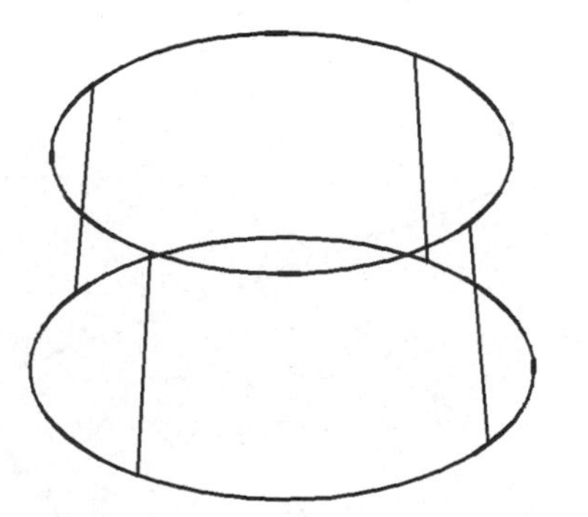

Figure 2.23

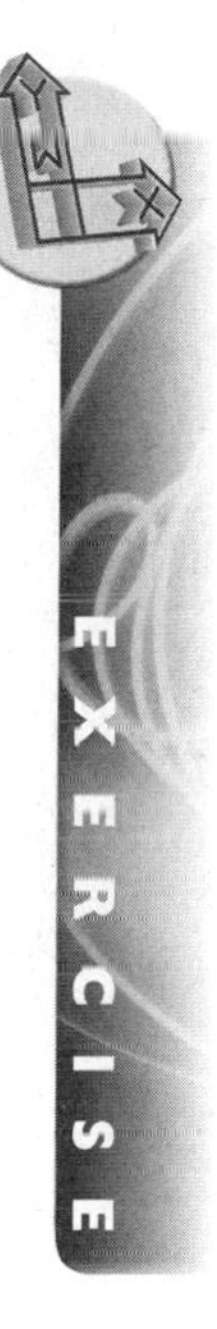

EXERCISES

The following section is intended to challenge you by providing open-ended problems. Like the problems you will encounter at work, each of the following problems has multiple solutions. To complete each exercise in this section, follow all steps, and save a file with the name of the drawing (e.g., \Md4book\Chapter2\Bracket.dwg). Your solutions should resemble the accompanying figures.

The extrusion direction for all exercises is the positive *Z* direction.

Exercise 2.1—Bracket

1. Open the file \Md4book\Chapter2\Bracket.dwg.
2. Extrude the bracket 3". Your drawing should resemble Figure 2.24.

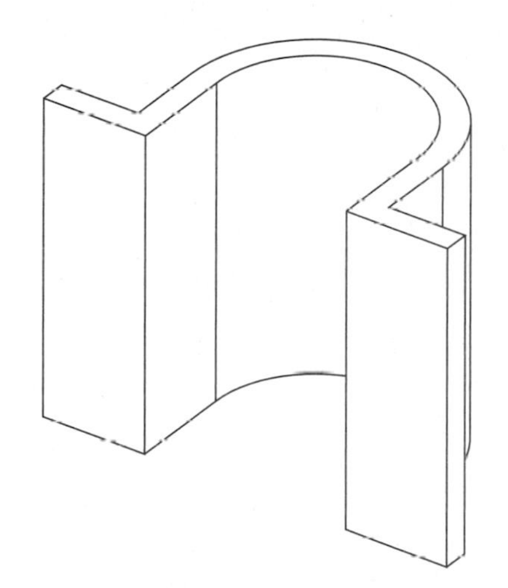

Figure 2.24

3. Save the file.

Exercise 2.2—Guide

1. Open the file \Md4book\Chapter2\Guide.dwg.
2. Extrude the profile 0.5".
3. Edit the feature so that the center-to-center distance is 4", as shown in Figure 2.38. (The 3" dimension is for reference only.) Remember—if you did not double click on the feature name in the browser to edit it, you will need to update the part after making the change.

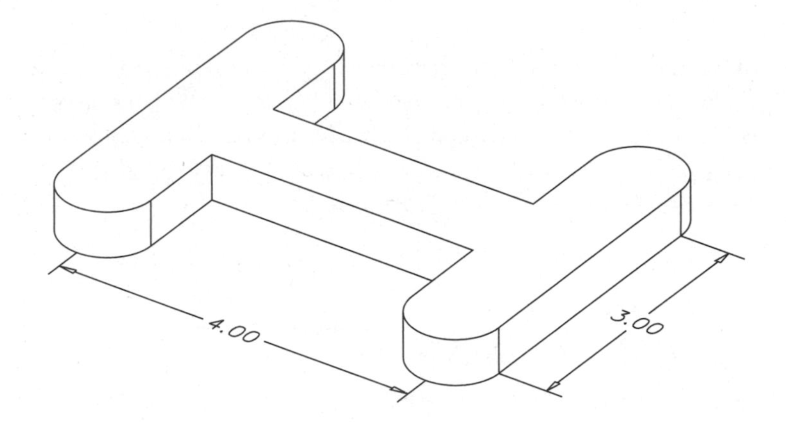

Figure 2.25

4. Save the file.

Exercise 2.3—Foot

1. Open the file \Md4book\Chapter2\Foot.dwg.
2. Extrude the profile 3".
3. Edit the feature's "sketch".
4. Delete the vertical constraint on the right line.
5. Add a dimension of 75° to the right vertical line, as shown in Figure 2.26 shown in plan view.
6. Delete the arc on the left side of the sketch.
7. Draw a line where the arc was.
8. Append the new line to the sketch.
9. Add a dimension of 30° to the upper left angled line as shown in Figure 2.26 shown in plan view.
10. Update the model. When complete, your screen should resemble Figure 2.27 shown in an isometric view with lines hidden.
11. Save the file.

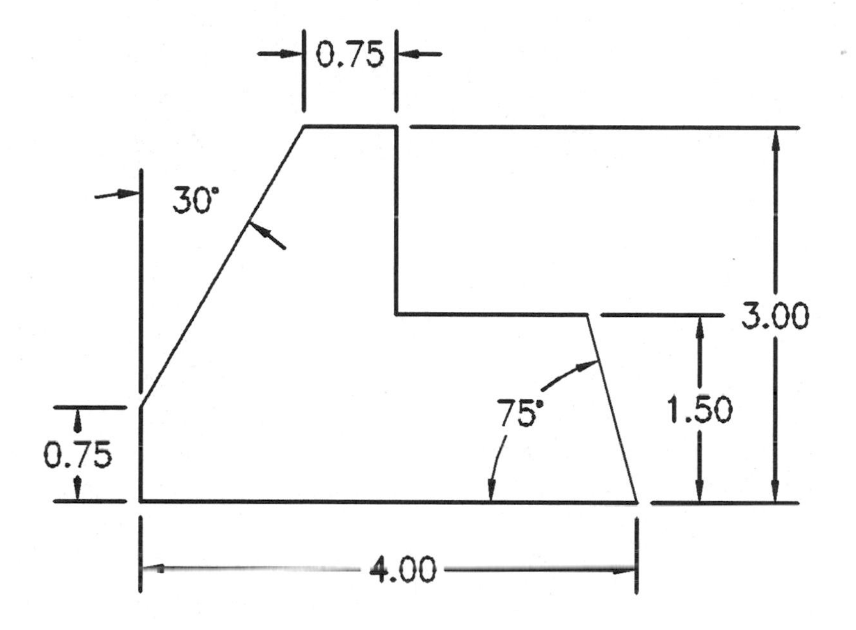

Figure 2.26

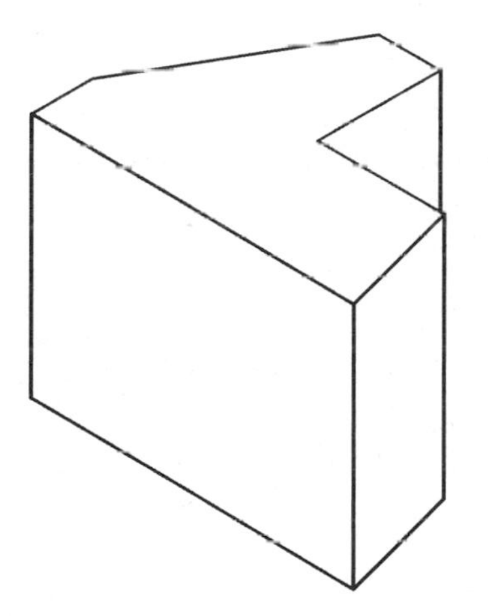

Figure 2.27

EXERCISE

Exercise 2.4—Lamp

1. Open the file \Md4book\Chapter2\Lamp.dwg.
2. Revolve the profile 360° about the inside vertical line.
3. Edit the feature's vertical dimensions, as shown in Figure 2.28.

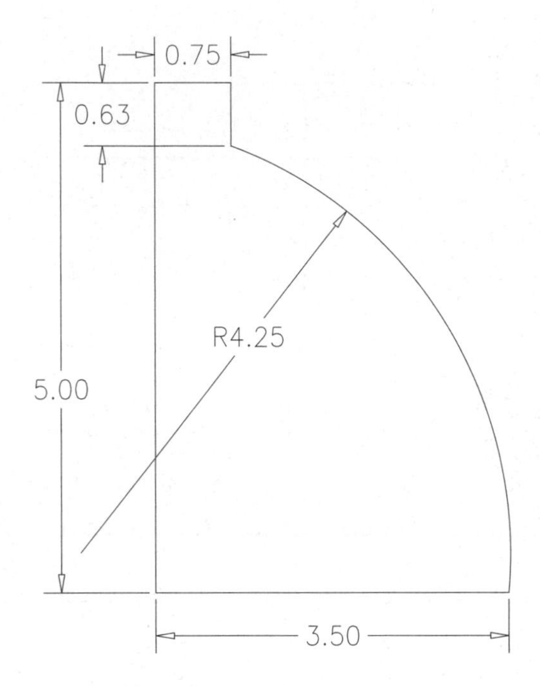

Figure 2.28

4. Update the part.
5. To better see the part, rotate it as required. Figure 2.29 shows the updated part in a rotated view.
6. Save the file.

Figure 2.29

REVIEW QUESTIONS

1. Name two methods for changing the viewpoint.
2. Explain three uses for the Mechanical Desktop browser.
3. The 3 key changes the screen to three viewports with a top, a front, and an isometric view. T or F?
4. A part can only be edited in wireframe mode. T of F?
5. A sketch does not require profiling before extruding. T or F?
6. What is a base feature?
7. What objects can be used as an axis of revolution?
8. Explain the main functionality of the 3D Orbit command.
9. A part cannot be saved in an edited state. T or F?
10. Once a profile is a base feature, you cannot delete or add constraints, dimensions or objects to the profile. T or F?

CHAPTER 3

Sketch Planes, Boolean Operations, Fillets, Chamfers, Holes and Arrays

In Chapter 2 you learned how to create a base feature. In this chapter you will learn how to make faces on a part become the active sketch plane, create profiles on these sketch planes and modify the part by adding or subtracting material from it. Then you will learn how to create placed features such as fillets, chamfers, holes and also how to array features. Before starting to create a part, think about how the finished part will look and break it down into simple shapes. Then create the part one shape at a time, adding and removing material as you go. If you are new to part creating, start with simple parts and work up to more complex parts.

AFTER COMPLETING THIS CHAPTER, YOU WILL BE ABLE TO:

- Explain what is meant by the term "feature".
- Make a planar face the active sketch plane.
- Create and modify existing parts using one of the four operations: Cut, Join, Intersect and Split.
- Use the browser to rename and delete features.
- Create four types of fillets.
- Create chamfers.
- Create holes.

MAKING A PLANE THE ACTIVE SKETCH PLANE

FACE HIGHLIGHTING AND UCS CYCLING

When adding sketched features to a part you may want to place them on a plane other than the one that is the active sketch plane. The active sketch plane is the plane on which a profile is drawn. There can only be one active sketch plane at a time. Any planar face (flat plane) on a part or work plane can be made the active sketch plane. If you are familiar with the UCS command, you will see some similarities to placing the active sketch plane. However, do not use the UCS command, because Mechanical Desktop keeps track of the sketch plane in which the profile was drawn, and this is not the case with the UCS command.

There are two requirements to make a plane on a part the active sketch plane: the part in which the plane will be placed must be a Mechanical Desktop part, and the part must contain a face or a work plane. The face does not need to have a straight edge. For example, a cylinder has two faces, one on the top and the other on the bottom of the part; neither has a straight edge. Issue the New Sketch Plane command **(AMSKPLN)** from the Part Modeling toolbar as shown in Figure 3.1 or right-click in the drawing area and from the pop-up menu, select New Sketch Plane. At the command line you will see options:

```
Select work plane, planar face or
[worldXy/worldYz/worldZx/Ucs]:
```

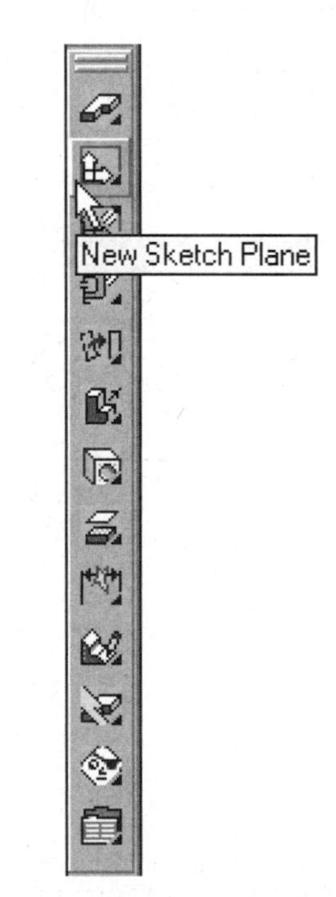

Figure 3.1

The default is to select a work plane or planar face. To help you select the correct face Mechanical Desktop has dynamic face and edge highlighting. Simply move the cursor over a given face and it will highlight. Keep moving the mouse, and different faces will highlight as the cursor passes over them. To cycle to a coincidental face, press the left mouse button and that face will highlight. A mouse icon will appear on the screen, as shown in Figure 3.2. The left mouse button is flashing to remind you how to cycle through the faces. Keep cycling through the faces until the correct face is highlighted, then press the right mouse button or ENTER to select the plane as the active sketch plane.

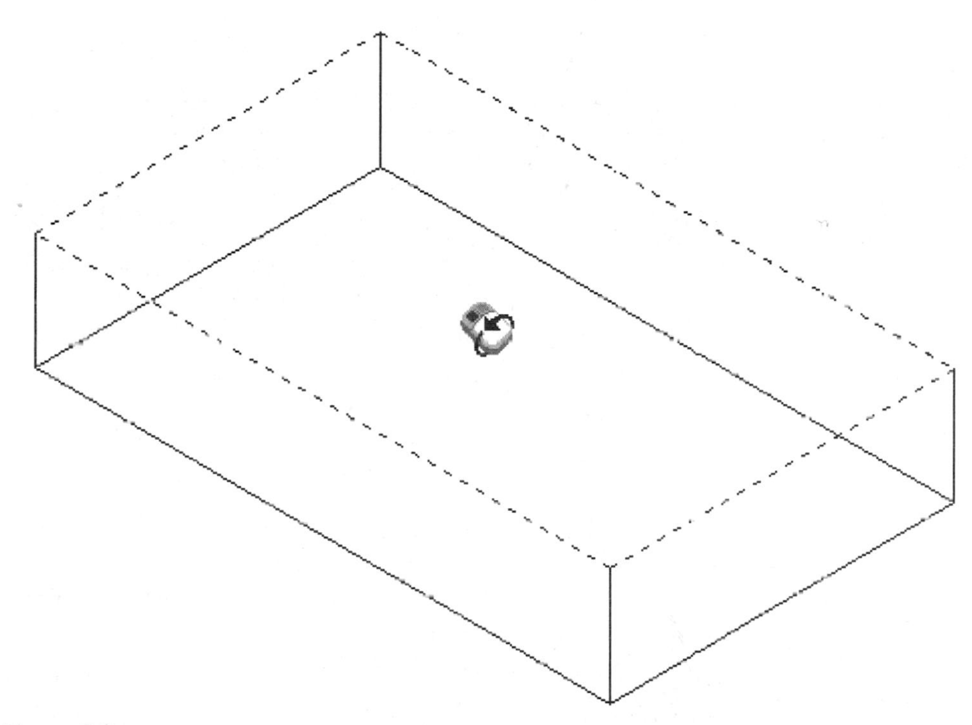

Figure 3.2

Two planes and an *X*, *Y* and *Z* arrow will appear, as shown in Figure 3.3, to identify the plane and UCS orientation. You can press ENTER to accept this orientation, select another edge to realign the *XY* or left click to rotate the *XY* in 90° increments. To flip the *Z* axis, select the *Z* arrow on the screen. To rotate the *XY* in 90° increments you could type R and ENTER or to flip the *Z* axis you could type Z and ENTER. When the orientation is correct press ENTER or right-click and select ENTER from the pop-up menu. The orientation of the *XY* axis will also be the orientation for the dimensions that will be placed on the sketch. The *XY* orientation is also used when rotating the current view to the plan view of the current sketch.

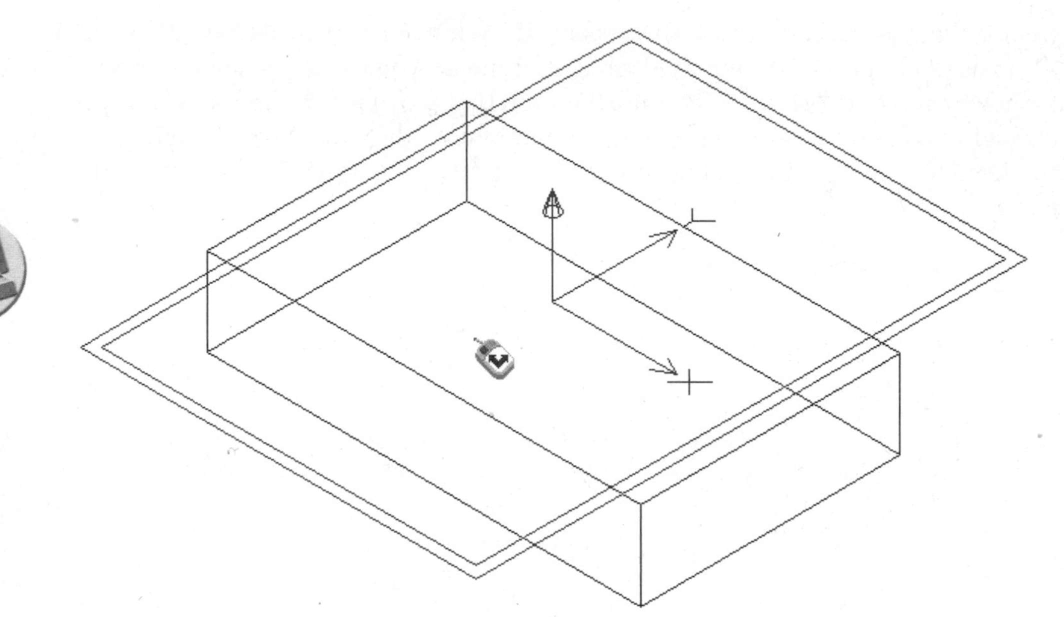

Figure 3.3

The last four options on the command line in brackets of the Create Sketch Plane command refer to the World coordinate system or current UCS. You will use these options if there is no geometry to select from on which to create a plane. Work planes that are also an option will be described in Chapter 5.

Tip: To see the active sketch plane in a plan view, press the 9 key.

TUTORIAL 3.1— FACE HIGHLIGHTING AND SKETCH PLANES

1. Start a new drawing from scratch.
2. Draw a square, profile it and place two "3" dimensions on it.
3. Change to an isometric view using the **8** key.
4. Extrude the profile "3" in the default direction with –15° of draft.
5. Make the top plane the active sketch plane by issuing the Create Sketch Plan command **(AMSKPLN)** and selecting in the middle of the top face.
6. Left-click (with the mouse left button) to cycle to the back face.
7. Left-click to cycle back to the top face, then right-click to accept the top face as your selection.

8. When the *XY* and *Z* arrows appear, rotate the *XY* arrows a few times by left-clicking four times to return it to its original position.
9. Flip the *Z* axis by selecting the *Z* arrow and select the *Z* arrow again to return it to its original position as shown in Figure 3.4.
10. Press ENTER to complete the command.
11. Draw a circle on this top face.
12. Make each of the other planar faces the active sketch plane and draw a circle on each plane. Rotate the part as needed to see it from a better point of view. When complete, your screen should resemble Figure 3.5 shown in a rotated view.
13. Rotate the part to verify that the circles are on each of the planes.
14. Save the file as \Md4book\Chapter3\TU3-1.dwg.

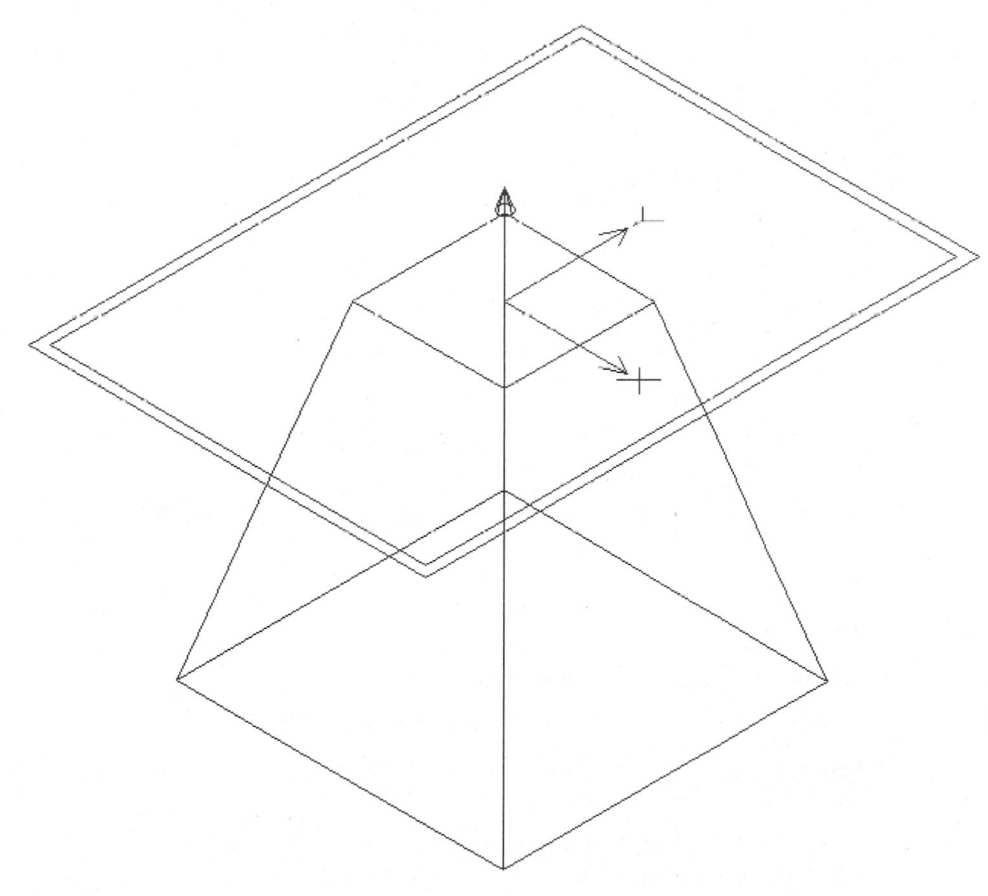

Figure 3.4

Figure 3.5

TUTORIAL 3.2— FACE HIGHLIGHTING AND SKETCH PLANES

1. Start a new drawing.
2. Draw, profile, dimension a 2" diameter circle and extrude it 3" with no draft angle.
3. Change to an isometric view with the **8** key.
4. Issue the Create Sketch Plan command **(AMSKPLN)**.
5. Select the top planar face of the cylinder. You will not cycle through faces since there was only one possible face.
6. When the *XY* and *Z* arrows appear, rotate the *XY* arrows a few times by left-clicking four times to return it to its original position and press ENTER to complete the command.

7. Draw a square on this top plane.
8. Make the bottom face the active sketch plane and then draw a triangle in the middle of it.
9. Rotate the part to verify that the square and triangle are on the correct plane.
10. When complete, your screen should resemble Figure 3.6.
11. Save the file as \Md4book\Chapter3\TU3-2.dwg.

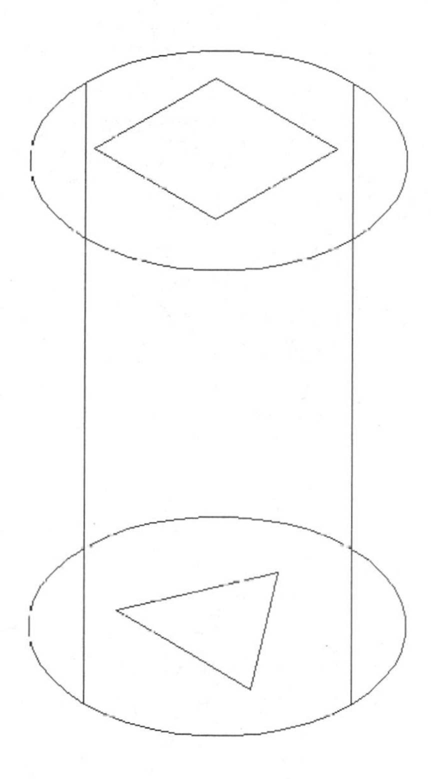

Figure 3.6

CUT, JOIN, INTERSECT AND THE SPLIT OPERATION

Up to now, the parts you created have not been complex in shape; the base part is usually derived from a simple extrude, or revolve. To create more complex parts, you can create sketch features on a plane then extrude, revolve or sweep them by adding or removing material to the part with one of four operations: cut, join, intersect or split.

There are seven basic steps that you will follow when creating sketched features on parts:

1. Create a base feature (part).
2. Create work planes as needed (work planes will be covered in Chapter 5).
3. Make a planar face or work plane your active sketch plane.
4. Create a sketch of the geometry to be profiled.
5. Profile the sketch.
6. Constrain and dimension the profile if desired.
7. Extrude, revolve or sweep the profile, removing from the part, adding to the part, keeping what is common to the base part or remove material from the part and create a new part from the removed material.

SKETCHED FEATURES

A sketched feature is a feature that you draw on a specified plane, profile, constrain, dimension and then perform a boolean operation either adding or removing material from the part. There are no limits to the number of sketched features that can be added to a part. After a plane has been made the active sketch plane it is sometimes easier to work in a plan view (looking straight at the current plane) of the current sketch plane. The 9 key is programmed to do this, as well as the Sketch View command (**AMVIEW** with the sketch option) as shown in Figure 3.7 or from the Mechanical View toolbar. You can apply dimensions and constraints exactly as you did with the first profile, with the addition of constraining and dimensioning the profile to the existing part. You may also place dimensions to geometry that does not lie on the current sketch plane, the dimensions will be placed on the current sketch plane. When you look at a part from different viewpoints, you will see arcs and circular edges appearing as lines. Remember that they are still circular edges; when you constrain or dimension them, the constraints and dimensions will go to their center points. After constraining and dimensioning the profile, extrude, revolve or sweep the profile. The sweep and loft commands will be explained in Chapter 5.

After you issue the command to extrude, revolve, sweep or loft a profile, a dialog box with options will appear. You will need to fill in information for the **Operation** type and **Termination** type. Descriptions follow for both Operation and Termination that will be needed for extruding, revolving, sweeping and lofting.

Operation: Select the type of operation you need. There are five options to choose from. The Base option is the default for the first part and will be grayed out after it has been created.

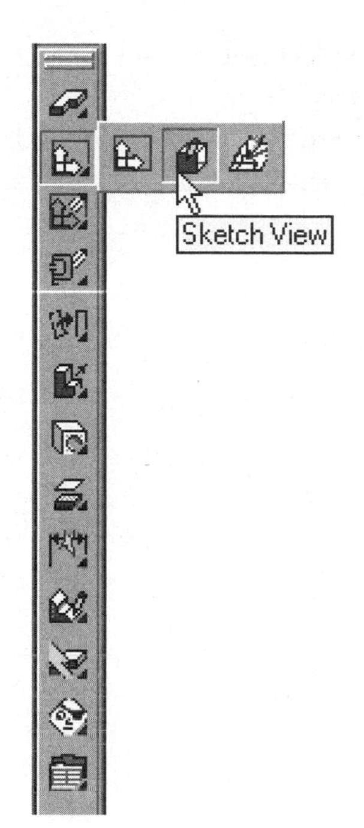

Figure 3.7

Option:	**Function:**
Base:	The first feature. After a part is created from a profile, this option will be grayed out.
Cut:	Removes material from the part.
Join:	Adds material to the part.
Intersect:	Keeps what is common to the part and this second feature.
Split:	Material will be removed from the part and the removed material will form a new part with the name of your choice. The new part has no relationship to the part it was created from unless global variables were used for the profile that was extruded, revolved, lofted or swept using the split operation. This new part also maintains the feature set that was used on the other part. They cannot be seen but they can be edited to modify this new part.

Figure 3-8 shows a pictorial representation of how the cut, join and intersect operations work. The top row represents a part on the left and a profile on the right. The bottom row shows the resulting material after this new feature is created. If you have used core solids in AutoCAD you will see a resemblance to the AutoCAD commands subtract, union and intersect to the cut, join and intersect commands that Mechanical Desktop uses. Even though they act the same, do not use the AutoCAD commands subtract, union and intersect on Mechanical Desktop parts.

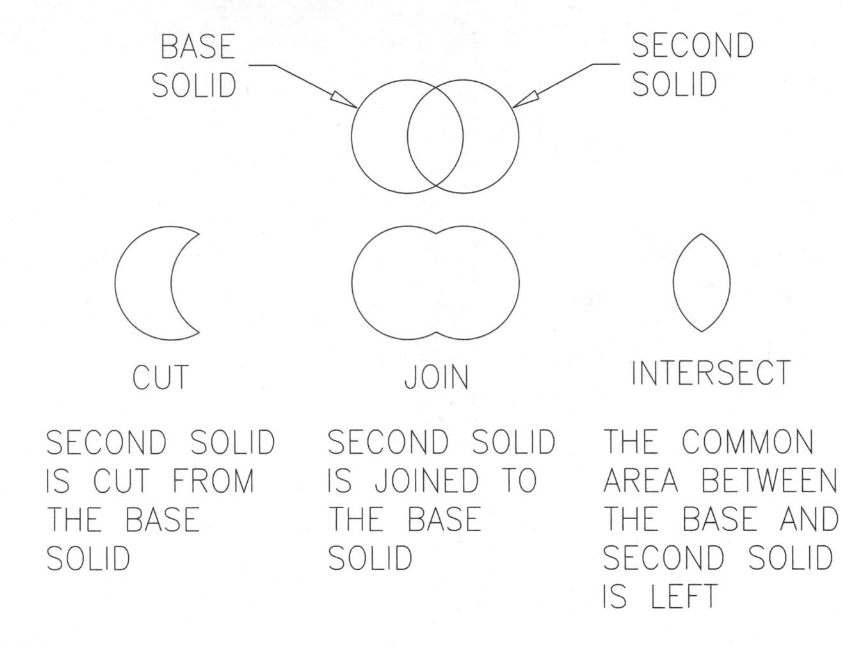

Figure 3.8

Termination

The Termination section of the Extrude, Revolve, Sweep and Loft dialog boxes have the following options that determines the boundary of the feature being created. After the function in parentheses is the command(s) that use that termination type.

Option:	Function:
Blind	Extrudes a specific distance in the positive or negative Z direction. (Extrude only)
Through	Goes all the way through the part in one direction. (Extrude only)
Mid Plane	Goes equal distances in the negative and positive directions. For example, if the extrusion distance is 2", the extrusion goes 1" in the negative and positive Z directions. (Extrude, Revolve only)

Mid Plane Through Goes through the part in both the negative and positive directions. (Extrude only)

To Face/Plane Continues until the profile reaches a specific face that is contoured or a plane that is flat. (Extrude, Revolve and Sweep)

From To Starts at one plane/face and stops at another. (Extrude, Revolve and Sweep)

By Angle Revolves the profile a specific number of degrees in the positive or negative Z direction. A arrow shows the direction in which to revolve. (Revolve only)

Path Only Causes the profile to follow the path from its start point to its end point. (Sweep only)

Note: There are no limits to the number of features that a part can have.

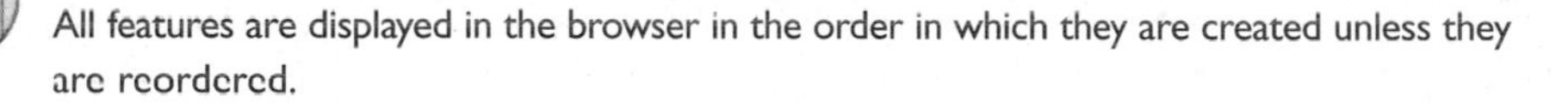
All features are displayed in the browser in the order in which they are created unless they are reordered.

Tip: Think about what the finished part will look like and create that part one feature at a time.

EDITING FEATURES AND FAILED FEATURES

To edit any feature, use the Edit Feature command **(AMEDITFEAT)** or the Desktop browser, which was covered in Chapter 2. When editing features, you can change the operation and termination types as well as the direction. If after editing a part and a red background appears behind the feature's name in the browser, Mechanical Desktop is telling you that the feature is in a failed state. This means that the new information was mathematically unable to be finished. Undo back and try different values or change the operation or termination types.

RENAMING FEATURES WITH THE BROWSER ASSEMBLY TAB

As previously explained, each feature and part is given a default name. You can set part name prefixes using the Desktop Options command **(AMOPTIONS)**. There are also two ways you can change the name after it has been created. For the first method, slowly click twice on the feature name but, instead of letting go on the second click, keep it depressed for a second longer. The name will be shaded blue and a box will appear around the name and the cursor will begin to blink. Rename the feature as you would using Microsoft Explorer. If the double click is too fast, you will edit the feature. The second method is to right-click on the feature or part name and choose rename from the pop-up menu.

Note: Renaming a part through the browser does not change the part definition name—only it changes the name that the browser uses. To change the name of a definition, use the Desktop catalog. The Desktop catalog will be covered in Chapter 8.

Each of the following tutorials guides you through a variety of methods for creating sketched features. Remember that there is no one way to generate a specific part. After completing the tutorial as shown, go back and try a different method.

TUTORIAL 3.3—EXTRUDING WITH CUT: THROUGH

1. Open the file \Md4book\Chapter3\TU3-3.dwg.
2. Make the front face the active sketch plane and orient the UCS as shown in Figure 3.9.

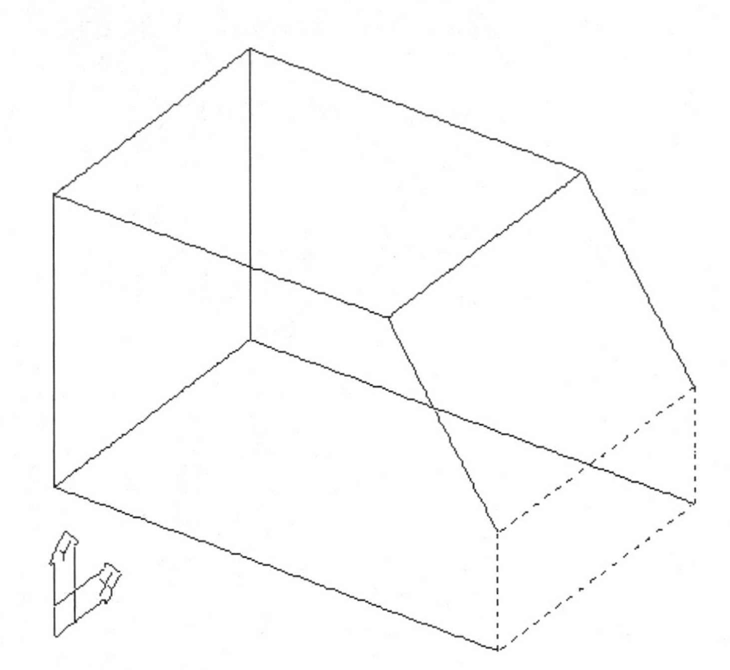

Figure 3.9

3. Change your view to the current sketch view (the **9** key).
4. Draw a rectangle and then profile, constrain (Collinear, top lines of the part and profile) and dimension the rectangle as shown in Figure 3.10.

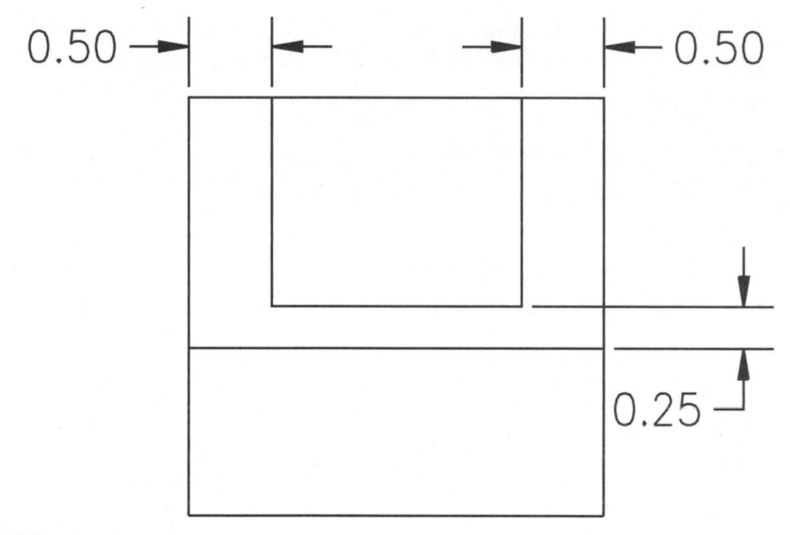

Figure 3.10

5. Switch to an isometric view (the **8** key).
6. Issue the Extrude command **(AMEXTRUDE)** and use the following settings:
   Operation = Cut
   Termination = Through
   Draft Angle = 0
   By default the extrusion direction should be directed into the part.
   Then select OK.
7. When complete, your screen should resemble Figure 3.11, shown with lines hidden.

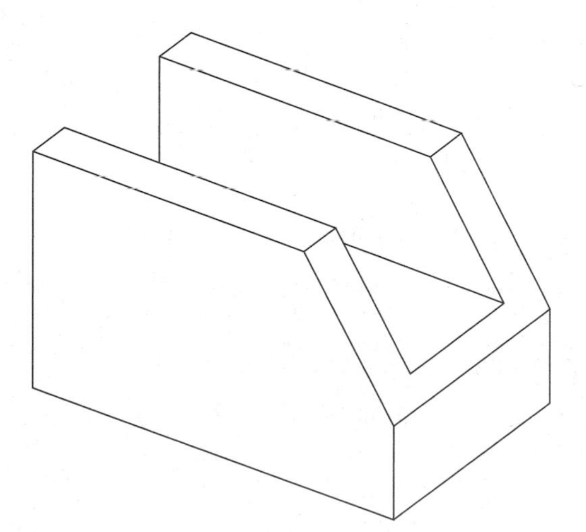

Figure 3.11

8. Edit the extrusion base length from "4" to "6" and update the part. The cut goes through the part even if the part's dimensions change.
9. Save the file.

TUTORIAL 3.4—EXTRUDING WITH CUT: BLIND

1. Open the file \Md4book\Chapter3\TU3-4.dwg.
2. Make the large outside back circle the active sketch plane and orient the UCS until your screen resembles Figure 3.12.

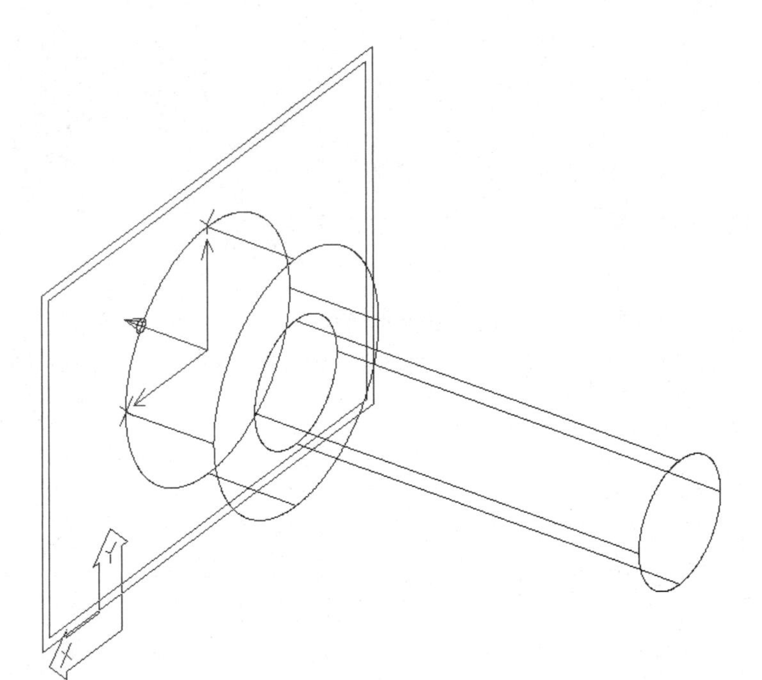

Figure 3.12

3. Change your view to the current sketch view (**9**).
4. Draw a rectangle and then profile, constrain (vertical lines tangent to the sides of the circle) and dimension the rectangle as shown in Figure 3.13.
5. Switch to a southwest isometric view (**88**).
6. Issue the Extrude command **(AMEXTRUDE)** and use the following settings:
   Operation = Cut
   Termination = Blind
   Distance = .25
   Draft Angle = 0

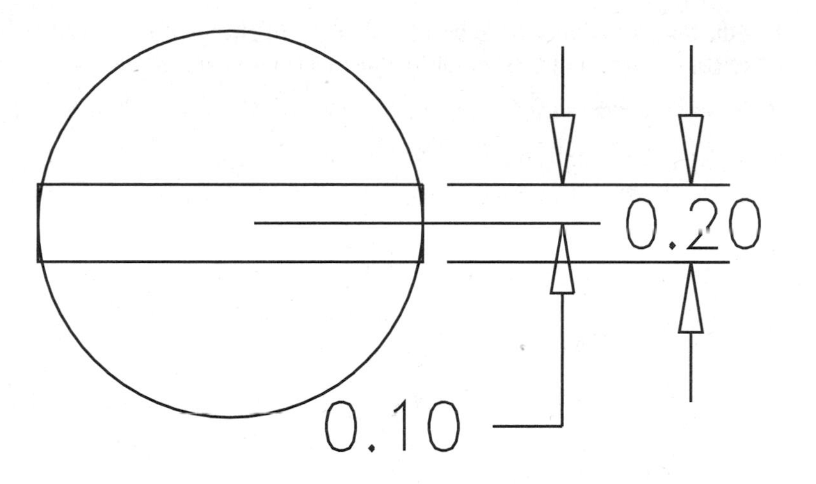

Figure 3.13

By default the extrusion direction should be directed into the part. Select OK.

7. When complete your screen should look like Figure 3.14, shown with lines hidden.

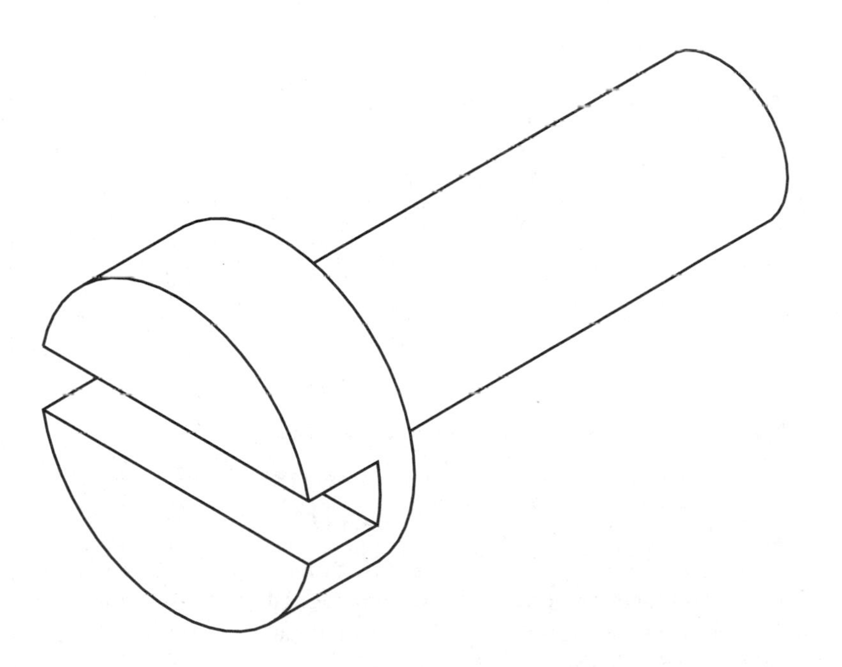

Figure 3.14

8. Edit the last extrusion "ExtrusionBlind1" and change the Operation to Join, Distance to .5 and Flip the direction so exits the end of the part.
9. If needed, update the part. When complete, your screen should look like Figure 3.15, shown with lines hidden.
10. Save the file.

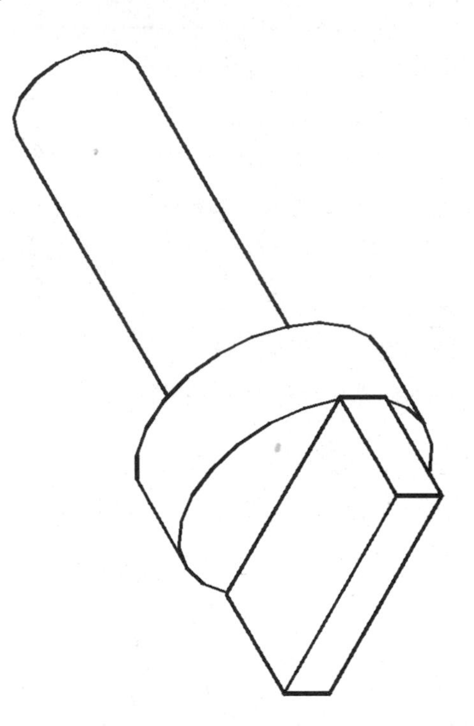

Figure 3.15

TUTORIAL 3.5 —REVOLVING WITH CUT

This file already has a work axis and a work plane that are invisible and a sketch that has been profiled.

1. Open the file \Md4book\Chapter3\TU3-5.dwg.
2. Change your view to the current sketch view (**9**).
3. Place the dimensions as shown in Figure 3.16.
4. Place a collinear constraint to the top horizontal line of the profile and the top of the cylinder and to the bottom horizontal line of the profile and the bottom of the cylinder. A collinear constraint can be applied between a line and an arc or circular edge of a feature.

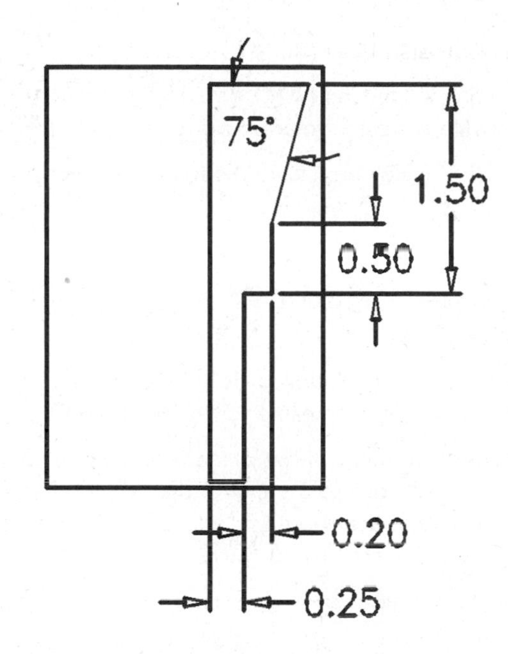

Figure 3.16

5. Place an XValue constraint between the inside vertical edge of the profile and one of the circular edges of the extrusion. This will fully constrain your part and when complete, your screen should look like Figure 3.17.

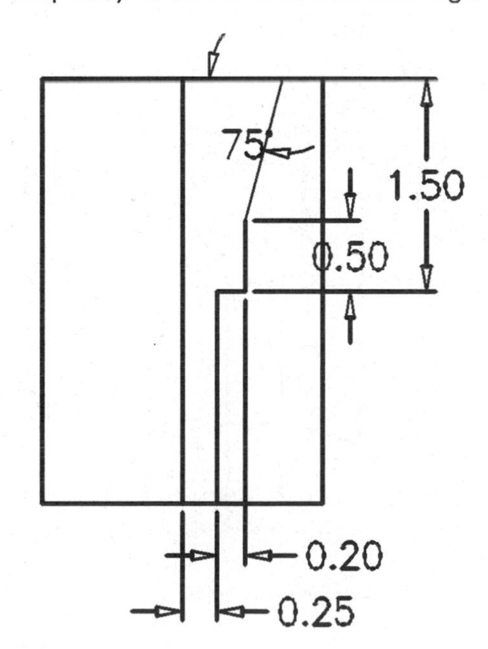

Figure 3.17

6. Switch to an isometric view (**8**).
7. Issue the Revolve command **(AMREVOLVE)** and select the inside vertical line of the profile for the revolution axis.
8. In the Revolution dialog box make the following settings:
   Operation = Cut
   Termination = By Angle
   Angle = 180
   Flip the direction
   Then select OK.
9. Rename the feature "RevolutionAngle1" by slowly selecting the name twice in the browser and change the name to "InsideRevolve".
10. Edit the extrusion distance of the cylinder to 3.5 and update the part. The revolved feature still goes to the end of the cylinder because the collinear constraint means that the two edges will always lie in the same plane, regardless of what changes occur to the part.
11. Edit the feature "InsideRevolve" and change the Angle to 360.
12. If needed, update the part. When complete your screen should look like Figure 3.18.
13. Save the file.

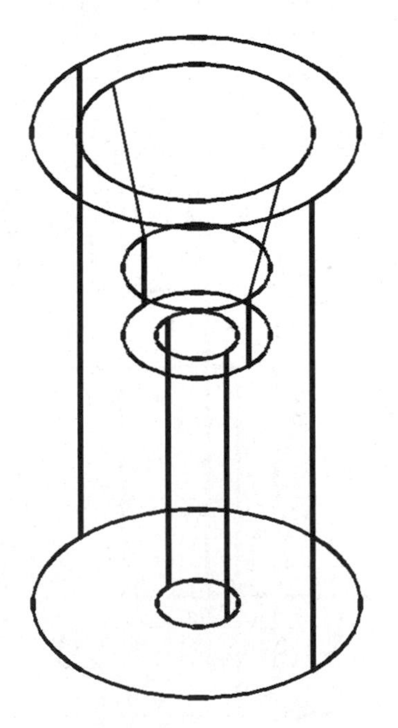

Figure 3.18

TUTORIAL 3.6 —EXTRUDING WITH JOIN: BLIND

1. Open the file \Md4book\Chapter3\TU3-6.dwg.
2. Make the right vertical face the active sketch plane and orient the UCS as shown in Figure 3.19.

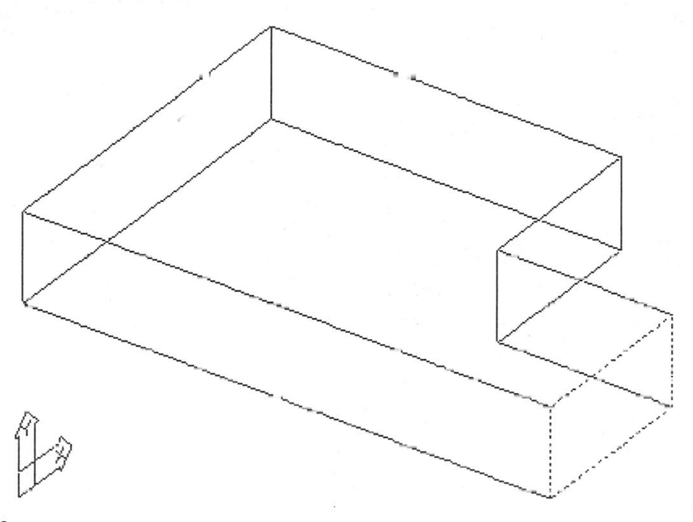

Figure 3.19

3. Change your view to the current sketch view (**9**).
4. Draw a tombstone shape and then profile, constrain and dimension the profile as shown in Figure 3.20. Hint: Apply a collinear constraint between the lines of the profile and to the edges of the extrusion.

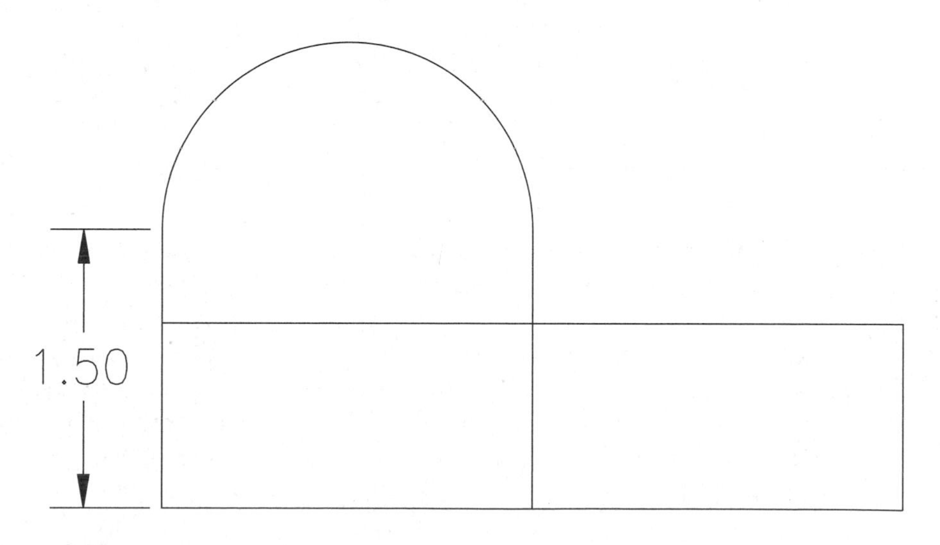

Figure 3.20

5. Switch to an isometric view (**8**).
6. Issue the Extrude command **(AMEXTRUDE)** and use the following settings:
   Operation = Join
   Termination = Blind
   Distance = 4
   Draft Angle = 0
   Flip the extrusion distance so the extrusion will go into the part.
   Then select OK to execute the command.
7. When complete, your screen should look like Figure 3.21, shown with lines hidden.
8. Save the file.

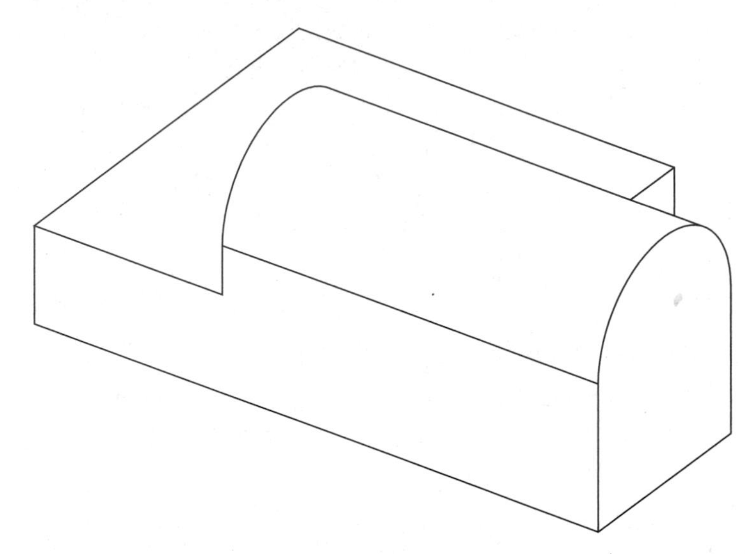

Figure 3.21

TUTORIAL 3.7—EXTRUDING WITH JOIN: TO FACE/PLANE

In this exercise you will use To Face/Plane as the Termination option. This allows you to stop the extrusion, revolve or sweep at a specified face or plane. Faces are not planar—they are contoured. When using the face option, the profile would need to be entirely enclosed within the selected face. After selecting the face, you will be prompted to cycle among the possible selection sets. Cycle through in the same manner as you did for making a planar face the active sketch plane. When the result on the screen is correct press ENTER to complete the command. The plane option allows you to select a planar face or a work plane for the place that the profile will stop at.

1. Open the file \Md4book\Chapter3\TU3-7.dwg.

2. Make the workplane on the right side the active sketch plane and orient the UCS, as shown in Figure 3.22
3. Turn off the visibility of the workplane.
4. Draw the tombstone shape and then, profile, constrain the bottom horizontal line (collinear with the front bottom edge of the extrusion) and dimension it as shown in Figure 3.23 (shown with lines hidden).

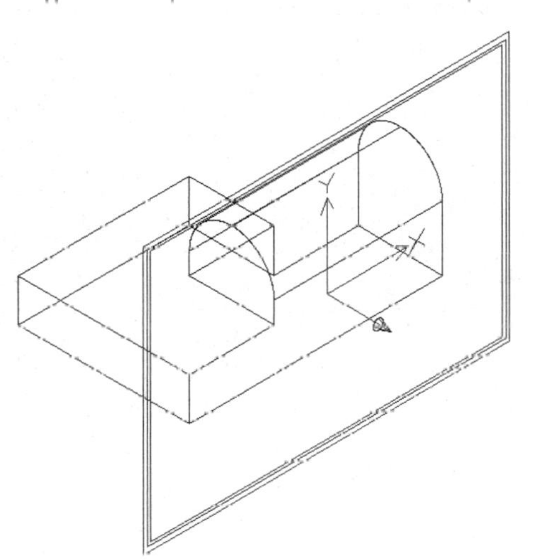

Figure 3.22

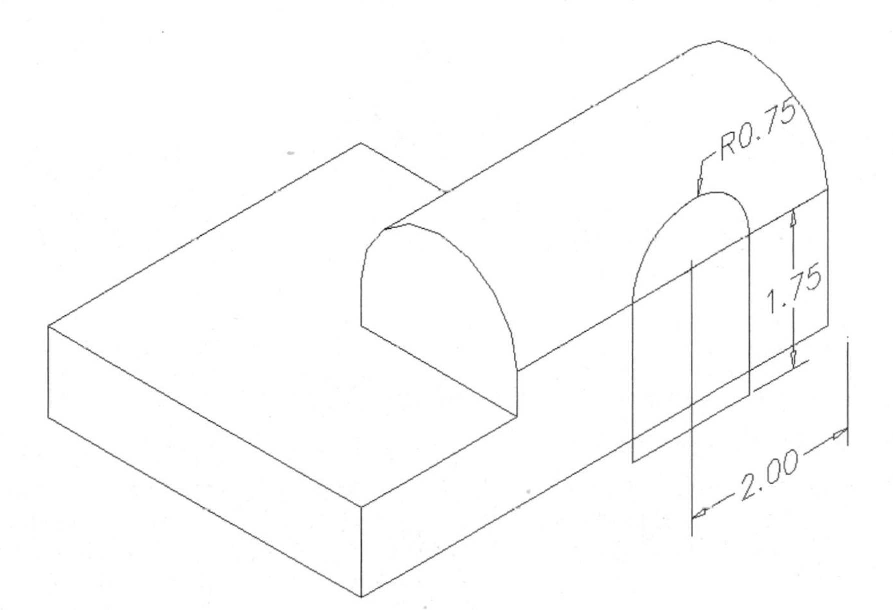

Figure 3.23

5. Issue the Extrude command **(AMEXTRUDE)** and use the following settings:
   Operation = Join
   Termination = To Face/Plane
   Draft Angle = 0
   Then select OK.
6. For the prompt "Select face or work plane": Select in the middle of the front face of the extruded arc as shown in Figure 3.24 with lines hidden.

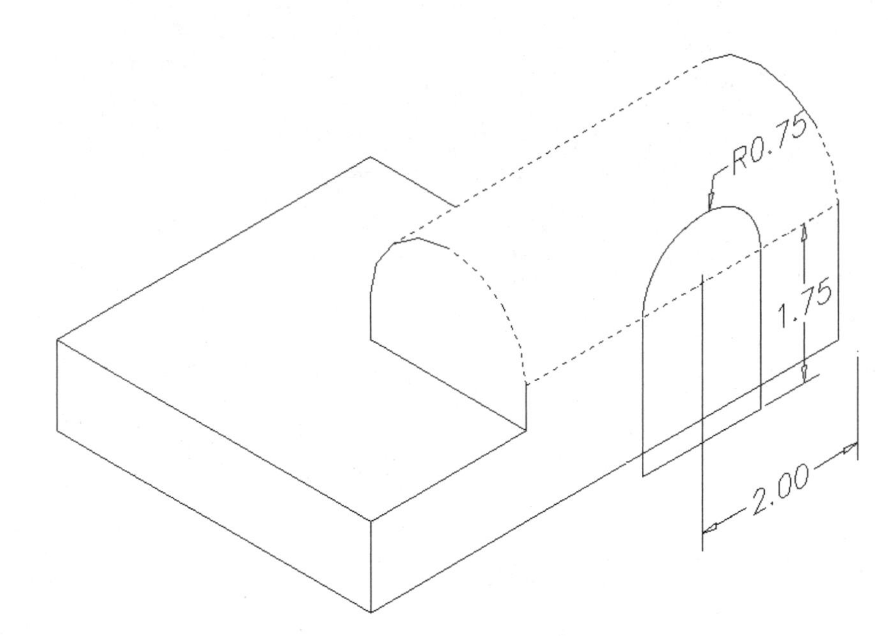

Figure 3.24

7. For the prompt "Select termination loop [Next/Accept] <Accept>": ENTER to accept the front face.
8. When complete, your screen should resemble Figure 3.25 shown with lines hidden.
9. Switch to a southeast isometric view (**8**).
10. Edit the last extrusion.
11. In the Extrusion dialog box select To Face/Plane again for the termination (this tells Mechanical Desktop that you want to change the face or plane that the profile stops at). Select OK in the dialog box to accept the other settings.

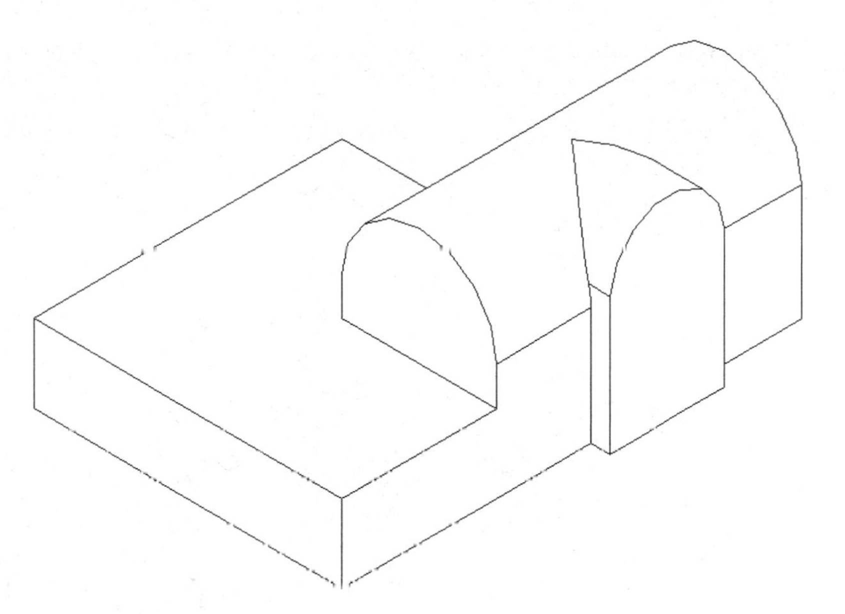

Figure 3.25

12. Then select the inside vertical face as shown in Figure 3.26 and when hightlighted, press ENTER to accept it.

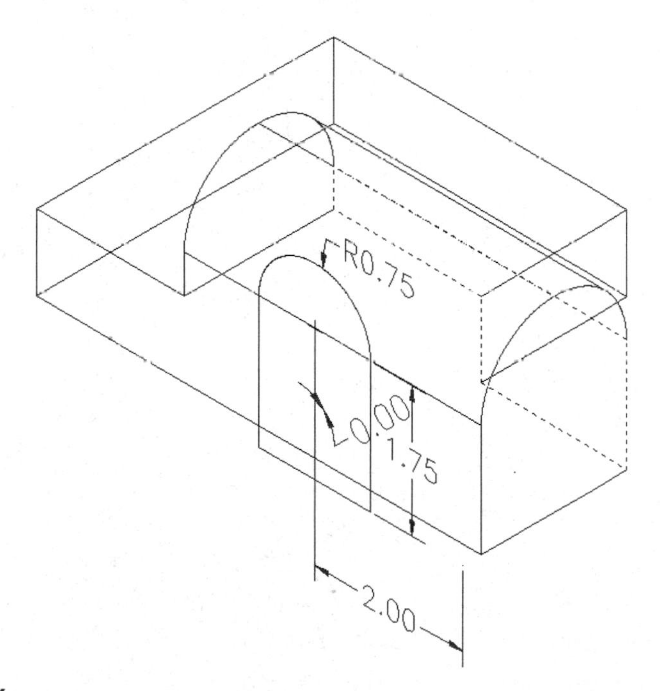

Figure 3.26

13. Then type P and ENTER or right click in the drawing area and select Plane from the pop-up menu.
14. Update the part if needed; when complete, your screen should resemble Figure 3.27 shown with lines hidden.
15. Rotate the part to better see that the profile did stop at the selected plane.
16. Save the file.

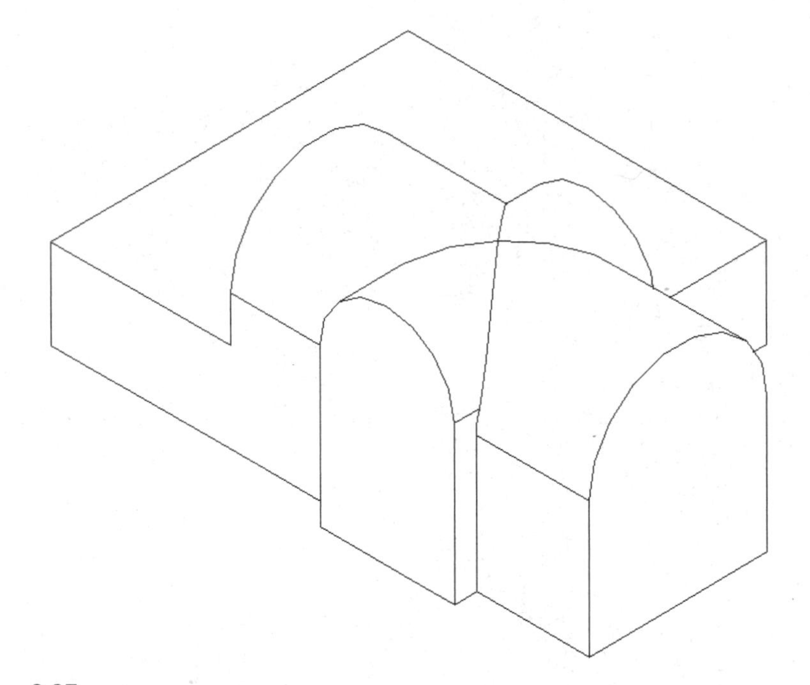

Figure 3.27

TUTORIAL 3.8—REVOLVING WITH JOIN: FROM-TO

In this exercise you will use From To as the Termination option. This allows you to start and stop the extrusion, and revolve or sweep at two different faces. Faces do not have to be planar, they can be contoured. In this tutorial, the profile (in magenta) has already been drawn and constrained. When performing a From To termination, zooming in closely will help you graphically isolate the face that is selected. You will have the option to cycle among the possible selection sets. You will select the first face and press ENTER and then select the second face and press ENTER to finish the command.

1. Open the file \Md4book\Chapter3\TU3-8.dwg.
2. Issue the Extrude command **(AMEXTRUDE)**.

3. In the Extrude dialog box make the following settings:
   Operation = Join
   Termination = From To
   Draft Angle = 0
   Then select OK.
4. For the first termination face/plane select the inside arced face as shown in Figure 3.28. You may need cycle through until the correct face is highlighted. Once highlighted press ENTER.

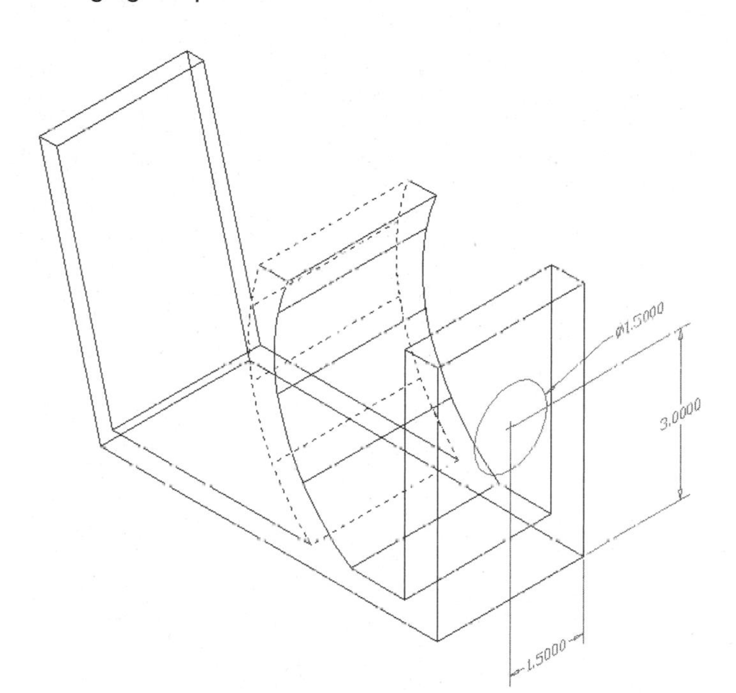

Figure 3.28

5. For the second termination face/plane, select the left inside slanted plane as shown in Figure 3.29. Once highlighted press ENTER.
6. Press ENTER to accept that the circle will stop at this plane.
7. When complete your screen should resemble Figure 3.29.
8. Rotate the part to see more clearly that the profile did start and stop at the selected faces.
9. Save the file.

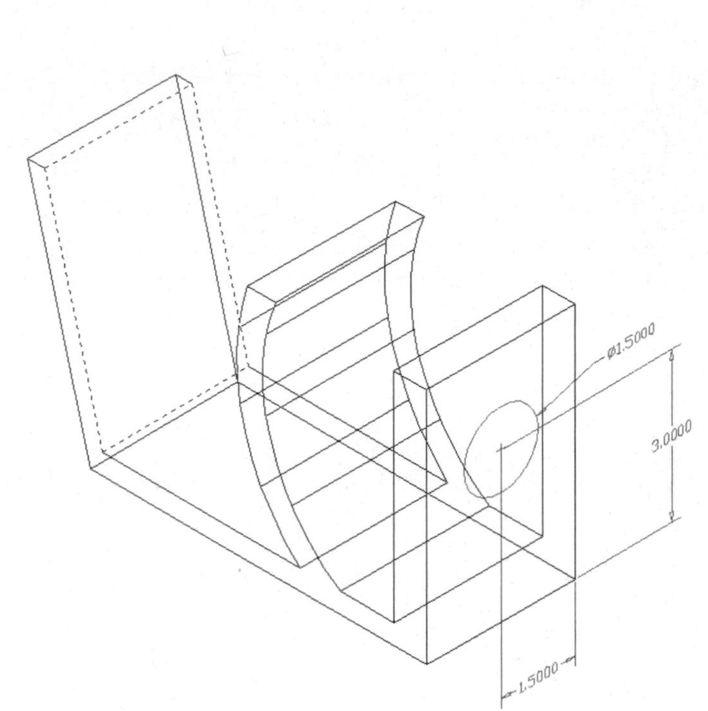

Figure 3.29

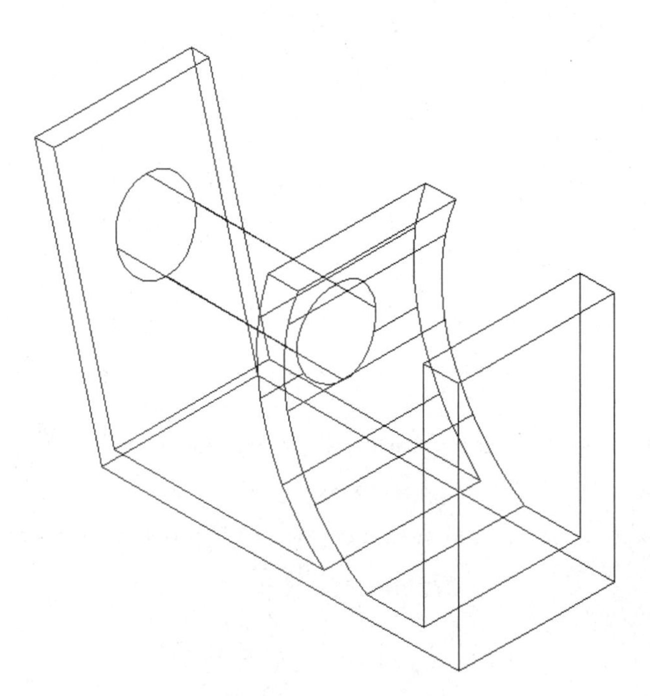

Figure 3.30

TUTORIAL 3.9 —REVOLVING WITH INTERSECT: FULL

In this tutorial, the sketch has already been profiled and constrained.

1. Open the file \Md4book\Chapter3\TU3-9.dwg.
2. Issue the Revolve command **(AMREVOLVE)** and for the revolution axis, select the green vertical line of the profile.
3. In the Revolution dialog box make the following settings:
   Operation = Intersect
   Termination = By Angle
   Angle = 360
   Then select OK.
4. When complete, your screen should resemble Figure 3.31.

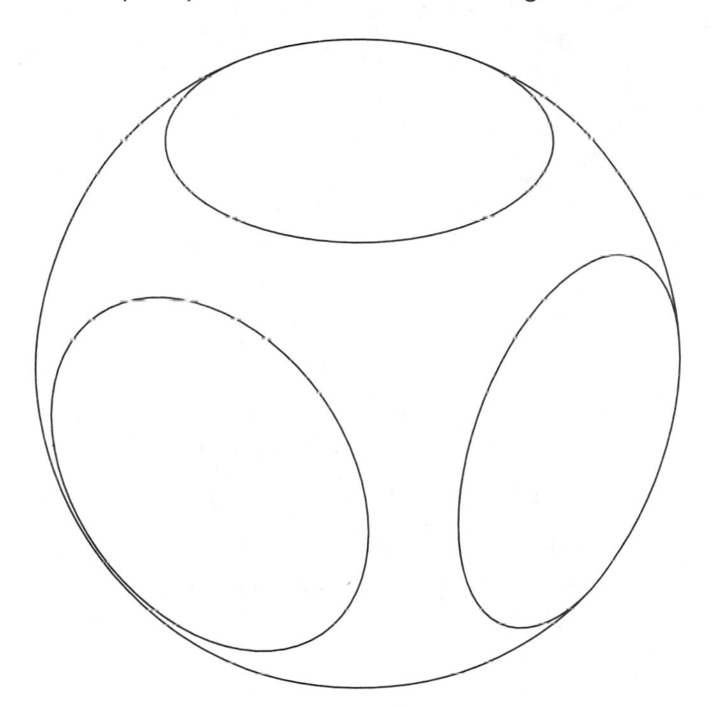

Figure 3.31

5. Issue the Edit Feature command **(AMEDITFEAT)** and edit the revolution.
6. Select OK in the revolution dialog box.
7. On the screen, change the radius from "1.25" to "1.5" and change the distance the arc is offset from the cube from ".25" to ".5" on both sides.
8. If needed, update the part; when complete, your screen should resemble Figure 3.32 shown with lines hidden.

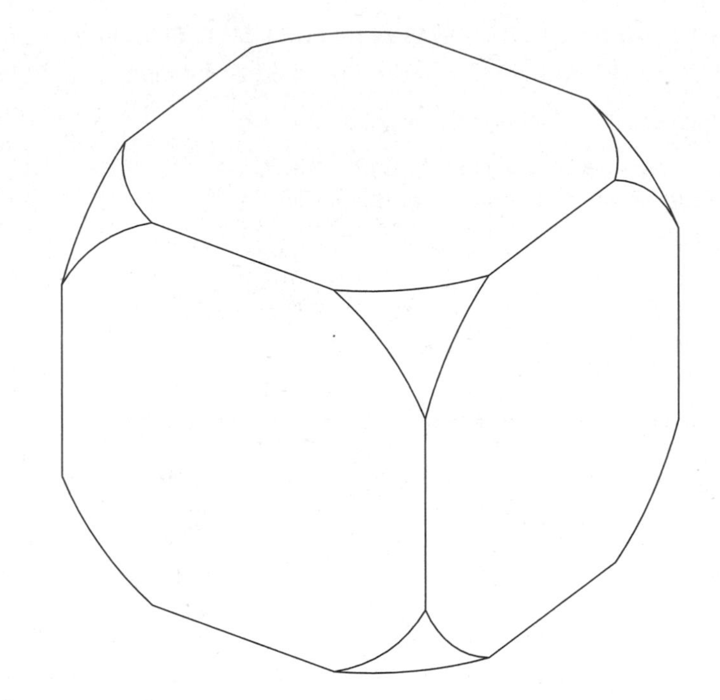

Figure 3.32

9. Save the file.

TUTORIAL 3.10 —SPLIT OPERATION

In this tutorial, a base feature and a profile have been created.

1. Open the file \Md4book\Chapter3\TU3-10.dwg.
2. Issue the Extrude command **(AMEXTRUDE)**.
3. In the Extrude dialog box make the following settings:
   Operation = Split
   Termination = Through
   Draft Angle = 0
   The direction should be down through the part.
   Then select OK.
4. For the new part name type "SPLIT" and then press ENTER to complete the command.
5. Move the part SPLIT_1 away from Part1.
6. Expand the feature list in the browser for both parts. All features and parametrics are maintained in both parts. If one part changes, the other will not unless global variables were used. Global variables will be covered in Chapter 6.

7. Make the part Split_1 the active part by double clicking on the part name in the browser or by double clicking on the Split_1 part in the drawing area. (Working with multiple parts will be covered in Chapter 7).
8. Edit ExtrusionThru1 of the part Split_1 and change the "3" dimension to "2".
10. If necessary, update the part. When complete, your screen should resemble Figure 3.33 shown with lines hidden.
11. Save the file.

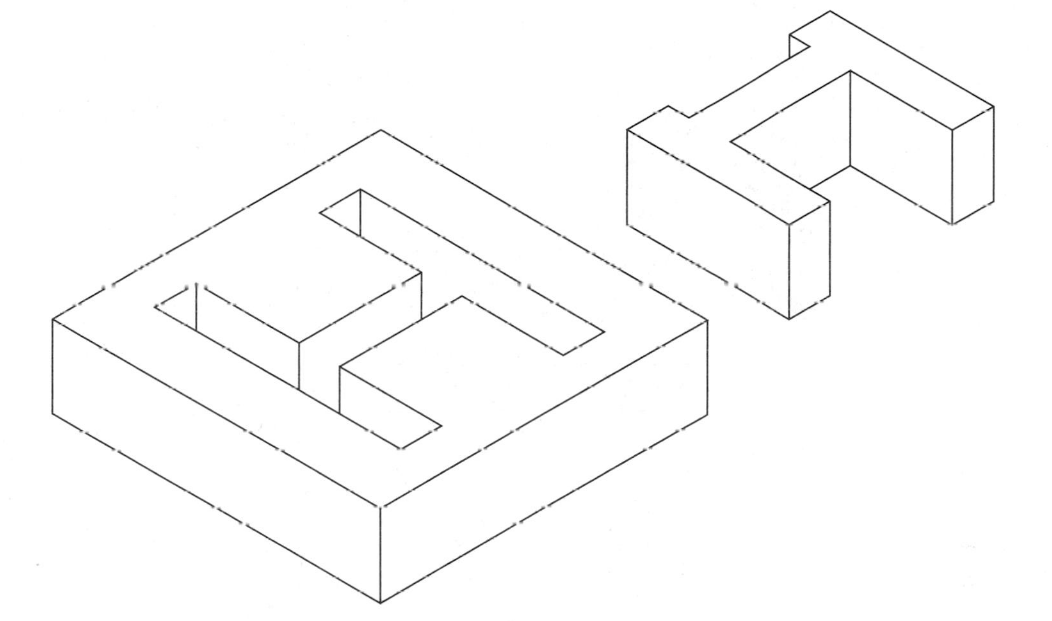

Figure 3.33

TUTORIAL

DELETING FEATURES

While working on a part you may need to delete a feature from the part. The delete feature command (AMDELFEAT) can be accessed in one of three methods. From the Part Modeling toolbar as shown in Figure 3.34, right-click in the drawing area and from the pop-up menu, select Delete from the Edit Features menu. Or you may select the feature name browser with the right mouse button and choose Delete from the pop-up menu. At the command line you will be prompted to either select a feature to delete or alerted that the "Highlighted features will be deleted. Continue? [Yes/No]" If you press enter the feature will be deleted. If you delete a parent feature that has a child feature dependent on it the child will also be deleted. Both features will be highlighted, alerting you that the child will also be deleted. An example of this would be a hole that was placed concentric to a fillet. If the fillet is deleted, the hole will be also.

If a feature was dimensioned to a feature that was deleted just the dimension will be deleted. To redimension the feature, return to its sketch and add a constraint or dimension. If directly after a feature is deleted you changed your mind you can use the AutoCAD UNDO command to bring it back.

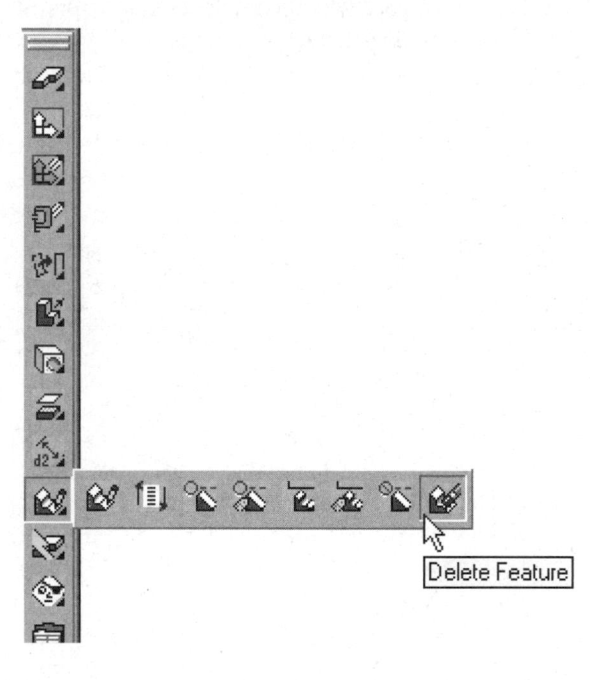

Figure 3.34

TUTORIAL 3.11 —DELETING FEATURES

1. Open the file \Md4book\Chapter3\TU3-11.dwg.
2. Expand all sections in the browser by selecting all the + signs.
3. Delete a few features using the browser right-click method.
4. Undo to bring all the features back.
5. Issue the delete feature command from the part modeling toolbar or right click menu, then delete a few features.
6. Undo to bring all the features back.
7. Delete the fillet feature on the front left face. The hole feature will also be highlighted and then deleted as well since it was placed concentric to the fillet.
8. Close the file without saving your changes.

PLACED FEATURES

In this next section you will learn how to create placed features. Placed features are features that are predefined except for specific values and just need to be placed. Placed features do not need to be placed on a sketch plane to be active except for From Hole option for hole placement. Placed features can be edited like sketched features. When a placed feature is edited, either a dialog box that was used to create it will pop up or features values will appear on the part. When creating your part it is usually better to have placed features instead of sketched features, whereever possible. For example, to make a through hole you could draw a circle, profile and dimension it, then extrude it through. Or from a dialog box select the type of hole and size and then place it. When drawing views are then generated, the type and size of the hole can be easily annotated and is automatically updated if the hole's type or value changes. For sketched features values would require more steps to annotate and would not change if hole size changes. Fillets, chamfers and holes will be covered in the next sections.

FILLETS

When creating fillets in 3D, you will select the edge that needs to be filleted and the fillet will be created between the two faces sharing this edge. This differs from filleting in 2D, because in 2D you select two objects and a fillet is created between them. When creating a part, it is usually good practice to create fillets and chamfers as one of last features in the part. Fillets add complexity to the part, which in turn adds to the file size and removes edges that may be needed to place other features. Issue the Desktop Fillet command **(AMFILLET)** from the Part Modeling toolbar as shown in Figure 3-35 or right-

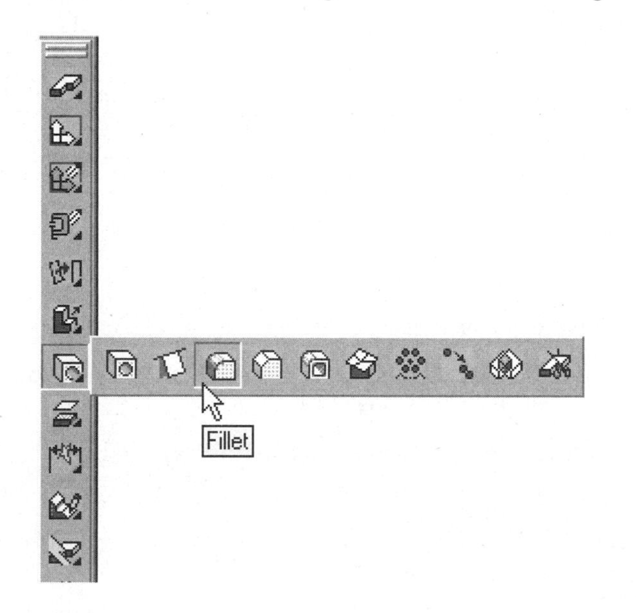

Figure 3.35

click in the drawing area and from the pop-up menu, select Fillet from the Placed Features menu. A Fillet a dialog box will appear. Select from four types of fillets (constant, fixed width, linear or cubic) and then type in a size if it is a constant or fixed width fillet. The four types of fillets will be described in the next section. To edit a fillet you can use either the Edit Feature command **(AMEDITFEAT)** or the browser, as described earlier in this chapter. Select the fillet to edit and the value of its radius will appear. Select the radius that you want to change and then type in a new value and press ENTER. Keep selecting the different radii that you want to change, pressing ENTER to exit the command, and then update the part.

TUTORIAL

CONSTANT FILLET

A constant fillet has the same radius from the beginning to the end of the fillet. Issue the Desktop Fillet command **(AMFILLET)** in the Fillet dialog box select Constant, type in a value for the radius, select Apply and then select the edge(s) that you want to fillet. There is no limit to the number of edges that can be filleted with a constant fillet. The order in which the edges are selected is not important. The edges that are to be filleted need to be picked individually, the use of window, crossing, etc. is not valid for creating fillets. When you are finished selecting the edge(s) to fillet, press ENTER. If you select multiple edges in the same command, they are linked together. If you edit a fillet that was selected as a group, they will all be highlighted. Change the value and they all change. Delete one fillet of the group and they are all deleted. If you think you would like to edit them independent of the group, check the box Individual Radii Override under Constant and then each fillet can have a different radius. In the Desktop browser, the fillets are grouped under one name, but when you edit them you can change their values independently.

At the bottom of the dialog box is a section Return to dialog box. If you want to create multiple fillets, one after another, check here and you will be returned to the dialog box after you create the fillet.

TUTORIAL 3.12 —CREATING A CONSTANT FILLET

1. Open the file \Md4book\Chapter3\TU3-12.dwg.
2. Issue the Desktop Fillet command **(AMFILLET).**
3. In the Fillet dialog box check Constant, Radius = .5 and select OK.
4. Select the front left vertical edge and press ENTER; when complete, your screen should resemble Figure 3.36.
5. Repeat the Desktop Fillet command **(AMFILLET).**
6. In the Fillet dialog box, change the radius to .125 and select OK.
7. Select all the remaining edges and press ENTER. When complete, your screen should resemble Figure 3.37.

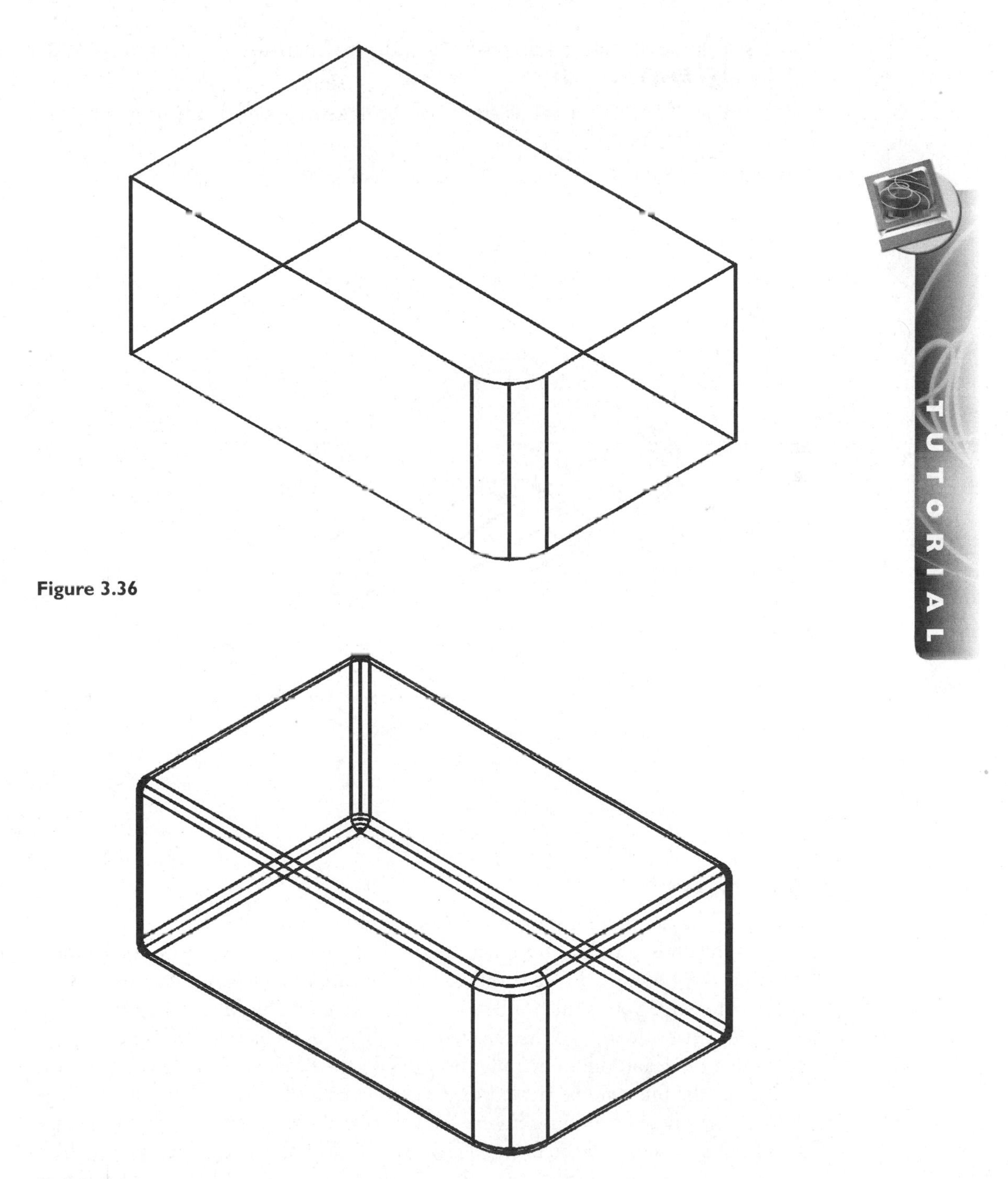

Figure 3.36

Figure 3.37

8. Edit the ".5" fillet (Fillet1) to ".125" using the Edit Feature command **(AMEDITFEAT)**.
9. Edit the ".125" (Fillet2) fillets to ".5" by double-clicking on the name Fillet2 in the browser.
10. Update the part if needed. When complete, your screen should resemble Figure 3.38.
11. Save the file.

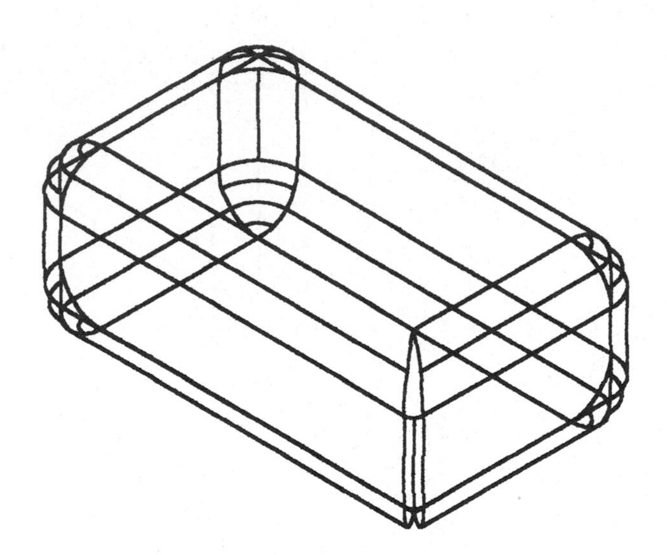

Figure 3.38

LINEAR FILLET

A linear fillet has a different starting and ending radius. The fillet will blend from the starting to the ending radius in a straight line. The value of zero is valid for a Linear fillet. Issue the Desktop Fillet command **(AMFILLET),** select Linear, select OK, and select the edge you want to fillet. Only one edge can be filleted at a time with Linear fillet. At both ends of the selected edge "R=0" will appear. Pick one of these and type in a value and press enter; do the same for the other side. After you enter the second value, the fillet will be created. A linear fillet cannot be created around an edge that is already filleted or round. Figure 3.39 shows a linear fillet with the back edge set to a radius of zero and the front edge set to .75. The front edge, marked with an *X,* could not be filleted with a linear fillet. A constant or cubic fillet could be placed on this edge instead.

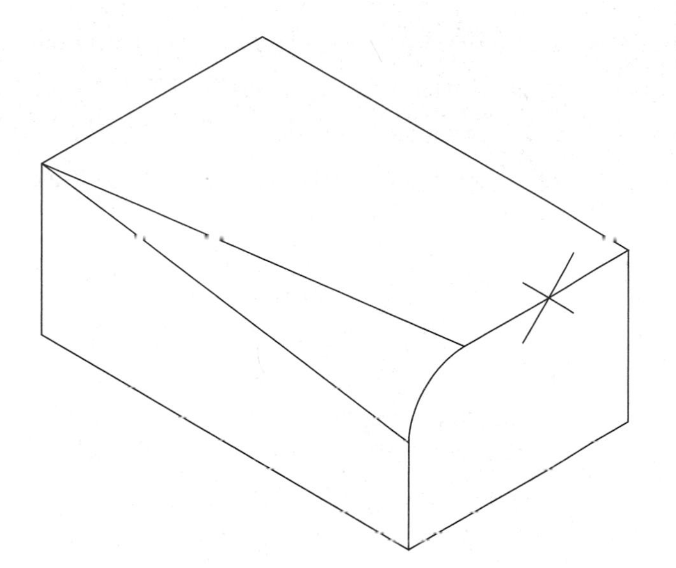

Figure 3.39

TUTORIAL 3.13 —CREATING A LINEAR FILLET

1. Open the file \Md4book\Chapter3\TU3-13.dwg.
2. Issue the Desktop Fillet command **(AMFILLET)**.
3. Select Linear and then select OK.
4. Select the front top left horizontal edge to fillet and then select the back R=0 and give it a value of .75. Press ENTER; your screen should resemble Figure 3.40.

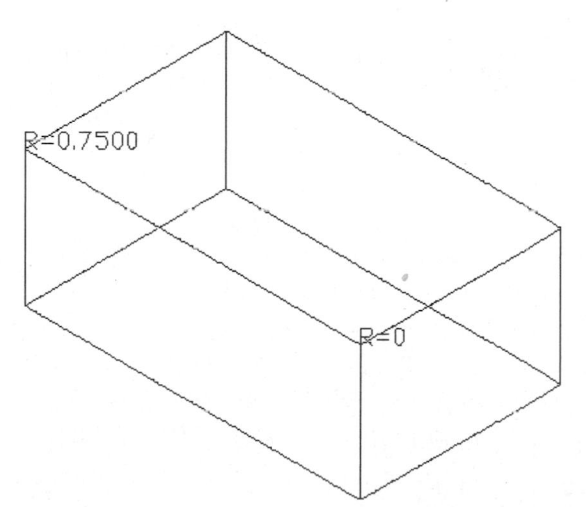

Figure 3.40

5. Press ENTER to accept the front R=0 value and complete the command.
6. Reissue the Desktop Fillet command **(AMFILLET)**.
7. Select OK to create another linear fillet.
8. Select the back top right horizontal edge. Pick the back R=0 and give it a value of .125 and change the front R=0 to .5, as shown in Figure 3.41.

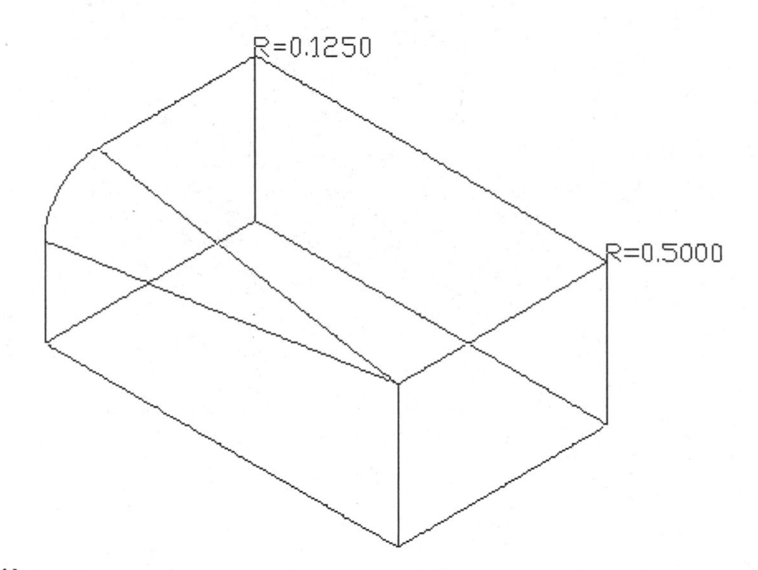

Figure 3.41

12. Press ENTER to accept these values. When complete, your screen should resemble Figure 3.42 shown with lines hidden.
13. Save the file.

CUBIC FILLET

A cubic fillet has a different starting and ending radius. The fillet will blend from the starting to the ending radius as a smooth transition, like a cubic spline. The value of zero is NOT valid for cubic fillet; the smallest fillet size is .0001. Issue the Desktop Fillet command **(AMFILLET)**, select Cubic, select OK, and select the edge you want to fillet. Only one edge can be filleted at a time. A "R=*" will appear at each vertex of the edge and at the command line you will see options:

```
Select radius or [Add vertex/Clear/Delete vertex]:
```

Add Vertex: Allows you to add a vertex and specify a radius. Select on the edge near where you want to add a different radius and then specify by a percentage value between the bounding vertices where you want this vertex added.

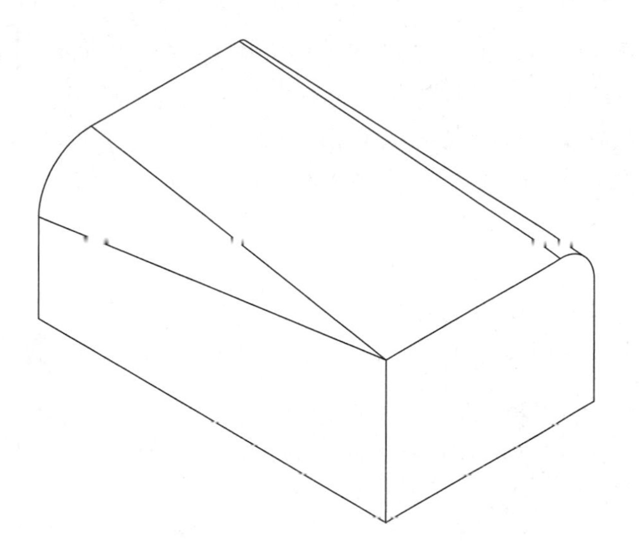

Figure 3.42

Clear: Removes the value of a selected fillet (and then you can select the "**R**=*" and type in a new value) or it will use the radius from the previous vertex.

Delete vertex: Removes the selected vertex from the part.

Select radius: To select a radius, pick one of the **R**=* and type in a value and press ENTER. Do the same for any vertex that was added. After entering all the values, press ENTER and the fillet will be created. If an **R**=* is not given a value, the value from the proceeding vertex will be used. Figure 4.41 shows a cubic fillet with the back edge set

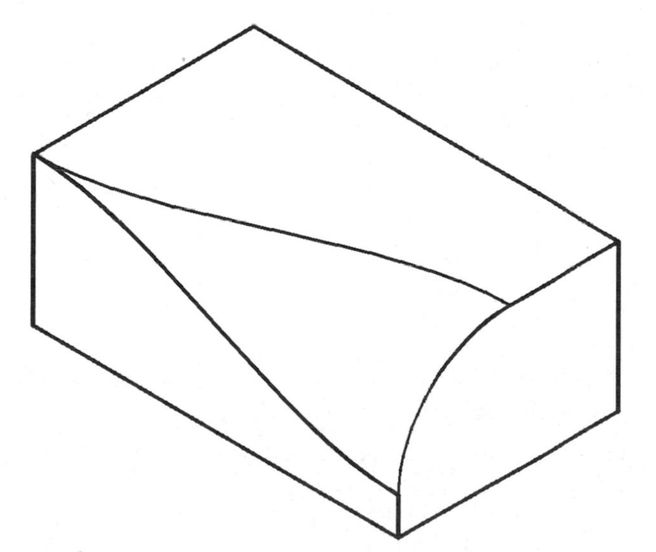

Figure 3.43

to a radius of .001 and the front edge set to .5. The front edge marked with an *X* could be filleted with a constant or cubic fillet.

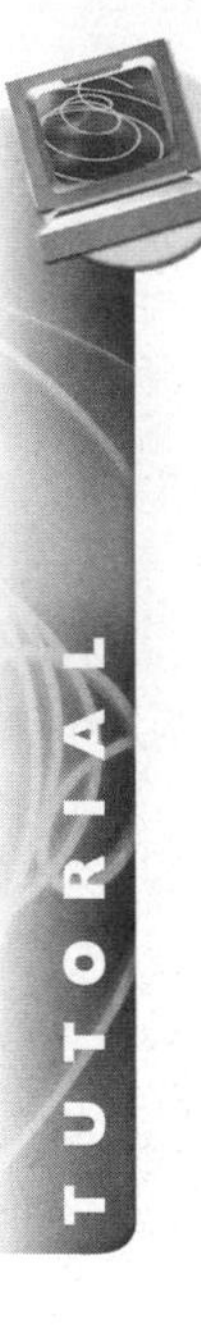

TUTORIAL 3.14 —CREATING A CUBIC FILLET

1. Open the file \Md4book\Chapter3\TU3-14.dwg.
2. Issue the Desktop Fillet command (**AMFILLET**).
3. Check Cubic and then pick OK.
4. Select the front left horizontal edge.
5. At the command line press A to add a vertex and press ENTER.
6. Select near the midpoint of the edge.
7. Type "50" so that the vertex is in the middle of the edge, and an R=* will appear at the midpoint of the edge.
8. Give a value of .125 for each end and .5 for the middle R=* as shown in Figure 3.44.

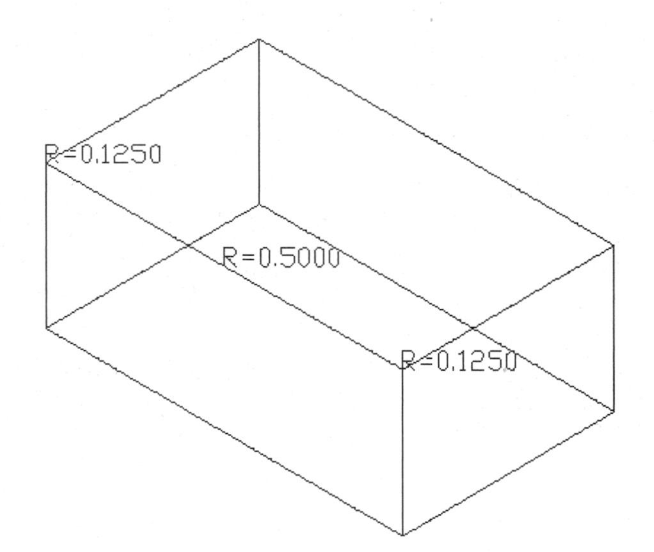

Figure 3.44

9. Press ENTER to accept these values. When complete, your screen should resemble Figure 3.45 shown with lines hidden.
10. Edit the fillet and give the vertices different values. When the values are entered, update the part if needed.
11. Save the file.

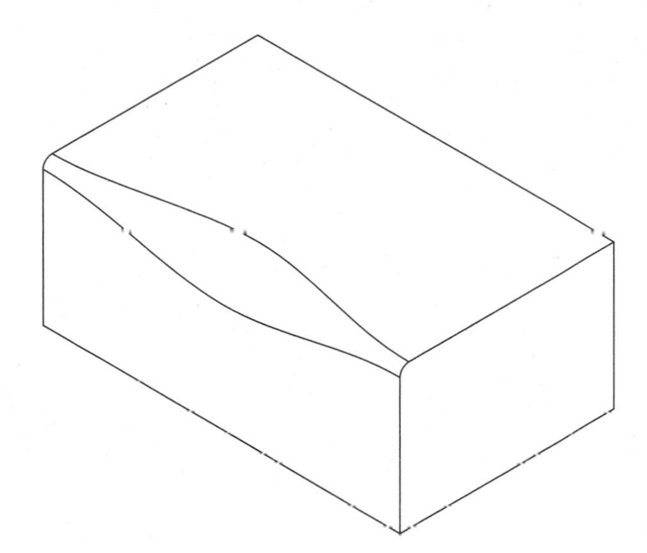

Figure 3.45

FIXED WIDTH FILLET

A fixed width fillet is used when a fillet is created between an angled face and an extrusion that is NOT perpendicular to that face. The fixed width is the chord length between the ends of the fillet. If you drew an arc, then found the distance between the two end points of the arc, this would be the chord length. The chord length will be the same as it goes around the edge. Issue the Desktop Fillet command **(AMFILLET)**, select Fixed Width, type in a value for Chord length, select OK, select the edge you want to fillet, and the fillet will be created. Only one edge can be filleted at a time.

TUTORIAL 3.15 —FIXED WIDTH FILLETING

1. Open the file \Md4book\Chapter3\TU3-15.dwg.
2. Issue the Desktop Fillet command **(AMFILLET)**.
3. Select Constant, type in a radius of .125 and then select OK.
4. Select the edge where the leftmost cylinder joins the angled plane. Press ENTER to accept this selection. When complete, your screen should resemble Figure 3.46 shown with lines hidden. Note that the chord length is not the same around the edge.
5. Change to a southeast isometric view by pressing **8** and ENTER.

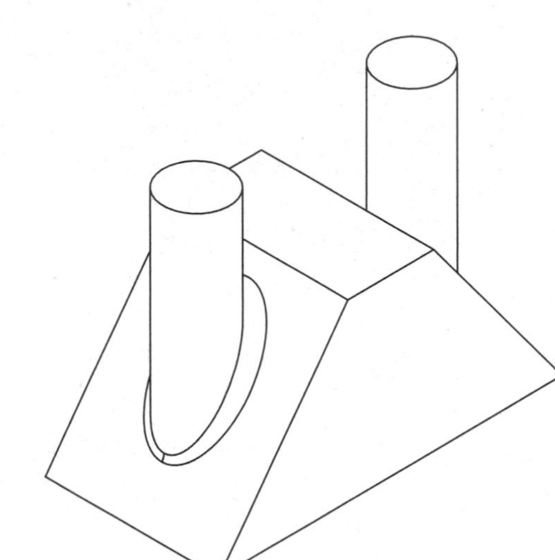

Figure 3.46

6. Issue the Desktop Fillet command **(AMFILLET)**.
7. Select Fixed Width, type in a Chord Length of .125 and then select Apply.
8. Select the edge where the right cylinder joins the angled plane. When complete, your screen should resemble Figure 3.47 shown with lines hidden. Note that the chord length is the same around the edge.
9. Save the file.

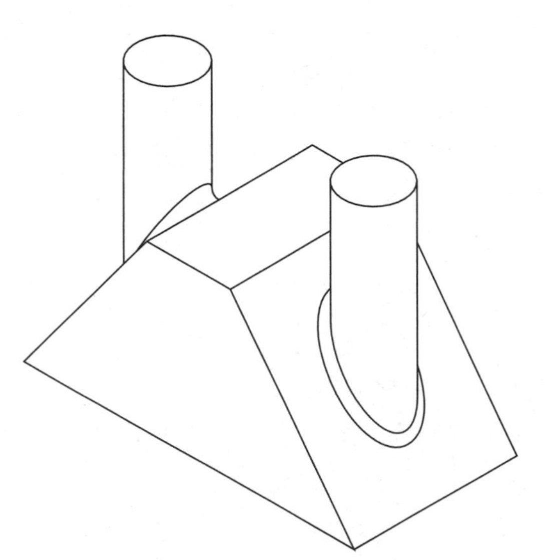

Figure 3.47

Tip: If you get an error when creating or editing a fillet, try to create it with a smaller radius.

After trying to use a smaller filler and you still get an error when creating the fillet, you can try to create the fillets in a different sequence or to create multiple fillets in the same command.

CHAMFERS

To create a 3D chamfer, you will select the common edge and the chamfer will be created between the two faces sharing this edge. Issue the Desktop Chamfer command **(AMCHAMFER)** from the Part Modeling toolbar as shown in Figure 3.48 or right-click in the drawing area and from the pop-up menu, select Chamfer from the Placed Features menu. The chamfer dialog box will appear; choose from three operations: Equal Distance, Two Distances and Distance x Angle. After selecting an operation, type in the required information and then select OK. You will be returned to the drawing where you will select the edge(s) that you want chamfered. A description of the three operations follows.

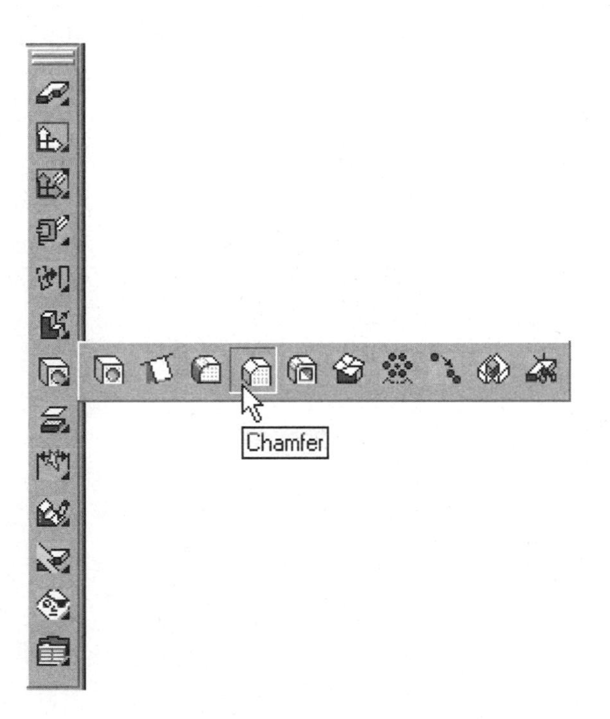

Figure 3.48

EQUAL DISTANCE CHAMFER

The equal distance option will create a 45° chamfer on the selected edge. The size of the chamfer is determined by the distance that you specify in the dialog box. The value is then offset in from the two common faces. From the dialog box, select Equal Distance, type in a value for the Distance1, and then select OK. Select the common edge and a 45° chamfer will be created. Multiple edges can be selected in a single command; these chamfers are linked together—if one changes or is deleted, they are all changed or deleted.

TUTORIAL 3.16 —CREATING A CHAMFER: EQUAL DISTANCE

1. Open the file \Md4book\Chapter3\TU3-16.dwg.
2. Issue the Desktop Chamfer command **(AMCHAMFER)**.
3. Select Equal Distance.
4. Give a value of .5 for Distance1.
5. Select OK.
6. Select the top left horizontal edge and then press ENTER. When complete, your screen should resemble Figure 3.49.

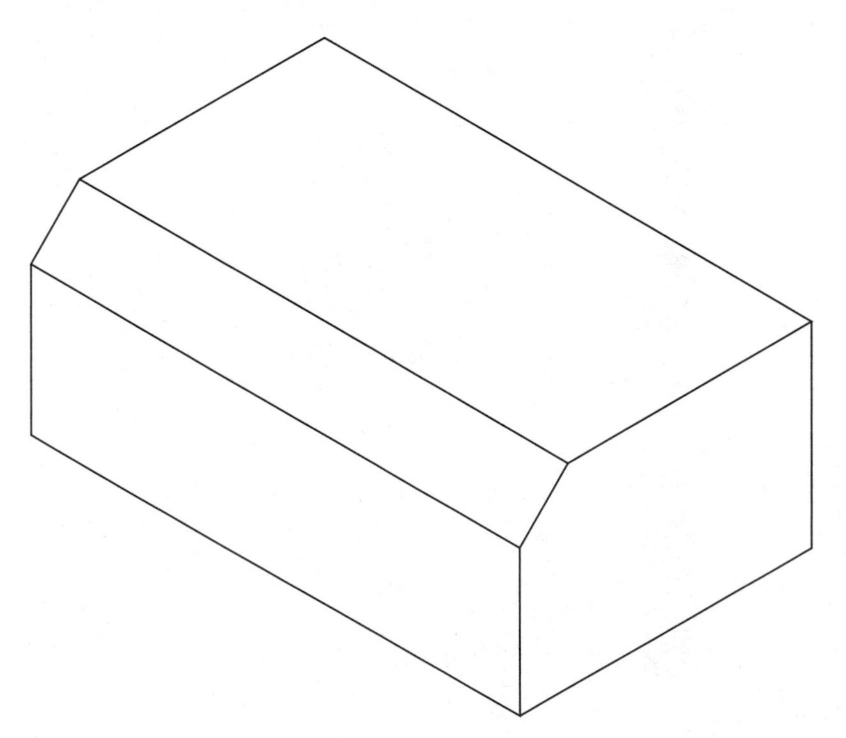

Figure 3.49

7. Repeat the Desktop Chamfer command.
8. Select Equal Distance.
9. Type in a value of .25 for Distance1.
10. Select OK.
11. Select all the remaining outside edges from the existing chamfer, including the 45° edges as shown in figure 3.50.

Figure 3.50

12. To complete the command press ENTER. When complete your screen should resemble Figure 3.51, which is shown with lines hidden and rotated to better show the chamfers.
13. Save the file.

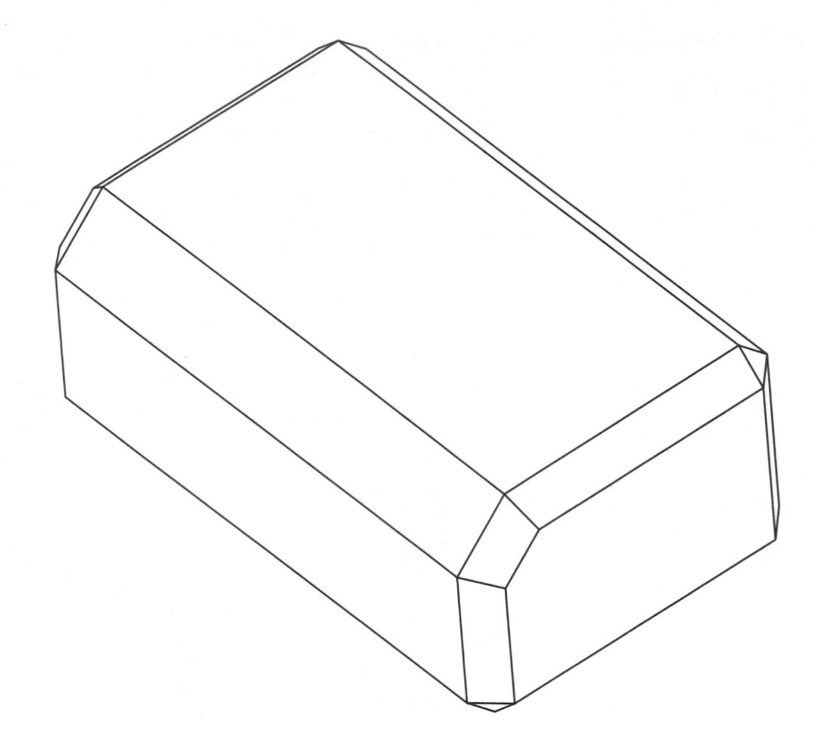

Figure 3.51

TWO DISTANCES CHAMFER

The Two Distances option will create a chamfer offset in from two faces, each the amount that you specify. In the dialog box, select Two Distances and give a value for both Distance1 and Distance2 (remember which distance is the first distance) and then select OK. Select the edge that you want to chamfer and a face will be highlighted. You have the option to accept this as the first face or press N and ENTER to highlight the other face. The highlighted face represents the first face that Distance1 will be applied to. The distance will be offset from this face. When the correct face is selected, press ENTER to create this chamfer. Only one edge can be chamfered at a time with this operation. After you create the chamfer, if it is the reverse of what you were expecting, edit the chamfer and the same dialog box will appear that was used to create it. Reverse the number in the dialog box and update the part.

TUTORIAL 3.17 —CREATING A CHAMFER: TWO DISTANCES CHAMFER

1. Open the file \Md4book\Chapter3\TU3-17.dwg.
2. Issue the Desktop Chamfer command **(AMCHAMFER)**.
3. Select Two Distances.

4. Give a value of ".125" for Distance1 and "1" for Distance2.
5. Select OK.
6. Select the top left horizontal edge as shown with an X in Figure 3.52, and press ENTER to accept this edge.
7. Press ENTER to accept the left vertical face as highlighted in Figure 3.52.

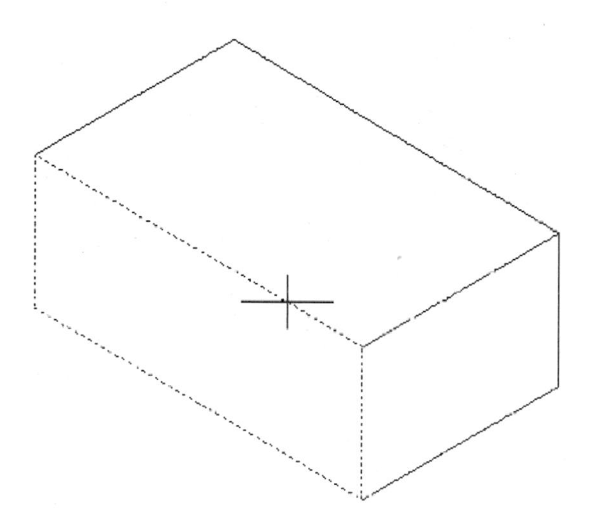

Figure 3.52

8. When complete, your screen should resemble Figure 3.53.

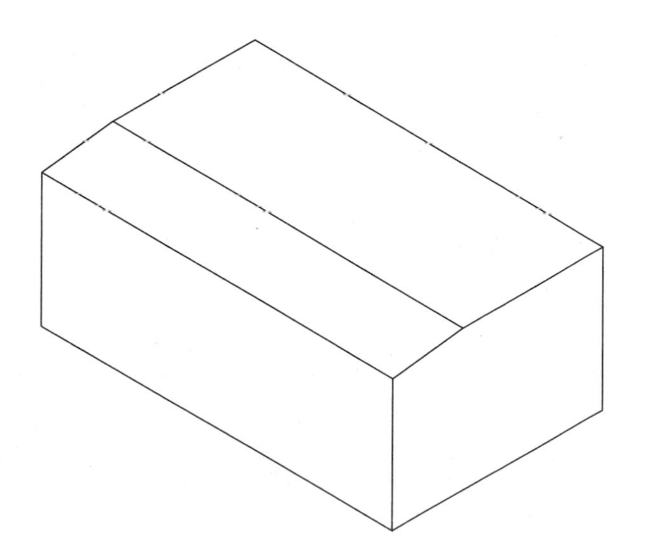

Figure 3.53

9. Create three identical chamfers for the three edges, as marked with an X in figure 3.54. The first face should be the vertical face, as highlighted in Figure 3.52 or 3.54. When complete, your screen should resemble Figure 3.55 shown with lines hidden.
10. Create a .25 equal distance chamfer around all eight front edges (they can all be selected in a single command). When complete, your part should look like Figure 3.56.
11. Save the file.

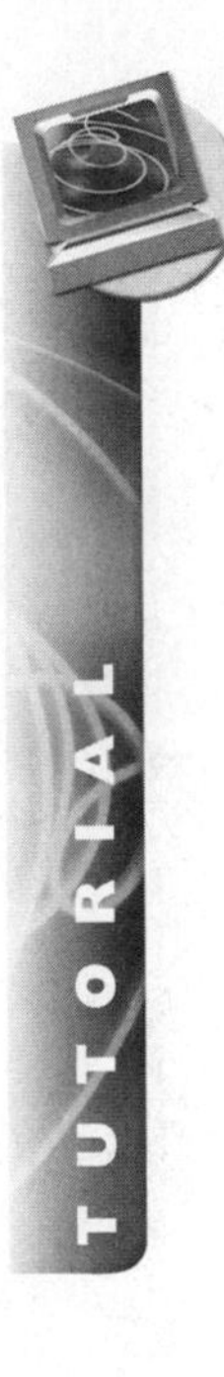

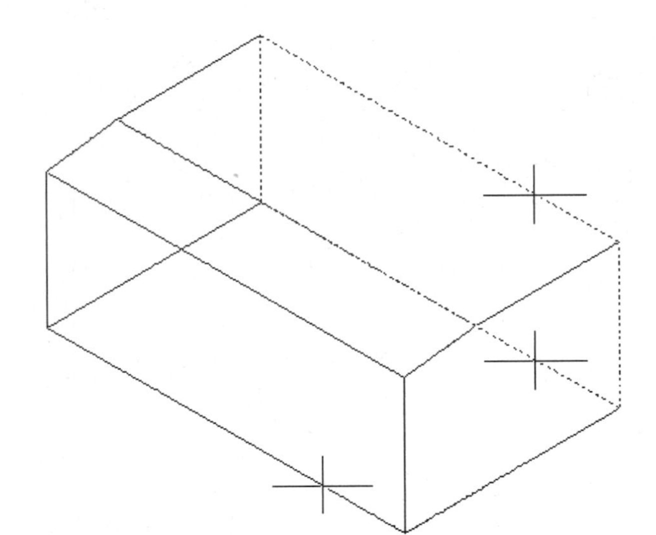

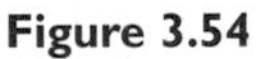

Figure 3.54

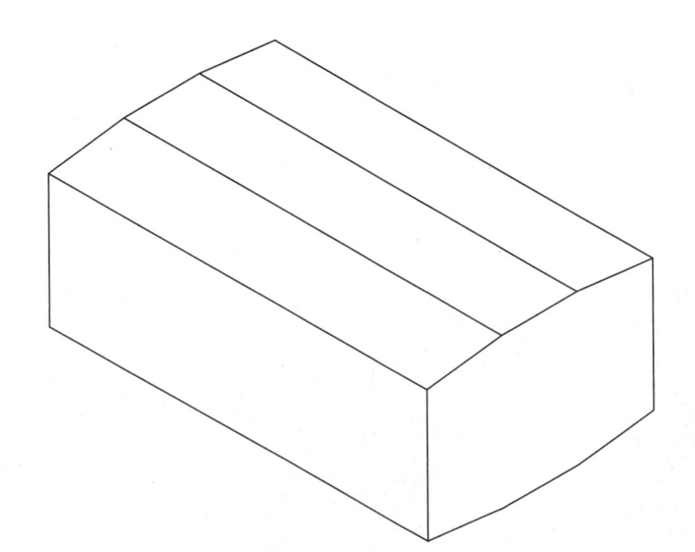

Figure 3.55

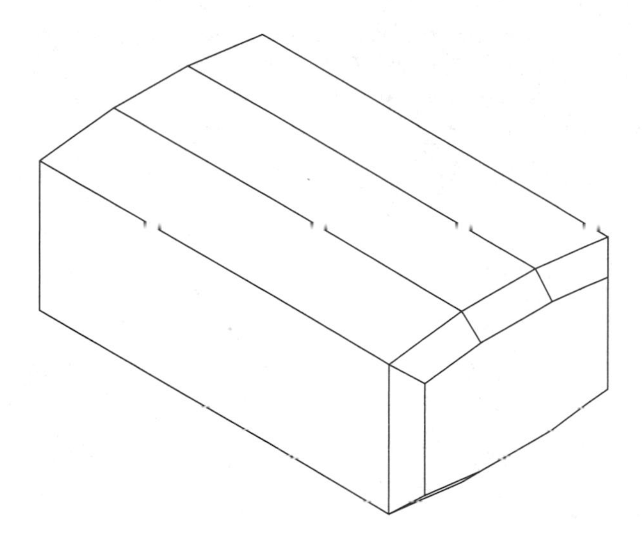

Figure 3.56

DISTANCE X ANGLE CHAMFER

The Distance x Angle option will create a chamfer offset down from a selected edge on a specified face and angled in from the number of degrees specified. In the dialog box, select Distance x Angle and give a value for both the Distance1 and the Angle and then select OK. Select the edge that you want to chamfer and press ENTER. Next a face will be highlighted. You have the option to accept this as the first face or cycle through with the left mouse button. When the correct face is highlighted press enter to create the chamfer. The highlighted face represents the first face that Distance1 will be applied to; the distance will be offset down this face. Only one edge can be chamfered at a time with this operation.

TUTORIAL 3.18 —CREATING A CHAMFER: DISTANCE X ANGLE

1. Open the file \Md4book\Chapter3\TU3-18.dwg.
2. Issue the Desktop Chamfer command **(AMCHAMFER)**.
3. Select Distance x Angle.
4. Give a value of 1 for Distance1 and 60 for the angle.
5. Select OK.
6. Select the top right horizontal edge as shown with an *X* in Figure 3.57, and press ENTER to accept this edge.
7. Press ENTER to accept the right vertical face as highlighted in Figure 3.57. When complete, your screen should resemble Figure 3.58 shown with lines hidden.

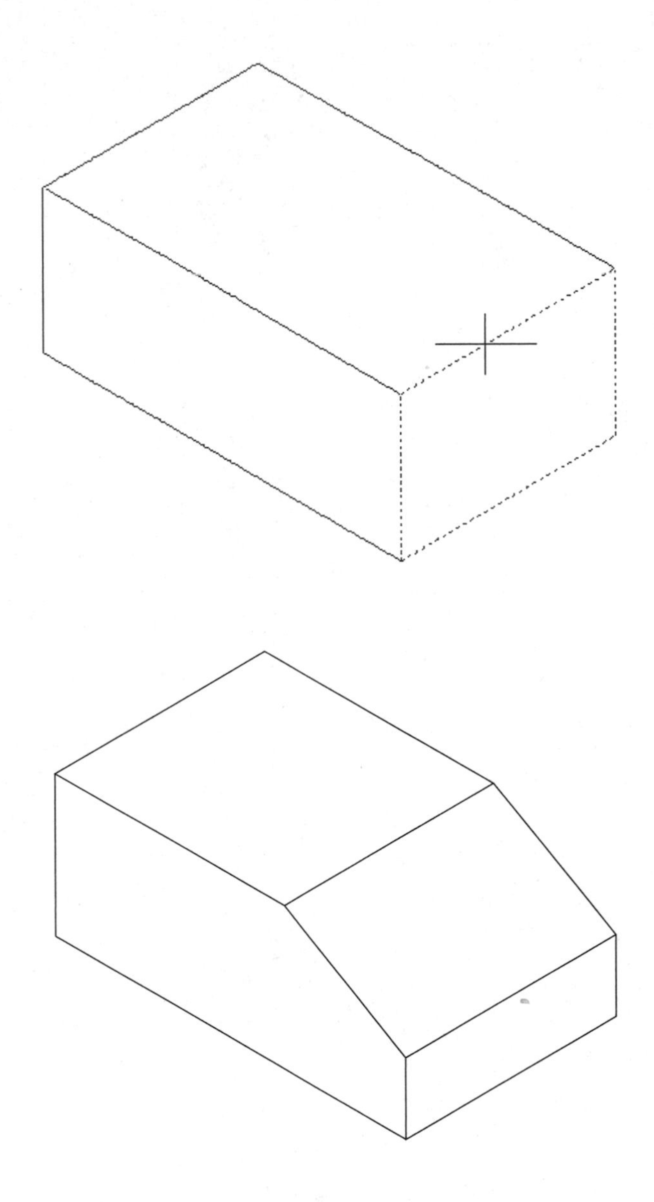

Figure 3.57

Figure 3.58

8. Repeat the Desktop Chamfer command **(AMCHAMFER)** and select OK to accept the previous settings.
9. Create a chamfer on the front bottom edge chamfer. The same front face will be used for the offset, as highlighted in Figure 3.57. When complete, your screen should resemble Figure 3.59 shown with lines hidden.
10. Save the file.

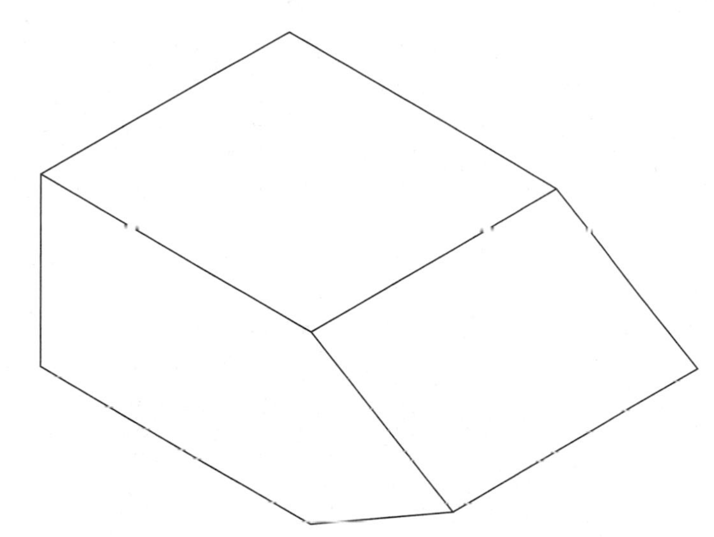

Figure 3.59

HOLES

There are thee basic types of holes that Mechanical Desktop can create: drilled, counter bore and counter sink. Mechanical Desktop can also simulate a tapped hole, but it will only show up in the drawing views, appearing as a drilled hole in the part. When editing a hole, use either the Edit Feature command **(AMEDITFEAT)** or the browser, and the Hole dialog box that created it will appear. Make any necessary changes and select Apply. If dimensions were used to place the hole, they will reappear on the screen. To change a dimension, select it, type in a new value, and press ENTER. Select another dimension to change or press ENTER to exit the command. Update the part if it did not automatically update. The information about the hole will be used when the hole is annotated.

To create a hole, issue the Hole command **(AMHOLE)** from the Part Modeling toolbar as shown in Figure 3.60. Or you may right-click in the drawing area and from the pop-up menu, select Hole from the Placed Features menu; the Hole dialog box will appear as shown in Figure 3.61. The hole dialog box is broken into six areas: Operation, Termination, Placement, Drill Size, C'Bore/Sunk Size and Tapped. A description of the different areas follows. After placing the first hole you can continue placing the same type hole without returning to the dialog box.

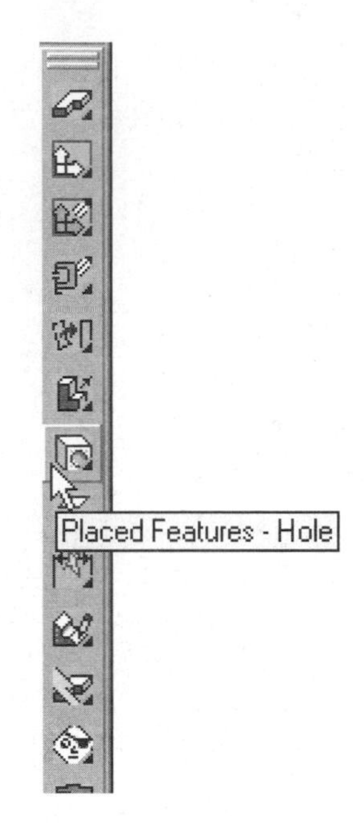

Figure 3.60

Hole

Copy Values

Operation: Drilled
Termination: Blind
Placement: 2 Edges

Drill Size
Dia: 0.5000
Depth: 1.5000
PT Angle: 118

Tapped
Major Dia: 0.6250
Full Thread Depth
Depth: 1.5000

C' Bore/Sunk Size
C' Dia: 0.8750
C' Depth: 0.3750
C' Angle: 45

OK Cancel Help <<

Return to Dialog

Figure 3.61

Copy Values

When selected, you will be prompted to select an existing hole and the values from the hole will populate the hole dialog box.

Operation

Option:	**Function:**
Drilled	Drilled hole.
C'Bore	Counter bore hole.
C'Sink	Counter sink hole.

Termination

Option:	**Function:**
Through	Will go all the way through the part.
Blind	Will stop at a specified depth.
To Plane	Will stop at a selected plane or planar face.

Placement

Option:	**Function:**
2 Edges	Select two edges that define a plane, then select a point near where the hole should be. You will then be prompted for the exact offset distance from each edge.
Concentric	With this option you will select a plane in which the hole should start and then select an arc or circular edge to which the hole will be concentric.
On Point	The hole will be placed on a selected work point.
From Hole	You will be prompted to select a plane in which to place a hole and then you will select a hole to offset for both the X and Y direction. The distances specified will be from center to center.

Drill Size

Option:	**Function:**
Dia	The diameter for the drilled hole.
Depth	The depth for a blind hole.
PT Angle	The point angle for a blind hole.

Note: To simulate a bore, use a point angle of 180 degrees.

C'Bore/Sunk Size

Option:	Function:
C'Dia	Diameter for the counter bore or sink.
C'Depth	Depth for the counter bore.
C'Angle	The angle for the counter sink.

Tapped

Tapped holes are represented in the part as drilled holes. When the drawing views are created they will be represented by the drafting standard such as ANSI, DIN and ISO that you select through the Desktop Options command. To create a tapped hole, first select Drilled hole, type in a diameter for the hole, select a termination and then select Tapped in the dialog box. Type in a value for the major diameter. Select Full Depth if the tap should be the same depth as the drilled hole; otherwise type in a value for the depth of the tap. Select OK to create the tapped hole.

TUTORIAL 3.19—CREATING A HOLE: DRILLED, THROUGH, BLIND, CONCENTRIC AND TWO EDGES

1. Open the file \Md4book\Chapter3\TU3-19.dwg.
2. Issue the Hole command **(AMHOLE)** and make the following changes:
   Operation = Drilled
   Termination = Blind
   Placement = Concentric
   Dia = .5
   Depth = .75
   PT Angle = 118
   Then select OK
3. For the prompt:
   `Select work plane or planar face`
   `[worldXy/worldYz/worldZx/Ucs]:`
   Select a point in the plane as highlighted in Figure 3.62.Figure 3.62
4. For the prompt:
   `Select concentric edge:`
   Select either the top or bottom circle defining the larger cylinder. When complete, your screen should resemble Figure 3.63.
5. To return to the Hole dialog box press ENTER twice.
6. Make the following settings:
   Operation = Drilled
   Termination = Through
   Placement = Concentric
   Dia = .25
   Then select OK

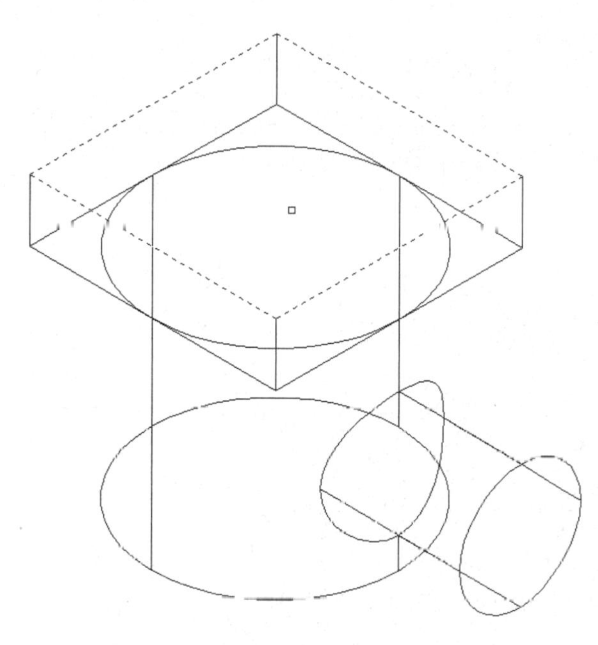

Figure 3.62

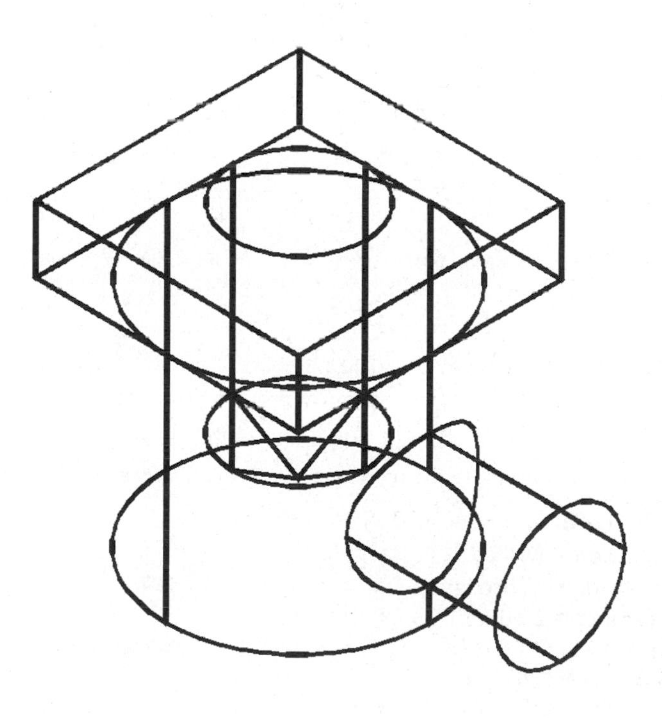

Figure 3.63

7. For the prompt:
   `Select work plane or planar face`
   `[worldXy/worldYz/worldZx/Ucs]:`
   Select on the circumference of the outside boss, as shown with an *X* in Figure 3.64 with the lines hidden for clarity.

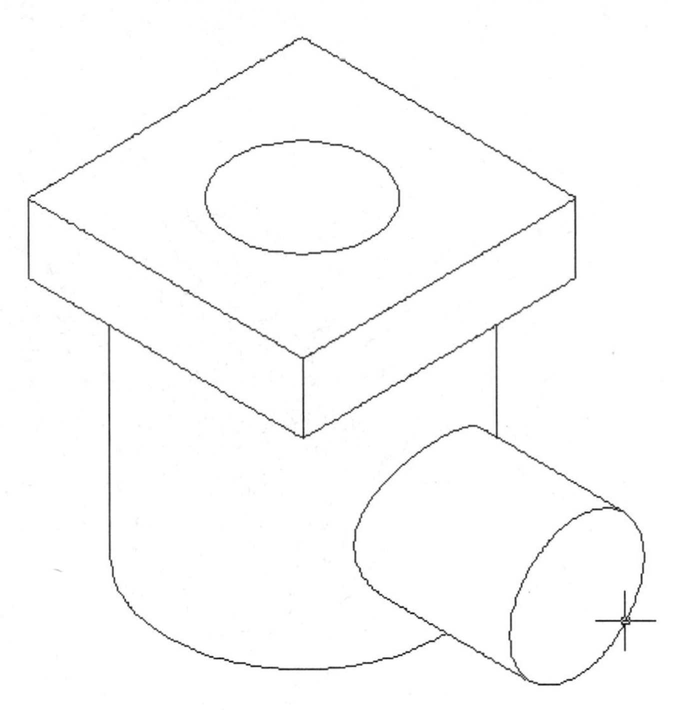

Figure 3.64

8. For the prompt:
   `Select concentric edge:`
   Select the same circle as in step 7 and when complete your screen should resemble Figure 3.65.
9. To return to the Hole dialog box press ENTER twice.
10. Make the following settings:
    Operation = Drilled
    Termination = Through
    Placement = 2 Edges
    Dia = .125
    Then select OK

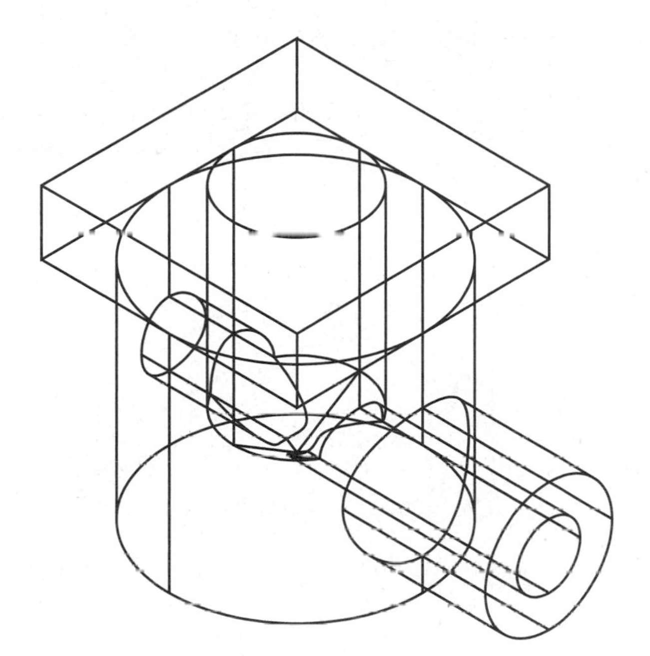

Figure 3.65

11. Select the two highlighted edges and then select a point where the two part lines intersect, as shown in Figure 3.66.

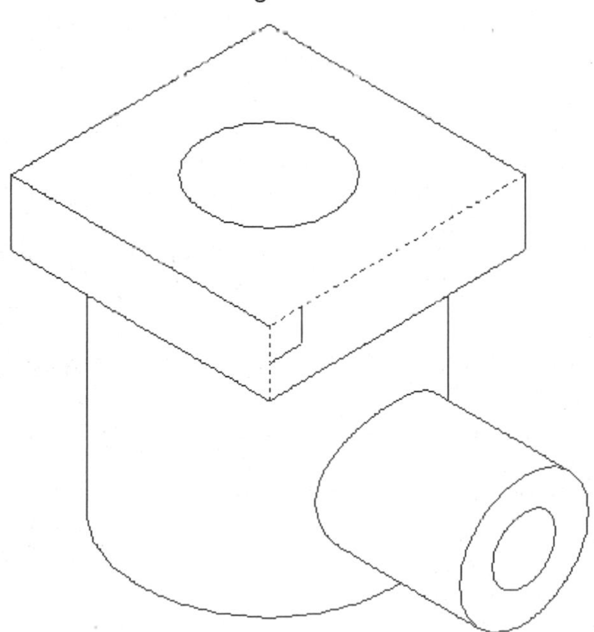

Figure 3.66

12. At the command line, type ".125" for both the distances from the first and the second edge.
13. While still in the command, select the two opposite edges on the same face and type a value of ".125" for each edge. When complete, your screen should resemble Figure 3.67.

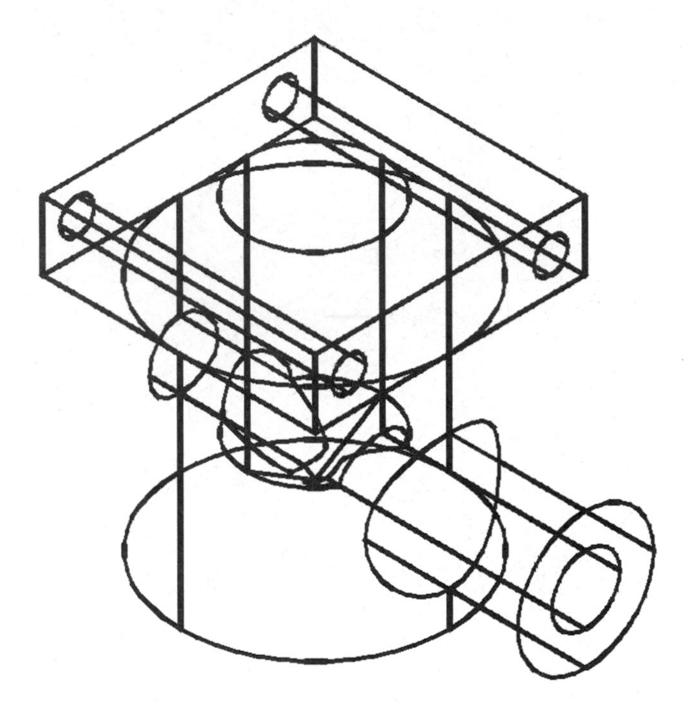

Figure 3.67

14. Save the file.

TUTORIAL 3.20 —CREATING A HOLE: COUNTER BORE, COUNTER SINK, THROUGH, CONCENTRIC AND ON POINT

In this tutorial, the work points (a work point is represented with three lines, one in each axis: *X*, *Y* and *Z*) have already been created and dimensioned. Work points will be covered in Chapter 4.

1. Open the file \Md4book\Chapter3\TU3-20.dwg.
2. Issue the Hole command **(AMHOLE)**.
3. Make the following settings:
   Operation = C' Bore
   Termination = Through
   Placement = On Point

Dia = .14
C' Dia = .25
C' Depth = .14
Then select OK

4. Select the work point on the bottom cylinder and then select the work point on the side of the inner cylinder. Press ENTER to accept the extrusion direction into the part. Press ENTER to exit the command. When complete, your screen should resemble Figure 3.68.

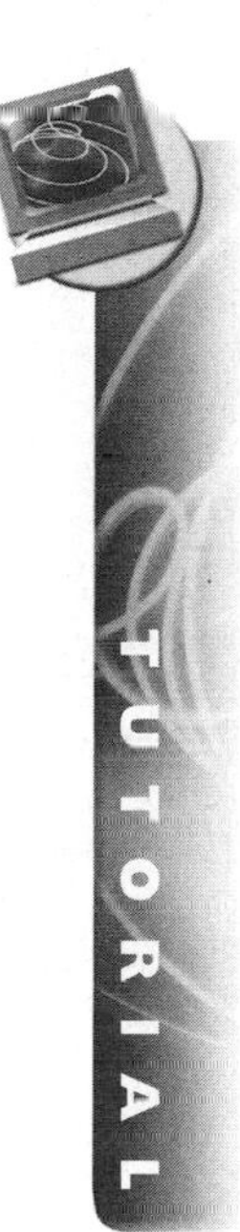

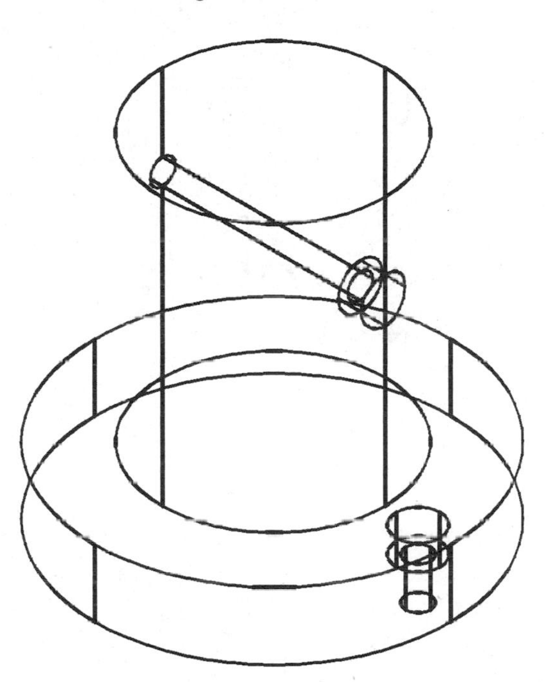

Figure 3.68

5. To return to the Hole dialog box press ENTER.
6. Make the following settings:
   Operation = C' Sink
   Termination = Through
   Placement = Concentric
   Dia = .5
   C' Dia = 1
   C' Angle = 45
   Then select OK
7. Select on the circumference of the top cylinder twice, press ENTER to finish the command. When complete your screen should resemble Figure 3.69.

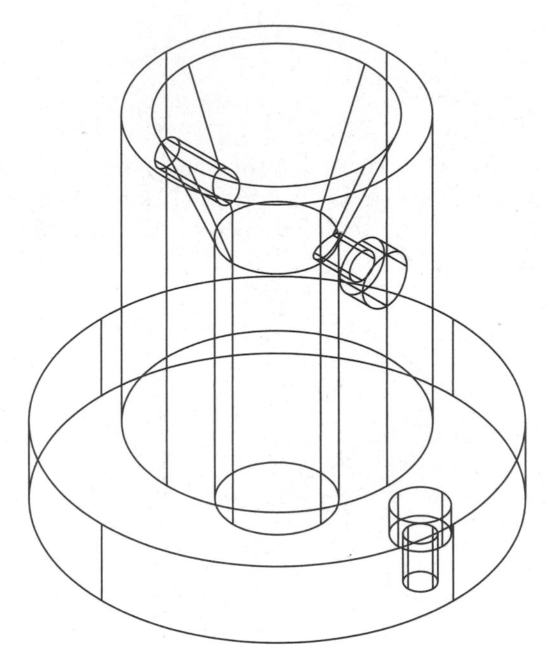

Figure 3.69

8. Save the file.

TUTORIAL 3.21 —CREATING A HOLE: DRILL, TWO EDGES, FROM POINT AND EDITING

1. Open the file \Md4book\Chapter3\TU3-21.dwg.
2. Issue the Hole command **(AMHOLE)** and in the Hole dialog box make the following settings:
   Operation = Drilled
   Termination = Through
   Placement = Two Edges
   Dia = .25
   Then select OK
3. Select the two highlighted edges and then select a point near the intersection of the two lines, as shown in Figure 3.70 with lines hidden for clarity.
4. At the command line, type in a value of .5 for distance from both edges and press ENTER after typing each value.
5. Press ENTER twice to return to the dialog box and in the Hole dialog box make the following settings:
   Operation = Drilled
   Termination = Through

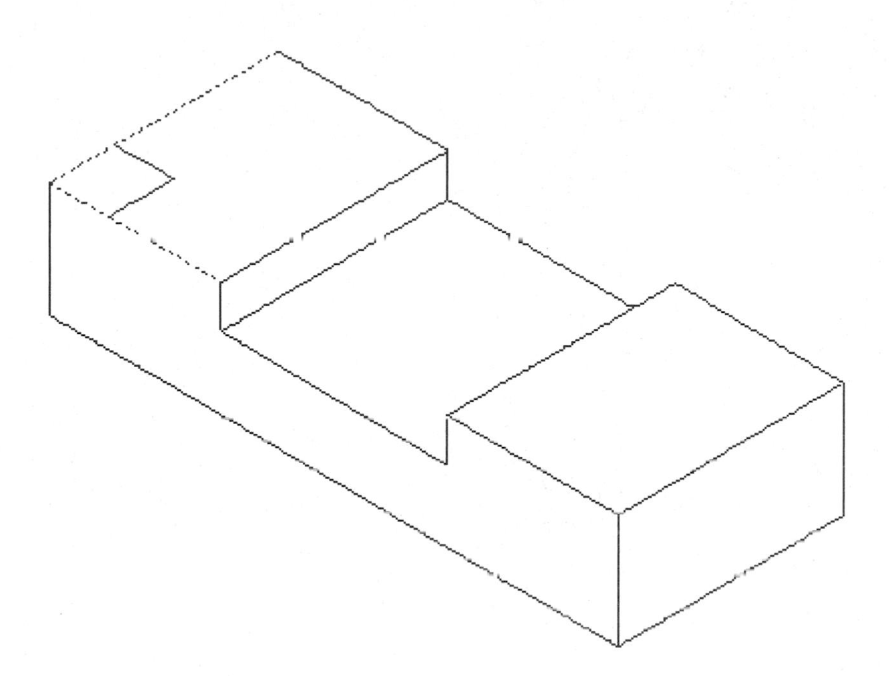

Figure 3.70

Placement = From Hole
Dia = .5
Then select OK

6. At the prompt:
   `Select work plane or planar face [worldXy/worldYz/worldZx/Ucs]:`
   Select the highlighted plane as shown in Figure 3.71 and then press ENTER.
7. At the prompt:
   Specify the orientation for the from hole placement.
   `Select edge to align X axis or [Rotate] <accept>:`
   Select the highlighted edge as shown in Figure 3.72 shown with lines hidden for clarity and press ENTER.
8. At the prompt:
   `Select hole for the X direction reference:`
   Select the hole that was just created.
9. At the prompt:
   `Select hole for the Y direction reference or <previous>:`
   Either select the same hole or press ENTER to use the previously selected hole.

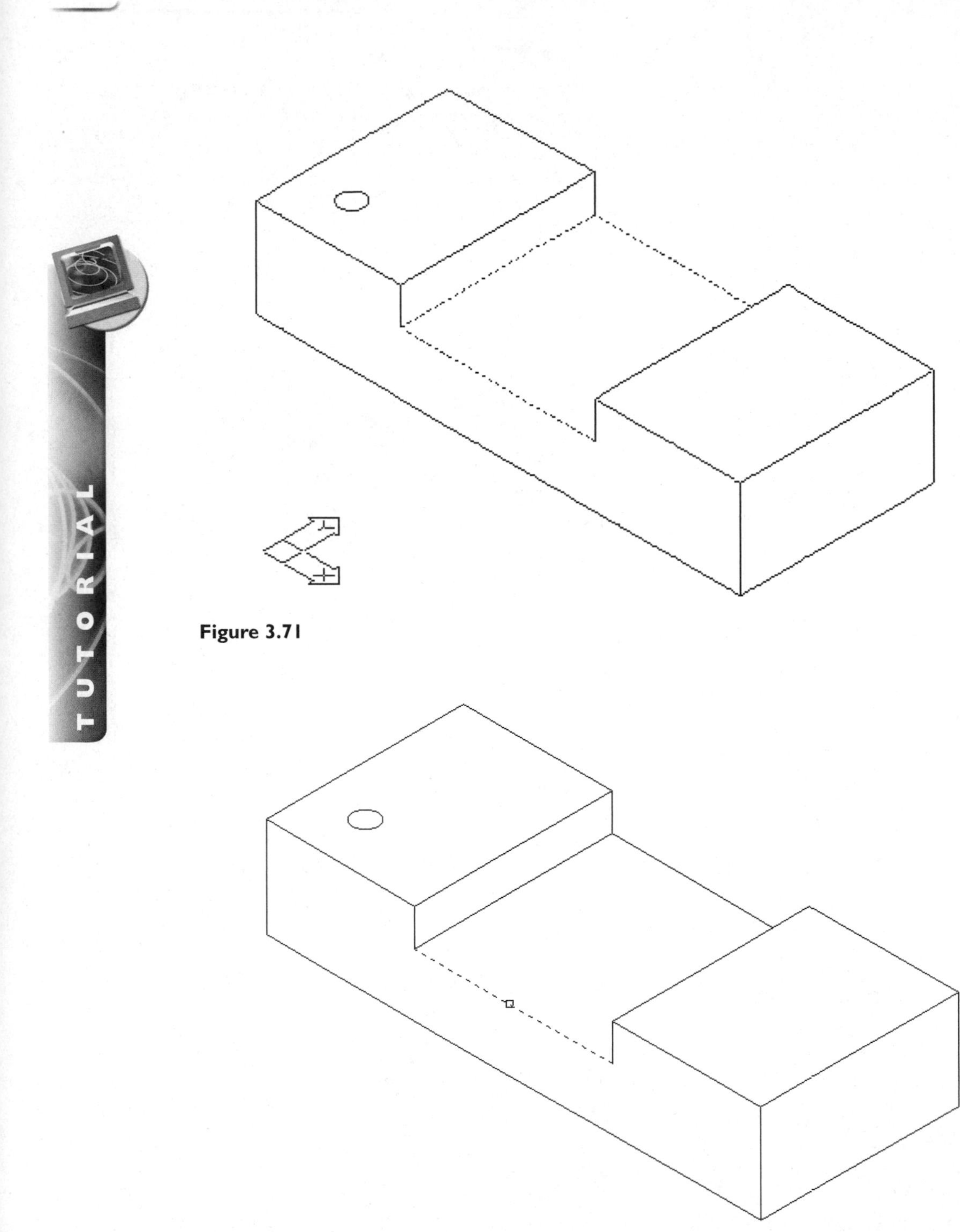

Figure 3.71

Figure 3.72

10. Select a point near the middle of the part, as shown in Figure 3.73 with lines hidden for clarity. As you move the mouse, you are graphically shown the approximate position of the hole.

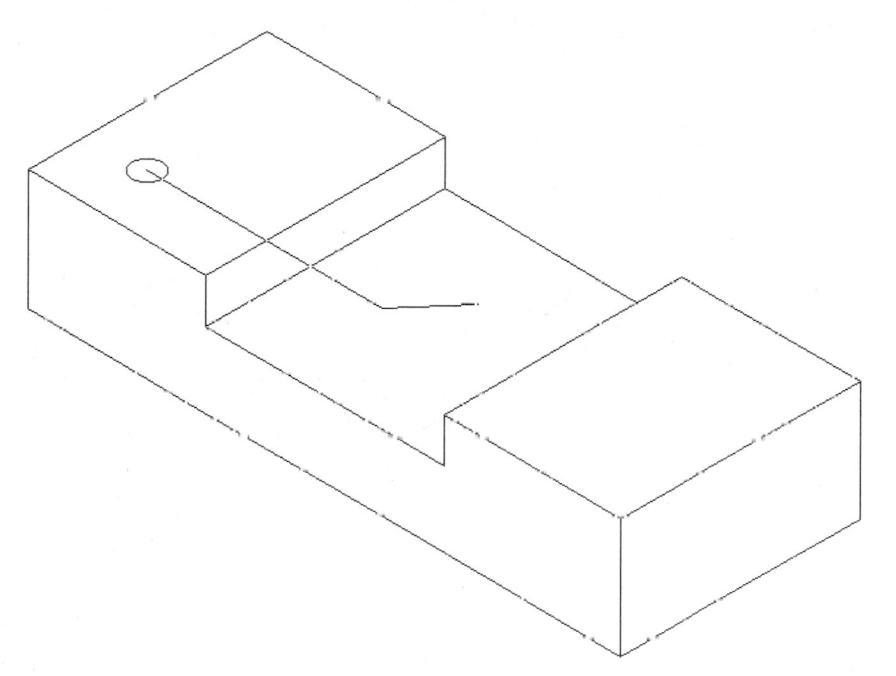

Figure 3.73

11. At the prompt:
    `Enter X distance <1.881393>:` (value will vary depending upon your selected point)
    Type "**2**" and press ENTER.
12. At the prompt:
    `Enter Y distance <0.560419>:` (value will vary depending upon your selected point)
    Type ".5" and press ENTER. When complete, your screen should resemble Figure 3.74 shown with lines hidden.
13. Press ENTER twice to return to the Hole dialog box and in the Hole dialog box make the following settings:
    Operation = Drilled
    Termination = Through
    Placement = From Hole
    Dia = .25
    Then select OK

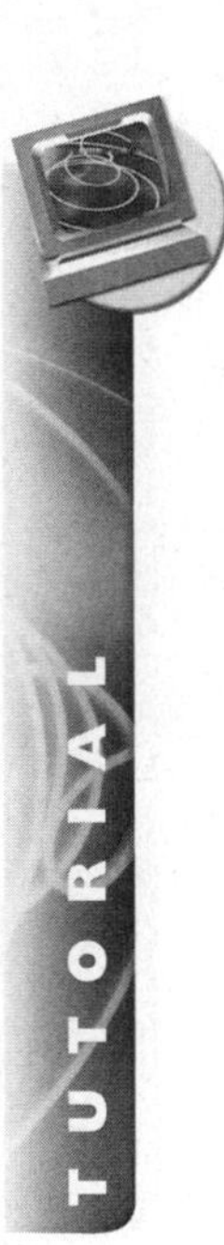

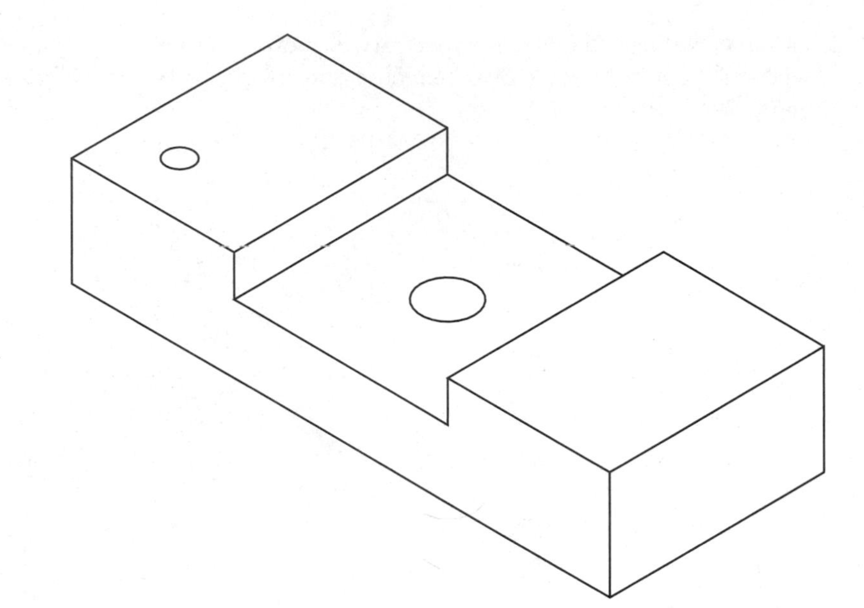

Figure 3.74

14. At the prompt:
    `Select work plane or planar face [worldXy/worldYz/worldZx/Ucs]:`
    Select the highlighted plane as shown in Figure 3.75 shown with lines hidden and then press ENTER.

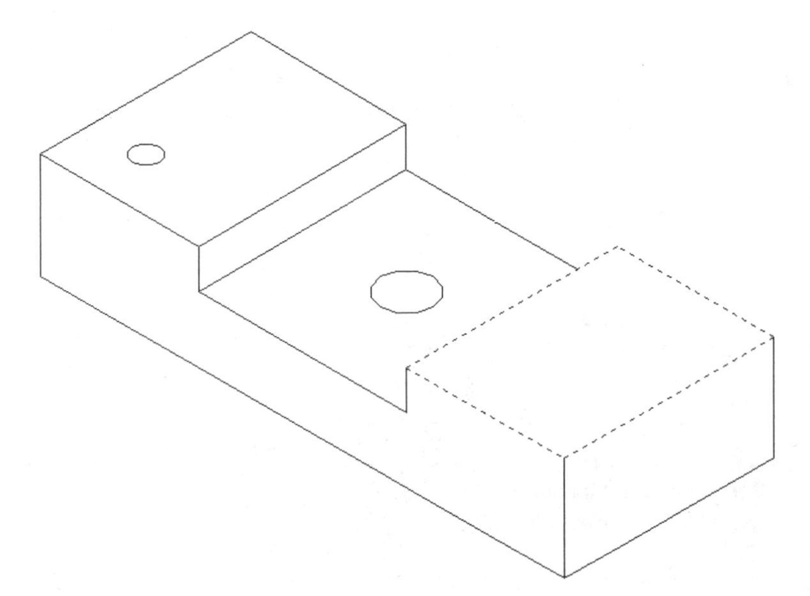

Figure 3.75

15. At the prompt:
    `Specify the orientation for the from hole placement.`
    Select edge to align X axis or [Rotate] <accept>:
    Select the highlighted edge as shown in Figure 3.76 shown with lines hidden and press ENTER.

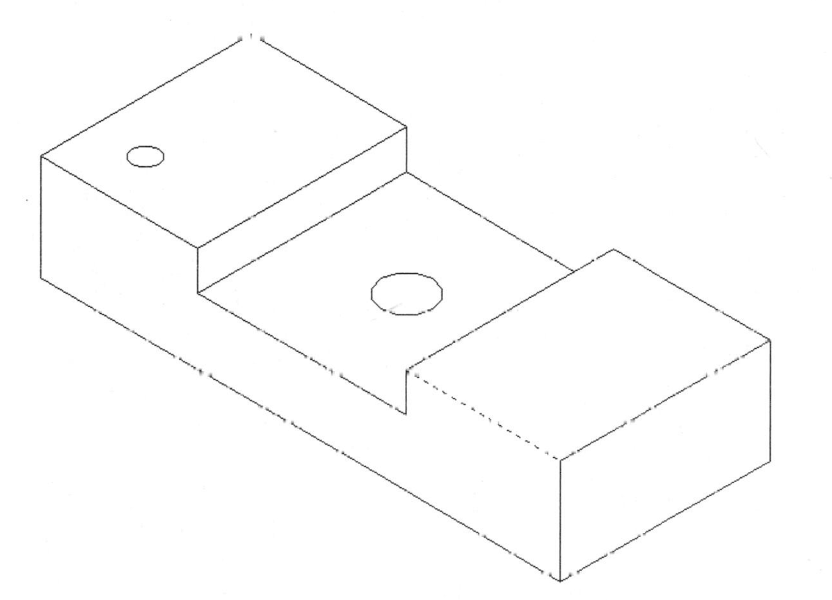

Figure 3.76

16. At the prompt:
    `Select hole for the X direction reference:`
    Select the hole in the lower left corner.
17. At the prompt:
    `Select hole for the Y direction reference or <previous>:`
    Select the hole in the middle of the part.
18. **Select a point near the top right corner of the part, as shown in Figure 3.77. As you move the mouse, you are graphically shown the approximate location of the hole.**
19. At the prompt:
    `Enter X distance <3.872356 >:` **(value will vary depending upon your selected point)**
    Type "4" and press ENTER.

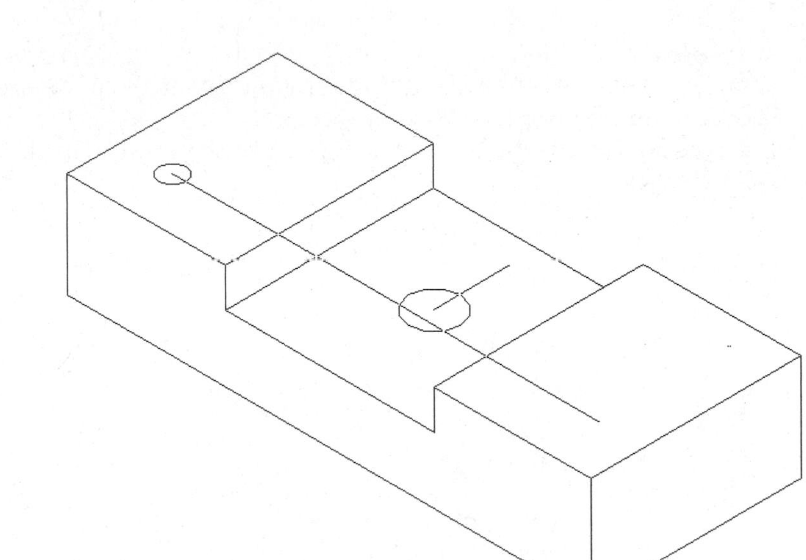

Figure 3.77

20. At the prompt:
    `Enter Y distance <0.480586>:` (value will vary depending upon your selected point)
    Type ".5" and press ENTER. When complete, your screen should resemble Figure 3.78 shown with lines hidden.

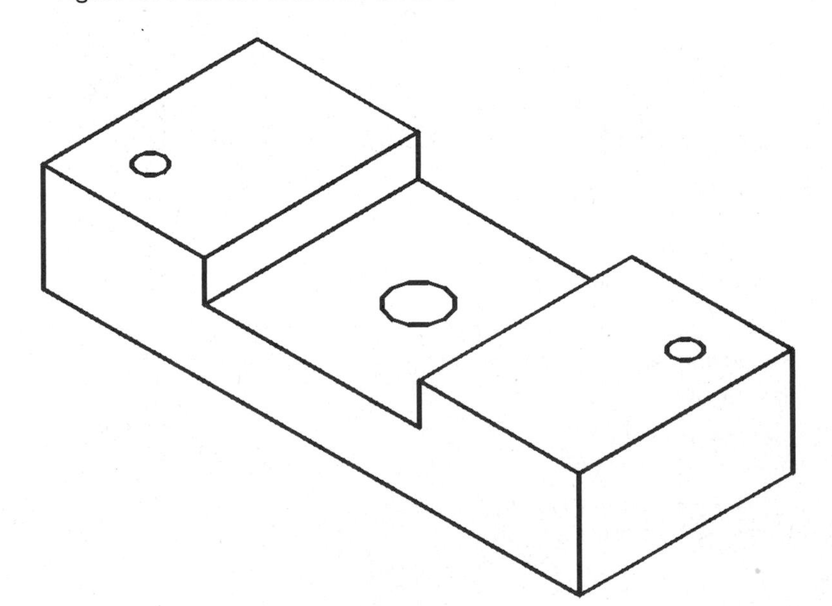

Figure 3.78

21. Edit the last hole created with the Edit Feature command **(AMEDITFEAT)** or use the browser to edit hole3. In the Hole dialog, change the drill diameter to .5 and select OK, change the dimensions from "4" to "3.5" and the ".5" to ".125" and press ENTER.

17. Update the part if necessary. When complete your screen should resemble Figure 3.79 shown with lines hidden.

22. Save the file.

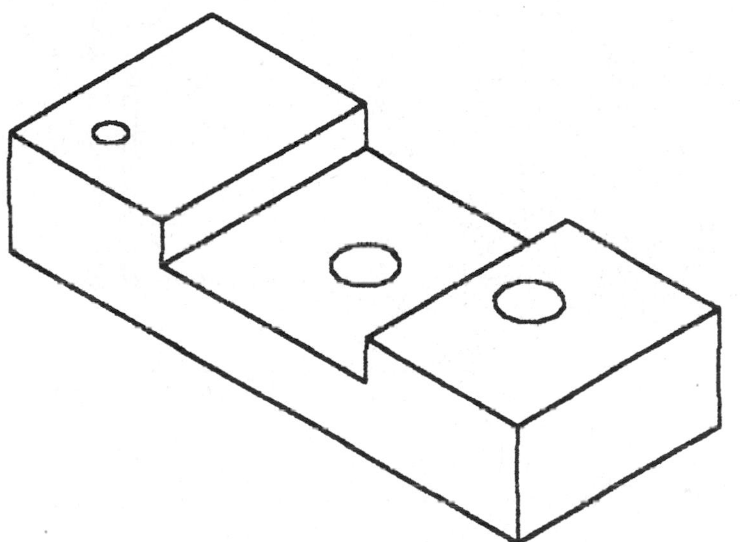

Figure 3.79

EXERCISES

Open each file and then create each part for the following three exercises, as shown for each. When each is complete, save the file. You will create drawing views and add annotations to the drawing views for the Chapter 4 exercises.

Exercise 3.1—Bracket

Open the file \Md4book\Chapter3\bracket.dwg.

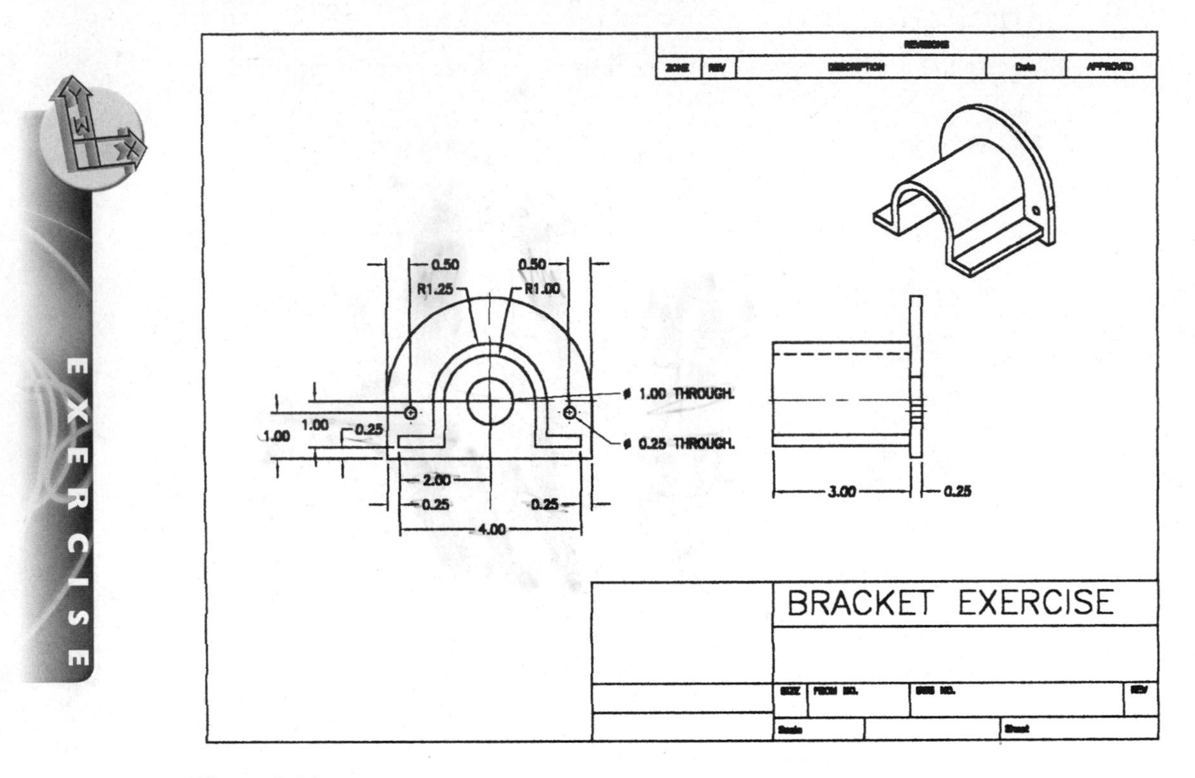

Figure 3.80

Exercise 3.2—Guide

Open the file \Md4book\Chapter3\Guide.dwg.

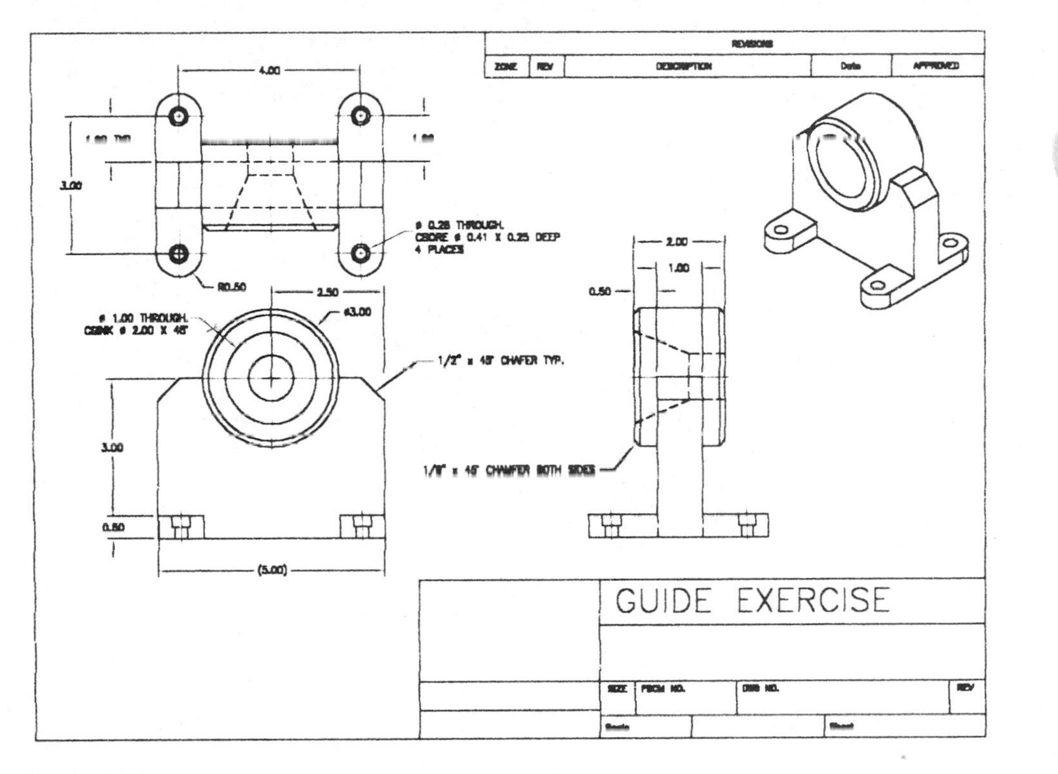

Figure 3.81

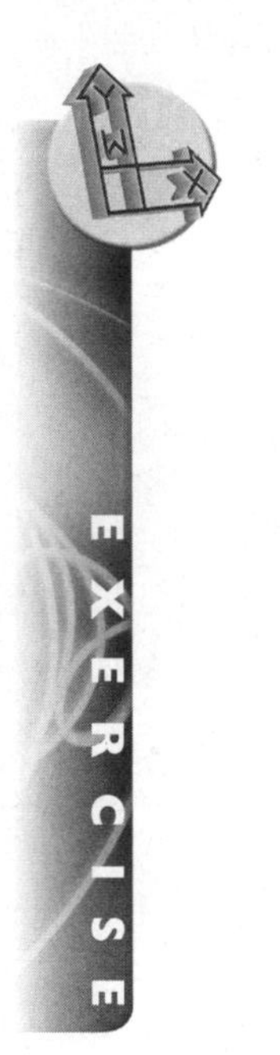

Exercise 3.3—Foot

Open the file \Md4book\Chapter3\Foot.dwg.

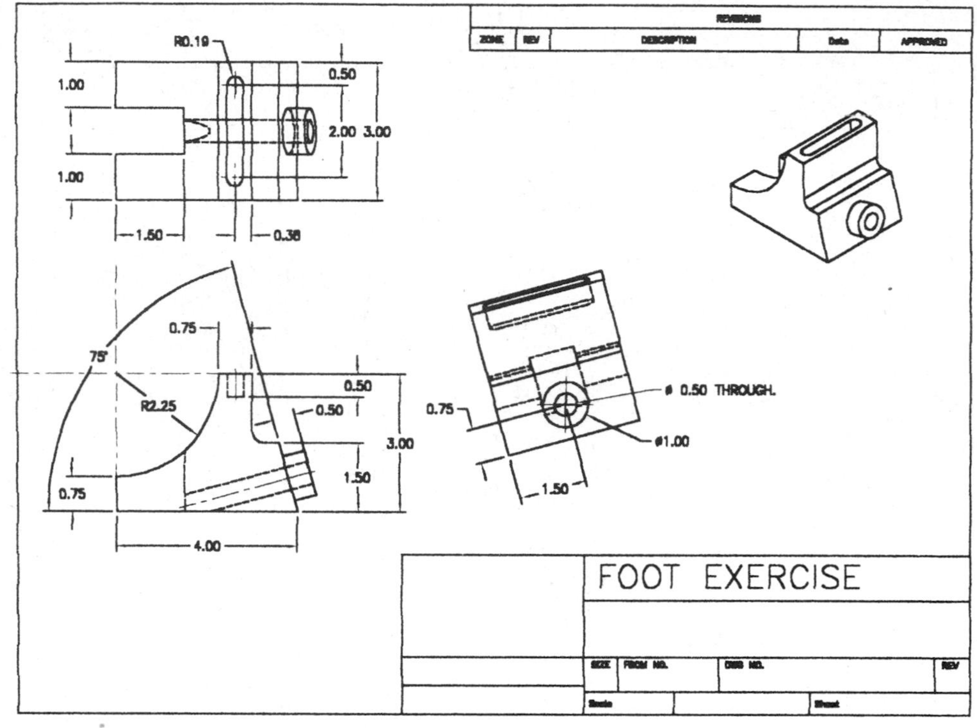

Figure 3.82

REVIEW QUESTIONS

1. What are the three different operation types used for sketched features?
2. A Cut operation can be performed without having a base feature (part). T or F?
3. When a feature is extruded Through a part and the depth of the part increases where the feature was extruded, the feature will no longer go through the part. T or F?
4. What is the difference between a sketch feature and a placed feature?
5. What is the active sketch plane?
6. What is the difference between a linear and a cubic fillet?
7. You can create a cubic fillet on multiple edges by picking the edges in a single command. T or F?
8. Name two ways to edit an existing feature.
9. Once a feature exists, its termination cannot be changed. T or F?
10. When deleting a feature that has a dimension from another feature (child feature) dimensioned to it, the child feature will also be deleted. T or F?

CHAPTER 4

Drawing Views and Annotations

After creating the part or assembly, the next step is to create drawing views. Drawing views can be created at any point after a part exists. The part does not need to be complete, since the part and drawing views are associative in both directions (bidirectional). This means that if the part changes, the drawing views will automatically be updated when you return to them. If a parametric dimension changes in a drawing view, the part will get updated before the drawing views get updated. This chapter will guide you through the steps for creating drawing views of a single part, cleaning up the dimensions and adding annotations. Chapter 9 will guide you through the steps for creating drawing views of assemblies.

AFTER COMPLETING THIS CHAPTER, YOU WILL BE ABLE TO:

- Understand Drawing Options.
- Create drawing views from a part.
- Edit and move drawing views.
- Move, hide, break, join, align, insert, and modify text properties of dimensions.
- Add reference dimensions.
- Add annotations such as geometric, surface, weld symbols, datums and centerlines.

DRAWING TAB

Before creating drawing views make any adjustments under the Drawing and Annotation tabs that can be accessed through the Desktop Options command **(AMOPTIONS)**. Figure 4.1 shows the Drawing option tab. There are five areas of the Drawing tab; each is explained next.

Suppress

Hidden Line Calculations:

When checked, hidden lines will not be calculated and the drawing views will appear like a wireframe model.

View Updates:

When checked, the drawing views will not automatically be updated after changes are made in the model and upon returning to the draw mode.

Drawing Viewport Borders:

When checked, the viewport borders will not be visible.

Use Part Color:

Visible Edges:

When checked, the part will use the parts color for visible (continuous lines) edges.

Hidden Edges:

When checked, the part will use the parts color for hidden (hidden lines) edges.

Hatch:

When checked, all hatch patterns for section views will use the parts color.

Hatch Pattern:

Default Hatch Pattern:

Selecting this button will take you to the hatch Pattern dialog box where you can set the hatch pattern to be used for section views.

One Layer for Hatch Patterns:

When selected, all hatch patterns will go to one layer named AM_8. Otherwise each hatch pattern for each section view will be placed on its own layer; the name is defined by the program at the time the hatch is created.

Display Hatch in ISO Views:

When checked, a pattern will be displayed in isometric views of section views.

Hide Obscure Hatch:

When Display Hatch in ISO Views is checked, this option will be available. When checked, hatch patterns in isometric views that should not be visible (because a visible edge is hiding the hatch pattern) the hatch pattern in this area will be suppressed.

Projection Type:

Third Angle:

When checked, ortho views will be calculated using third angle projection.

First Angle:

When checked, ortho views will be calculated using first angle projection.

Layer Object Settings:

Selecting this button will bring up the Layer/Object Setting dialog box. This is where you can set the layers on which each object type will be created, as well as layer color and line weight

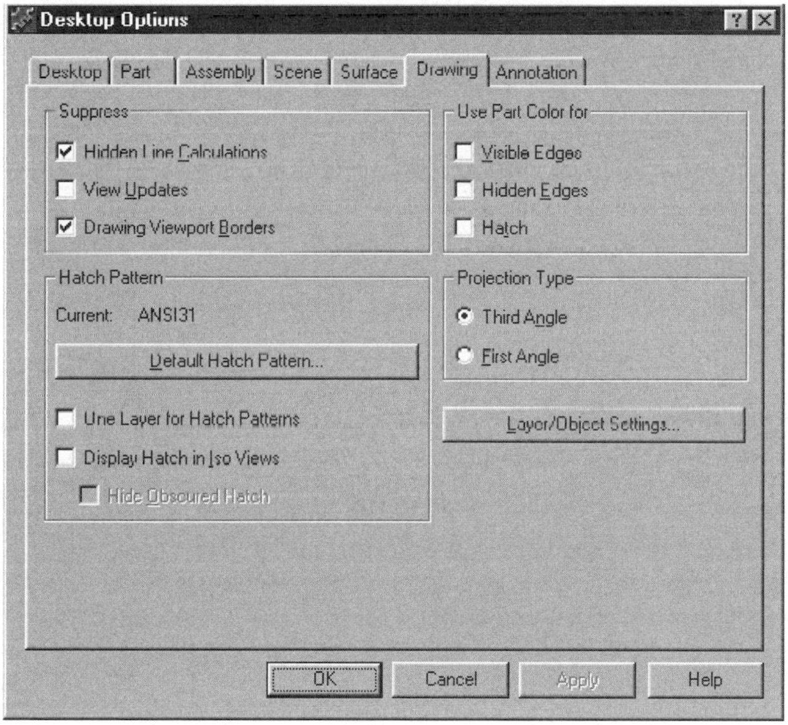

Figure 4.1

ANNOTATION TAB

Parametric Dimension Display:

Active Part:

When checked, parametric dimensions of the active part will be displayed in the appropriate drawing views. When unchecked the visibility of the parametric dimension will be off.

Scenes, Groups, and Selected Objects:

When checked, parametric dimensions of the active part will be displayed for the data sets of scenes, groups and selected objects.

Section Views:

When checked, parametric dimensions of the active part will be displayed in section views.

Hide Zero-Length Parametric Dimensions:

When checked, all dimensions that have a value of "0" will have their visibility turned off.

Automatically Arrange Dimensions:

When checked, the dimensions will be evenly spaced in the drawing views according to the current dimension style regardless of their placement.

On External Parts as Reference Dimensions:

When checked in a file that has parts referenced into it, the parametric dimensions will appear as reference dimensions in the drawing views.

- Drafting Standards:

When this button is selected, the Drafting Standards dialog box will appear. You can set the standard type to be used for Taped Holes, Detail Views and Section Views with htis button. The standards to choose from are: ANSI, DIN, GB, ISO, JIS or None.

Centerline Settings:

When this button is selected, the Centerlines dialog box will appear. It allows you to set whether or not the centerlines should automatically resize if the object size changes, the size of the centerlines and if the cenerlines should automatically be placed on Holes, Fillets and Circular Edges. You can also set the type of view in which these centerlines should automatically appear.

Display Tapped Holes:

When checked, lines representing tapped hole will be generated in the appropriate drawing view. When unchecked, tapped holes will appear as drilled holes.

Save Orphan Annotations in Group:

When checked, and an annotation that was attached to a specific edge in a drawing view is deleted, the resulting annotation will be saved in a group named Orphan. When an annotation is orphaned, a dialog box will appear to alert you. The orphaned annotation can then be deleted or reattatched.

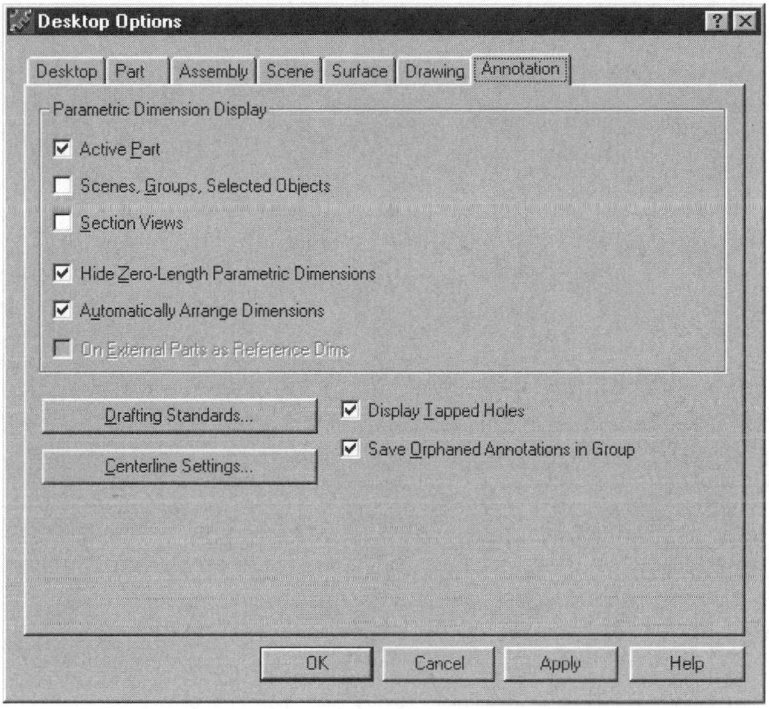

Figure 4.2

CREATING DRAWING VIEWS

Before creating drawing views, you should have created a part or an assembly. Then, to create a drawing view for the current part, issue the New View command **(AMDWGVIEW)** from the Drawing Layout toolbar, as shown in Figure 4.3, or right-click in the layout and select New View from the pop-up menu. You will automatically be placed in drawing mode or drawing manager. Drawing mode does switch to paperspace, but it is recommended that you switch to this mode with the Drawing tab because Mechanical Desktop manages the objects created here. All drawing views are created in paperspace. Insert your title block in paperspace at full scale, because you will plot at full scale (1=1). All layer creation and the choice of layer that objects are placed on that were set through the Layer/Object Settings from the Drawing tab of the Desktop options command **(AMOPTIONS)** are automatically handled by Mechanical Desktop. Then a dialog box will appear, as shown in Figure 4.4. The dialog box is divided into several areas: view type, data set, layout, hidden lines, section views, calculate hidden lines and properties.

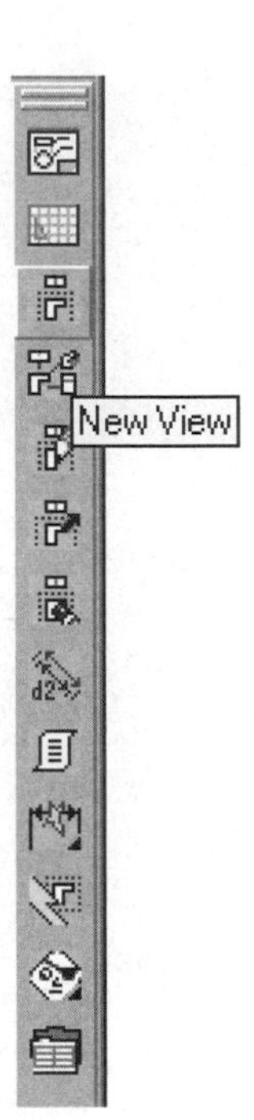

Figure 4.3

Figure 4.4

VIEW TYPE

The first three options: Base, Multiple and Broken are the only three options that are available until a base view exists.

Base: The first drawing view from a part or assembly, the base view is required before other views can be created. It is up to you to decide which view is the base view: top, front or side.

Multiple: You can place a base view, ortho views and isometric views by selecting points on the screen.Press ENTER to place a view (will appear as a wireframe) and continue in this manner until all the views are in place and then press ENTER and the hidden line calculations will be run.

Broken: A view in which the middle of the part is removed and the ends remain. The overall dimension will reflect the correct length of the part.

Ortho: A drawing view that is projected horizontally or vertically from another view.

Aux: A drawing view that is perpendicular to a selected edge of another view.

Iso: A 30° isometric view generated from any view. An isometric view can be projected to any of the four quadrants.

Detail: A selected area of an existing view will be generated at a specified scale.

Depending on whether or not there are existing drawing views in a file, some options may be grayed out.

DATA SET

Active Part: Only the active part will appear in the drawing view.

Scene: You will specify the scene in which the drawing view will be regenerated. This is used for creating views of an assembly.

Select: Only the selected parts and objects will appear in the drawing view.

Group: After you create a group in AutoCAD, this will be a valid option. The parts in the selected groups will appear in the drawing view.

LAYOUT

From the drop down list, specify the layout in which the drawing view(s) should be created.

HIDDEN LINES TAB

Calculate Hidden Lines: When checked, hidden lines will be calculated. Otherwise, all the lines in the view will be a continuous linetype, appearing as a wireframe part.

Display Hidden Lines: When checked, the hidden lines will appear in the drawing views.

Display Tangencies: When checked, tangency geometry will be displayed.

Remove Coincident Edges: When checked, duplicated edges, when found in a drawing, will be erased.

Display As: When calculate hidden lines is unchecked, this area will become available.

Wireframe: All lines in the view will appear as a continuous linetype and look like a wireframe model.

Wireframe with Sihouettes: All lines in the view will be appear as a continuous linetype and look like a wireframe model plus the exterior shape will appear as a continuous linetype.

SECTION VIEWS TAB

After selecting this tab, you will have a choice of creating the section views detailed below, along with the specified hatch pattern and symbol. A section view can be a base view or an ortho view. In Mechanical Desktop, you can also create a sectioned iso view.

None: No section view will be created.

Full: Creates a section view that will go straight through a specified point or selected work plane. When you select a point, the point will go to the center of a selected arc or circle or to the nearest endpoint of a selected line.

Half: Creates a section view that is perpendicular to the specified point. You will be prompted to flip the direction of the cut.

Offset: Creates a section view based on the selected cutting line.

Aligned: An aligned view requires that a cut line exist before an aligned section view is created. An aligned section view has one line that needs to be perpendicular to the view it is being projected to and the other line can be at an angle. Cut lines will be covered later in this chapter.

Breakout: A breakout view requires that a breakout line exist before a breakout section view is created. A breakout line can be a line, arc, circle or spline. Breakout lines will be covered later in this chapter.

Radial: A radial view is a view that will rotated a specified angle and projected perpendicular to the angle.

Symbol: Specify the symbol to be used in the section view.

View Label: Specify the text to be used to label the section view.

Hatch: Check this area if you want the view hatched and to select the pattern.

Hide Obscure Hatch: When checked, hatch patterns in isometric views that should not be visible (because a visible edge is hiding the hatch pattern), the hatch pattern in this area will be suppressed.

PROPERTIES

Scale: Specify the scale for the created view. When you create an Ortho or an Aux view, the scale will be grayed out because the scale is dependent on the view from which it is projected.

Relative to Parent: When checked for isometric views the specified scale will be relative to the view from which it is projected.

To create a base view after issuing the New View command **(AMDWGVIEW)**, select Base for the View Type and type in a scale. Select OK and you will be returned to the part or assembly. Then select a plane, face, work plane or UCS to align the view and orient the *XY* axis for the base view. This sequence will be similar to making a plane or work plane the active sketch plane. After the correct orientation is selected, press ENTER and then select and edge or type and axis to align the view to and then press ENTER. You will be returned to paperspace, where you can select a point to place the view. Keep selecting a point until the location is correct and then press ENTER. The view will then be created at this location. If the scale or location is incorrect, you can edit or move the views later. To create a view based on another view, issue the New Drawing View command **(AMDWGVIEW)** Select the type of view (Ortho, Iso etc.) and select OK. You will be prompted to select a parent view if you issued the command from the icon. If the command was issue from the browser, the parent view is already identified. Select a point in the parent view from which to create this new view. This parent view does not have to be the base view. After selecting the parent view, you will be prompted for a location for the view. Select a point to place the new view. After the location is correct, press ENTER and the view will be created.

TUTORIAL 4.1—CREATING DRAWING VIEWS FROM A PART

1. Open the file \Md4book\Chapter4\TU4-1.dwg.
2. Click on the Drawing tab or Drawing Layout icon on the express toolbar.
3. To create a base view, issue the New Drawing View command **(AMDWGVIEW)**. In the Create Drawing View dialog box, change the scale to ".25" and the other values will be the defaults. Then select OK. You will be returned to part mode.
4. Select in the middle of the highlighted face as shown in Figure 4.5 and ENTER.

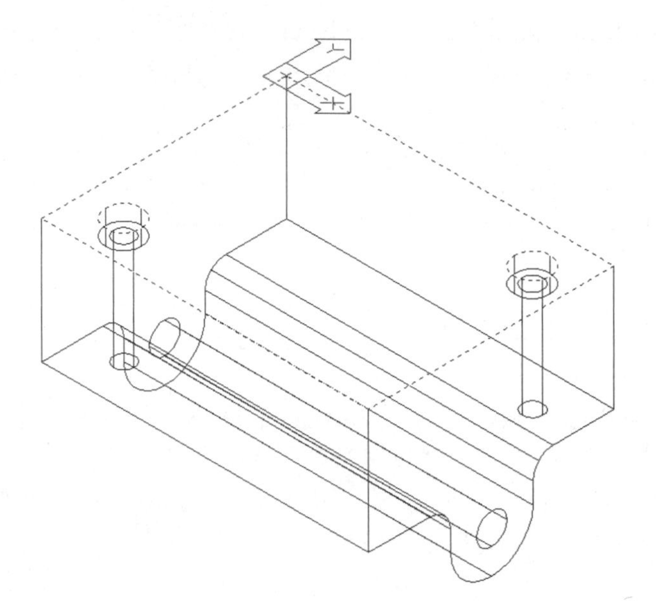

Figure 4.5

5. For the prompt
   `Select work axis, straight edge or [worldX/worldY/worldZ]:`
   Select the top horizontal edge of the previously highlighted plane as defined by the "X" axis of the ucs icon. (This edge will define the orientation of the X axis of the view.)
6. An arrow will appear, and the direction that the arrow is pointing will define the positive *Z* axis. Press ENTER to accept the orientation out of the part.
7. You will be returned to the layout; select a point in the upper left corner of the existing border and press ENTER to create a drawing view that resembles the view shown in Figure 4.6.
8. Press ENTER or right click and select Repeat New View from the pop-up menu to repeat the New Drawing View command.
9. Select Ortho as the View Type and then select OK,. Select a point in the top (parent) view and then select a point to the bottom to the view and press ENTER.
10. Press ENTER or right click and select Repeat New View from the pop-up menu to repeat the New Drawing View command. Select OK, select a point in the front view and then select a point to the right of the side view. When complete, your screen should resemble Figure 4.7.

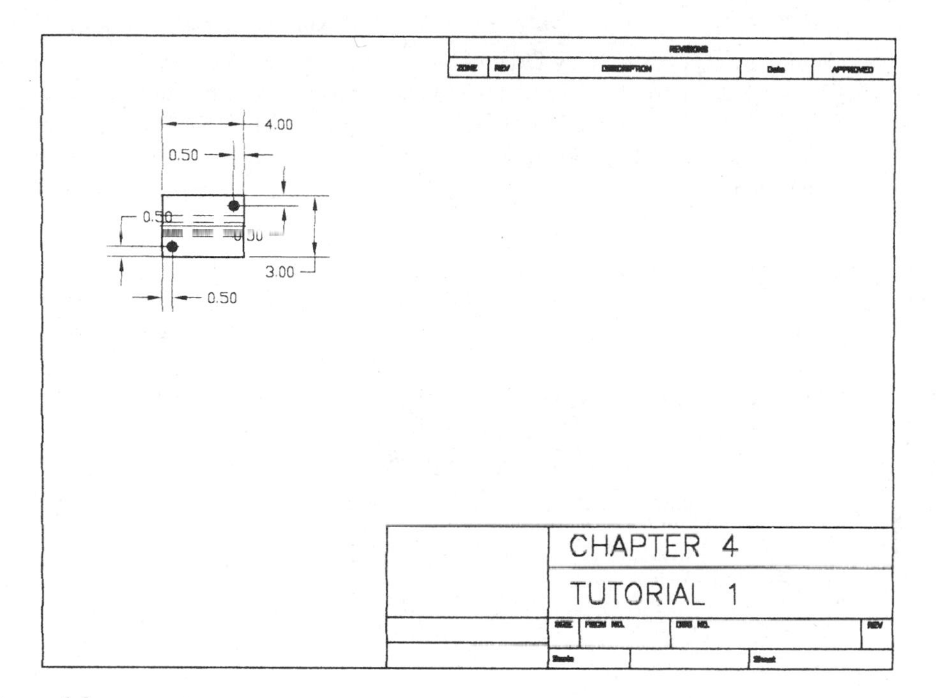

Figure 4.6

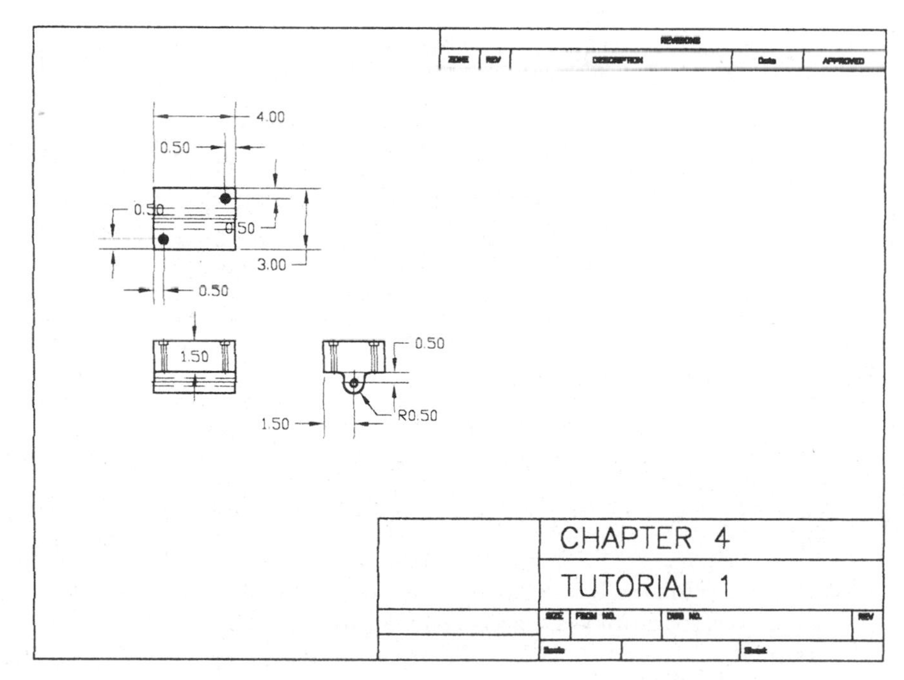

Figure 4.7

TUTORIAL

11. Press ENTER or right click and select Repeat New View from the pop-up menu to repeat the New Drawing View command. Select Section View tab and select Full for the Type. Then select OK to accept the other defaults. Select any point in the right side view, then pick a point to the right so that the section view will be located to the right of the view and press ENTER to accept this location.

12. For the prompt:
    Enter section through option
    `[Point/Ucs/Work plane] <Work plane>:`
    press P and press ENTER. Select a point on the bottom arc (this is the line where the section line is drawn). When complete, your drawing should look like Figure 4.8.

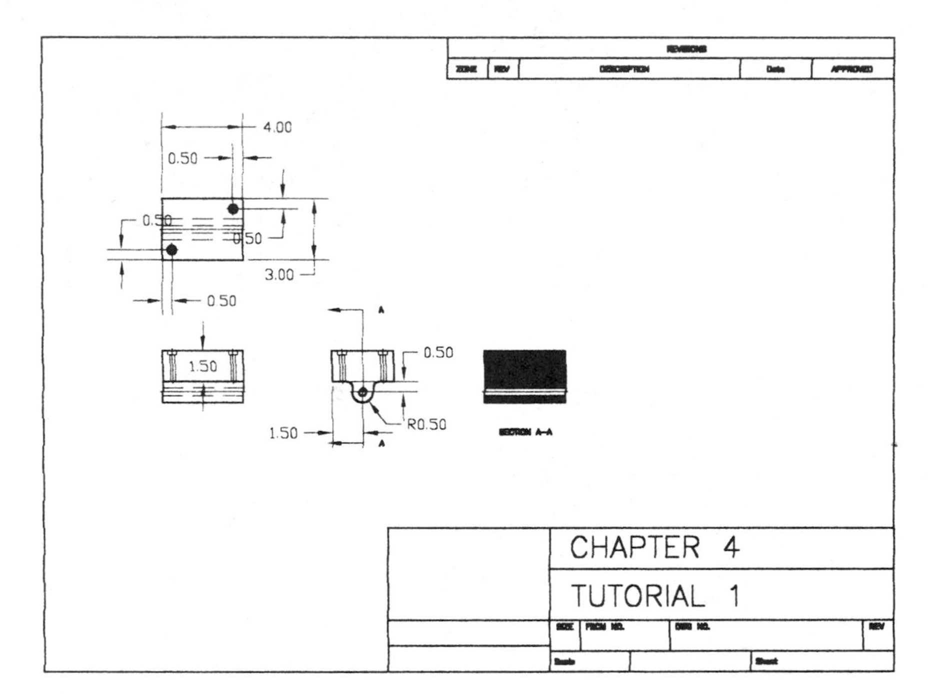

Figure 4.8

13. Press ENTER to repeat the Create Drawing View command. Select Iso for the View type, change the scale to ".75", check relative to parent and then select OK.

14. Select a point in the front view and then locate the isometric view in the upper right corner of the drawing. When complete, your drawing should resemble Figure 4.9.

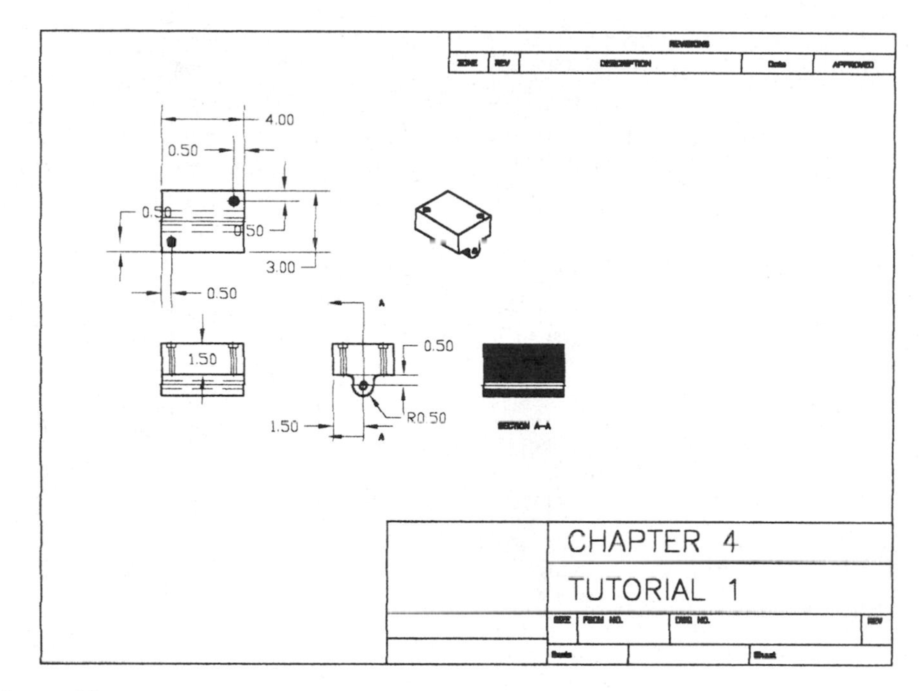

Figure 4.9

15. Save the file.

In tutorial 4.7 you will edit this drawing so it better fits the paper.

MULTIPLE VIEWS

To create multiple views within the same command without returning to the dialog box either issue the New Drawing View command **(AMDWGVIEW)** and select Multiple for the view type or select the Multi Views icon as shown in Figure 4.10. The first view you create will be the base view and then select points where the ortho or isometric view(s) will be projected from this base view. After selecting the location for the base view press ENTER then select the location for the projected view; a highlighted image will appear to show the outcome. When the location is correct press N and then ENTER. Continue selecting locations for the views in this manner and then press ENTER to complete the command. Then all the views will be calculated in one operation. Section views will need to be created after the multiple views are placed.

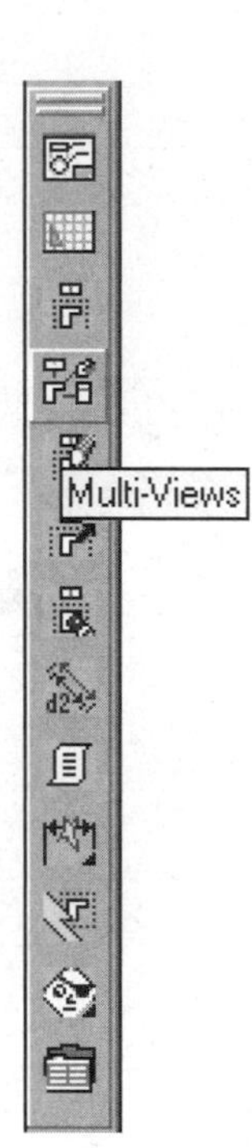

Figure 4.10

TUTORIAL 4.2—CREATING DRAWING VIEWS WITH THE MULTIPLE OPTION

1. Open the file \Md4book\Chapter4\TU4-2.dwg.
2. Click on the Drawing tab or Drawing Layout icon on the express toolbar.
3. To create multiple views in single command string, issue the New Drawing View command **(AMDWGVIEW)** and select Multiple for the View Type or select the Multi Views icon as shown in Figure 4.10. In the Create Drawing View dialog box, change the scale to ".5" and the other values will be the defaults. Then select OK. You will be returned to part mode.
4. Select in the middle of the highlighted face as shown in Figure 4-11 and ENTER.
5. For the prompt.
   `Select work axis, straight edge or [worldX/worldY/worldZ]:`
   Select the bottom horizontal edge of the previously highlighted plane as defined by the "X" axis of the ucs icon. (This edge will define the orientation of the *X* axis of the view.)

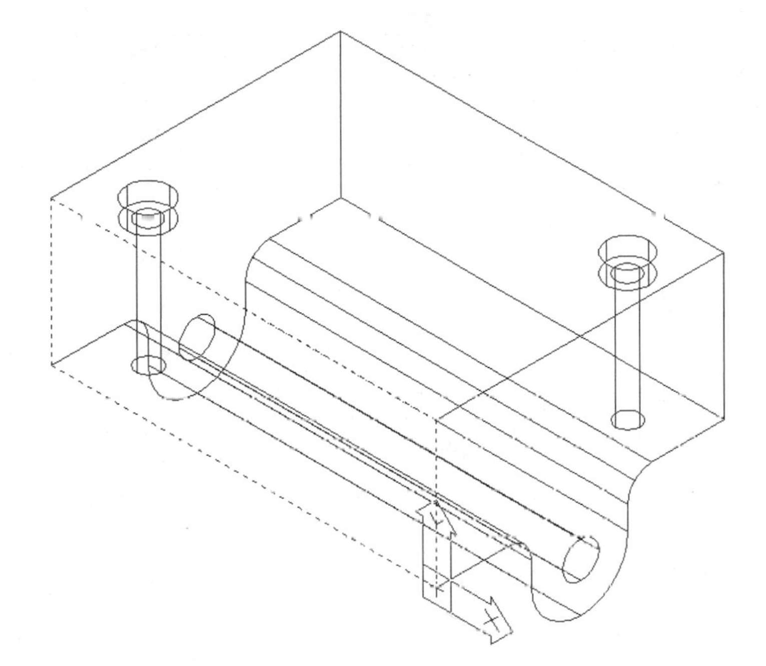

Figure 4.11

TUTORIAL

6. An arrow will appear, the direction that the arrow is pointing will define the positive Z axis. Press ENTER to accept the orientation out of the part.
7. You will be returned to the layout; select a point in the lower left corner of the existing border and press ENTER.
8. To create the right side view, select a location to the right of the front view and press the N and then ENTER.
9. To create a top view, select a location to the top of the front view and press the N and then ENTER
10. To create an isometric view: turn off ortho mode (F8), then select a location inside the top right corner of the border and then press ENTER. When complete, your screen should resemble Figure 4.12.
11. Save the file.

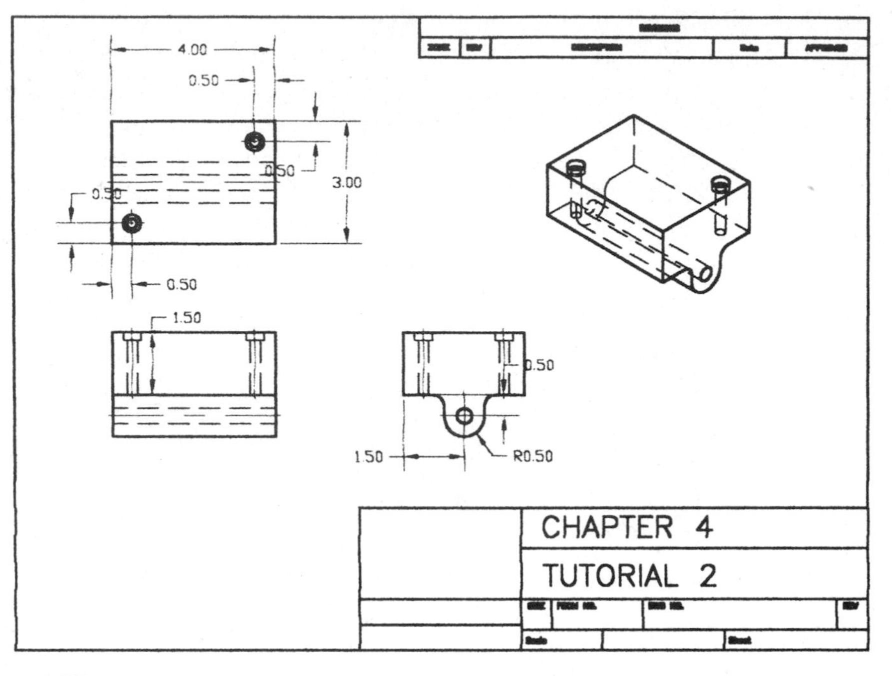

Figure 4.12

SECTION VIEWS

There are three types of section views that require you to create objetcs that define the section line. The three type are: offset, aligned and breakout section views. Offset and aligned section views require a cutline and a broken section view requires breakline. Each tells Mechanical Desktop what the objects are going to be used for, similar to the profile command.

OFFSET SECTION VIEWS

An offset section view is cut out of the part using horizontal and vertical line segments. Before creating an offset section view you will draw either lines or polylines on the current sketch plane. The lines must be horizontal or vertical—no angled lines are permitted. The lines must at least touch or exceed the outside edge of the part. The lines that enter and exit the part must be either horizontal or vertical, parallel to each other and perpendicular to the view being projected. After sketching the objects issue the Cut Line command **(AMCUTLINE)** from the part Modeling toolbar as shown in Figure 4.13 or right-click in the drawing area and from the pop-up menu, select Cut Line from the Sketch Solving menu. Select the objects and then constrain and dimension them if desired. The same constraining and dimensioning rules apply to cut lines as well as to profiles.

ALIGNED SECTION VIEWS

An aligned section view is cut out of the part using two lines; horizontal, vertical or angled line segments. Before creating an aligned section view you will draw either lines or polylines on the current sketch plane. There can only be two lines and they may be angled. The lines must at least touch or exceed the outside edge of the part. After sketching the objects issue the Cut Line command **(AMCUTLINE)** from the part Modeling toolbar as shown in Figure 4.13 or right-click in the drawing area and from the pop-up menu, select Cut Line from the Sketch Solving menu. Select the objects and then constrain and dimension them if desired. The same constraining and dimensioning rules apply to cut lines as well as to profiles.

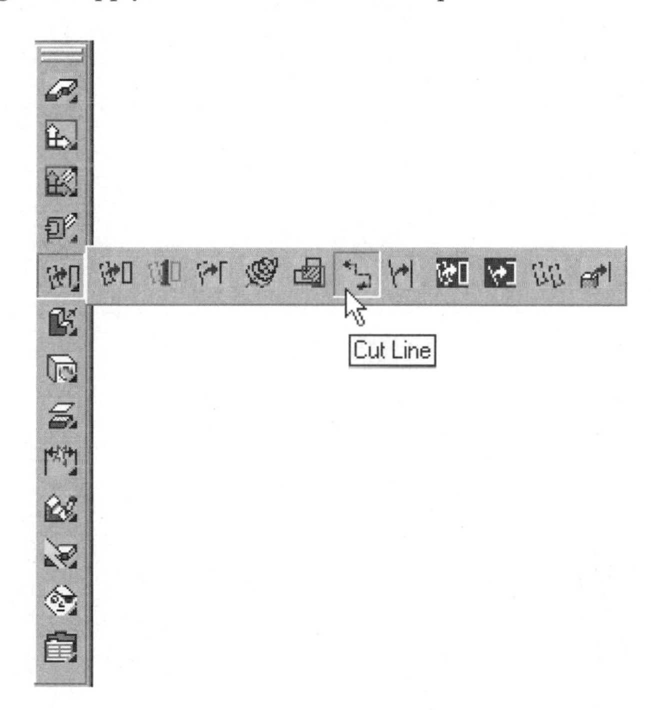

Figure 4.13

BREAKOUT SECTION VIEWS

A breakout section view is used to remove material from a part to see what is inside it. Before creating a broken out section view you will draw either lines, arcs, a circle, polyline or a spline on the current sketch plane. The break line needs to be enclosed or use the part edges to close the edges. The lines must at least touch or exceed the outside edge of the part. After sketching the objects, issue the Break Line command **(AMBREAKLINE)** from the part Modeling toolbar as shown in Figure 4.14 or right-click in the drawing area and from the pop-up menu, select Break Line from the Sketch Solving menu. Select the objects and then constrain and dimension them if desired. The same constraining and dimensioning rules apply to break lines as well as to profiles. A break out section can be a base or an ortho view. For creating an ortho view you will select the view from which to project and in what direction. Then you will be prompted to select the break line. The next prompt "Specified work plane will project the break line to this plane. Select a work plane for section depth or <use break-line plane>:" here you can use the plane in which the break line was drawn or select a work plane to project the break line onto. After pressing ENTER to use the plane that the break line is on or selecting a work plane, an arrow will appear showing the side in which material will be removed. Either accept the direction or flip it.

Define portion to cut away [Flip/Accept] <Accept>:

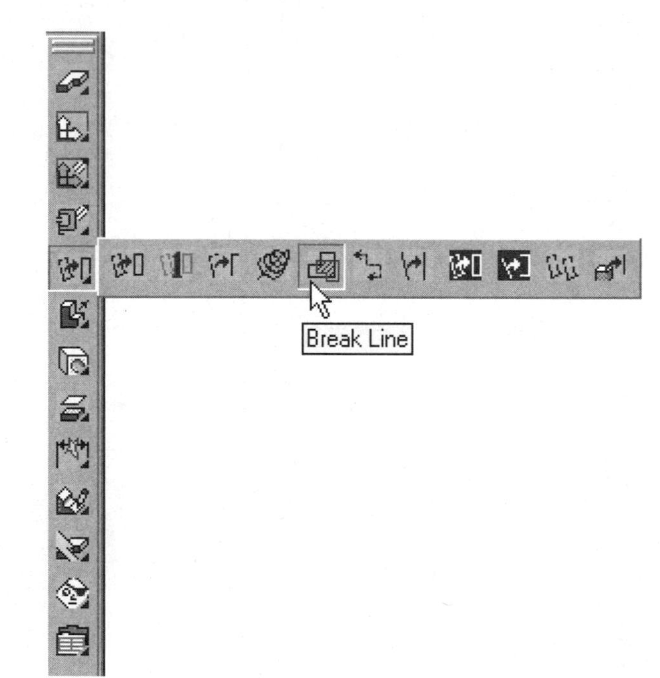

Figure 4.14

TUTORIAL 4.3—CREATING AN OFFSET SECTIONED DRAWING VIEW IN THIS TUTORIAL A BASE VIEW HAS ALREADY BEEN CREATED.

1. Open the file \Md4book\Chapter4\TU4-3.dwg.
2. Change to the sketch view using the **9**.
3. Draw three lines or a polyline, as shown in Figure 4.15. (They are shown with thickness for clarity only.)

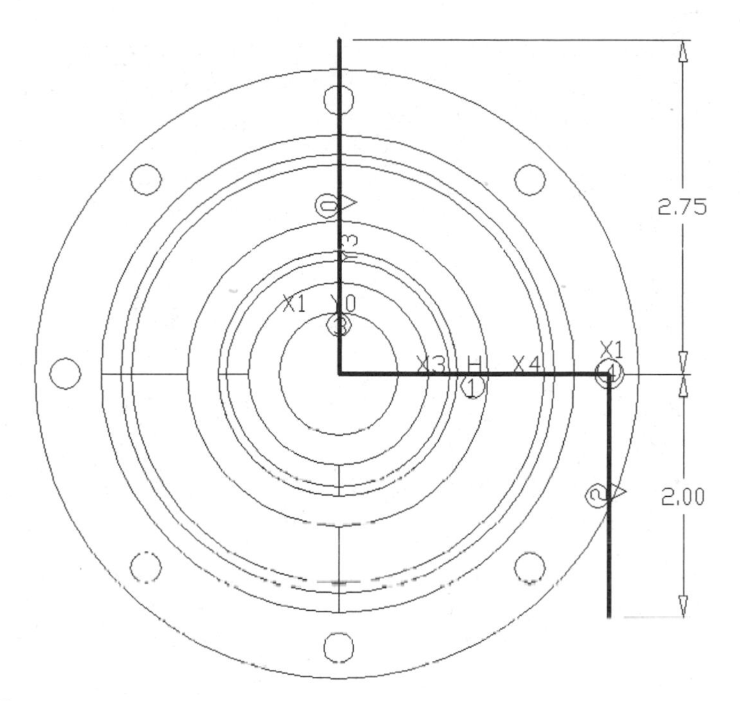

Figure 4.15

4. Issue the Cut Line command **(AMCUTLINE)** and select the lines or polyline.
5. Constrain and dimension the cut line as shown in Figure 4.15 (constraints are shown for clarity). Apply an *X* and *Y* value constraint to the center circle and the intersection of the lines in the middle of the part.
6. Click on the Drawing tab or Drawing Layout icon on the express toolbar.
7. Issue the New Drawing View command **(AMDWGVIEW)**. In the Create Drawing View dialog box, change the View Type to Ortho and the section Type to Offset and the other values will be the defaults. Then select OK.
8. For the prompt to select the parent view, select any point in the base view and then select a point to the right and press ENTER.
9. You will be returned to part mode and prompted to select the cutting line sketch. Select the cut line.
10. Create a "1" scale relative to the parent isometric view of the section view. Place the isometric view in the upper right corner of the drawing. When complete, your screen should resemble Figure 4.16.
11. Save the file.

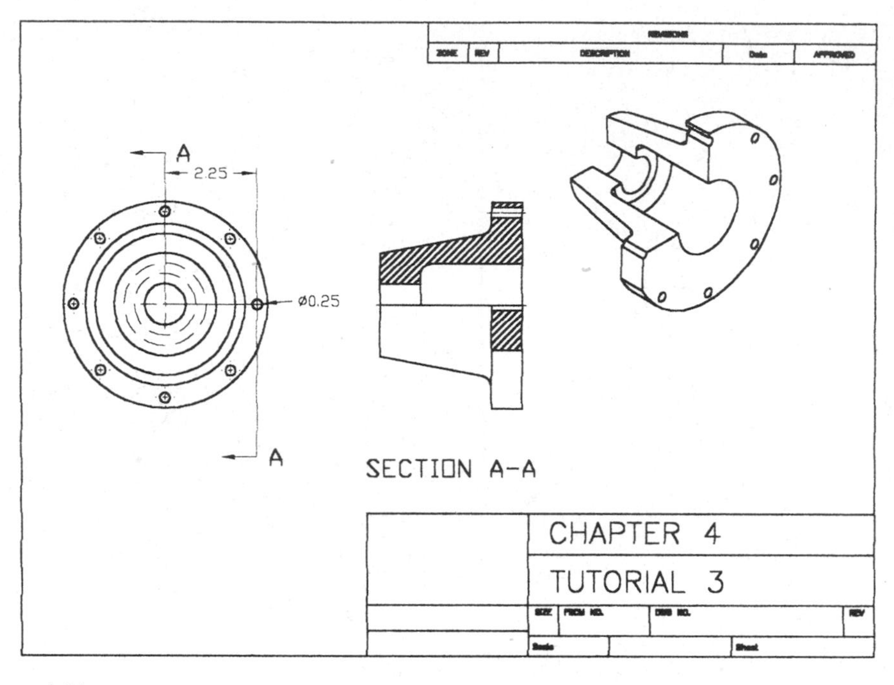

Figure 4.16

TUTORIAL 4.4—CREATING AN ALIGNED SECTIONED DRAWING VIEW

In this tutorial a base view has already been created.

1. Open the file \Md4book\Chapter4\TU4-4.dwg.
2. Change to the sketch view using **9**.
3. Draw two lines or a polyline, as shown in Figure 4.17. (They are shown with thickness for clarity only.)
4. Issue the Cut Line command **(AMCUTLINE)** and select the lines or polyline.
5. Constrain and dimension the cut line as shown in Figure 4.17 (constraints are shown for clarity). Apply an *X* and *Y* value constraint to the center circle and the intersection of the lines.
6. Click on the Drawing tab or Drawing Layout icon on the express toolbar.
7. Issue the New Drawing View command **(AMDWGVIEW)**. In the Create Drawing View dialog box, change the View Type to Ortho and the section Type to Aligned and the other values will be the defaults. Then select OK.
8. For the prompt to select the parent view select a point in the base view and then select a point to the right and press ENTER.

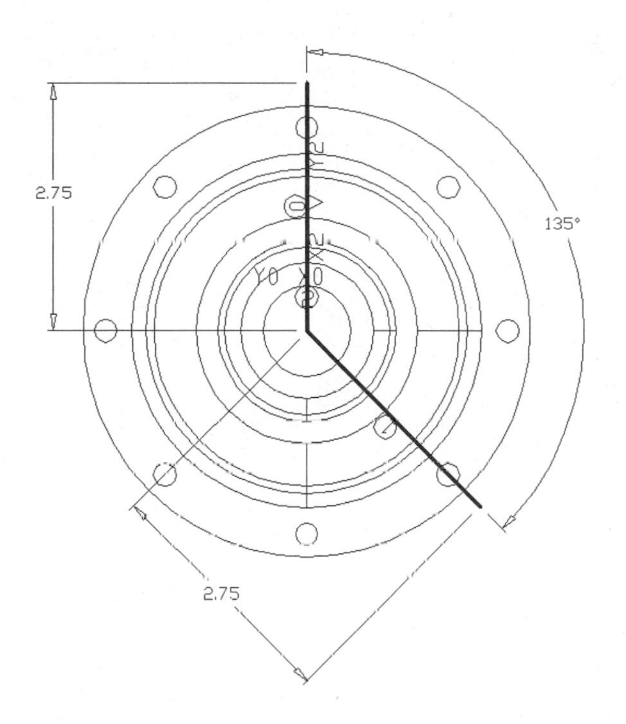

Figure 4.17

9. You will be returned to part mode and prompted to select the cutting line sketch. Select the cut line.
10. Create a "1" scale relative to the parent isometric view of the section view. Place the isometric view in the upper right corner of the drawing. When complete, your screen should resemble Figure 4.18.
11. Save the file.

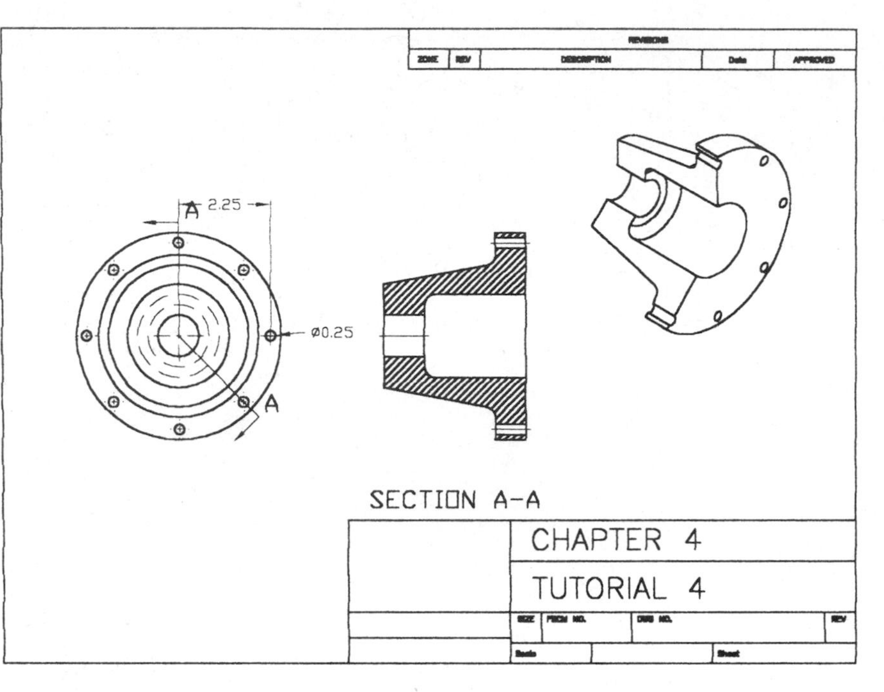

Figure 4.18

TUTORIAL

TUTORIAL 4.5—CREATING A BREAKOUT SECTIONED DRAWING VIEW

In this tutorial a base view has already been created.

1. Open the file \Md4book\Chapter4\TU4-5.dwg.
2. Change to the sketch view using **9**.
3. Draw a polyline that resembles the shape that is shown in Figure 4.19 (the polyline is shown with thickness and the workplane is turned off for clarity).
4. Issue the Break Line command **(AMBREAKLINE)** and select the polyline.
5. For this tutorial you will not constrain the break line. Break lines do not need to be constrained.
6. Click on the Drawing tab or Drawing Layout icon on the express toolbar.
7. Issue the New Drawing View command **(AMDWGVIEW)**. In the Create Drawing View dialog box, change the View Type to Ortho and the section Type to Breakout and the other values will be the defaults. Then select OK.

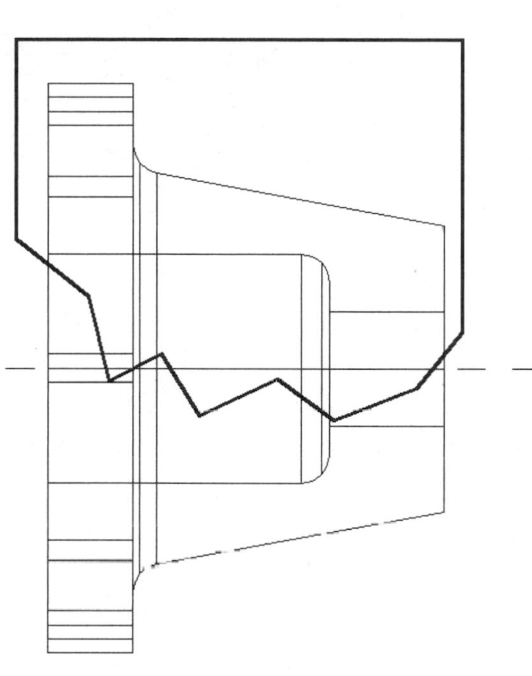

Figure 4.19

8. For the prompt to select the parent view, select a point in the base view and then select a point to the right and press ENTER.
9. You will be returned to part mode and prompted to select the break line sketch. Select the break line. Then ENTER twice to accept the break line location and the direction of the cut.
10. Create a isometric view that is "1" scale and relative to the parent view. Place the isometric view in the upper left corner of the drawing. When complete, your screen should resemble Figure 4.20.
11. Save the file.

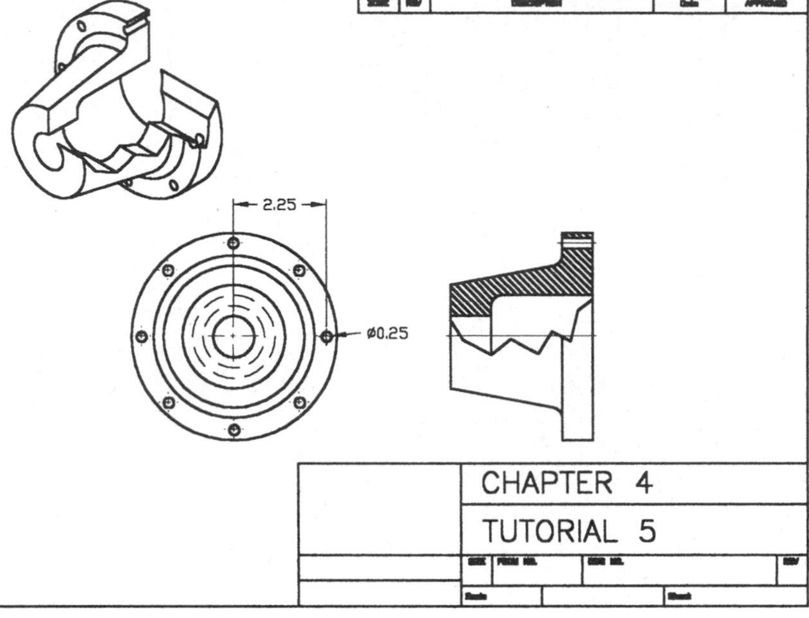

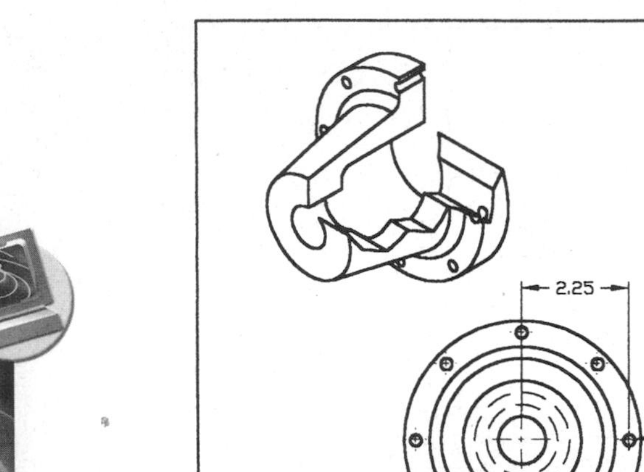

Figure 4.20

RADIAL SECTION VIEWS

Another type of a section view is radial section view. A radial section view rotates an auxiliary section view about a selected point in the view and then it is projected like an orthographic section view. A work plane needs to be created to represent the angle for the auxiliary view. Then you will select a point to rotate the view about. This can be a selected point or the midpoint of the section. If an arc or circle is selected, the center of the arc or circle will be used. The view type needs to be ortho and the section type set to radial.

TUTORIAL 4.6—CREATING A RADIAL SECTIONED DRAWING VIEW

In this tutorial a base view and a work plane have already been created.

1. Open the file \Md4book\Chapter4\TU4-6.dwg.
2. Click on the Drawing tab or Drawing Layout icon on the express toolbar.
3. Issue the New Drawing View command **(AMDWGVIEW)**. In the Create Drawing View dialog box, change the View Type to Ortho and the section Type to Radial and the other values will be the defaults. Then select OK.
4. For the prompt to select the parent view, select a point in the base view and then select a point to the right and press ENTER.

5. For the prompt "Select workplane", in the parent view, select the blue angled line (work plane).
6. Press ENTER to rotate about the middle of the section.
7. Move the text "SECTION A-A" up and out of the border.
8. Create a "1" scale relative to the parent isometric view of the section view. Place the isometric view in the upper right corner of the drawing. When complete, your screen should resemble Figure 4.21.
9. Save the file.

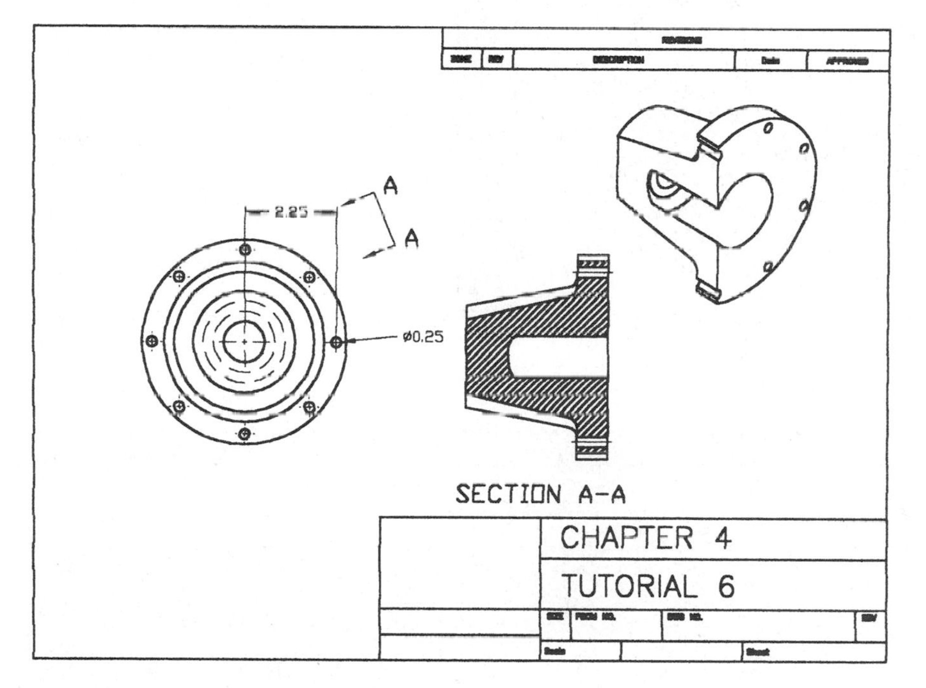

Figure 4.21

EDITING DRAWING VIEWS

After creating the drawing views, you may need to edit the properties of a view. To edit a view's properties, issue the Edit Drawing command **(AMEDITVIEW)** from the Drawing Layout toolbar, as shown in Figure 4.22. Right-click in the layout and select Edit View from the pop-up menu or double-click on the view name in the browser under the Drawing tab. If the command was issued from the toolbar or right click menu in the layout, select a point in the view to edit and the Edit Drawing View dialog box that is appropriate for the selected view will appear. Figure 4.23 shows the Edit Drawing View dialog box for an ortho-sectioned view. Depending on the view

Figure 4.22

Figure 4.23

that is selected, certain options may be missing or grayed out from the dialog box. For example, if a non-sectioned ortho view is selected, the scale will be grayed out because it is dependent on the view from which it was projected and the section tab will be missing because it is not a section view. Most of the items in the dialog box are the same as when it was created with the exception of the following: When the scale is changed in the base view all the dependent views will be scaled as well.

On the bottom left of the Edit Drawing view dialogue box is a check box named Placement for ortho views that has the following options:

Placement

Move with Parent: Available on orthogonal and auxiliary views. When checked, the view will be aligned with its parent view. If unchecked, the view can move freely and then the options below will be available. This option does not apply to base views.

Horizontal: When selected, you can realign a view along the horizontal axis that was moved away from its parent view. Use object snaps to realign the two views.

No Alignment: When Move with Parent is unchecked, this option will be checked. This shows that the view can move freely.

Vertical: When selected, you can realign a view along the vertical axis that was moved away from its parent view. Use object snaps to realign the two views.

Under the Display tab is a button named Edge Properties that has the following options:

Edge Properties: With this option you can change the properties of selected objects. The properties that can be changed are; hide them, color, layer, linetype, linetype scale or unhide the objects that were turned off. After issuing the option you will be prompted to either "Remove all/Select/Unhide all". To change the properties of visible objects press enter, select the objetcs and press enter. An Edge Properties dialog box will appear. Change the properties that need to be changed and then select OK. To Unhide all edges that have been hidden with this command type u and then enter. Then select OK in the original dialog box and the changes will take effect in the view.

Under the Section tab is a option named **Hatch**.

Hatch

When checked, the hatch pattern will be visible in the view; when unchecked, the hatch pattern will be invisible in the view. This switch can be used in ortho, auxiliary and isometric section views.

MOVING DRAWING VIEWS

To move a drawing view, issue the Move Drawing View command **(AMMOVE-VIEW)** from the Drawing Layout toolbar as shown in Figure 4.24, right-click in the layout and select Move View from the pop-up menu or right click on the view name in the browser and select Move View from the pop-up menu. If the command is issued from the toolbar or right click menu in the layout, select a point in the view to move and left-click to select a point to move the view to. Keep selecting points until you are happy with the view placement and then press ENTER, and the view will be moved to this new location. If you select a parent view, the children or dependent views will also move with it. Orthogonal views can only be moved along the axis in which they were created, unless you uncheck Move with Parent in the Edit Drawing View dialog box. Detail and isometric views can be moved freely.

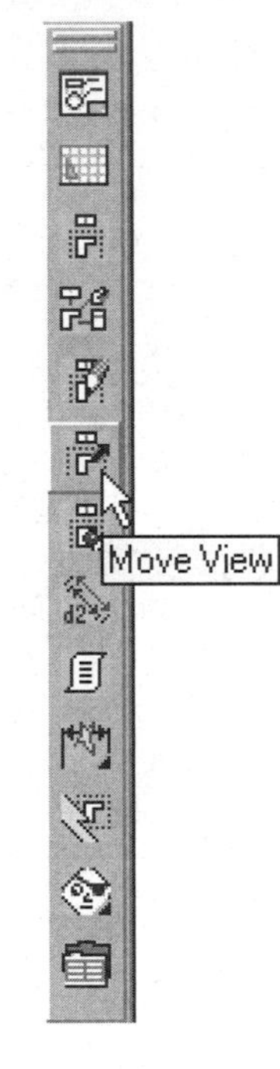

Figure 4.24

DELETING DRAWING VIEWS

To delete a drawing view, issue the Delete Drawing View command **(AMDELVIEW)** from the Drawing layout toolbar, as shown in Figure 4.25, right-click in the layout and select Delete View from the pop-up menu or right click on the view name in the browser and select Delete View from the pop-up menu. If the command is issued from the toolbar or right click menu in the layout, select a point in the view to delete. If the selected view has a view that is dependent on it, you will be prompted: to "Delete dependent views?". If you select Yes, the dependent view will be deleted. If you select No, only the selected view will be deleted. If you select Cancel, the command will be cancelled and no view will be deleted. If the selected view has no dependent views, it will be erased without an alert box.

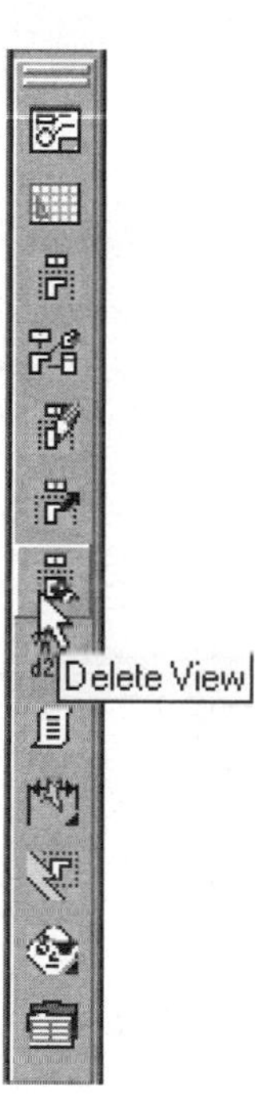

Figure 4.25

TUTORIAL

TUTORIAL 4.7— EDITING, MOVING AND DELETING DRAWING VIEWS

1. Open the file \Md4book\Chapter4\TU4-1.dwg.
2. If the drawing mode isn't current, click on the Drawing tab or Drawing Layout icon on the express toolbar.
3. Issue the Edit Drawing command **(AMEDITVIEW)** or, in the browser, double-click on the name Base under the Drawing tab.
4. If the command was selected from the toolbar or right click menu in the layout select a point in the top view. Change the scale to ".5" and then select OK.
5. Repeat the Edit Drawing command and edit the scale of the isometric view to "1", check on the Display Hidden Lines and then select OK. When complete, your screen should resemble Figure 4.26.

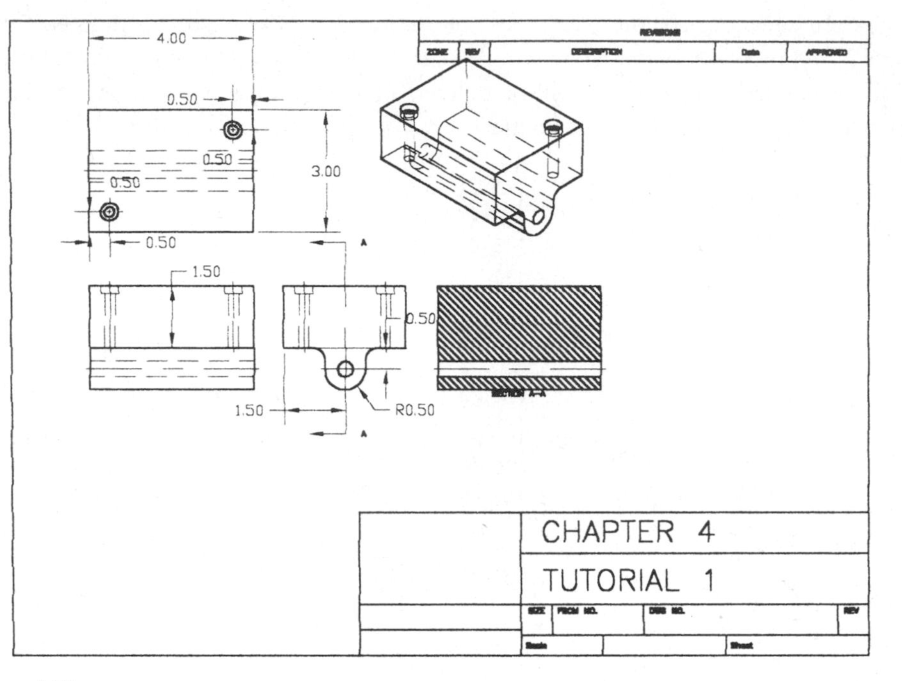

Figure 4.26

6. Repeat the Edit Drawing command and in the isometric view uncheck the Display Hidden Lines, then select OK.

7. Repeat the Edit Drawing command to edit the isometric view, select Edge Properties from the Display tab and press ENTER to select geometry. Select a few objects of your choice by selecting, windowing or crossing them and then press ENTER to continue. Select Hide Edges and then select OK in both dialog boxes. The selected geometry should have disappeared from the view.

8. Repeat the Edit Drawing command to edit the isometric view. Select Edge Properties from the Display tab and type U and ENTER to unhide all hidden geometry. Select OK to complete the command and the geometry will reappear.

9. Issue the Move Drawing View command **(AMMOVEVIEW)** and practice moving the views around. First select the top view, which is the base view, then try moving the front view and then the isometric views. After practicing moving the views, move the views so they better fit the border.

10. Create an isometric view of the sectioned ortho view at a scale of "1" relative to the parent.

11. Edit the sectioned isometric view and under the Section tab check the hatch option. The hatch pattern will appear in the isometric view.
12. Issue the Delete Drawing View command **(AMDELVIEW)** and select the ortho section view. In the Delete Dependent Views dialog box select Yes. All associated views should have been erased.
13. Use the Undo command to bring back the drawing views, issue the Delete Drawing View command and select the ortho section view. In the Delete Dependent Views dialog box select No and only the ortho section view will be erased.
14. Save the file.

EDITING DIMENSIONS

When drawing views are created, the dimensions do not always appear in the correct location—there may be dimensions that are not required in the views or they may have the wrong value. The dimensions are placed where they were created in the profile unless through the Desktop Options command and under the Annoation tab, Automatically Arrange Dimensions is checked then they will be spaced per your dimension style. In AutoCAD there are many ways to edit dimensions, and the same is true with Mechanical Desktop. In the following section you will learn how to change a dimensions value inside a drawing as well as hide, change the location, align, join, insert, break and modify the text properties.

CHANGE DIMENSIONS

The dimensions that automatically appear in the drawing views are parametric dimensions and they can be changed in the same way as you would a parametric dimension on a 2D sketch. To change a dimension's value issue the Change Dimension command **(AMMODDIM)** from the Drawing Layout toolbar as shown in Figure 4.27 Then update the part with the Update Drawing View command **(AMUPDATE)** from the Drawing Layout toolbar, as shown in Figure 4.28 or right-click in the layout and select Update View from the pop-up menu. Another method for changing a dimension value is to double click on the dimension and type in a new value for the expression. When updating the drawing views you will be prompted:

```
Update part now? [Yes/No] <Yes>:
```

Press ENTER to update the part. The next prompt will give you the option to update the current layout, all layouts or a specified view.

```
Select view to update or [All layouts/current Layout]:
```

Type L and ENTER to update all the views in the current layout.

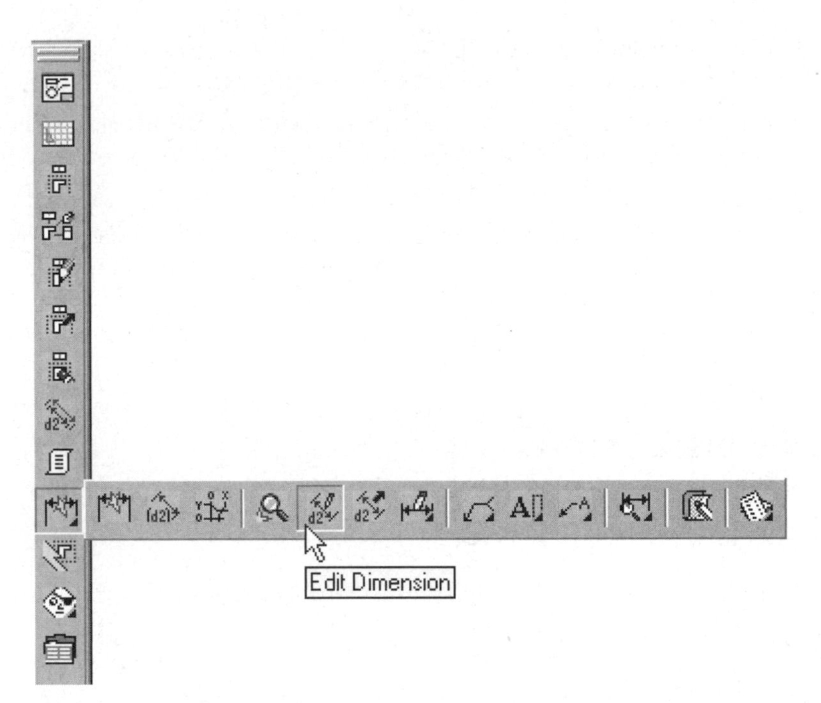

Figure 4.27

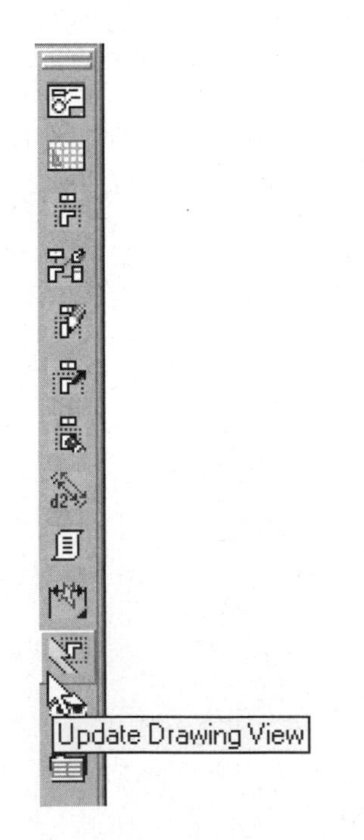

Figure 4.28

HIDE DIMENSIONS

When drawing views are created, not all the dimensions that appear will be needed for the actual drawing. These dimensions are important to the integrity of the part, so do not erase them. Instead, hide them in the drawing view. If you have zero length dimensions appearing in the drawing views and you want to turn off the visibility of all them, check Hide Zero-Length Parametric Dimensions through the Desktop Options command and under the Annoation tab. Do this before creating the drawing view that the zero-length dimensions in which will appear. To hide a dimension, issue the Drawing Visibility command **(AMVISIBLE)** from the Drawing Layout toolbar as shown in Figure 4.29. A Desktop Visibility dialog box will appear. From the drawing tab, check Hide and then pick Select. You will be returned to the drawing views, where you can select the dimensions to hide. To hide all parametric dimensions, check Hide from the Drawing tab and then check Parametric Dims. To bring back a specific hidden dimension, issue the same command, but select Unhide. The dimensions that were hidden will reappear on the screen; select the dimension(s) that you want to unhide. To bring back all hidden dimensions, issue the same command, but select Unhide and then check Parametric Dims. The dimensions that were hidden will reappear on the screen.

Drawing Visibility

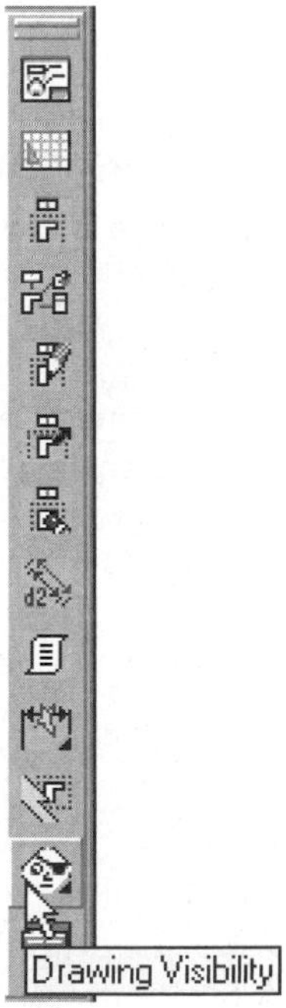

Figure 4.29

TUTORIAL 4.8— CHANGING AND HIDING DIMENSIONS

1. Open the file \Md4book\Chapter4\TU4-8.dwg.
2. Issue the Change Dimension command **(AMMODDIM)** and change the "4.00" dimension in the top view to "5.00" and press ENTER, change the "3.00" dimension in the top view to "3.50" and press the ENTER twice to complete the command.
3. Update the drawing views.
4. For the prompt: `Update part now? [Yes/No]` <Yes>: ENTER
5. For the prompt: `Select view to update or [All layouts/current Layout]:` Type L and press ENTER.
6. Repeat Steps 2 to 5 and change the "5.00" dimension to "4.50".
7. Issue the Drawing Visibility command **(AMVISIBLE)**.
8. Pick the Select button and pick a few dimensions of your choice and then press the ENTER, then pick the OK button in the dialog box to complete the command.
9. Repeat the Drawing Visibility command **(AMVISIBLE)**.
10. Pick the Unhide radio button, then pick the Select button and pick the dimensions that you hid in step 8. Press ENTER to return to the dialog box then pick the OK to complete the command.
11. Go to the part mode by selecting either the Model tab in the browser or picking the part modeling icon of the express toolbar.
12. Edit the ExtrusionBlind1 and change the extrusion distance "1.50" to "2" and the "3.50" width dimension to "3.25".
13. Update the part if necessary.
14. Return to the drawing mode by either picking the Drawing tab on the browser or picking the Drawing Layout icon on the express toolbar. The drawing views should automatically be updated to reflect the changes.
15. Save the file.

MOVE AND REATTACH DIMENSIONS

If the dimensions are not in the correct location, you have many ways to move the dimensions to a different position. If you are familiar with grips, you can use them in exactly the same way as you would in regular AutoCAD. To use grips while not in any command, select the dimension that you want to relocate with the left mouse button and grip points will appear. Do not select the grips that are located at the points where the dimension meets the geometry. This location is the definition point and will alter the value of the dimension. However, when the drawing view is next updated the dimension will revert back to its original location. Select the grip on the dimension's text and select a new location. The dimension will move to the new point. To exit grip editing, you can either press ENTER or press the escape key twice. A second method is to use AutoCAD's **STRETCH** command. Again, only change the dimensions location.

A third method is to use a Mechanical Desktop command called Move Dimension **(AMMOVEDIM)**. This command allows you to flip, reattach or move Mechanical Desktop dimensions within the view or to another view. After issuing the command, press ENTER to move a dimension, select a dimension to move and then select a point in the view for its new location. The view does not need to be the view in which it now appears. However, the view does need to represent the same set of geometry. Then select a new point and the dimension will follow to that point. Continue selecting a point until the dimension is located in the correct position. The other two options are to flip and reattach a dimension. The flip option will change the text of a vertical dimension going from the upper right to the lower left or vice versa. The reattach option is used for dimensions that have their extension lines going through the geometry. Issue the Move Dimension command **(AMMOVEDIM)** from the Drawing Layout toolbar, as shown in Figure 4.30 then press R and ENTER. Select the dimension to reattach and then select the extension line to reattach, select a new location. The same rules apply for reattaching as for creating dimensions. The point will go to the nearest endpoint of a line or center of an arc or circle. If the point that you need

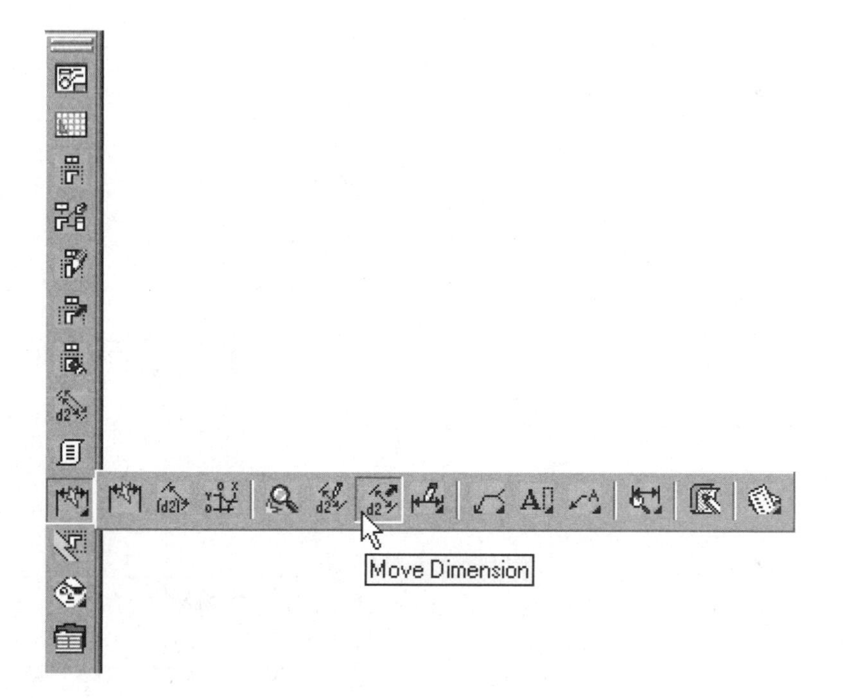

Figure 4.30

to select is under the extension line, you can cycle through the geometry by holding down the control key and selecting the point until the correct object is highlighted. Then press ENTER. Repeat the sequence until the dimensions are correctly reattached.

TUTORIAL 4.9—MOVING DIMENSIONS

1. Open the file \Md4book\Chapter4\TU4-8.dwg if it is not the current file.
2. Practice repositioning a few of the dimensions using grips.
3. Practice repositioning a few of the dimensions using AutoCAD's STRETCH command.
4. Practice repositioning a few of the dimensions using the **(AMMOVEDIM)** command.
5. Issue the Move Dimension command and ENTER to move a dimension. Select the "2.00" dimension in the front view. Then select a point in the right side view and position the dimension on the right side of the view. When complete, your screen should resemble Figure 4.31.

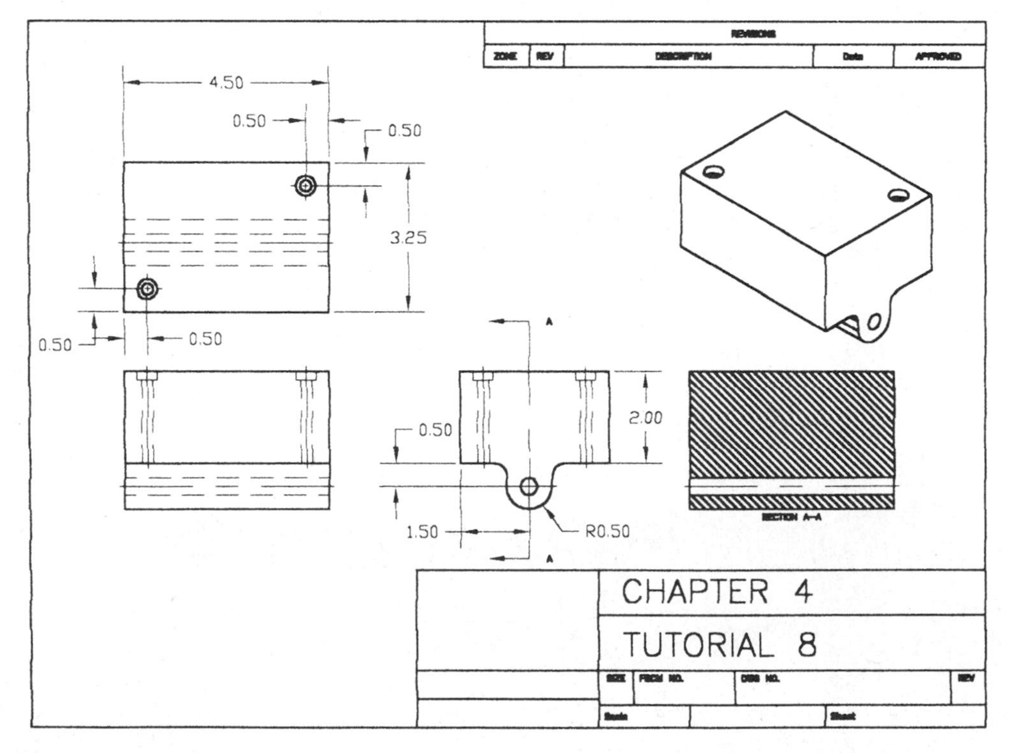

Figure 4.31

6. Repeat the Move Dimension command and press R and ENTER to reattach a dimension.
7. For the prompt "Select extension line:" select the bottom horizontal extension line of the "2.00" vertical dimension that you just moved to the right side view.
8. For the prompt "Select attachment point:" select near the bottom of the out side vertical line, as shown in Figure 4.32 with the pickbox.

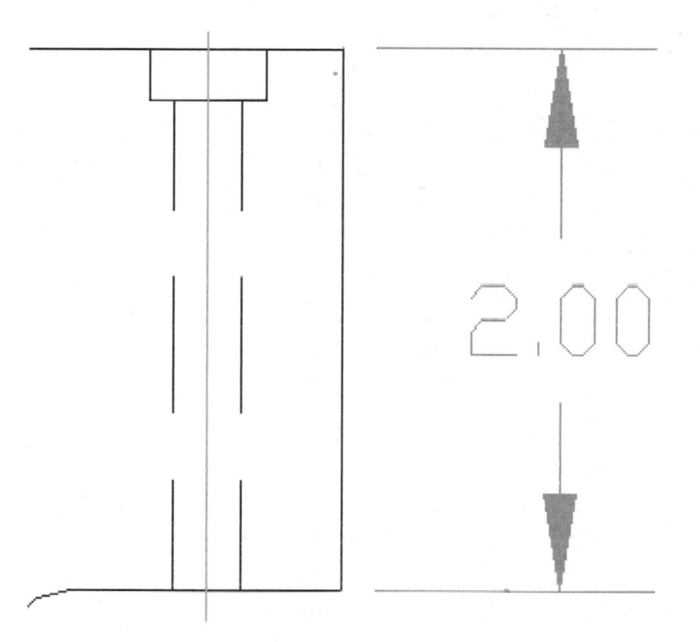

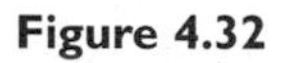

Figure 4.32

9. Then press ENTER to complete the command. When done, the "2.00" dimension should resemble Figure 4.33.
10. Save the file.

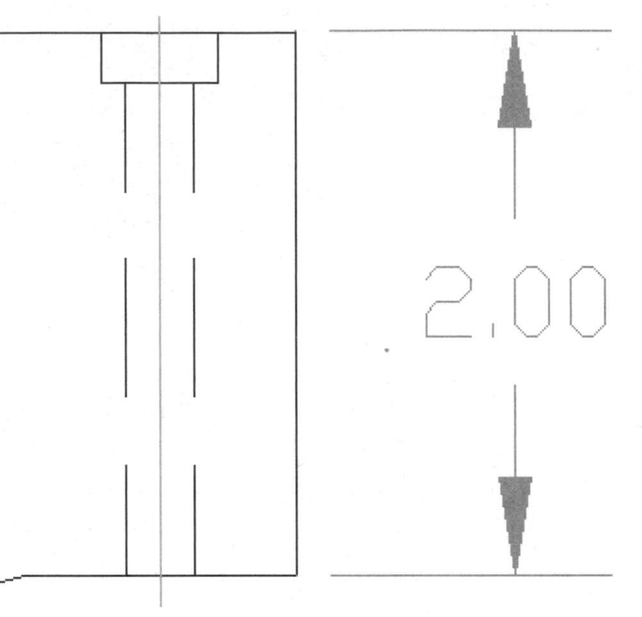

Figure 4.33

REFERENCE DIMENSIONS

After laying out the drawing views, you may find that another dimension is required to better define the part. You could go back and add the dimension to the part if the part was unconstrained by the missing dimension, edit the sketch and modify it or you could add a reference dimension. A reference dimension is not a parametric dimension, but it is associative and reflects the length of the geometry being dimensioned. After a reference dimension is created and the views change, the reference dimension will get updated to reflect any changes. A reference dimension is added with the Reference Dimensions command **(AMREFDIM)** from the Drawing Layout toolbar, as shown in Figure 4.34. After issuing the command, create a reference dimension in the same manner as you would a parametric dimension. The reference dimension will be placed on the AMREFDIM layer and will take on the color of that layer. Therefore, if you want to change the color of all the reference dimensions, change the color of the AMREFDIM layer. Reference dimensions can also be placed to the end points of centerlines.

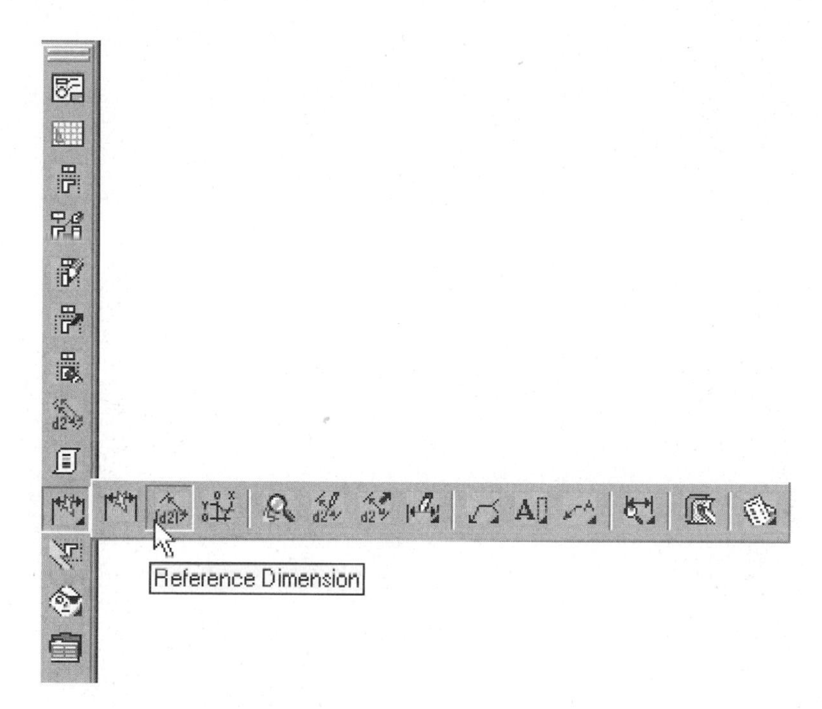

Figure 4.34

HOLE NOTES

Before you create a hole note, a hole feature must exist as well as a drawing view that shows the hole. You can also modify a dimension style to represent the way the hole note(s) should appear: leader style, text height, unit type (decimal, fraction and so on). The hole note will not reflect the number of holes or tap information; you can add this information by using the same Hole Note command with the Edit option or by using the AutoCAD DDEDIT command. Issue the Hole Note command **(AMHOLENOTE)** from the Drawing Layout toolbar, as shown in Figure 4.35. Press ENTER to create a new hole note or press E and ENTER to edit an existing hole note. Then select a hole to annotate or an existing hole note to edit. A Create Holenote dialog box will appear, as shown in Figure 4.36. Select the template to use for the selected hole and make other changes as needed. If a hole type is changed you will need to edit the hole note and select a hole template to use. The Create holenote dialog box is broken into the following areas.

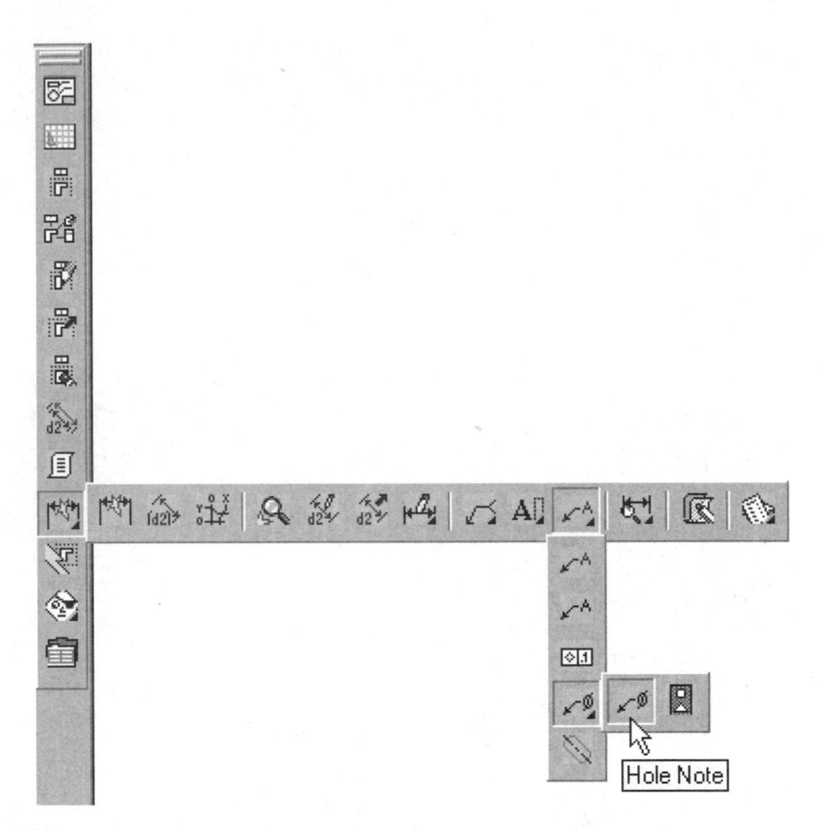

Figure 4.35

Create Holenote

Template Name

THRU_CBORE
THRU_DRILL

Leader Justification

Middle of Top Line
Middle of All Text
Middle of Bottom Line

Edit Template

Dimension Style Overrides

Primary Dimension Style

Apply Dimension Style toText Height Only

Standard

OK Cancel Help

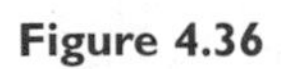
Figure 4.36

Template Name: A listing of possible template names will appear for the selected hole. If more than one template name appears, select the one to be used. Templates can be created and edited with the Template command **AMTEMPLATE.**

Leader Justification: Select one of the three justifications for the text: Middle of Top Line, Middle of All Text or Middle of Bottom Line.

Edit Template: Allows you to modify the format of the hole note. When the changes are complete, a new name will appear in the Template Name area called "EDITED" and it will be the current template.

Dimension Style Override: If this button is selected, a Holenote Dimension Style Override dialog box will appear. The Expression cannot be changed, but you can change the others by selecting in the text area. A drop-down list will appear. Select the entry you want.

Primary Dimension Style:

> **Apply Dimension Style to Text Height Only:** This should be checked if you want only the text height of the selected dimension style to be used. Otherwise, the entire format of the chosen dimension style will be used.
>
> From the drop-down list, select the dimension style to be used.

TUTORIAL

TUTORIAL 4.10— ADDING REFERENCE DIMENSIONS AND HOLE NOTES

1. Open the file \Md4book\Chapter4\TU4-8.dwg if it is not the current file.
2. Issue the Reference Dimensions command **(AMREFDIM)**.
3. Place a vertical reference dimension on the left side of the front view. When complete, your screen should resemble Figure 4.37.
4. Issue the Change Dimension command **(AMMODDIM)** and change the "4.50" dimension in the top view to "3.50" and press ENTER twice to complete the command.
5. Update the part and the drawing views. The reference dimension should now be "2.50".
6. Practice adding reference dimensions to the part and between the centerlines of the holes.
7. Issue the Hole Note command **(AMHOLENOTE)** and add a hole note to one of the counter bore holes.
8. Go to the part mode by selecting either the Model tab in the browser or picking the part modeling icon of the express toolbar.
9. Edit both counter bore holes and change them to .50 diameter through holes.

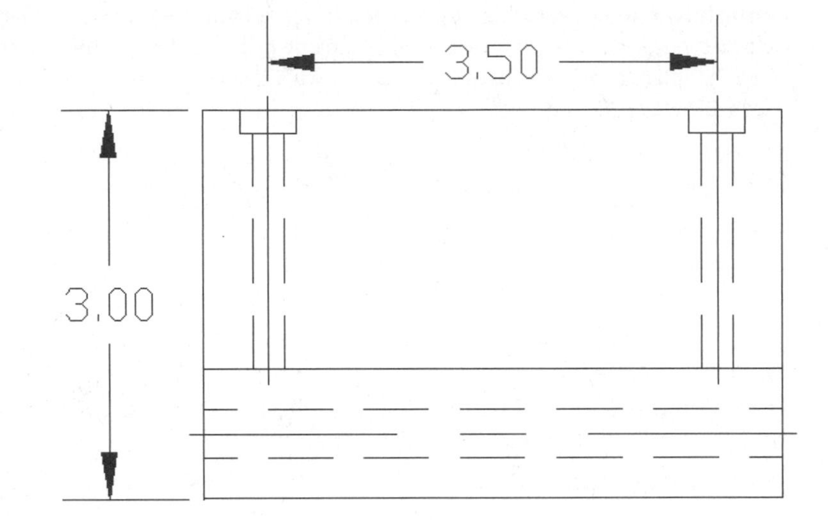

Figure 4.37

10. Update the part if necessary.
11. Return to the drawing mode by either picking the Drawing tab on the browser or picking the Drawing Layout icon on the express toolbar.
12. Issue the Hole Note command **(AMHOLENOTE)** and select THRU_DRILL as the template to use. Select OK in the dialog box to complete the command and the hole note should reflect the new hole specifications.
13. Save the file.

EDIT, ALIGN, JOIN, INSERT AND BREAK DIMENSIONS

In this section you will learn how to change the style of a dimension and how to align, join, insert and break them as well.

The dimension style of a dimension that Mechanical Desktop creates can be edited like any other AutoCAD dimension or you can use the Edit Format command **(AMDIMFORMAT)**, as shown in Figure 4.38, to modify the format of the dimension. After issuing the command, select the dimension to modify and a dialog box will appear. As you make changes in the dialog box, the change will appear on the dimension.

To align dimensions, issue the Align Dimension command **(AMDIMALIGN)**, as shown in Figure 4.38. Select the base dimension that you want the others aligned to, select the dimension(s) that you want to align to first and press ENTER. The dimensions will be aligned to the first selected dimension. After selecting the first dimension multiple dimension can be selected using any AutoCAD selection method.

The Join Dimensions command **(AMDIMJOIN)**, as shown in Figure 4.38, will join two dimensions into one. The two selected dimensions will be hidden and a reference dimension will be created that joins the two selected dimensions. The reference dimension will be placed at the location of the extents of the selected dimensions.

The Insert Dimension command **(AMDIMINSERT)**, as shown in Figure 4.38, will insert a second dimension where there was only one. Issue the command, select a dimension and then select a target location with object snaps for the inserted dimension. The selected dimension will be hidden and two reference dimensions will be created.

The Break Dimension command **(AMDIMBREAK)**, as shown in Figure 4.38, can break a dimension by selecting two spots where you want it broken. When you break dimensions, it is recommended that you turn off object snaps. The broken dimension will remain parametric and associative. If the dimension is changed or stretched, the broken area will revert back to a solid line.

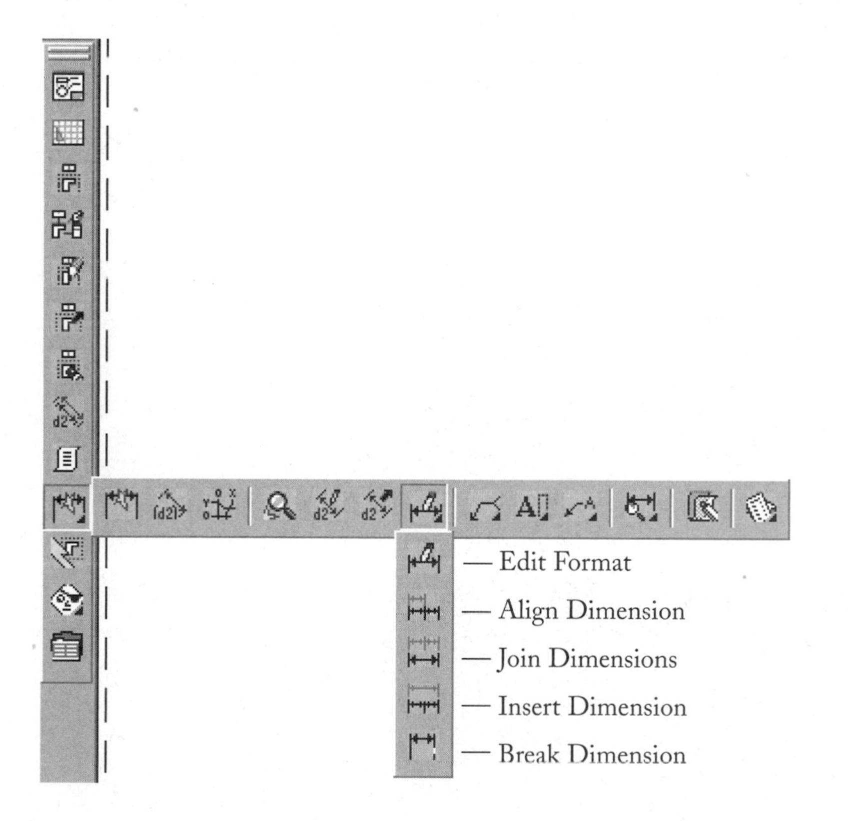

Figure 4.38

TUTORIAL 4.11— ALIGNING, JOINING, INSERTING AND BREAKING DIMENSIONS

1. Open the file \Md4book\Chapter4\TU4-11.dwg.
2. Issue the Insert Dimension command **(AMDIMINSERT)** and select the "4.50" horizontal dimension in the top view and then select the top of the centerline of the lower left hole as shown with the pickbox over it in Figure 4-39.

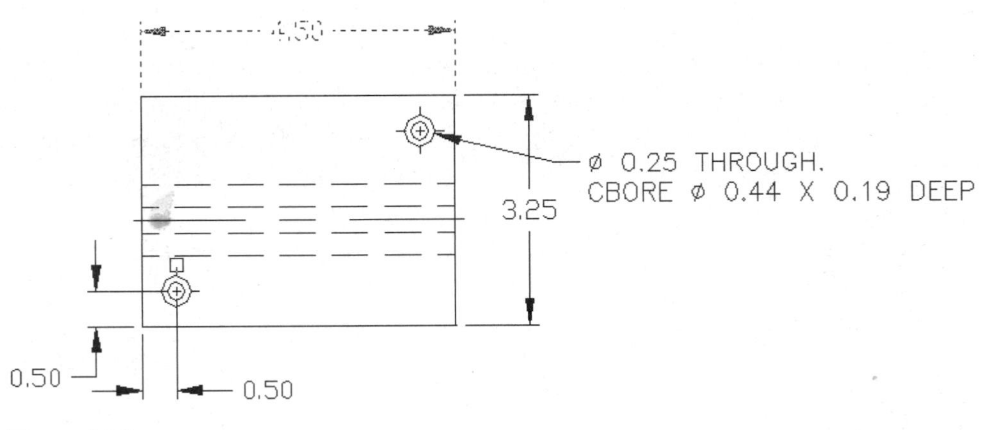

Figure 4.39

3. Grip edit the "0.50" dimension that was just inserted in the top view until your screen resembles Figure 4.40.

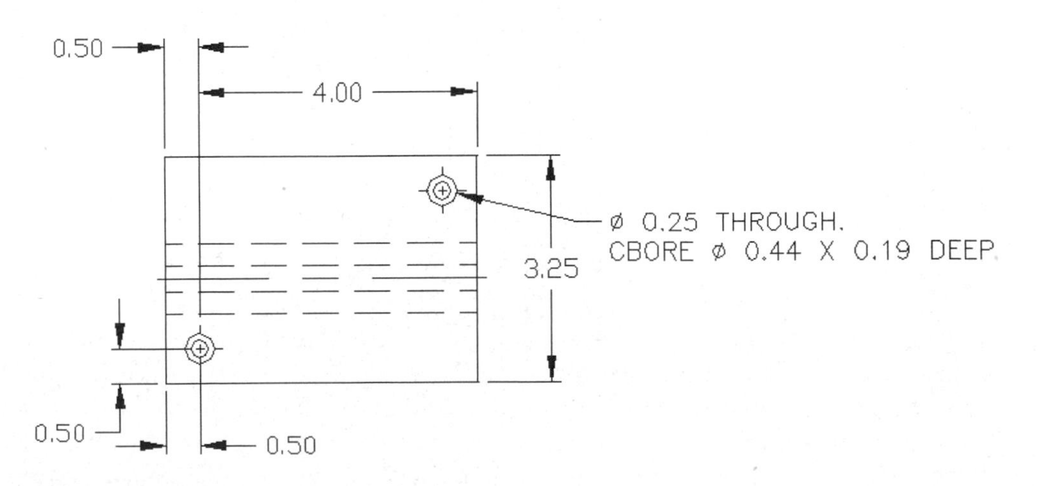

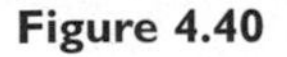
Figure 4.40

4. Issue the Align Dimension command **(AMDIMALIGN)** and select the "4.00" and "0.50" dimensions in the top view and then press ENTER. Both dimensions will be inline with one another. The second one picked will move to the new position.
5. Issue the Join Dimensions command **(AMDIMJOIN)** and select both the "4.00" and "0.50" top horizontal dimensions in the top view. A "4.50" reference dimension will be created.
6. In the top view the hole note is going through the "3.25" dimension. To break out a section of the "3.25" dimension, issue the Break Dimension command **(AMDIMBREAK)** and break the "3.25" vertical dimension in the top view by picking a point just above and below the angled line of the hole note. When complete, your screen should resemble Figure 4.41.

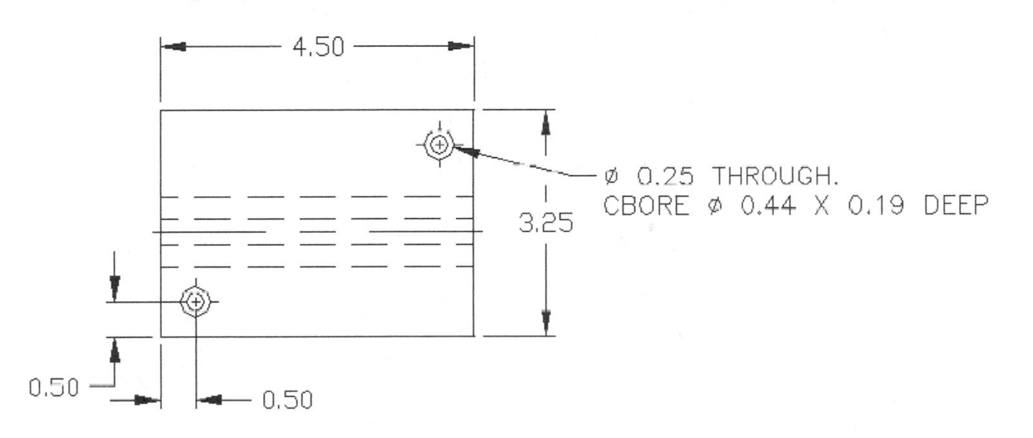

Figure 4.41

7. Issue the Edit Format command AMDIMFORMAT and make changes to the "4.50" dimension in the top view by trying a few different operations under each tab in the Dimension Formatter dialog box. When complete, apply your changes.
8. Save the file.

CENTER MARKS AND CENTERLINES

Center lines can automatically be created for holes, fillets and circular edges by checking the specific options under the Centerline Settings button under the Annotation tab of the Desktop Options command. These setting should be done before creating drawing views. To manually add centerlines to arcs, circles or in between two lines (circular edges), issue the Centerline command **(AMCENLINE)** from the Drawing Layout toolbar, as shown in Figure 4.42. Then select an arc or circle and press ENTER and a center mark will be placed. To place a centerline between two parallel lines, issue the Centerline command, select the two parallel lines, and then select a start and an end point for the centerline. When creating a centerline of a tapped hole, in the side view select the two lines that represent the drill. The lines representing the tap are annotations and cannot be used for centerline placement. These centerlines are attached to the geometry; if the view moves or changes, the centerlines will also move. The distance the centerline goes past the arc or circle can be controlled through Desktop Options (under Annotation tab select Centerline Settings).

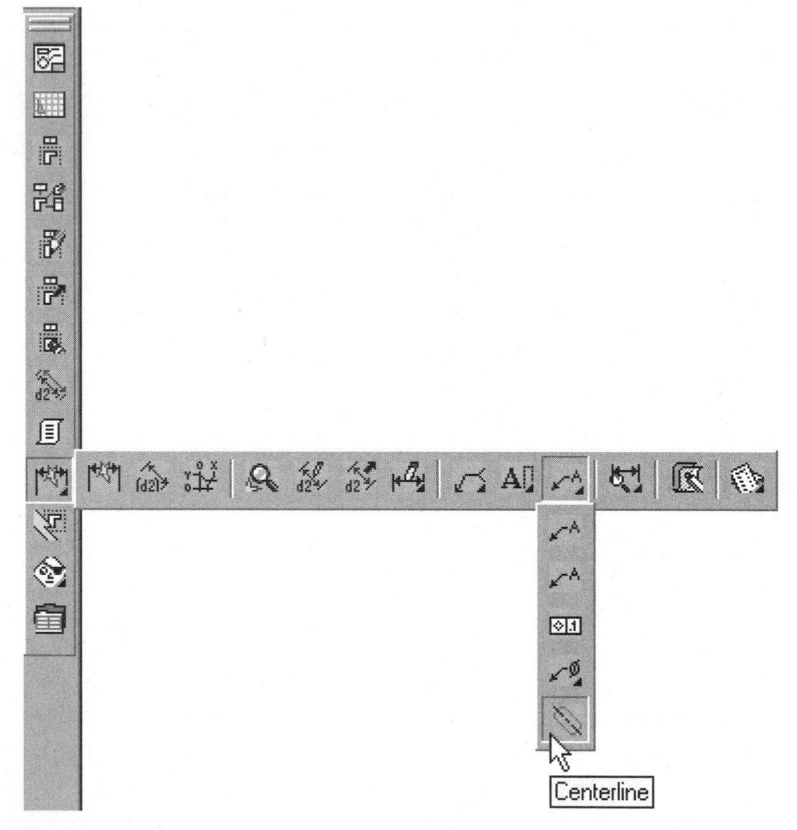

Figure 4.42

ANNOTATIONS

As you create drawing views, you will find it necessary to add your own 2D geometry, text, etc. to a drawing view. Once the information is in the correct position, you can attach it to a point in a view with the Annotation command **(AMANNOTE)** from the Drawing Layout toolbar, as shown in Figure 4.43. Once the information is attached, it will maintain a distance relationship to the selected point. If the view moves or changes, so will the annotation. After issuing the Annotation command, press ENTER to create an annotation, select the geometry or text that will be part of the annotation and press ENTER. Then select a point in a view to attach to. The point selected will go to the nearest endpoint of a line or center of an arc or circle. Once an annotation is created, you can add geometry to it with the Add option. You can delete an annotation with the Delete option and reposition the annotation with the Move option.

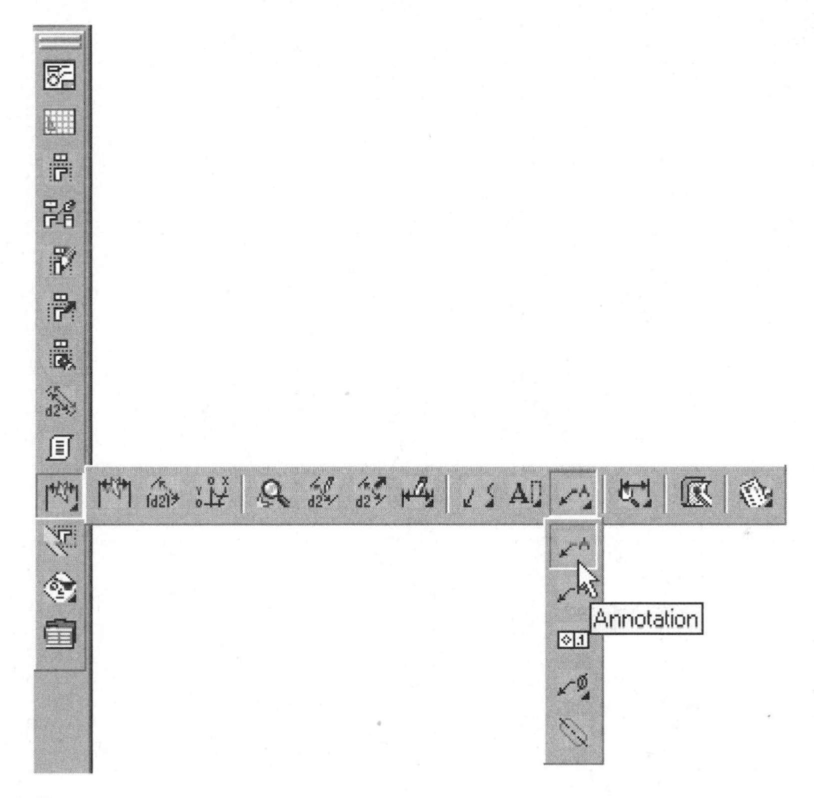

Figure 4.43

SYMBOLS

Mechanical Desktop also has standard symbols such as weld, surface texture, GD&T, datum identifier, datum target, feature identifier, and control frames can be created. You can also create your own standard based on ANSI, BSI, DIN, JIS and ISO, by issuing the Symbol Standard command **(AMSYMSTD)** from the Drawing Layout toolbar, as shown in Figure 4.44. You can copy a standard and then customize the copied standard to meet your requirements. The method for creating all the symbols is the same: issue the command, select a location that the symbol will be attached to (the distance between the symbol and this point will be maintained) then press ENTER or pick points to create a leader and then press ENTER. As the information in the dialog box is filled in, it will appear or when you select in another field. To edit a placed symbol double-click on it and the same dialog box will appear that it was created with.

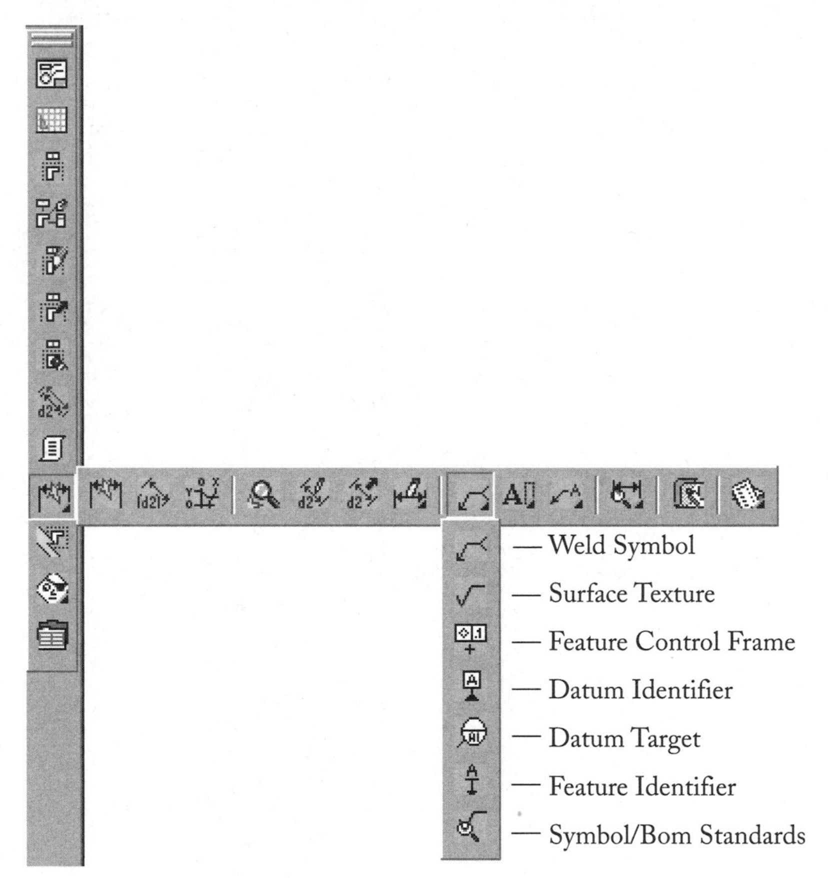

Figure 4.44

TUTORIAL 4.12—CENTER MARKS, CENTER LINES AND SYMBOLS

1. Open the file \Md4book\Chapter4\TU4-12.dwg.
2. Add a surface finish symbol to the RIGHT SIDE view AS SHOWN IN Figure 4.45
3. Add the note to the right side of the surface finish symbol using MTEXT.
4. Issue the Annotation command **(AMANNOTE)**.

5. Press ENTER to create a annotation.
6. Select the note and then select a point near the top right corner of the right side view. (An X will appear noting the attached location.)
7. Move the right side view to verify that the surface finish symbol and annotation also move with the view.
8. Save the file.

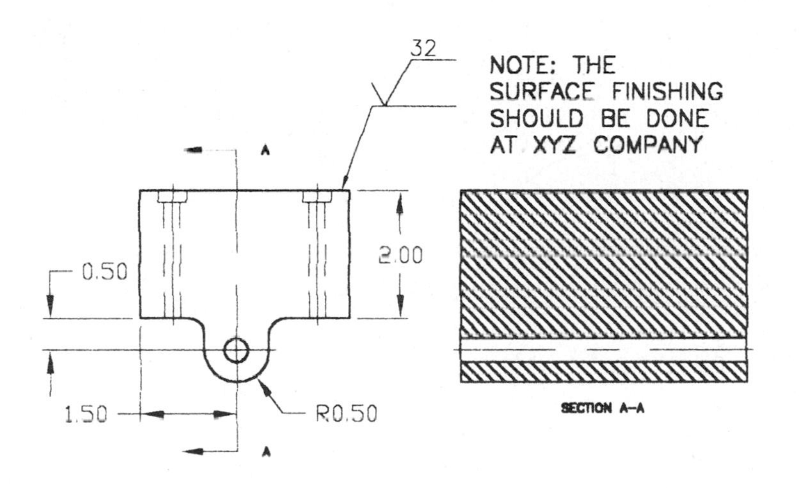

Figure 4.45

EXERCISES—DRAWING VIEWS WITH ANNOTATIONS

For the following exercises, follow the instructions before each.

TUTORIAL

Exercise 4.1—Bracket

Open the file \Md4book\Chapter3\bracket.dwg. Create a base and a ortho drawing view with a scale of 1=2 and the isometric view at .75 relative to the parent, reposition dimensions and add hole notes as shown in Figure 4.46.

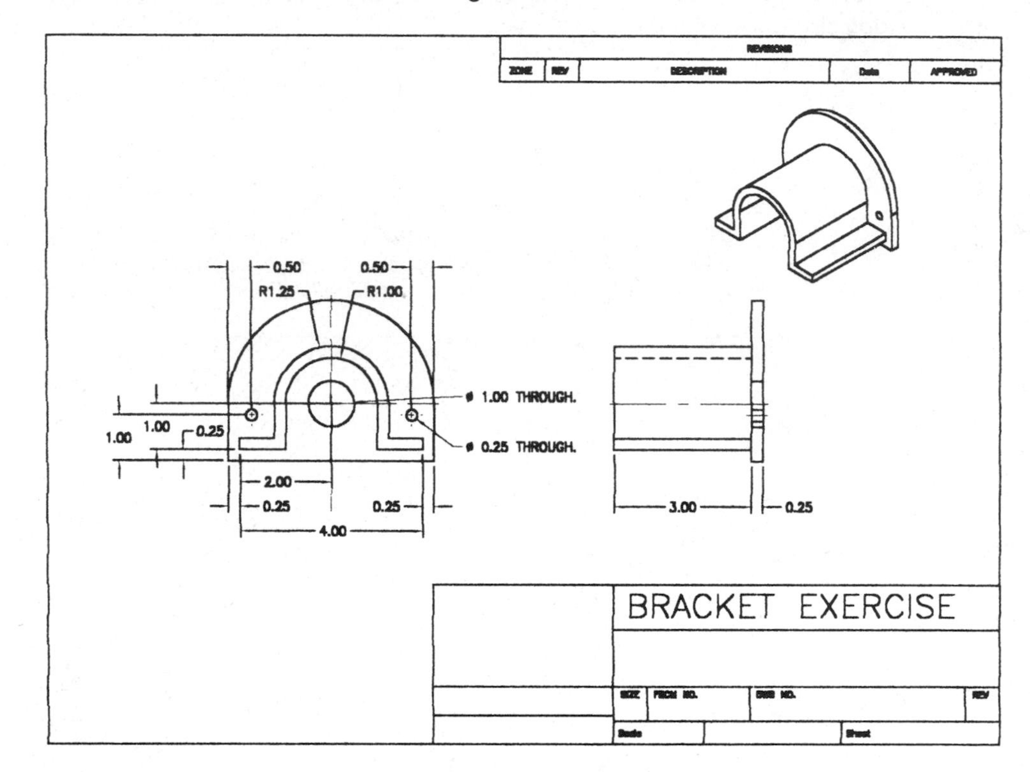

Figure 4.46

Exercise 4.2—Guide

Open the file \Md4book\Chapter3\Guide.dwg. Create a base and a ortho drawing view with a scale of 1=2 and the isometric view at .75 relative to the parent, reposition dimensions and add hole notes as shown in Figure 4.47.

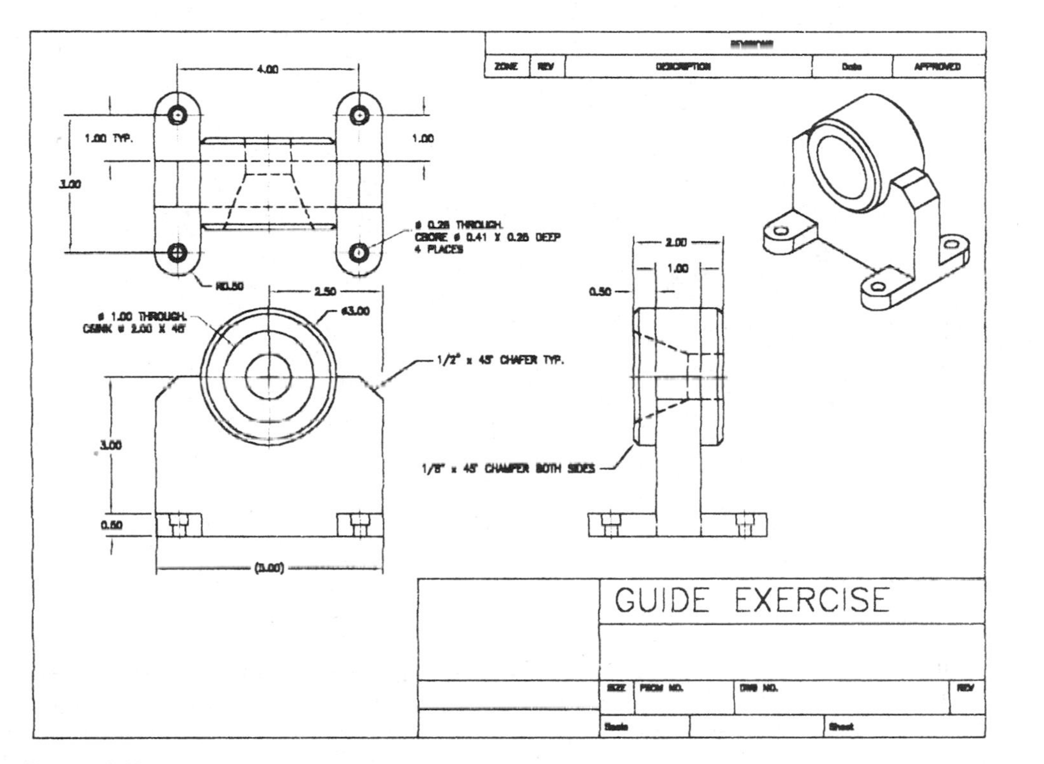

Figure 4.47

Exercise 4.3—Foot

Open the file \Md4book\Chapter3\Foot.dwg. Create a base, ortho and an auxiliary drawing view with a scale of 1=2 and the isometric view at .75 relative to the parent, reposition dimensions and add hole notes as shown in Figure 4.48.

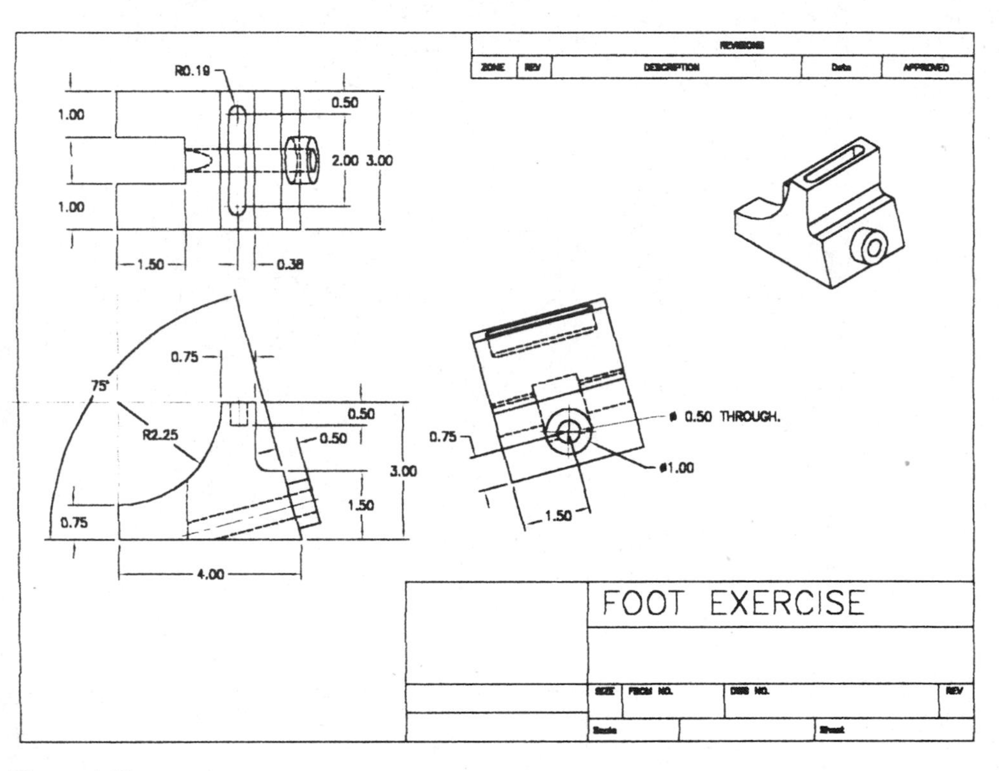

Figure 4.48

REVIEW QUESTIONS

1. There can be only one base view in a drawing. T or F?
2. The color and linetype of layers that Mechanical Desktop information is placed on cannot be changed. T or F?
3. Why is it not recommended to use the Select option when creating drawing views?
4. The only way to create an isometric view is to first create a base view and then create an isometric view from it. T or F?
5. When plotting a drawing with Mechanical Desktop views, you always plot at 1=1. T or F?
6. Explain how a dimension can be moved from one view to another and also how to reattach a dimension.
7. Every center line needs to be placed manually. T or F?
8. Reference dimensions can be changed to modify the part. T of F?
9. Explain how to remove a line section from a dimension.
10. Explain what an annotation is in reference to Mechanical Desktop drawing views.

CHAPTER 5

Work Axis, Work Planes, Work Points, 3D Path (sweep), Loft and Visibility

Up to now, you have created sketches on planes that lie on the part (active sketch plane). On that plane you created your sketch, constraints, dimensions and then extruded or revolved that profile into a part. There are times that you will need to create a sketch or place features on a plane that does not exist on the part. To do this you will create work planes, work axis and work points that are parametrically tied to the selected objects. In this chapter you will learn how to create work features that will extend the modeling capabilities and then apply those work features. The creation of sweeps and lofts will also be introduced.

AFTER COMPLETING THIS CHAPTER, YOU WILL BE ABLE TO:

- Create a work axis.
- Create work planes.
- Create work points.
- Control the visibility of objects.
- Create sweeps.
- Create lofts.

CREATING A WORK AXIS

A work axis is a line that goes through the center of an arc or circular edge of a part or can be sketched on a plane. A work axis has three purposes:

1. It can be used to align the UCS of a selected sketch plane or work plane.
2. It is used to create a work plane that goes through the center of a cylindrical shaped part.
3. It is used as the axis of revolution in a polar array.

To create a work axis, issue the Work Axis command **(AMWORKAXIS)** from the Part Modeling toolbar as shown in Figure 5.1 or right-click in the drawing area and from the pop-up menu select Work Axis from the Sketched & Work Features menu. Select an arc or circular edge or Press S and ENTER to create the work axis on the current sketch plane and then pick two points to locate the axis. A work axis will be created through the center of the selected arc or circular edge extending beyond both sides of the part. If the arc or circular edge is present on both sides of the part, it does not matter which side of the geometry is selected. For example, to place a work axis in a cylinder, you could select either the top or bottom circular edge and you would get the same result. The work axis is parametrically tied to the arc or circular edge that was selected. If the location of the arc or circular edge changes, the work axis will also move. For work axes that are placed on the current sketch plane the work axis will snap horizontally or vertically if it falls within the Angular Tolerance of horizontal and vertical—set through the Desktop options command **(AMOPTIONS)**. To guarantee a horizontal or vertical work axis, turn ortho on. To add a different constraint or dimension to the work axis, follow these steps.

1. Issue the Edit Feature command **(AMEDITFEAT)** and select the work axis.
2. The work axis will appear between the two points that were selected to create the work axis.
3. Delete constraints if needed.

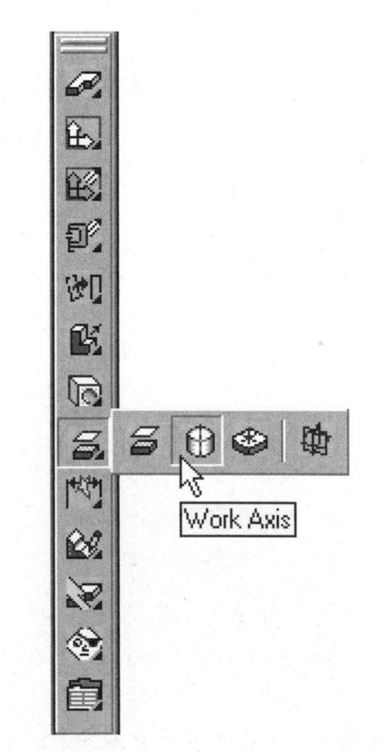

Figure 5.1

4. Add dimensions and constraints to the work axis.
5. When finished, update the part.

Note: A work axis does not need to be fully constrained.

TUTORIAL 5.1—WORK AXIS

1. Open the file \Md4book\Chapter5\TU5-1.
2. Issue the Work Axis command **(AMWORKAXIS)** and select one of the arcs on the right side of the part.
3. Make the left side face the current sketch plane as highlighted in Figure 5.2. Accept the default orientation for the XY axis.4. Issue the Work Axis command **(AMWORKAXIS)** and press S and ENTER to place the work axis on the current sketch plane. Select two points that are nearly vertical to one another near the middle of the work plane. When complete you should have a work axis in the middle of the sketch plane.

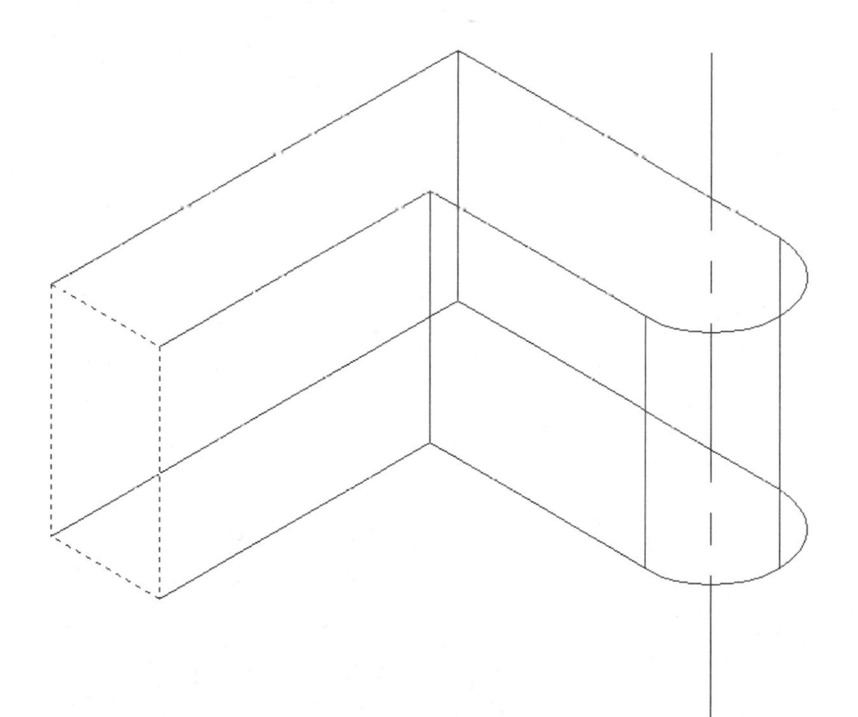

Figure 5.2

TUTORIAL

5. Edit the work axis and add a "0.50" dimension to the work axis and the end of the part as shown in Figure 5.3. The work axis will still require two dimensions: one for the location of each end point. However, those dimensions will not affect the work axis location so you will not add those dimensions.

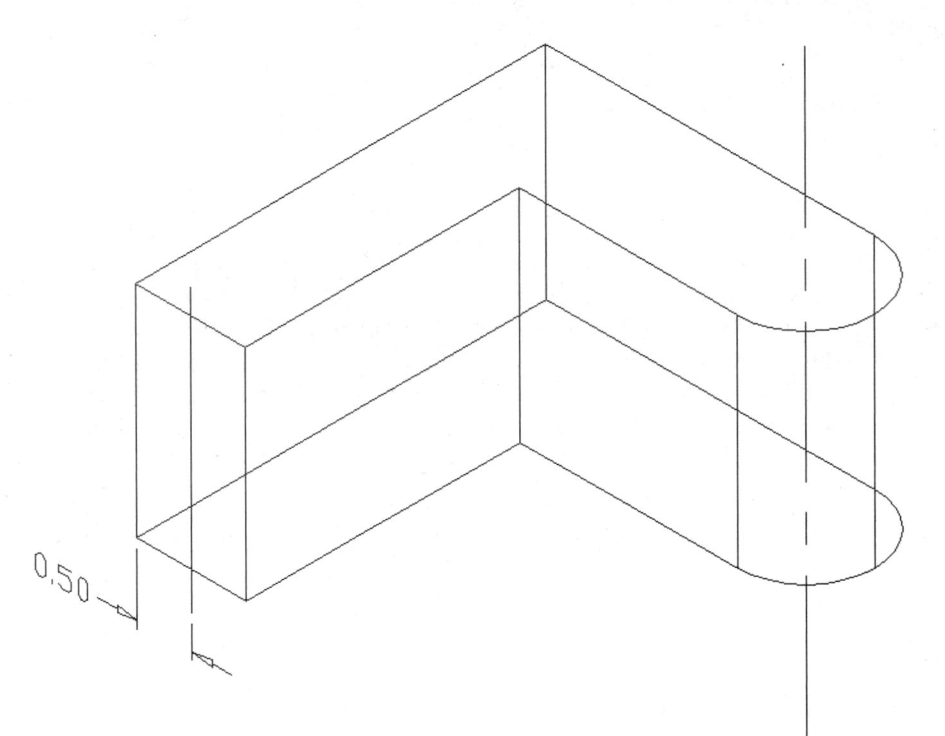

Figure 5.3

6. Update the part if needed. When updated, the dimension will disappear.
7. Issue the Work Plane command **(AMWORKPLN)** and check On Edge/Axis for both the first and second modifier. Check Create Sketch Plane and then select OK to return to the part.
8. Select each work axis—the order is not important.
9. The work plane will appear between each work axis. Rotate the *X* , *Y* and *Z* axis so the *X* is pointing into the screen and the *Y* axis is pointing upward press ENTER to accept this orientation.
10. Save the file.

CREATING A WORK PLANE

Before you consider work planes, it is important that you understand when you need to create a work plane. If there is a face on the part that can be used as the active sketch plane, make it the active sketch plane. If there is no face where you need one, then you create a work plane. A work plane can be made the active sketch plane and used as the plane for extruding, revolving or sweeping. A work plane can also be used as a plane to stop or start an extrusion, revolve, sweep or loft. It can also be used for positioning section cuts in drawing views.

A work plane is a rectangular plane that, ideally, is parametrically tied to the part and that will always be larger than the part. If made parametrically, if the part moves, the work plane will also move. For example, if a work plane is tangent to the outside face of a 1"-diameter cylinder and the cylinder diameter changes to 2", the work plane moves with the outside face of the cylinder. You can create as many work planes on a part as needed, and any work plane can be made the active sketch plane with the Create Sketch Plan command **(AMSKPLN)**. A work plane is similar to moving the UCS, but the trouble with this is that the UCS is not parametrically tied to the part and could give you unexpected results.

A work plane is a feature and must be edited and deleted as a feature. For example, to edit an angle or offset distance of a work plane, use Edit Feature command **(AMEDITFEAT)** or double-click on the work plane's name in the browser. To delete work planes, use the Delete Feature command **(AMDELFEAT)**. Do not use the erase command, because this will delete the entire part.

Later in this chapter you will learn how to control the visibility of work planes. Before creating a work plane, ask yourself where this work plane needs to exist and what you know about this location. For example, you might want a plane to be tangent to a given face and parallel to another plane, or to go through the center of two arcs. Once you know what you want, select from the appropriate options and create a work plane. There are times when you may need to create an intermediate (construction) work plane before creating the final work plane. An example of this need: creating a work plane that is at 30° and tangent to a cylindrical face. First create a work plane that is at a 30° angle and located at the center of the cylinder, and then create a work plane parallel to the angled work plane that is also tangent to the cylinder.

When work planes are created, it is best if they are created from a Mechanical Desktop part. They can also be created with the World Coordinate system. However, if you rotate the part in 3-D space you may get unexpected results because the work plane is not following the geometry. Also, if you place a work plane on a face of the part using the UCS, and that face moves, the work plane will not move with the face.

To create a work plane, issue the Work Plane command **(AMWORKPLN)** from the Part Modeling toolbar as shown in Figure 5.4 or right-click in the drawing area and from the pop-up menu select Work Plane from the Sketched & Work Features menu. A Work Plane Feature dialog box will appear on the screen as shown in Figure 5.5. This dialog box consists of four sections: 1st Modifier, 2nd Modifier, Create Sketch Plane and an area for offset distance and angle. The 1st and 2nd Modifiers will determine how the work plane is tied to the part. As different options for the 1st Modifier are selected, there will be corresponding options that are grayed out in the 2nd Modifier area. It does not matter which modifier is selected for the first or second modifier, except for the Sweep Profile option, which does not require a second modifier. The Create Sketch Plane area can be checked if you want to make the new work plane the active sketch plane. After creating the work plane, you will be prompted to align the UCS to that plane. The last section is an area for specifying an offset distance or an angle.

Figure 5.4

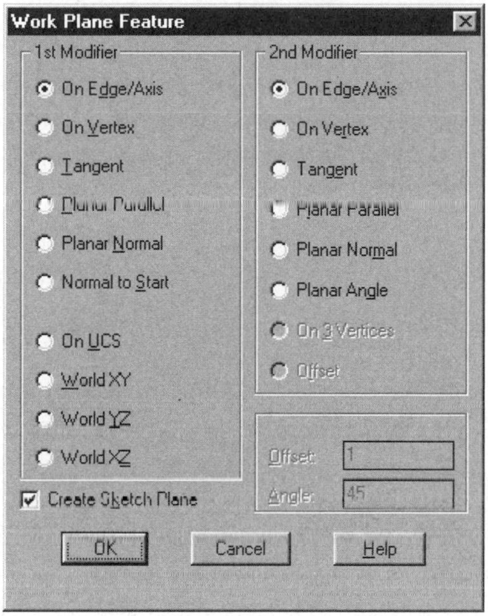

Figure 5.5

Notes:

- When making a work plane the active sketch plane and aligning the UCS, follow the same steps as you did for making a planar face the active sketch plane.
- When the modifiers for creating a work plane are selected, it does not matter which option is the first or second modifier, except for Sweep Profile, which does not require a second modifier.
- Work planes are tied to the part—if the part changes, the work plane will maintain the original modifiers used when it was created.
- Work planes and work axes will always extend beyond the part; if the part gets larger, the work plane will re-size.
- The color of the work features is controlled by the color of the layer "AM_WORK".

The best way get an understanding of work planes is to create them. In the tutorials that follow, you will create the most common types of work planes. Before each tutorial, there is a description of the type of work plane that will be created.

WORK PLANES ON EDGE/AXIS–ON EDGE/AXIS

In this exercise you will create two work axes and then create a work plane that passes through each axis. The options that you will use are On Edge/Axis and On Edge/Axis. An edge is any line in the part, and an axis refers to a work axis. With the modifiers On Edge/Axis and On Edge/Axis, you can create a work plane using these combinations: two edges, two work axes or an edge and a work axis. The screen has been split in two by using the 2 key.

TUTORIAL 5.2—CREATING A WORK PLANE ON EDGE/AXIS–ON EDGE/AXIS

1. Open the file \Md4book\Chapter5\TU5-2.
2. Issue the Work Axis command **(AMWORKAXIS)** and create a work axis through the two large arcs.
3. Issue the Work Plane command **(AMWORKPLN)** to create a work plane though each work axis. Use On Edge/Axis for both the 1st and 2nd Modifiers, check Create Sketch Plane and then select OK.
4. Select both work axes; it does not matter which axis is selected first.
5. Rotate the UCS until your screen resembles Figure 5.6.

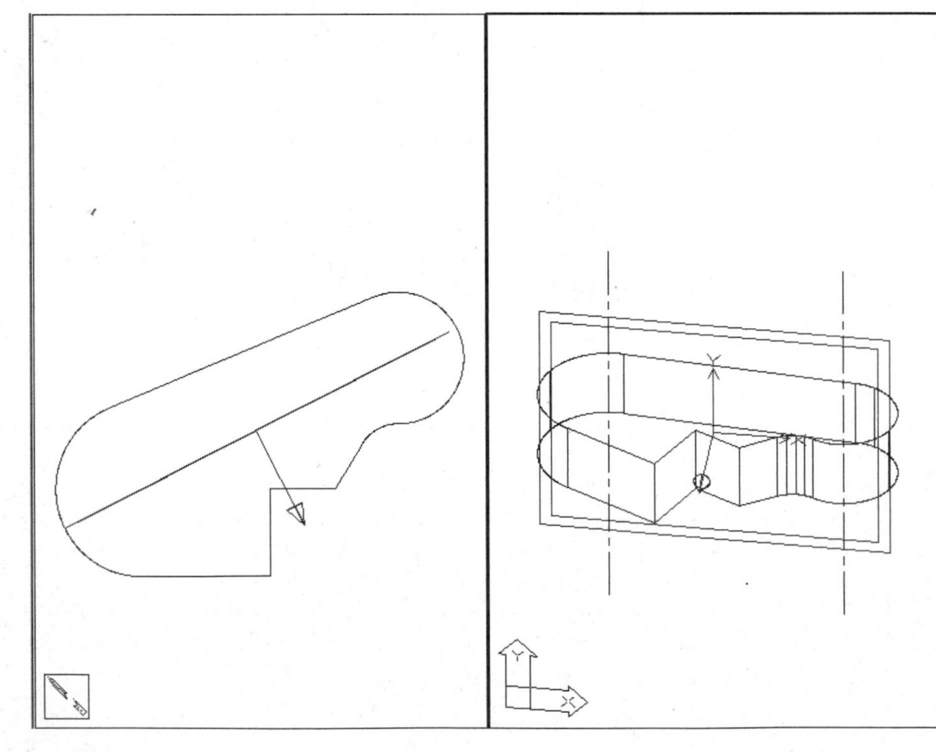

Figure 5.6

6. Repeat the Work Plane command and create a work plane though the two front edges. Again set both the 1st and 2nd Modifiers to On Edge/Axis, check Create Sketch Plane and then select OK. The two Xs in Figure 5.7 show the edges to select; the order is not important.
7. Rotate the UCS so the X axis is pointing into the screen as shown in Figure 5.7.
8. Save the file.

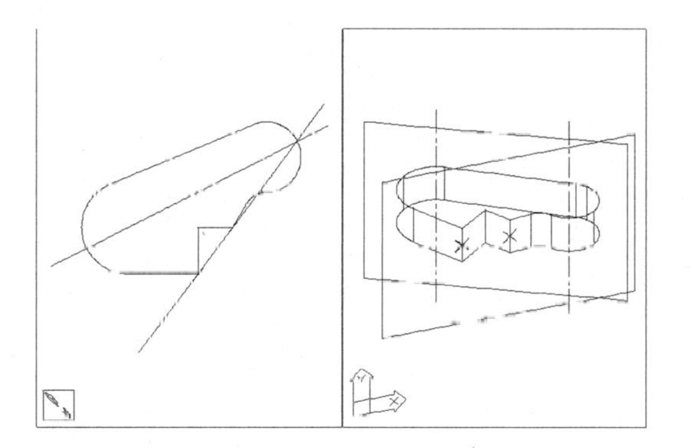

Figure 5.7

WORK PLANES ON EDGE/AXIS – PLANAR ANGLE

In this tutorial you will create a work plane that is angled: you will select a work axis, or straight edge, that will act as the center of rotation and specify a plane as the reference for the angle. After the plane is selected, an arrow will appear showing the direction. You can either accept or flip the direction.

TUTORIAL 5.3—CREATING A WORK PLANE ON EDGE/AXIS – PLANAR ANGLE

1. Open the file \Md4book\Chapter5\TU5-3.
2. Issue the Work Axis command **(AMWORKAXIS)** and create a work axis through the left arc.
3. Issue the Work Plane command **(AMWORKPLN)** to create an angled work plane. Set the 1st Modifier to On Edge/Axis and the 2nd Modifier to Planar Angle, type in "60" for the Angle, check Create Sketch Plane and then select OK.
4. For the prompt:
   ```
   Select work axis, straight edge or
   [worldX/worldY/worldZ]:
   ```
   Select the work axis.

5. For the prompt:
   ```
   Select work plane, planar face or
   [worldXy/worldYz/worldZx/Ucs]:
   ```
   Select the front plane as shown in Figure 5.8. When the front face is highlighted, press ENTER.

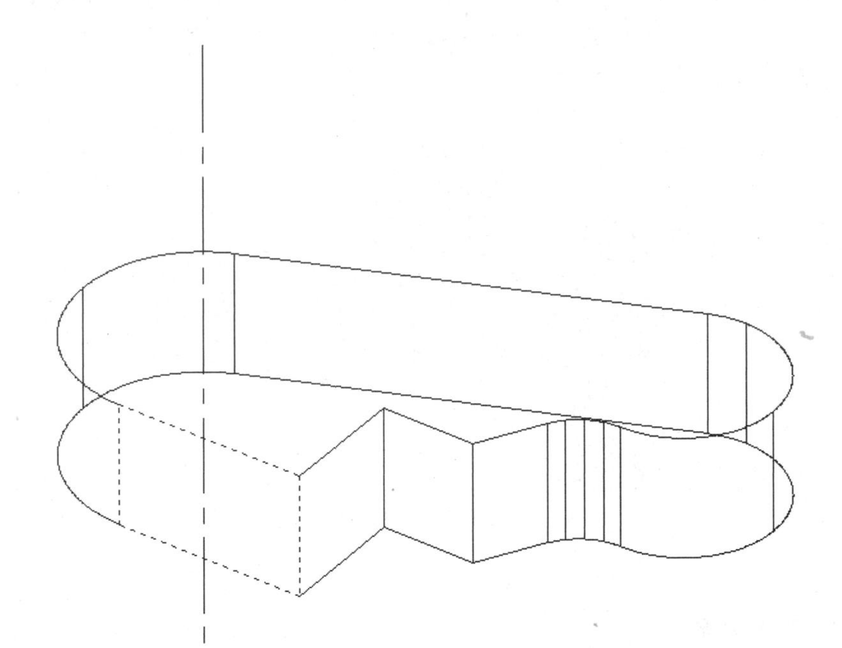

Figure 5.8

6. The work plane will appear on the screen. Flip the direction by clicking the left mouse button. Click the left mouse button again to return to the original position. Press ENTER to accept this position.
7. Rotate the *XY* axis until your screen resembles Figure 5.9. When complete press the ENTER.
8. Change the angle of the work plane to 45° with the Edit Feature command **(AMEDITFEAT)**: Issue the command, select the work plane or double-click on the name WorkPlane1 in the browser, select the 60° dimension and change it to 45. Then update the part with the Update Part command **(AMUPDATE)**.
9. Save the file.

TUTORIAL

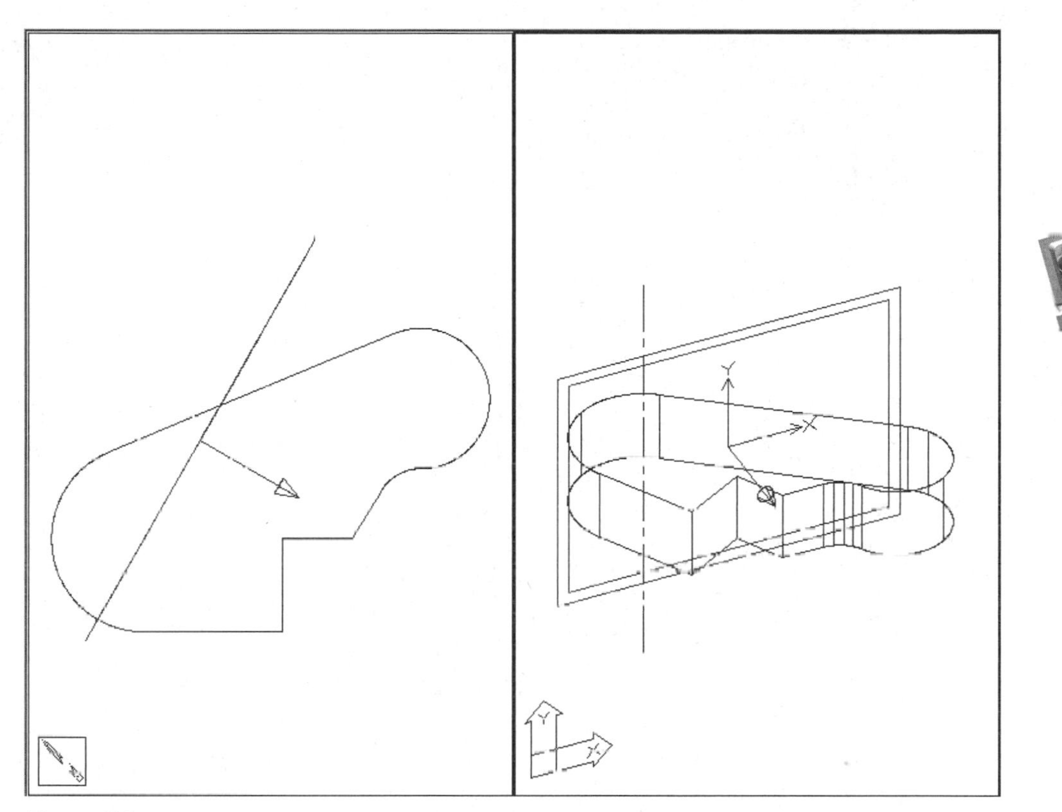

Figure 5.9

WORK PLANES: TANGENT – PLANAR PARALLEL

In this tutorial you will create a work plane that is tangent to an arc and parallel to another work plane. To create a tangent work plane, you will select an arc or circular edge to be tangent to; it does not matter what side of the extrusion you select. Then you will select a plane to which you want the work plane to be parallel. After the work plane appears on the screen, you will prompted to either accept or flip the side to which the work plane will be tangent.

TUTORIAL 5.4—CREATING A WORK PLANE: TANGENT – PLANAR PARALLEL

1. Open the file \Md4book\Chapter5\TU5-4.
2. Issue the Work Plane command **(AMWORKPLN)** to create a Tangent work plane. Set the 1st Modifier to Tangent and the 2nd Modifier to Planar Parallel, check Create Sketch Plane and then select OK.
3. For the prompt:
   `Select cylindrical or conical face:`
   Select the leftmost arc.

4. For the prompt:
   `Select work plane, planar face or [worldXy/worldYz/worldZx/Ucs]:`
   Select the existing work plane.
5. Flip the direction for the tangency by clicking the left mouse button and then press ENTER to accept this direction.
6. Rotate the *XY* axis until your screen resembles Figure 5.10. When complete, press ENTER.

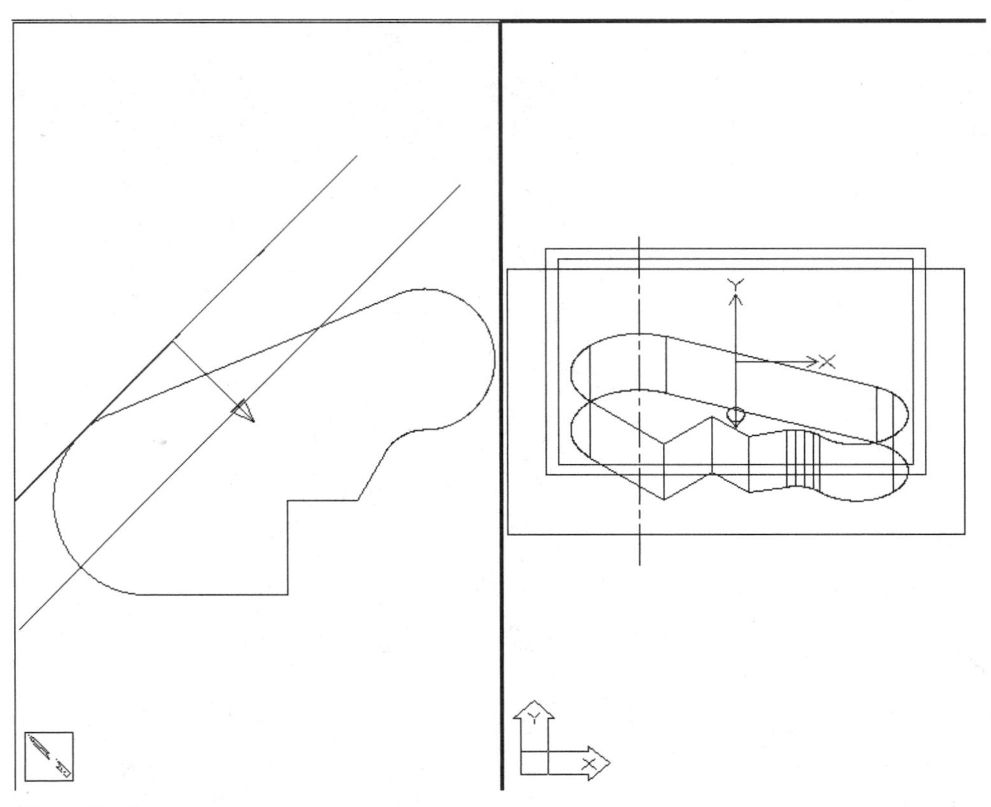

Figure 5.10

7. Repeat the Work Plane command, set the 1st Modifier to Tangent and the 2nd Modifier to Planar Parallel, check Create Sketch Plane and select OK.
8. For the prompt:
   `Select cylindrical or conical face:`
   Select the rightmost arc.

9. For the prompt"
   ```
   Select work plane, planar face or
   [worldXy/worldYz/worldZx/Ucs]:
   ```
   Select the highlghted face as shown in Figure 5.11

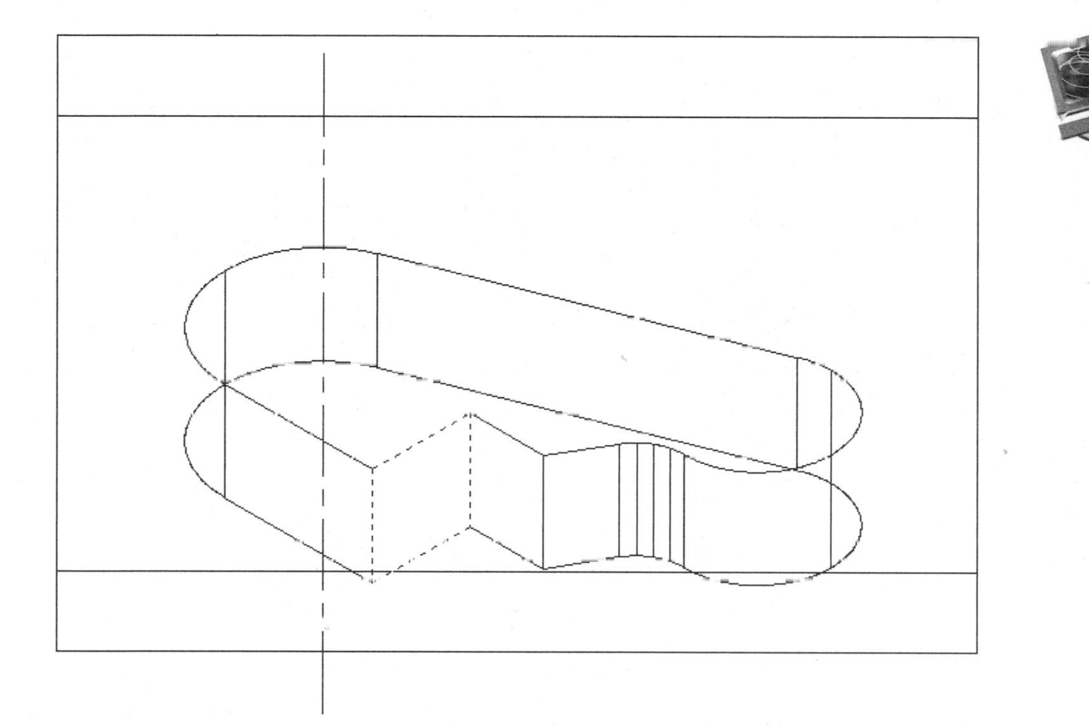

Figure 5.11

TUTORIAL

10. Press ENTER to accept the default direction of the tangency.
11. Press ENTER to accept the default orientation of the *XY* axis as shown in figure 5.12.
12. Save the file.

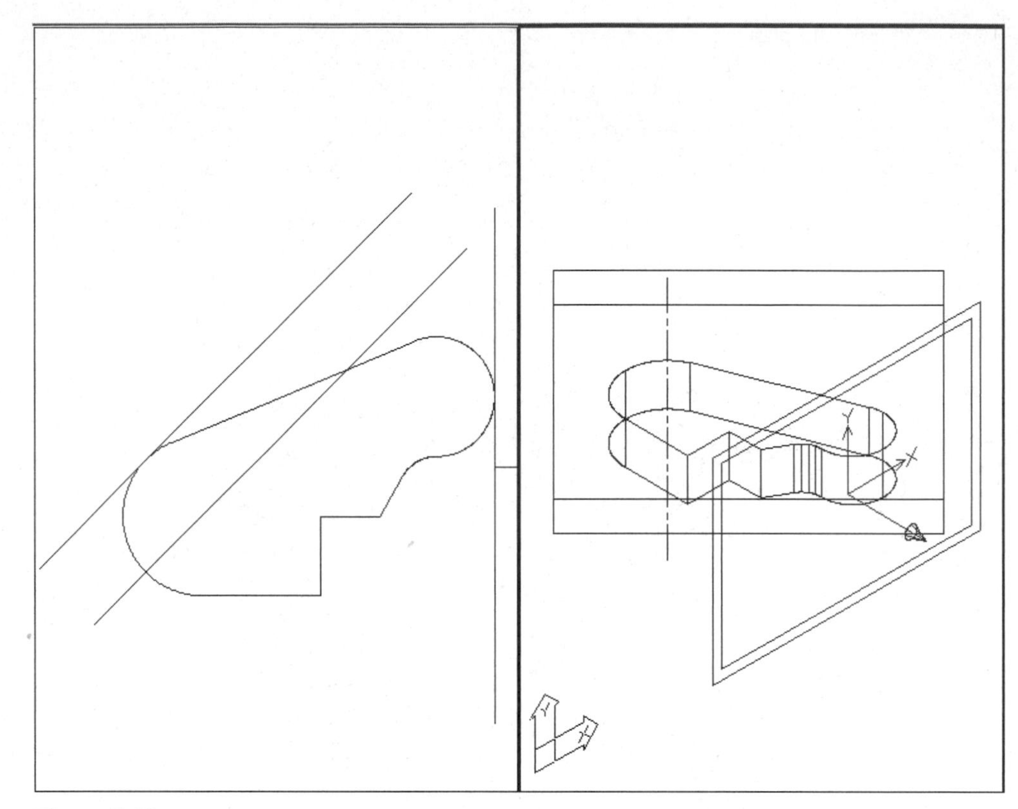

Figure 5.12

TUTORIAL 5.5—CREATING A WORK PLANE TANGENT – PLANAR PARALLEL

1. Start a new drawing from scratch.
2. Create a 2" diameter circle and extrude it 2".
3. Change to an isometric view using the **8** key.
4. Create a work plane with the 1st Modifier set to Tangent and the 2nd Modifier to Planar Parallel, check Create Sketch Plane and then click OK.
5. For the prompt:
   `Select cylindrical or conical face:`
   Select either the top or bottom circle (it does not matter where you select).
6. For the prompt:
   `Select work plane, planar face or [worldXy/worldYz/worldZx/Ucs]:`
   Since there is no axis or edge available as a parallel plane, you must refer back to the world coordinate system. Press Y and ENTER to orient it to the world *YZ* plane.

7. Press ENTER to accept the default direction.
8. Press ENTER to accept the default orientation of the *XY* axis as shown in Figure 5-13.
9. Save the file as \Md4book\Chapter5\TU5-5. You will use this file later in the chapter.

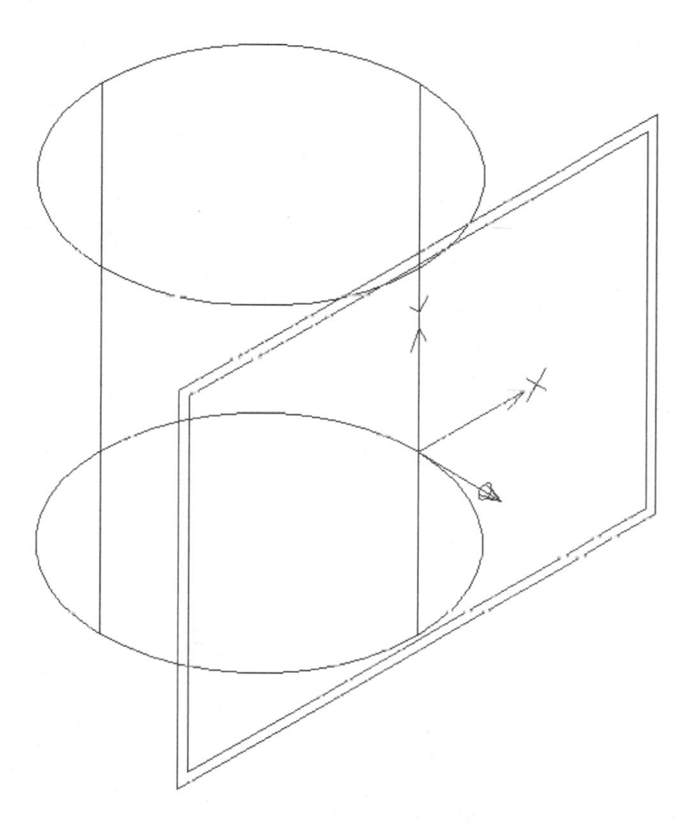

Figure 5.13

TUTORIAL

WORK PLANES: TANGENT – ON EDGE/AXIS

In this tutorial you will create a work plane that is tangent to an arc and attached to an edge. To create a work plane that is tangent to an arc and tied to an edge, select on the side of the arc to which you want the work plane to be tangent.

TUTORIAL 5.6—CREATING A WORK PLANE TANGENT – ON EDGE/AXIS

1. Open the file \Md4book\Chapter5\TU5-6.
2. Create a work plane with the 1st Modifier set to Tangent and the 2nd Modifier to On Edge/Axis, check Create Sketch Plane then select OK.

3. For the prompt:
   `Select cylindrical or conical face:`
   Select near the bottom of the right arc, as shown with an X in Figure 5.14.
4. For the prompt:
   `Select work axis, straight edge or [worldX/worldY/worldZ]:`
   Select the vertical edge, as shown with an X in Figure 5.14.

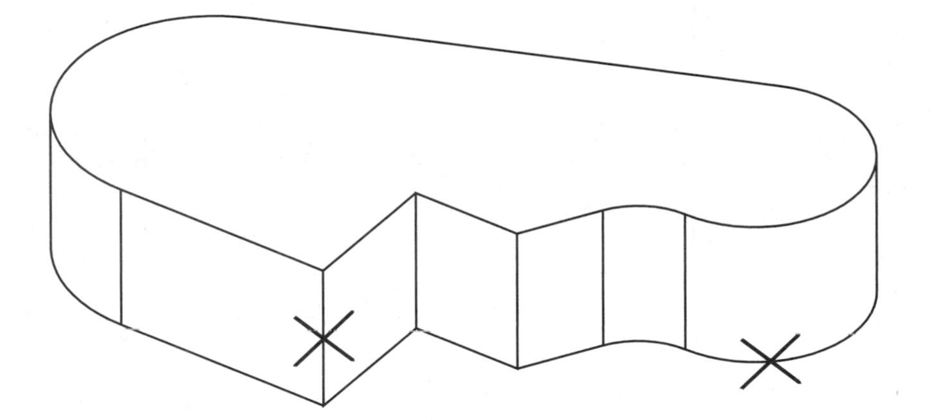

Figure 5.14

5. Rotate the *XY* axis until your screen resembles Figure 5.15. When complete, press ENTER.
6. Save the file.

WORK PLANES: PLANAR PARALLEL

When creating a parallel work plane, you can create it from another work plane, a plane on the part, or parallel to the world coordinate system. The plane will be parallel to the selected plane and will be placed at the location specified by the second modifier.

TUTORIAL 5.7—CREATING A WORK PLANE PLANAR PARALLEL – OFFSET

1. Open the file \Md4book\Chapter5\TU5-6 if it is not the current file.
2. Create a Work Plane with the 1st Modifier set to Planar Parallel and the 2nd Modifier set to Offset, check the Create Sketch Plane, type in "2" for the Offset distance and then select OK.
3. For the prompt:
   `Select work plane, planar face or [worldXy/worldYz/worldZx/Ucs]:`
   Select the work plane that was created in Tutorial 5.6.

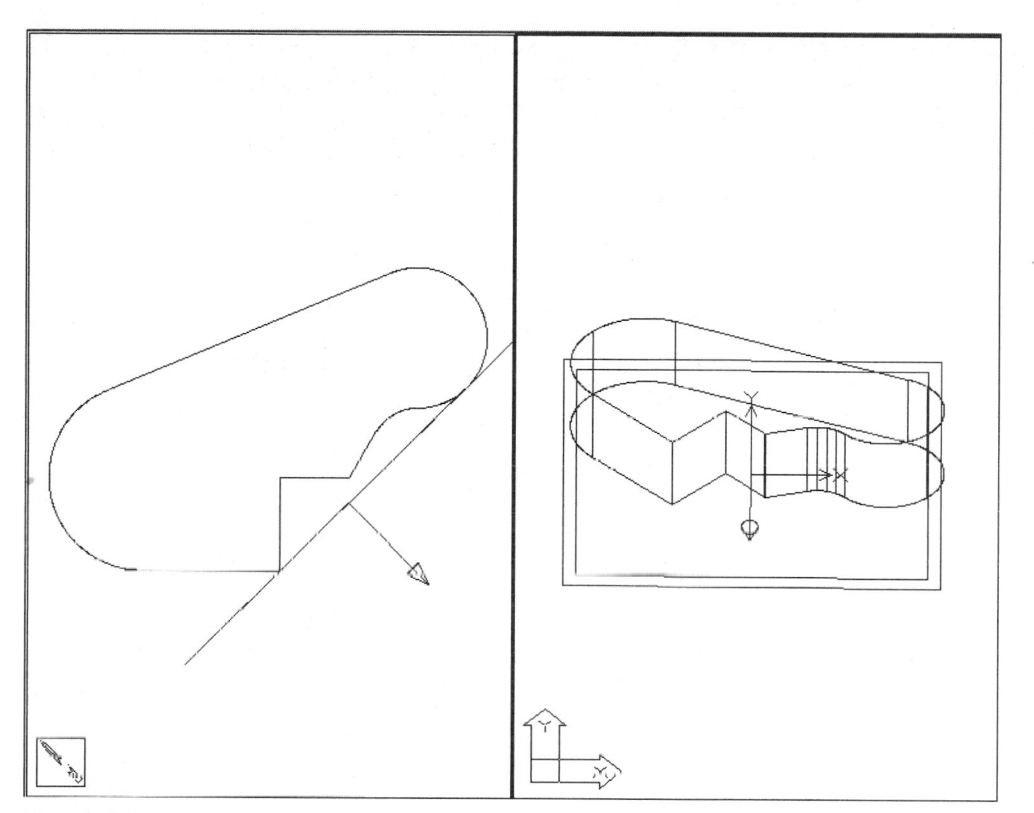

Figure 5.15

4. Press ENTER to accept the default direction.
5. Press ENTER to accept the default orientation of the *XY* axis.
6. Delete the first work plane and the second work plane will also be deleted since it is dependant on the first work plane.
7. Practice offsetting, editing and deleting work planes from planar faces on the part.
8. When finished practicing save the file.

WORK PLANES: TANGENT TO TWO CIRCULAR EDGES

Work planes can also be tangent to two circular edges on the same part. Issue the Work Plane command **(AMWORKPLN)**. In the Work Plane Feature dialog box, select Tangent for the first and second modifiers and then select OK. On the part, pick the first and then the second circular edge. Pick close to where you want the work plane's point of tangency. The work plane will be created near the selected points and an arrow will appear in each of the corners. Mechanical Desktop looks at the circular edges as though they were circles and to cycle through the four combinations of tangencies even though you may have selected an arc. Cycle through by clicking the left mouse button until the correct tangency is shown then press ENTER to create the work plane. If Create Sketch Plane was checked in the Work Plane Feature dialog box, you will orient the *X*, *Y* and *Z* axes on the work plane. This work plane is parametrically tied to the chosen tangencies—if the size of the circular edge changes, the work plane will also adjust to the new size.

TUTORIAL

TUTORIAL 5.8—CREATING A WORK PLANE TANGENT – TANGENT

1. Open the file \Md4book\Chapter5\TU5-8.
2. Create a Work Plane with the 1st Modifier set to Tangent and the 2nd Modifier set to Tangent, check the Create Sketch Plane and then select OK.
3. For the prompt:
   `Select first cylinder or sphere:`
   Select one of the circular edges.
4. For the prompt:
   `Select second cylinder or sphere:`
   Select the other circular edge.
5. Cycle through the possible combinations until your screen resembles Figure 5.16.
6. Then press ENTER twice to accept this location and default *XY* axis orientation.
7. Save the file.

WORK PLANES: BASIC WORK PLANES

If you prefer to start a drawing that has a work plane in all three planes this can be accomplished by issuing the Create Basic Workplanes command **(AMBASICPLANES)** from the Part Modeling toolbar as shown in Figure 5.17. You will be prompted to select an origin for the work planes. After selecting a location, three work planes and a work point at the intersection of the three planes will be created. Before sketching on one of the work planes make it the active sketch plane with the Create Sketch Plan command **(AMSKPLN)**.

CREATING WORK POINTS

A work point is a feature consisting of three short lines, one in each axis (*X*, *Y*, and *Z*), that intersect at their midpoints. A work point can be used as the axis of rotation for polar arrays, for placing hole features and for locating points for a sweep path. To create a work point, issue the Work Point command **(AMWORKPT)** from the Part Modeling toolbar as shown in Figure 5.18. Or you can right-click in the drawing area and from the pop-up menu select Work Point from the Sketched & Work Features menu. Then select a point on the active sketch plane. Because the work point is a fea-

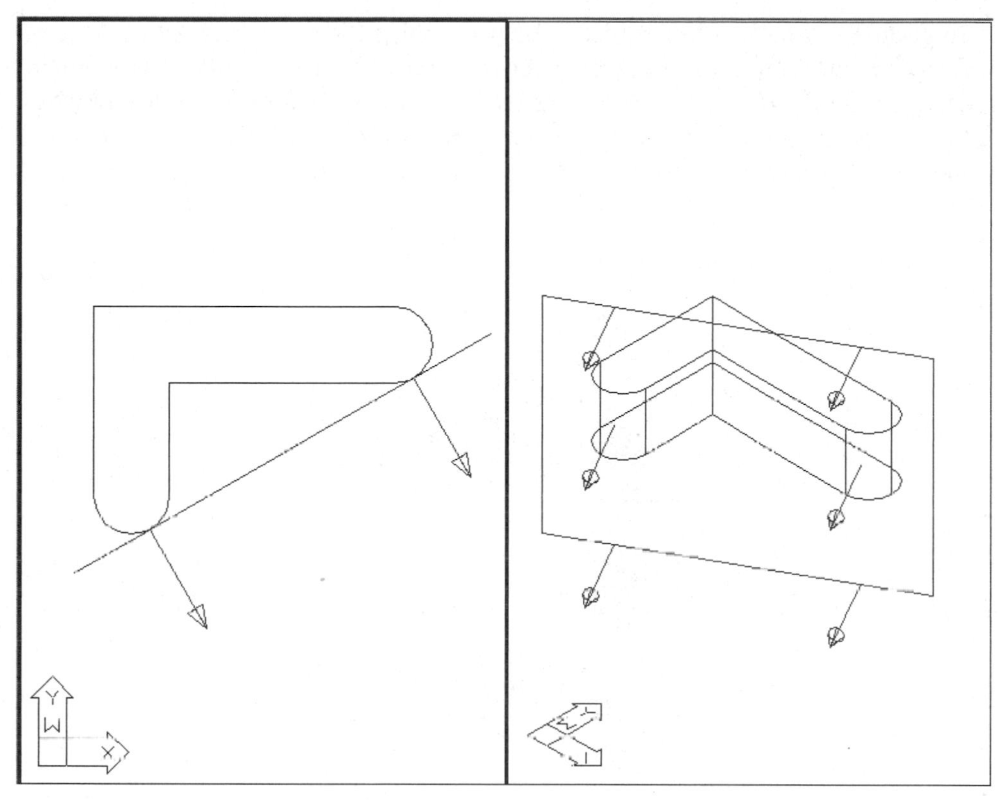

Figure 5.16

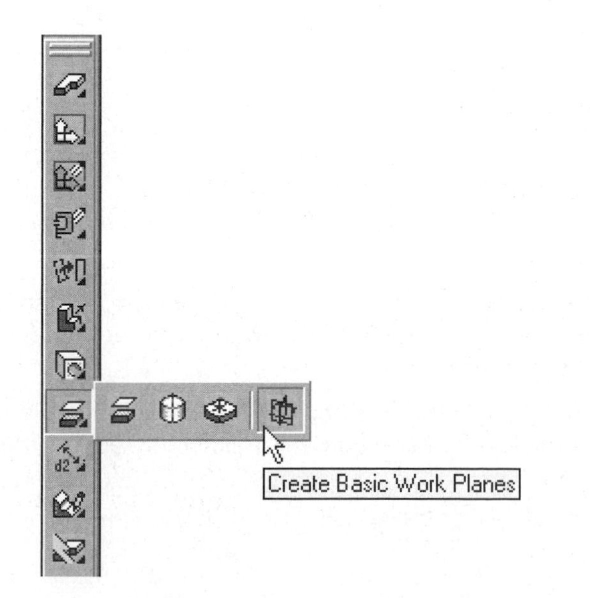

Figure 5.17

ture, it does not need to be profiled. The work point will not automatically be constrained; it will require two dimensions or constraints. When a work point is dimensioned, the dimension will go to the intersection of the three lines, regardless of where you select on the work point. After the work point is dimensioned, the dimensions will appear on the screen until the work point is used for an array, hole or sweep. A concentric constraint can be applied to constrain a work point to the center of an arc or circle edge. To delete a work point, use the Delete Feature command

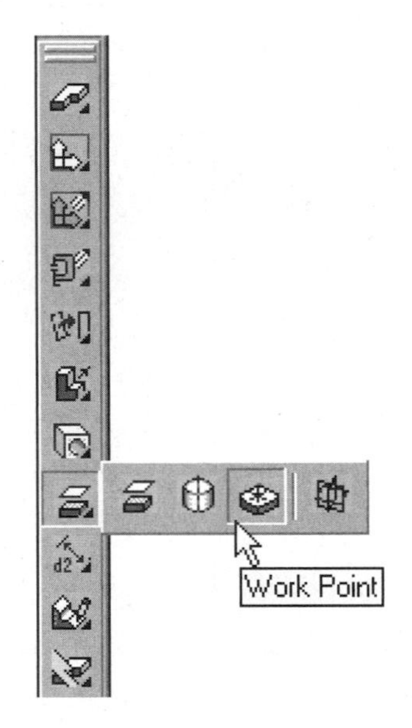

Figure 5.18

(AMDELFEAT).

TUTORIAL 5.8—CREATING WORK POINTS

1. Open the file \Md4book\Chapter5\TU5-5 that you worked on earlier in this chapter.
2. If the work plane is not the active sketch plane make it active. (Note: Use the Show Active command **(AMSHOWACT)** to determine if the workplane is also the current sketch plane.)
3. Issue the Work Point command **(AMWORKPT)** and place a work point to the right side of center on the *X* axis and near center in the *Y* axis.
4. Add a ".25" and "1.00" dimension to the work point and the cylinder as shown in Figure 5.19. The UCS Icon is off for clarity.

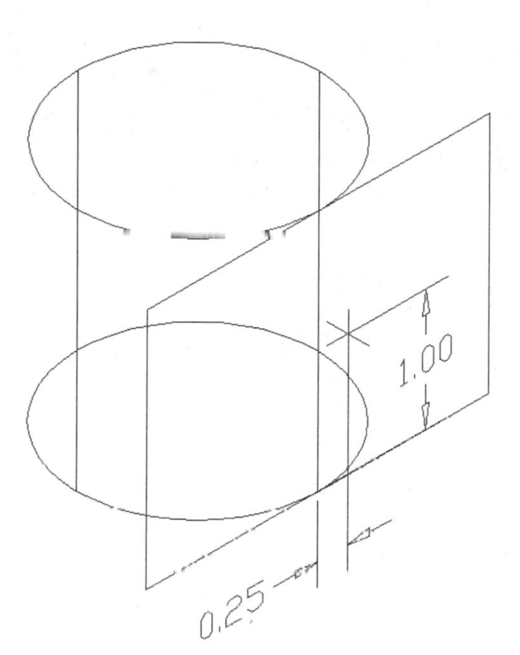

Figure 5.19

5. Make the top circular edge of the cylinder the active sketch plane.
6. Issue the Work Point command **(AMWORKPT)** and place a work point to the right side of center on the *X* axis and to the bottom of center in the Y axis.
7. Add two ".50" dimensions to the work point and the cylinder as shown in Figure 5.20. The UCS Icon is off for clarity.
8. Issue the Hole command **(AMHOLE)** and make the following settings:
   Operation = Drilled
   Termination = Through
   Placement = On Point
   Dia = .5
   Then select OK
9. Select the top work point and the hole will go through the part.
10. Select the work point on the side of the part and ENTER to accept the arrow into the part.
11. Press ENTER to complete the command.
12. Rotate and shade the part to make sure that both holes went through the part.

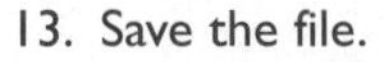

13. Save the file.

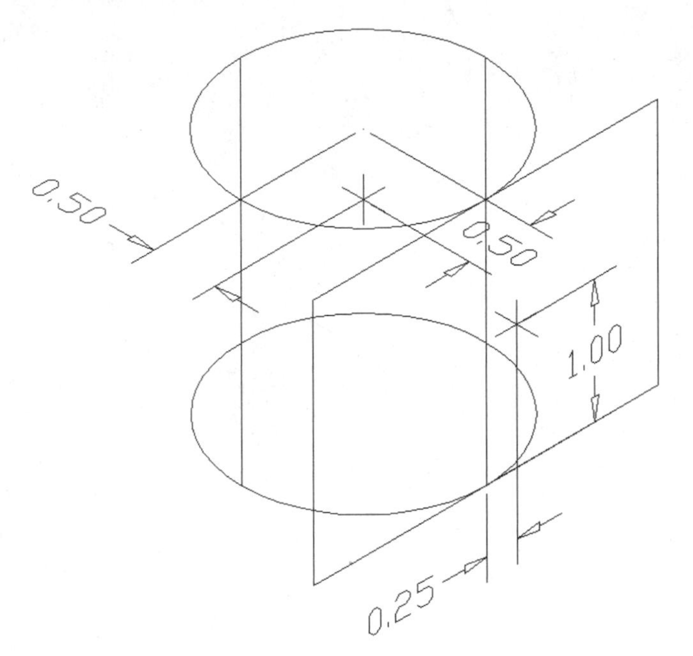

Figure 5.20

TUTORIAL 5.9—WORK FEATURES

1. Start a new drawing from scratch.
2. Sketch, profile and dimension a 4" square.
3. Extrude it 2" in the default direction.
4. Make the left vertical face, as highlighted in Figure 5.21, the active sketch plane. Accept the default orientation of the *XY* axis.
5. Place a work axis vertically in the middle of the plane.
6. Edit the work axis and add a 2" horizontal dimension as shown in Figure 5.22
7. Update the part.
8. Issue the Work Plane command **(AMWORKPLN)** to create an angled work plane. Set the 1st Modifier to On Edge/Axis and the 2nd Modifier to Planar Angle, type in "60" for the Angle, check Create Sketch Plane and then select OK.
9. For the prompt:
   ```
   Select work axis, straight edge or
   [worldX/worldY/worldZ]:
   ```
   Select the work axis.

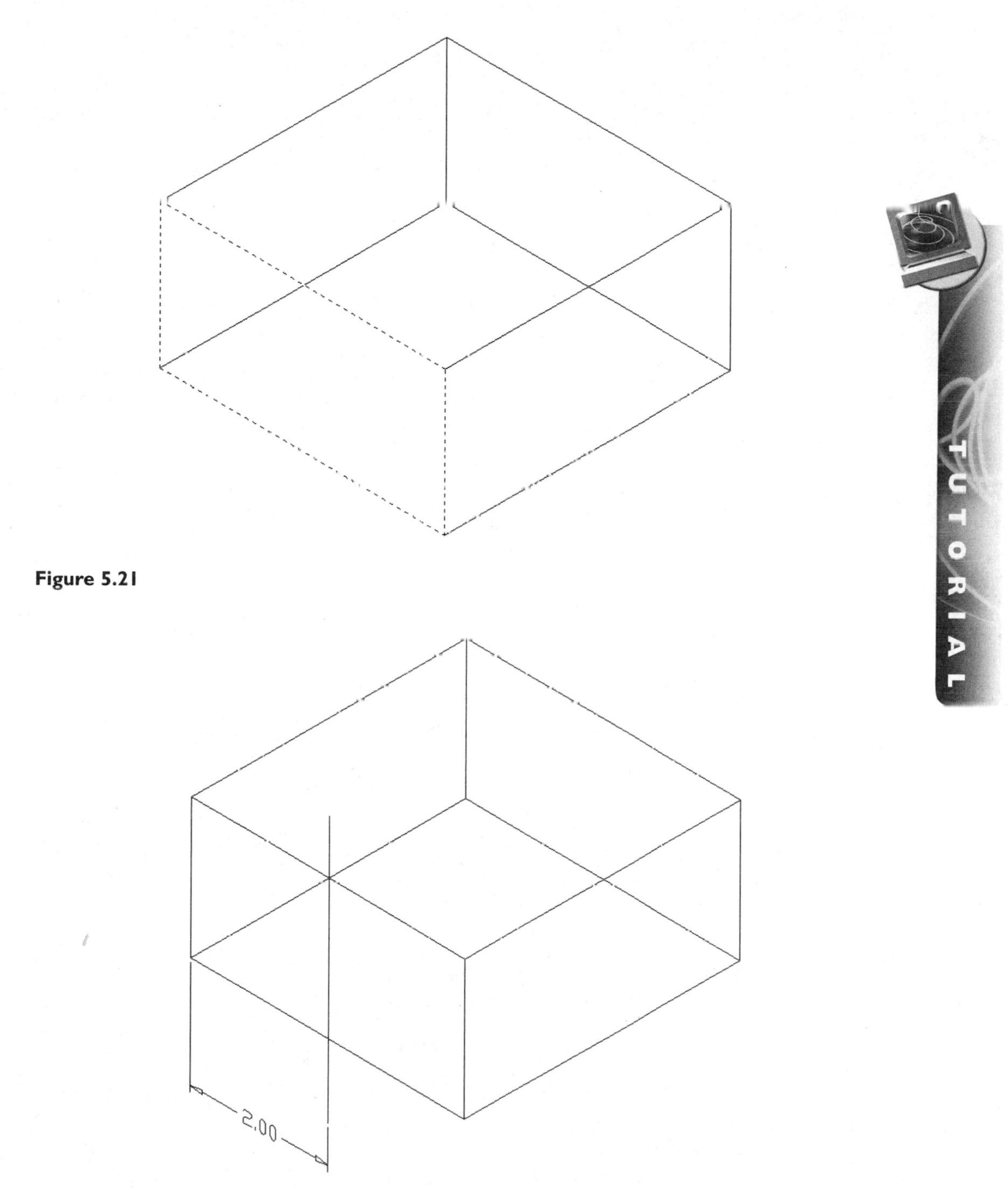

Figure 5.21

Figure 5.22

10. For the prompt:
    `Select work plane, planar face or [worldXy/worldYz/worldZx/Ucs]:`
    Select the same face that is highlighted in Figure 5.21. When the correct face is highlighted, press ENTER.

11. Rotate the *XY* axis until your screen resembles Figure 5.23 and then press ENTER.

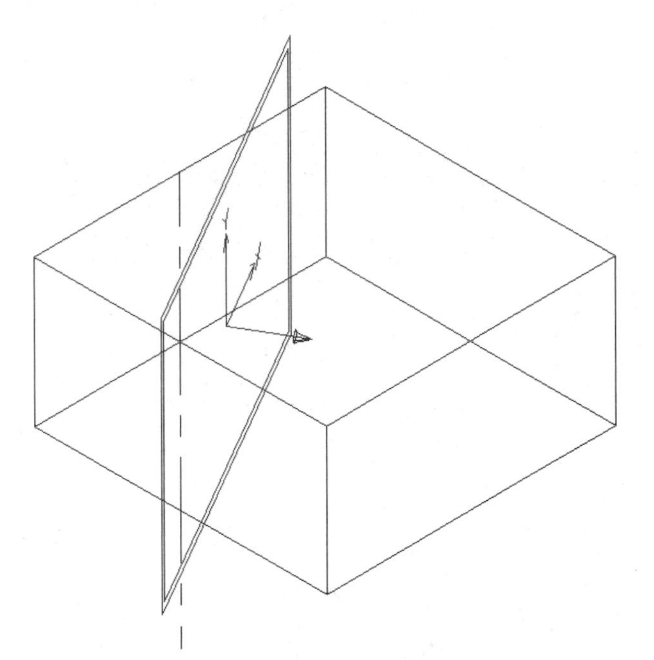

Figure 5.23

12. Make the right vertical face as highlighted in Figure 5.24 the active sketch plane. Accept the default orientation of the *XY* axis.

13. Sketch, profile and dimension a 1" diameter circle. Add two 1" dimensions to locate the circle as shown in Figure 5.25.

14. Issue the Revolve command **(AMREVOLVE).**

15. Select the work axis as the revolution axis.

16. The Revolution dialog box appears.

17. Input the following data:
    Operation = Cut
    Termination = To Face/Plane
    Then select OK.

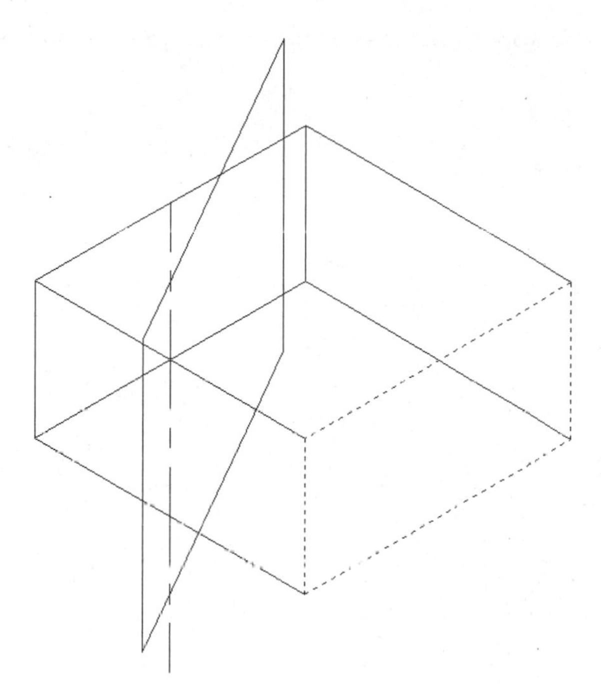

Figure 5.24

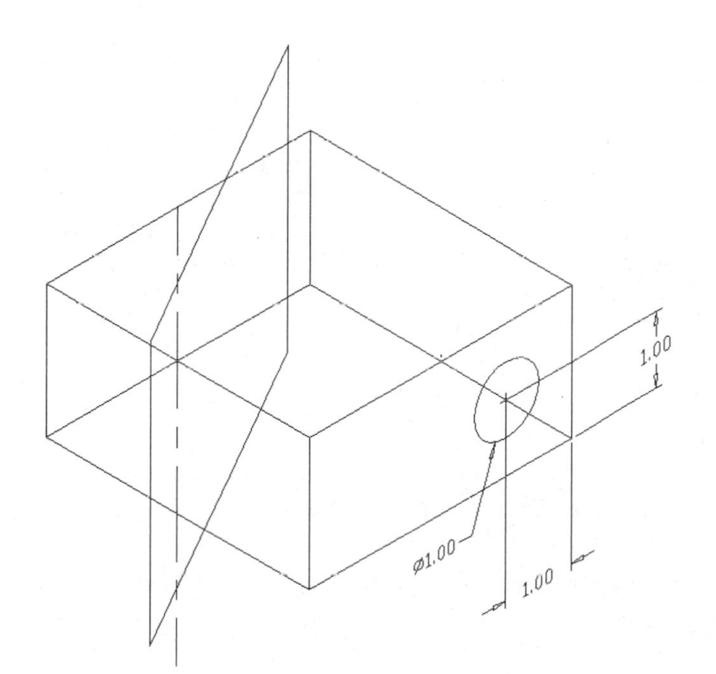

Figure 5.25

18. Select the work plane. When complete, your screen should resemble Figure 5.26.

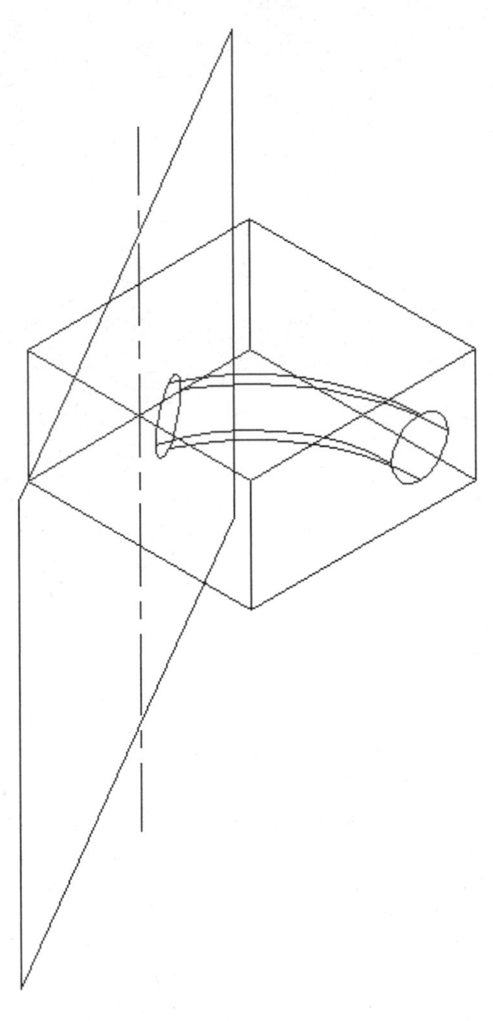

Figure 5.26

19. Create a Work Plane with the 1st Modifier set to Planar Parallel and the 2nd Modifier set to Offset, check the Create Sketch Plane, type in "1" for the Offset distance and then select OK.
20. Select the highlighted plane as shown in Figure 5-27 and then press ENTER.

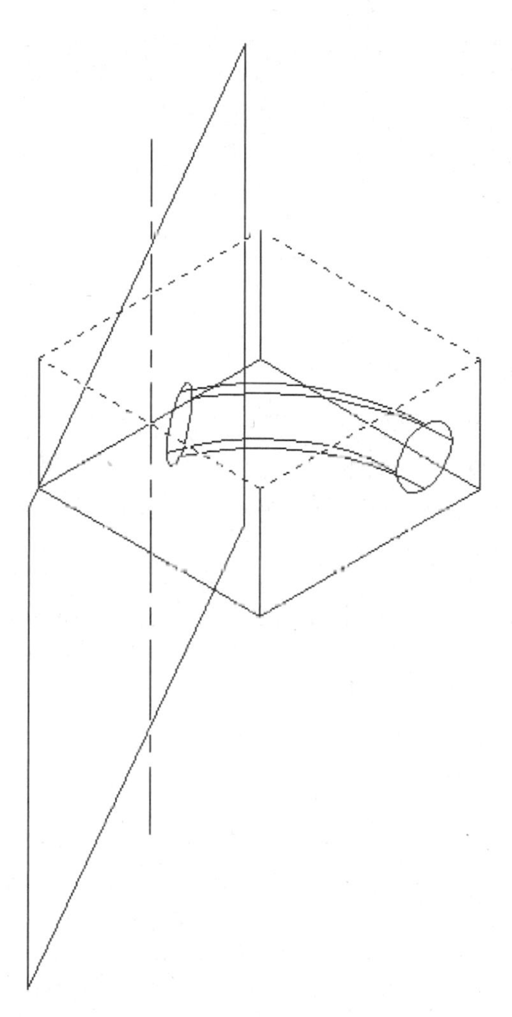

Figure 5.27

21. Flip the offset direction so it goes into the part and ENTER twice to accept this direction and the default *X Y* orientation.
22. Click on the Drawing tab or Drawing Layout icon on the express toolbar.
23. To create a base view, issue the New Drawing View command **(AMD-WGVIEW)**. In the Create Drawing View dialog box, change the scale to ".5" and the other values will be the defaults. Then select OK. You will be returned to part mode.
24. Select in the middle of the highlighted face as shown in Figure 5.21 and then press ENTER.

25. To align the X axis, select one of the horizontal edges of the selected face in step 24. Press ENTER to accept the arrow pointing out of the part.
26. For the view location select a point in the lower left corner of the drawing layout.
27. To create a ortho section view, issue the New Drawing View command **(AMDWGVIEW)**. In the Create Drawing View dialog box, change the view type to Ortho, the section type to Full and the other values will be the defaults. Then select OK.
28. Select a point in the front view, then select a point to the top of it and press ENTER.
29. Press ENTER again to cut through using a work plane.
30. Select the work plane.
31. Create an iso view of the cross section view, place it in the upper right corner of the drawing.
32. Edit the iso view and turn on the hatch pattern (hint: you will find the hatch option under the Section tab of the Edit View dialog box). When complete, your screen should resemble Figure 5.28. The "2.00" horizontal dimension is for the placement of the work axis.
33. As time permits, add reference dimensions to complete the drawing.
34. Save the file as \Md4book\Chapter5\TU5-9.

TUTORIAL

CONTROLLING THE VISIBILITY OF OBJECTS

When working in AutoCAD, you can turn layers on and off to control the visibility of all objects on a particular layer. To control the visibility of object types or selected objects independent of the layer they are on, issue the Part Visibility command **(AMVISIBLE)** from the Part Modeling toolbar as shown in Figure 5.29. (Objects refer to work axes, work planes, work points, lines, arcs, circles etc.) After the command is issued, the Desktop Visibility dialog box appears, as shown in Figure 5.30. There are six possible tabs along the top: you may not see all the tabs (part, assembly, scene or drawing)depending on if there is more than one part and if scenes exit.

- Part modeling mode: where a part is created.
- Assembly modeling mode: where parts are assembled.
- Scene mode: where a scene is created for use with assemblies.
- Drawing mode: where drawing views are created.
- Object mode: For AutoCAD objects.
- Toolbody mode: When toolbodies are present.

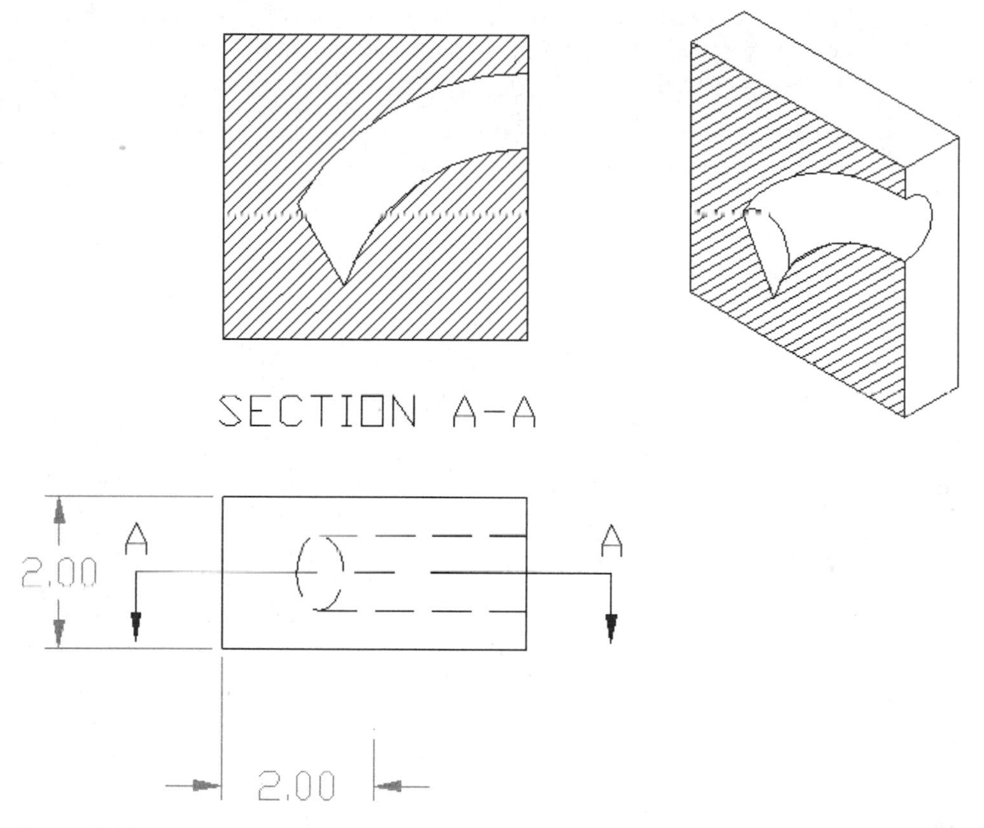

Figure 5.28

To hide or unhide objects, select the Hide or Unhide button. The All box will hide/unhide all objects. To hide/unhide all objects except those selected, select the All button and the Except button. You will then be returned to AutoCAD to select objects. To hide an entire object group, check the box before the group. To hide specific object(s) select the Select option and you will be returned to Mechanical Desktop to select the object(s). Select Apply to see the results without exiting the command or select OK to exit the command and see the results.

Under the Objects tab, you can control the visibility of AutoCAD objects including lines, arcs, circles, splines, parts, surfaces, etc. Another method for toggling the visibility of a work plane, work axis, work points and parts is to right click on its name in the browser and select Visibility from the pop-up menu.

Note: Not all of the tabs will be available at any one time.

The scene tab is only visible when you are working in a scene.

If a Mechanical Desktop drawing contains objects whose visibility has been turned off and it is opened in regular AutoCAD, those objects cannot be made visible through regular AutoCAD commands.

Tip: On the fly-out of the Part Visibility command you will find icons for controlling the visibility of specific objects.

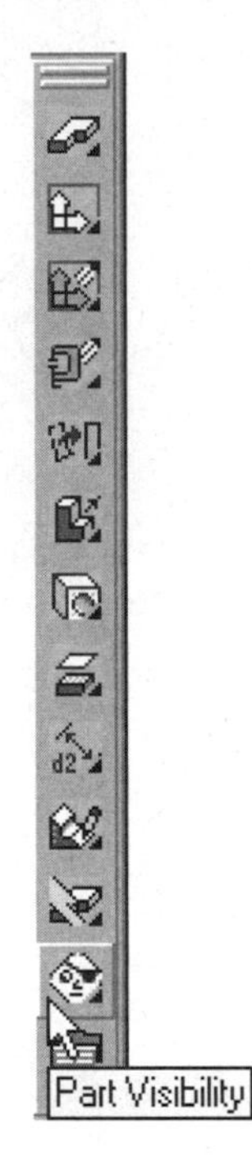

Figure 5.29

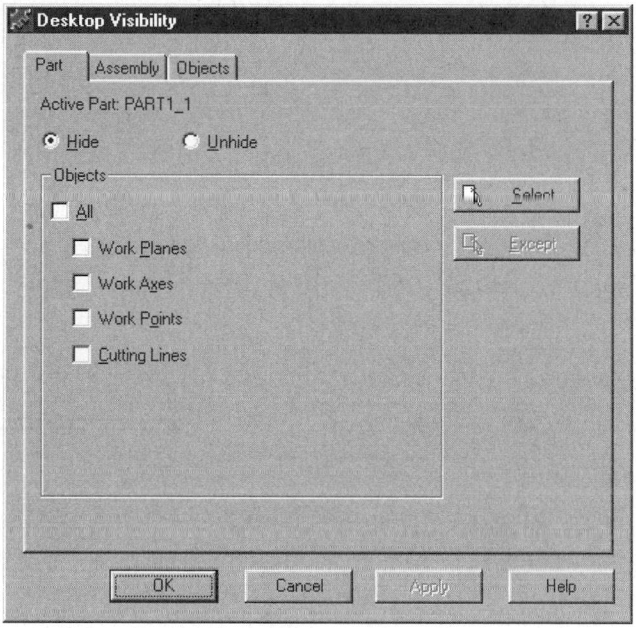

Figure 5.30

TUTORIAL

TUTORIAL 5.10—USING DESKTOP VISIBILITY OPTIONS

1. Open the file \Md4book\Chapter5\TU5-10.
2. Issue the Part Visibility command **(AMVISIBLE)**.
3. Hide all work planes and then select Apply.
4. Hide all work axes and then select Apply.
5. Unhide All and select Apply.
6. Hide one work plane, one work axis and one work point.
7. Unhide the one work plane using the Unhide select method.
8. From the Object tab, hide all circles and then select Apply.
9. Hide one of the lines with the Select option.
10. Continue hiding and unhiding objects until you feel comfortable with the command.
11. Save the file when finished.

SWEEPING THE PROFILE

Another method for creating or modifying a part is sweeping. In this section you will learn how to perform a 2-D sweep and four methods performing 3-D sweeps; using an edge, helical, 3D polyline (pipe) and a spline. Each method will be covered in its own section. A sweep is similar to an extrusion, except that a sweep follows a path. In contrast, an extrusion extrudes the profile in one direction only, in the positive or negative *Z* axis.

2-D SWEEP

Before performing a 2-D sweep, there are six possible steps to create a 2D sweep: create the path, constrain the path, create a work plane, sketch a profile, constrain the profile (not required) and sweep the profile along the path.

Create a Path: Step 1

1. Sketch a 2-D path using lines, arcs, polylines or spline. This path can be open or closed, but it must lie in one plane.
2. Issue the 2-D Path command **(AM2DPATH)** command from the Part Modeling toolbar as shown in Figure 5.31 or right-click in the drawing area and from the pop-up menu, select 2D Path from the Sketch Solving menu.
3. Select the sketched path; Mechanical Desktop analyzes the sketch like a profile and prompts you to select a start point from which to draw the profile.
4. For the start point choose a point near the geometry (do not use object snaps).
5. You will be prompted:
   `Create a profile plane perpendicular to the path? [Yes/No] <Yes>:`
   It is recommended that you allow Mechanical Desktop to create this work plane, if not you will need to manually create a work plane at the start of the path.
6. If the work plane was created, you can rotate the *XY* axis as needed.
7. The number of constraints and/or dimensions required to constrain the path appears on the command line.

Constrain the Path if desired: Step 2

In this step, make the plane that the path is on the active sketch plane then add constraints and dimensions to the path, as you did with profiles. You do not need to constrain the path, but it is recommended to do so.

Create Work Plane: Step 3

The third step in sweeping is to make the previously created work plane, the active sketch plane or create a work plane that is normal (perpendicular) to the path's start point. Normal is 90° from a given point. For example, if you stand a pencil straight upright on a desk, the pencil is normal to the desk.

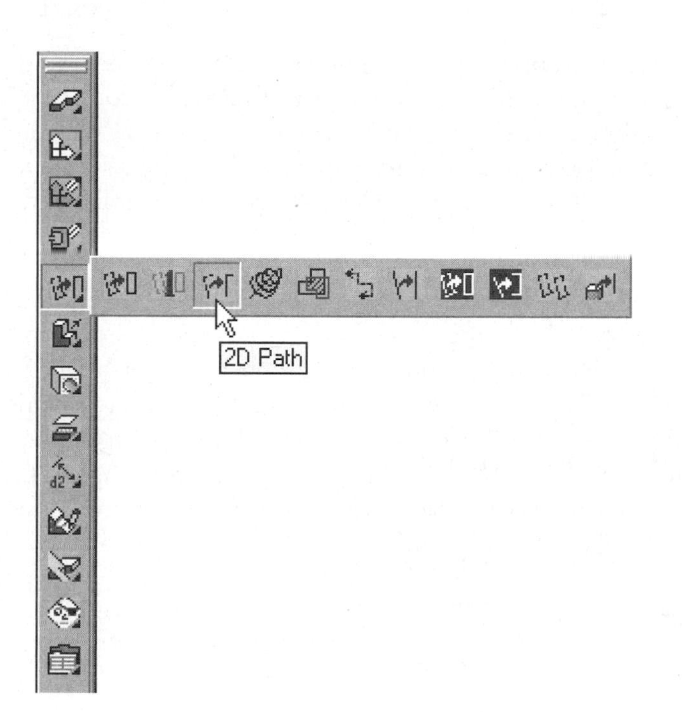

Figure 5.31

To create a work plane:

1. Issue the Work Plane command **(AMWORKPLN),** and for the 1st Modifier select Sweep Profile.
2. Below the 1st Modifier section, choose Create Sketch Plane.
3. Select OK. A work plane appears at the path's start point. Mechanical Desktop then prompts you to rotate the *XY* orientation.
4. A work point will be created at the start of the path. The work point will be used to dimension the profile to the path.

Sketch the Profile: Step 4

The fourth step is to draw the profile to be swept along the path. This profile is the same as a profile you would extrude or revolve.

Constrain the Profile: Step 5

The fifth step is to constrain and dimension the profile, then constrain it to the end of the path. You can use the work point to constrain the profile to the end of the path. When you select the work point, it snaps to the intersection of the three lines.

When you dimension a base feature (the first feature), the command prompt does not reflect the fact that profile requires two more dimensions to constrain it to the end of the path. For example, when you sweep a circle along a path, the command prompt reflects the fact that the circle requires only one dimension. While the prompt says "one," it accepts the two dimensions needed to constrain the profile in the *X* and *Y* axes.

Sweep the Profile: Step 6

The final step in this process is to sweep the profile along the path. Issue the Sweep command **(AMSWEEP)** on the Part Modeling toolbar as shown in Figure 5.32 or right-click in the drawing area and from the pop-up menu, select Sweep from the Sketched & Work Features menu. The Sweep dialog box appears and has four sections: Operation, Body Type, Termination, and Draft Angle. Operation and draft angle have the same options introduced in Chapter 2. Body type and another termination type will be covered next. After making changes to the dialog box, select OK. The profile sweeps along the path as specified.

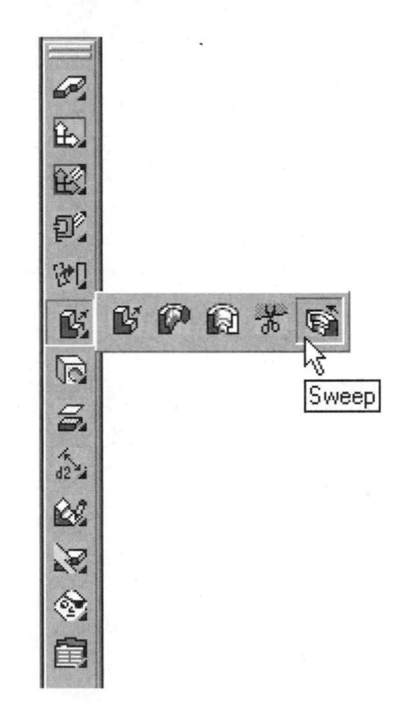

Figure 5.32

Termination: Has one new option called **Path Only.** The Path Only option will cause the profile to follow the path from its start point to its end point adding or removing material.

Body Type. In this section, you determine the way in which the profile should travel along the path. You have two options:

Normal	The profile will travel 90° all the way around the path (see Figure 5.33)
Parallel	The profile, as it travels around the path, will be parallel to the original profile in the start position (see Figure 5.33)

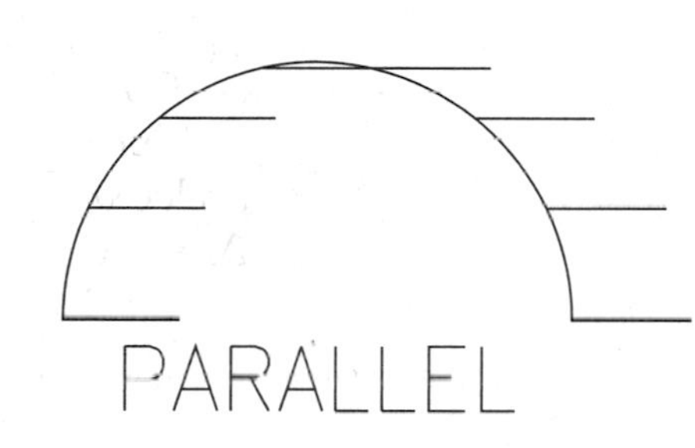

Figure 5.33 *Comparing normal and parallel*

TUTORIAL 5.11—SWEEPING AN OPEN PATH

1. Start a new drawing.
2. Draw a line and an arc or polyline as shown in Figure 5.34.

Figure 5.34

3. Issue the Path command **(AM2DPATH)**.
4. Select the geometry.
5. For the start point, select near the left end of the horizontal line.

6. Press ENTER twice to create a work plane and to accept the default orientation.
7. Issue the New Sketch plane command.
8. Make the world *XY* axis current by typing the X and press ENTER twice to accept the default orientation.
9. Add the dimensions shown in Figure 5.35.

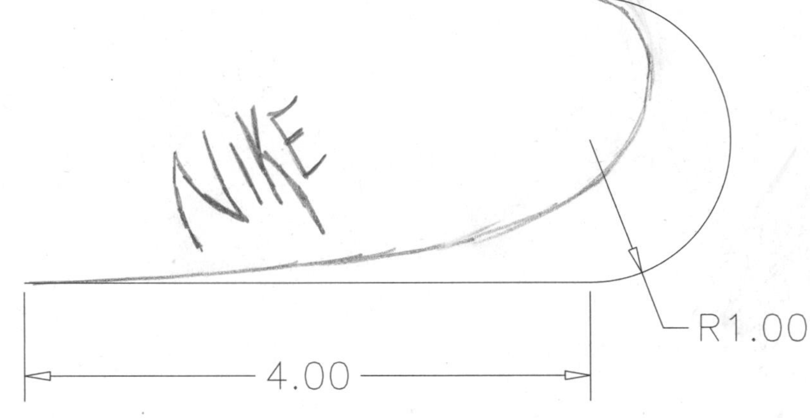

Figure 5.35

10. Make the work plane the current sketch plane. Accept the default orientation.
11. Draw a circle near the end of the line, but do not use object snaps.
12. Profile the circle.
13. Delete the fix constraint on the circle.
14. Add a concentric constraint, to the circle and then to the work point.
15. Add a ".5" diameter dimension to the circle.
16. Issue the Sweep command **(AMSWEEP)**. The Sweep dialog box appears.
17. Select OK. The profile sweeps along the path. Your part should resemble Figure 5.36.
18. Save your file as \Md4book\Chapter5\TU5-11.

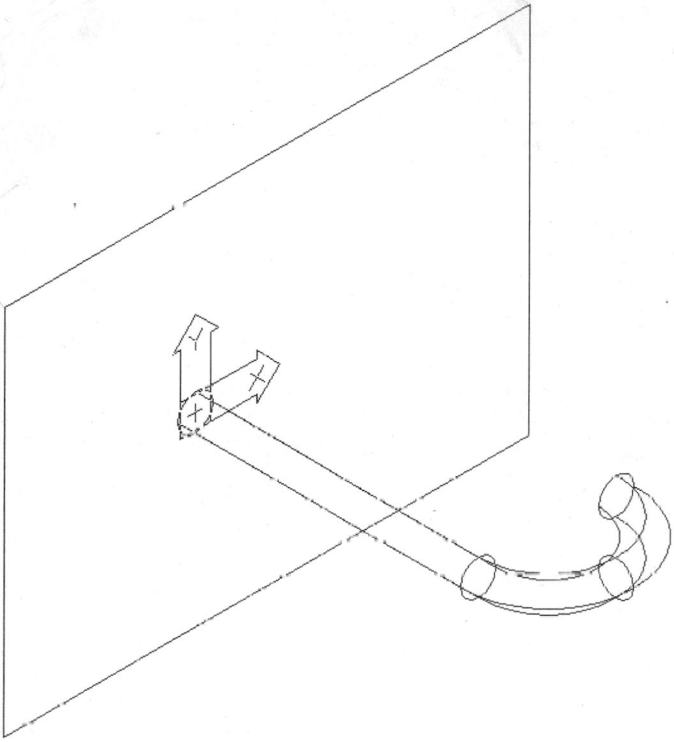

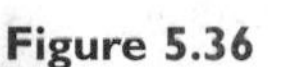

Figure 5.36

TUTORIAL 5.12—SWEEPING A CLOSED PATH

For this tutorial a base feature has been created as well as a polyline that will be used for a path.

1. Open the file \Md4book\Chapter5\TU5-12.
2. Issue the Path command **(AM2DPATH)**.
3. Select the blue polyline.
4. For the start point select near the top endpoint of the rightmost vertical line on the path.
5. Press ENTER twice to create a work plane and to accept the default orientation.
6. Issue the New Sketch plane command.
7. Make the top plane that the path is on the active sketch plane. Press ENTER twice to accept the default orientation.
8. Add the dimensions as shown in Figure 5.37. The lines are hidden and the UCS Icon and workplane are off for clarity.

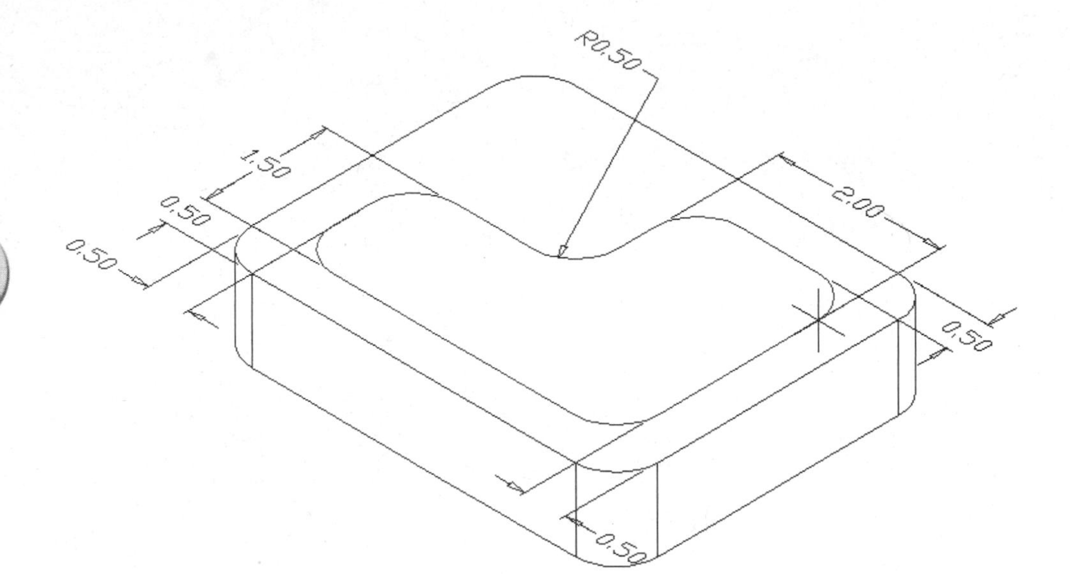

Figure 5.37

9. Make the work plane the current sketch plane. Accept the default orientation.
10. Draw a circle near the end of the line, but do not use object snaps.
11. Profile the circle.
12. Add a concentric constraint first to the circle and then to the work point.
13. Add a ".5" diameter dimension to the circle.
14. Issue the Sweep command **(AMSWEEP)**. The Sweep dialog box appears.
15. Change the Operation to Join and then select OK to accept the other defaults. The profile sweeps along the path.
16. Turn off the visibility of the workplane. When complete, your screen should resemble Figure 5.38.
17. Save the file. You will use it in the next tutorial.

3-D PATH

In the next sections, using four methods, you will learn how to create a 3-D sweep (where the path can have changes in the Z axis). Using an edge as a path, helical, 3D polyline (pipe) and a spline will each be covered in their own sections. To create one of the four 3-D paths, issue the 3-D path **(AM3DPATH)** command from the Part Modeling toolbar as shown in Figure 5.39 or right-click in the drawing area and from the pop-up menu, select corresponding 3D path option from the Sketch Solving menu.

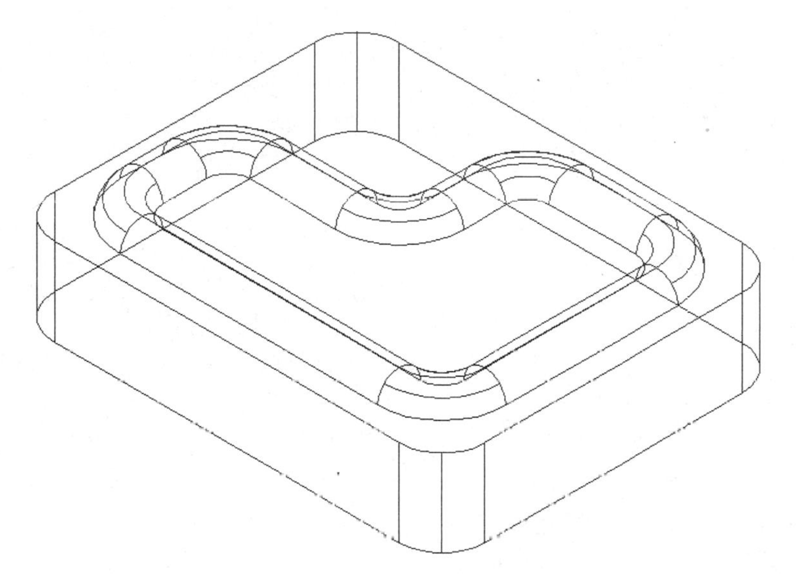

Figure 5.38

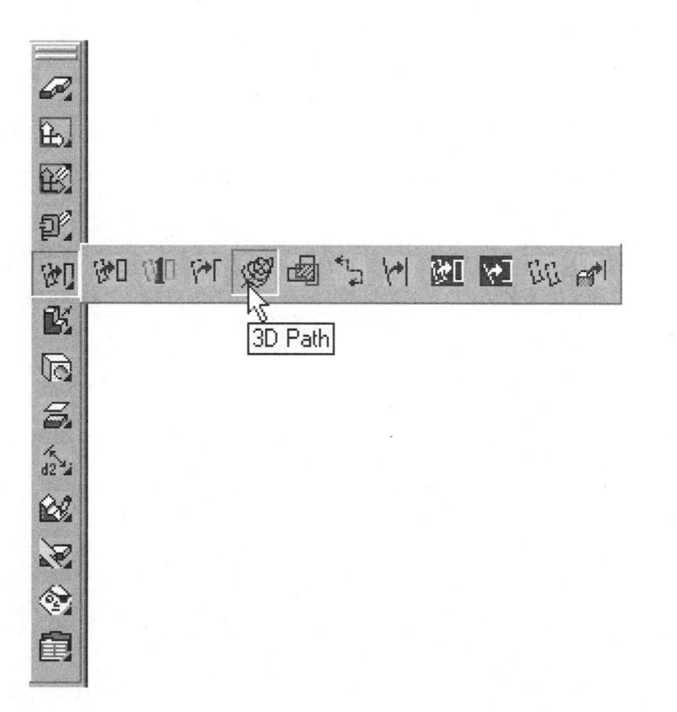

Figure 5.39

Edge:

The easiest method for creating a 3D path is to use an existing edge of a part for the path. The path can change *Z* values as it goes but the edges must be tangent. The path is parametrically constrained to the part, if the part edges change shape so will the path. After issuing the 3-D path **(AM3DPATH)** command type E and ENTER to select the edges to use as the path. Then individually pick the edges to use as the path. When finished selecting the edges, press ENTER. You will be prompted to specify a start point; select were the profile will be drawn. Then you will be prompted to create a workplane. Press ENTER to create a work plane at the start point of the path. Orientate the *X Y* axis as needed. Create your profile, constrain it and then issue the Sweep command **(AMSWEEP)** on the Part Modeling toolbar as shown in Figure 5.32. Or you may right-click in the drawing area and from the pop-up menu, select Sweep from the Sketched & Work Features menu. The same Sweep dialog box and options are used for both a 2D and 3D sweep.

TUTORIAL 5.13—SWEEPING ALONG AN EDGE

1. Open the file \Md4book\Chapter5\TU5-12 if it is not currently open.
2. Issue the 3-D path **(AM3DPATH)** command with the edge option.
3. Select the highlighted edges as shown in Figure 5-40 and then press ENTER to accept these edges.

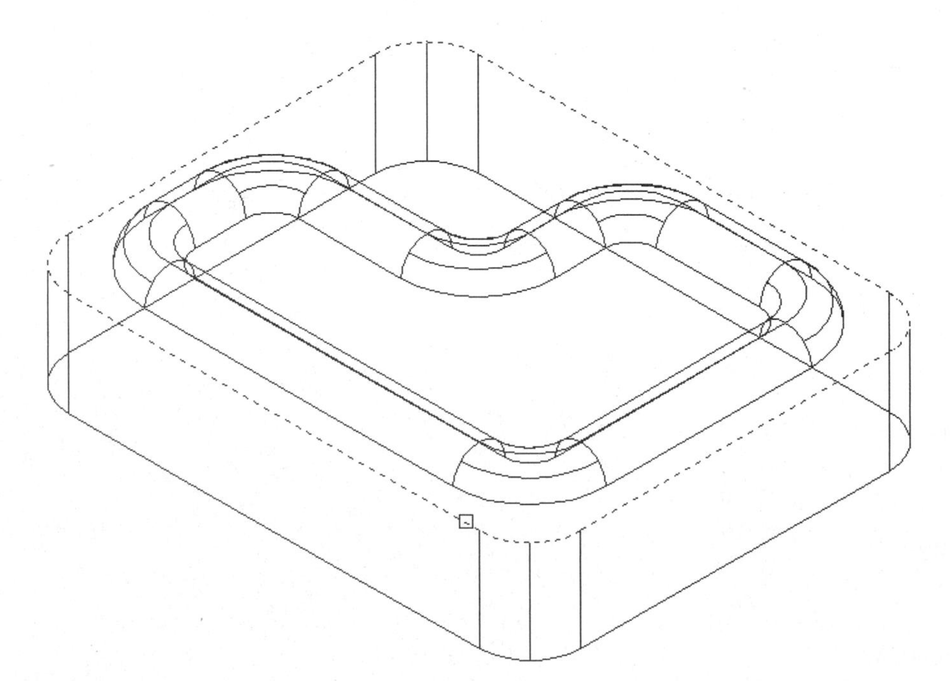

Figure 5.40

4. For the start point, select near the bottom of the end point of the top horizontal line. The pickbox is shown over this point in Figure 5.40.
5. For the prompt, Create a Workplane. Press ENTER to create it.
6. Press ENTER to accept the default orientation of the *XY* axis.
7. Change to the sketch view by using the 9 key.
8. Turn off the visibility of the workplane.
9. Sketch and profile a small square in the upper left corner of the part.
10. Add two ".125" dimensions to the square as well as two collinear constraints to position the top and left side of the square to the outside edges of the part. Figure 5.41 shows how the constrained and dimensioned profile should look.

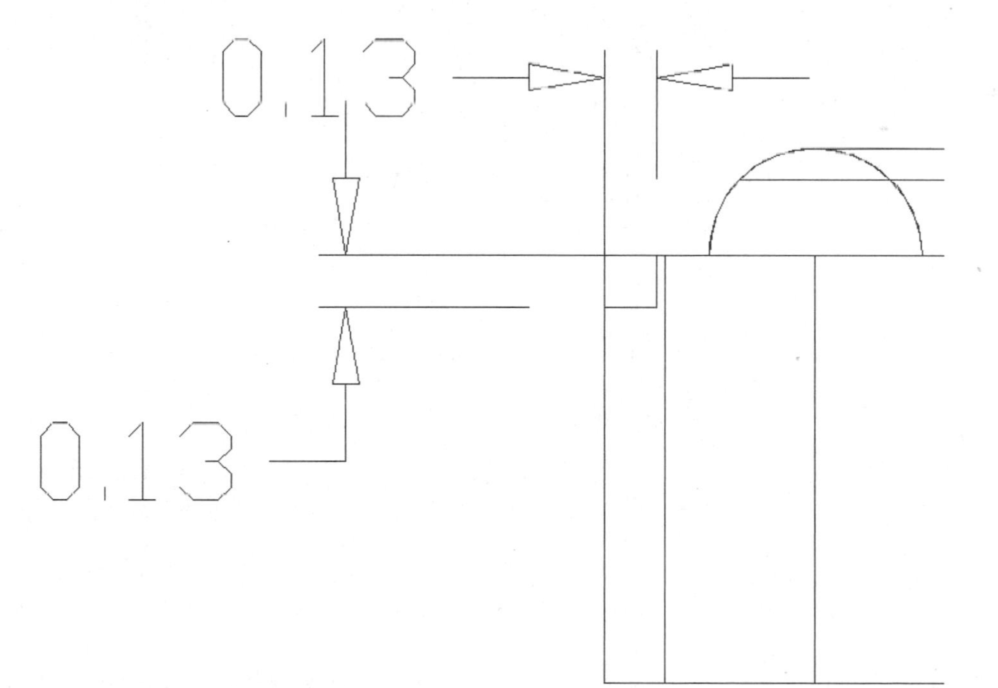

Figure 5.41

11. Change back to an isometric view using the **8** key.
12. Issue the Sweep command **(AMSWEEP)**. The Sweep dialog box appears.
13. Change the Operation to Cut and then select OK to accept the other defaults. The profile will sweep along the path and when complete, your screen should resemble Figure 5.42 shown with lines hidden.

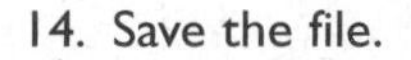

14. Save the file.

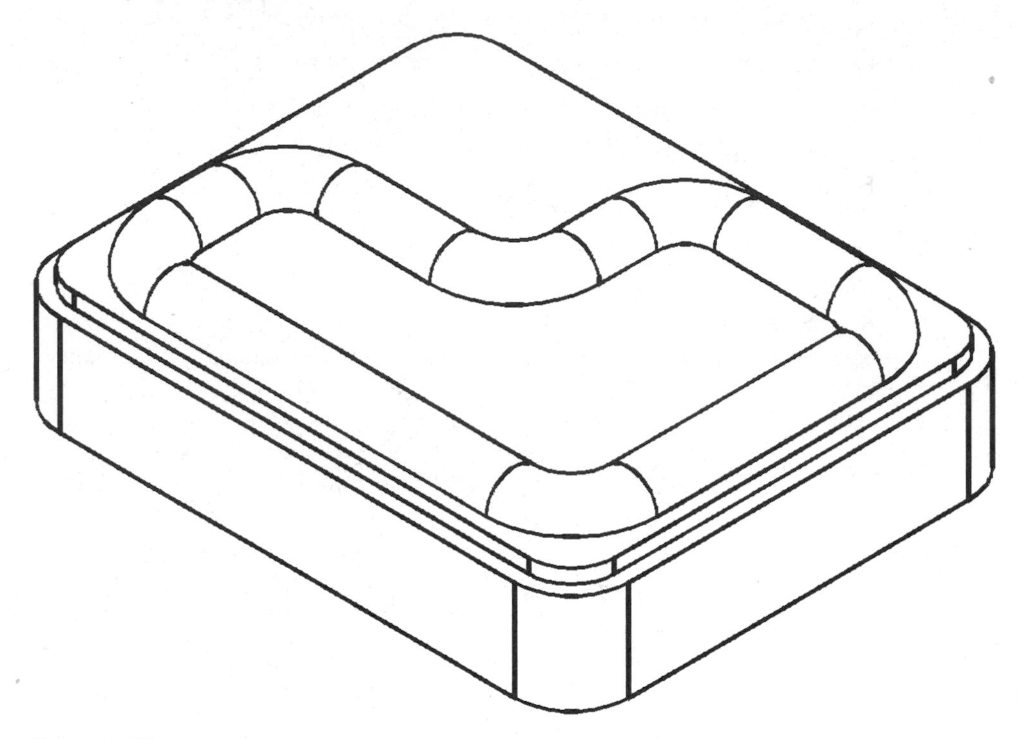

Figure 5.42

Helical:

One of the most requested options is the ability generate a spring or create threads. The helical option of the 3-D path **(AM3DPATH)** command will allow you to create a 3D helical path that can be used to create a spring or to create threads. Before you create a 3D path, there must first be a work axis or a cylindrical face. If you want to generate a spring as a separate part, create a work plane and make it the active sketch plane. Next, create a work axis with the sketch option (select two points with ortho on) and create a work plane that is perpendicular to the work axis. Issue the 3D Path (Helix) command **(AM3DPATH)** and the Helix dialog box will appear. As you input data, the objects on the screen will update in real time so that you can see the results before you have created anything. The dialog box is divided into four sections.

Type: Specify what type of a helix you will generate.

> **Pitch and Revolution:** With this option, you specify the pitch (distance) between each of the revolutions and how many revolutions there will be in the helix.
>
> **Revolution and Height:** With this option, you specify how many revolutions

fit within a given height.

Height and Pitch: With this option, you specify the overall height of the helix and the pitch (distance) between each revolution.

Spiral: With this option, you create a 2D spiral; you control the number of revolutions and the pitch (distance) between each revolution.

For Revolutions, Pitch and Height, you input the values for the helix. Some of the options may be grayed out, depending on the option you choose. You can either type in a value or select the up and down arrow. The incremental value each arrow goes up or down is specified through the Desktop Preferences command under the Desktop tab: in the Spinner Controls section, set angle (degree) and Length.

Orientation: Determine if the helix should go clockwise or counter-clockwise and where the helix should start in relation to the axis's sketch plane.

Shape: Determine if the helix will be circular or elliptical as well as the diameter and taper angle. Taper angle will determine if the revolutions go straight, in or out.

Profile Plane: You determine if a work plane will be created for the profile to be swept along the path and also how is should be orientated. If a work plane is to be created, it will be generated after the helix is created.

None: No work plane will be created.

Center Axis Path: A work plane will be created on the center of the axis of the helix.

Normal to Path: A work plane will be created normal (perpendicular) to the start point of the helix.

At the bottom of the dialog box, is a Flip and a double arrow button. The flip button will flip the helix 180° about the current sketch plane. The double arrow will turn the pictorial section of the dialog box on and off.

TUTORIAL 5.14—3D PATH SPRING

1. Start a new drawing.
2. Create three work planes using the Create Basic Workplanes command **(AMBASICPLANES)** as shown in Figure 5.17. Select a location in the middle of the screen.
3. Create a vertical work axis on the active sketch plane (World XY) by issuing the Work Axis command **(AMWORKAXIS)**. Press S and ENTER then select a point near the middle of the bottom of the work plane and then pick another point near the middle of the top of the work plane.

TUTORIAL

4. Change to an isometric view using the **8** key.
5. Make the workplane that lies in the world *ZX* plane the active sketch plane and press ENTER to accept the default orientation of the *XY* axis as shown in Figure 5.43.

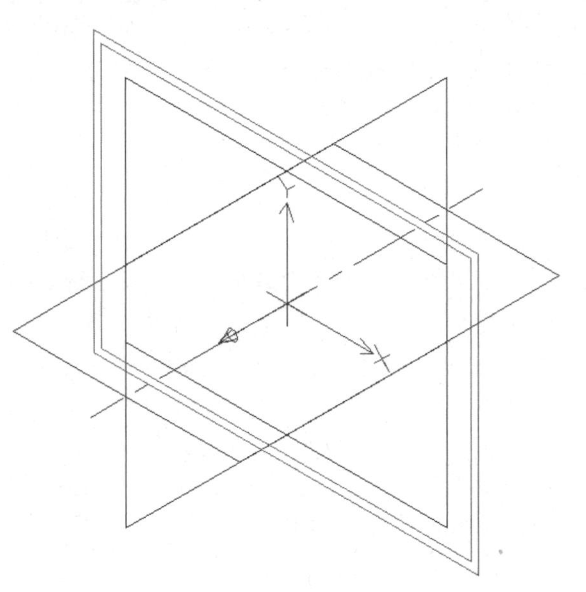

Figure 5.43

6. Create a helical path by issuing the 3D Path command **(AM3DPATH)** with the Helix option and then select the work axis as the center of the helix.
7. In the Helix dialog box make the following settings: (as you make changes in the dialog box the helix will update to reflect the new values).
   Type to: Revolution and Height.
   Change the number of revolutions to: 5.
   Change the Height to: 4.
   Change the Diameter to: 1.50.
   Change the Profile Plane to: Normal to Path.
   Select the Flip button to change the Z direction of the helix.
   Select OK to create the helix.
8. Accept the defaults for the orientation of the *X*, *Y* and *Z* axes.
9. For better clarity, turn off all workplanes; when complete, your screen should resemble Figure 5.44.
10. Draw a circle near the work point at the start of the helix.
11. Profile the circle and constrain it to the work point using the concentric constraint. Then apply a ".375" diameter dimension to the circle.

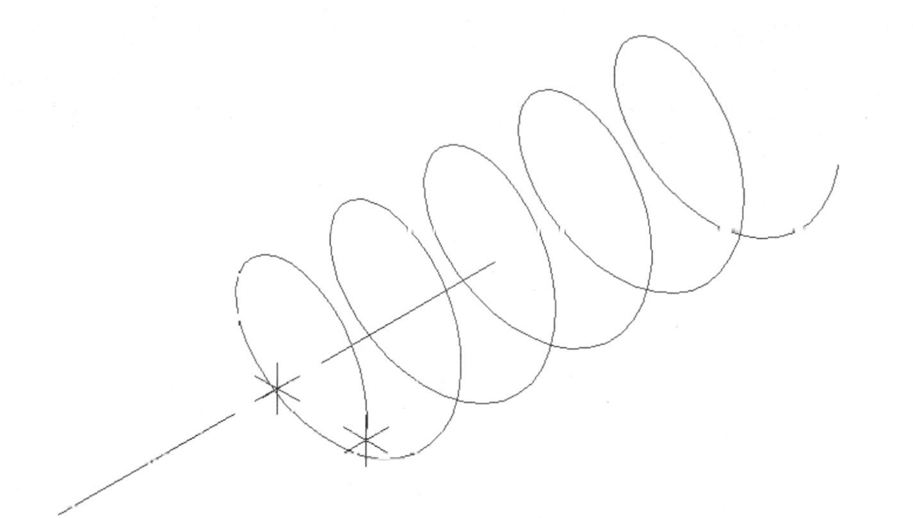

Figure 5.44

12. Sweep the circle along the path using the Sweep **(AMSWEEP)** command accepting the defaults.
13. To better see the part, shade it, and when complete, your screen should resemble Figure 5.45 shown in shaded mode with the work axis off.
14. Save the file as \Md4book\Chapter5\TU5-14.

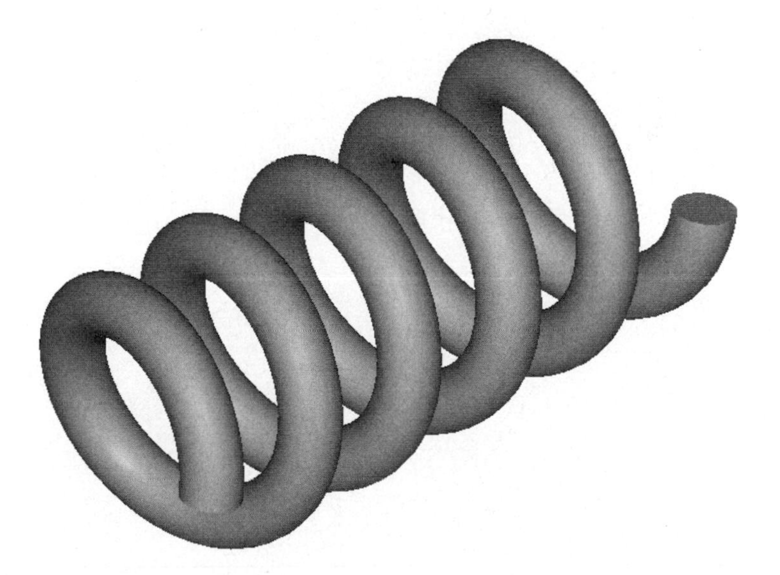

Figure 5.45

TUTORIAL 5.15—3D PATH THREAD

1. Start a new drawing.
2. Draw a circle, profile it and give it a "2" diameter dimension.
3. Change to an isometric view using the **8** key.
4. Extrude the profile "4" in the default direction.
5. Issue the 3D Path command **(AM3DPATH)** with the helix optio. Then, to center the helix, select the bottom circular face of the extrusion.
6. In the Helix dialog box make the following settings:
   Change the Type to: Height and Pitch.
   Change the height to: 4.
   Change the Pitch to: .5.
   Change the Diameter to: 2.
   Change the Profile Plane to: Center Axis/Path.
   Select OK to create the helix.
7. Rotate the *XY* axis so that *X* is pointing down toward the bottom of the screen as shown in Figure 5.46.

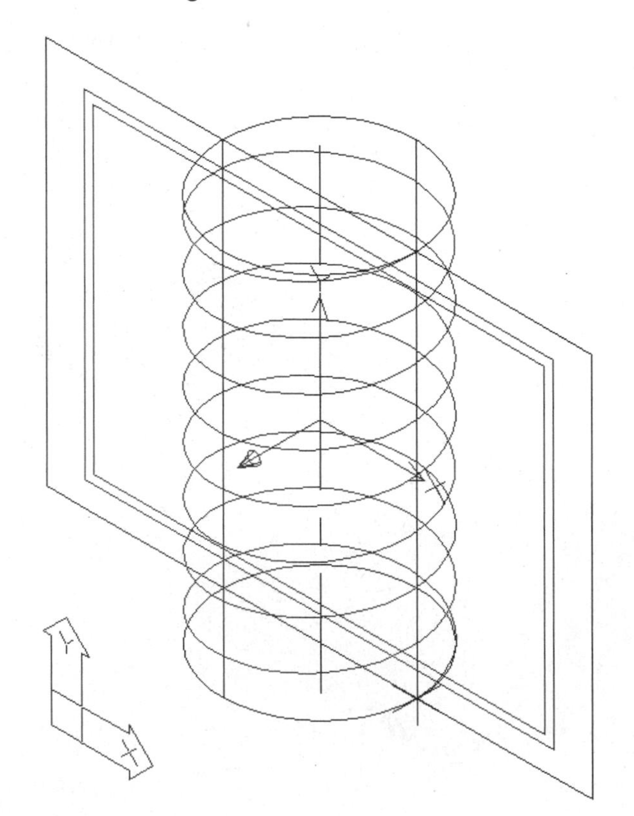

Figure 5.46

8. For clarity turn off the work plane.
9. Change to the sketch view using the **9** key.
10. Draw a triangle, profile it and then constrain and dimension it as follows:
    – Add an equal-length constraint to the two angled lines.
    – Add a Y Value constaint to the intersection of the angled lines and the bottom circular edge of the extrusion.
    – Add the three dimensions as shown in Figure 5.47. The "0.01" dimension is dimensioning the vertical line of the triangle to the outside of the work point. If this was at "0", the operation would fail.

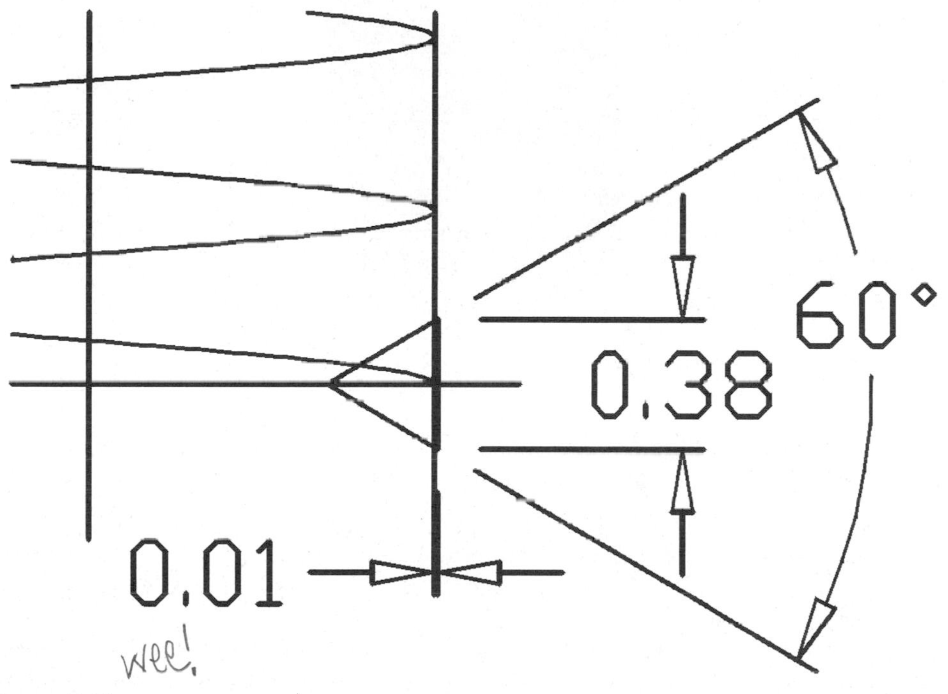

Figure 5.47

11. Change to an isometric view using the **8** key.
12. Sweep the circle along the path using the Sweep **(AMSWEEP)** command. Accept the defaults to cut away material and follow the path. When complete, your screen should resemble Figure 5.48, shown in shaded mode.

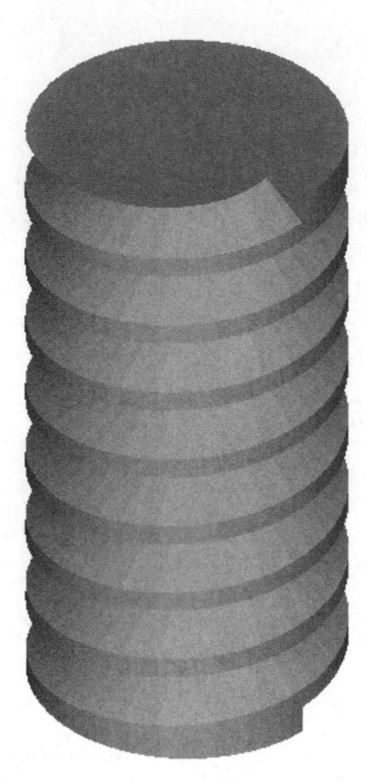

Figure 5.48

13. To have the thread go beyond the part, edit the 3D path and change the height to "4.5". Update the part. When complete, your screen should resemble Figure 5.49, shown in shaded mode.

14. Save the file as \Md4book\Chapter5\TU5-15.

Pipe:

Another option of the 3-D path **(AM3DPATH)** command is the Pipe option—this option will assist in creating a pipe run. After issuing the 3-D path **(AM3DPATH)** command with the Pipe option, you will be prompted to "Select polyline path source:" Now you have two options to use the command.

1. Select a 3D objetc that consists of any one or a combination of: 3D polylines, lines or arcs that that share the same start and end points. You will be prompted to select a start point. Then a Pipe Path dialog box like the one shown in Figure 5.50 will appear. Note: if arc segments or fillets exists they must be tangent to the adjacent object(s,) then the arc data will be imported into the 3D Pipe Path dialog box. If no arc segments exist, they can be created in the 3D Pipe Path dialog box.

2. Or, at the prompt, press ENTER and you can input *X Y* and *Z* point and arc data into the 3D Pipe Path dialog box to define the centerline of a path.

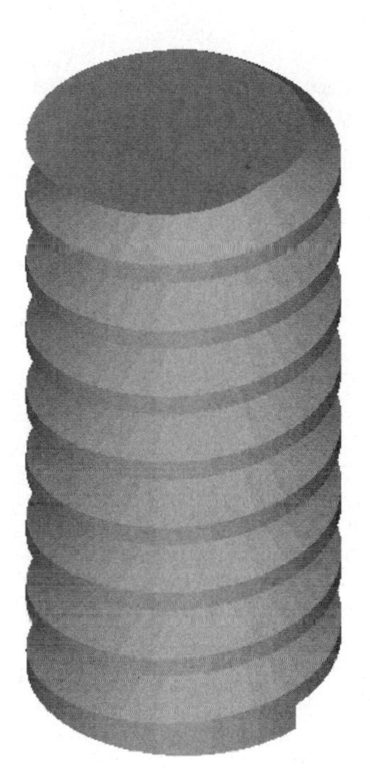

Figure 5.49

The 3D Pipe Path dialog box has ten columns. Each is described below.

No.
Specifies the point number from the start point.

3D Pipe Path

No.	C	From	Delta X	Delta Y	Delta Z	Length	Angle XY	Angle Z	Radius
1			1.8533	3.6016	0.0000	3.8486	0	0	
2		1	3.8486	0.0000	0.0000	2.1751	90	0	0.0000
3		2	0.0000	2.1751	0.0000	2.0000	0	90	0.0000
4		3	0.0000	0.0000	2.0000	2.9338	0	0	0.0000
5		4	2.9338	0.0000	0.0000	2.9317	-90	0	0.0000
6		5	0.0000	-2.9317	0.0000	3.1623	180	-72	0.0000
7		6	-1.0000	0.0000	-3.0000	1.0718	-90	0	0.0000
8		7	0.0000	-1.0718	0.0000				
9									

☑ Create Work Plane ☐ Closed OK Cancel Help

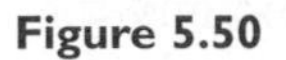

Figure 5.50

C

Specifies whether or not the point is constrained. If constrained, the icon will change to a small rectangle with a work point in it. If not constrained, the icon will look like a magnet, indicating that it is constrained.

From

Specifies the number the current point is Constrained From.

Delta X

Specifies the change in the *X* coordinate from the last point in the path, or from 0,0,0 of the current UCS for the first point in the path.

Delta Y

Specifies the change in the *Y* coordinate from the last point in the path, or from 0,0,0 of the current UCS for the first point in the path.

Delta Z

Specifies the change in the *Z* coordinate from the last point in the path, or from 0,0,0 of the current UCS for the first point in the path.

Length

Specifies the length of the line segment. If an arc exists, the length will be calculated to the theoretical corner (without the fillet).

Angle XY

Specifies the angle in the *XY* plane in relation to the current sketch plane.

Angle Z

Specifies the angle in the *Z* plane in relation to the current sketch plane.

Radius

Specifies the radius at the point. If the point has no radius, a value of zero will be shown.

To change a value in the 3D Pipe Path dialog box, click in a field and type in a new value. Entire columns can also be changed to the same value. To highlight a column: put the cursor in the first (or last) field in the column, hold the mouse button down, and drag down (or up) the column to highlight the the portion you wish to change. Depress the mouse button. When the fields are highlighted, type in a new value and press ENTER. When values in the 3D Pipe Path dialog box change, the path on the screen will update to reflect the change. To constrain a point in the path, a work point needs to exist. Right click on the constrain icon for the point to constrain and then select Constrain to Work Point from the pop-up menu. Then select the work point to constrain the point to. When finished making changes in the dialog box, select OK and a work plane will be created normal to the start point. Create a sketch, profile and constrain it as needed and then issue the Sweep command **(AMSWEEP)** on the Part Modeling toolbar as shown in Figure 5.32. Or you may right-click in the drawing area, and from the pop-up menu, select Sweep from the Sketched & Work Features menu. The same Sweep dialog box and options are used for both a 2D and 3D sweep. The sweep can be edited with the edit feature command **(AMEDITFEAT)** and the originating dialog box will appear. Make changes as you would to any other feature.

For the following exercises a 3D polyline will be used to sketch the paths.

TUTORIAL 5.16—3D PATH PIPE UNCONSTRAINED

1. Start a new drawing.
2. Draw a 3D polyline with the following coordinates using the AutoCAD 3DPOLY command.
   0,0,0
   @2,0
   @0,2
   @0,0,2
   @3,0
   @2,-2,-2
   @2,0
   Then press ENTER to complete the command.
3. Change to an isometric view using the **8** key.
4. Issue the 3D Path command **(AM3DPATH)** with the Pipe option. Then select the 3D polyline for the polyline path source and then press ENTER.
5. For the start point, select near the end of the line at the 0,0 coordinate.
6. In the 3D Pipe Path dialog box change the Radius for points 2 through 6 to ".75" as shown in Figure 5.51. Then select the OK button to exit the 3D Pipe Path dialog box.
7. Press ENTER to accept the default orientation of the *XY* axis. When complete, your screen should resemble Figure 5.52.
8. Sketch a circle near the start of the path.
9. Profile the circle, then apply a concentric constraint to the circle and the work point. Add a ".5" diameter dimension to the circle.
10. Sweep the circle along the path using the Sweep **(AMSWEEP)** command. Accept the defaults to follow the path.
11. Turn off the visibility of the work plane.
12. Shade the part and when complete, your screen should resemble Figure 5.53.

3D Pipe Path

No.	C	From	Delta X	Delta Y	Delta Z	Length	Angle XY	Angle Z	Radius
1			0.0000	0.0000	0.0000	2.0000	0	0	
2		1	2.0000	0.0000	0.0000	2.0000	90	0	0.7500
3		2	0.0000	2.0000	0.0000	2.0000	0	90	0.7500
4		3	0.0000	0.0000	2.0000	3.0000	0	0	0.7500
5		4	3.0000	0.0000	0.0000	3.4641	-45	-35	0.7500
6		5	2.0000	-2.0000	-2.0000	2.0000	0	0	0.7500
7		6	2.0000	0.0000	0.0000				
8									

☑ Create Work Plane ☐ Closed

OK Cancel Help

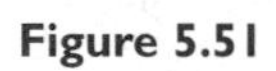

Figure 5.51

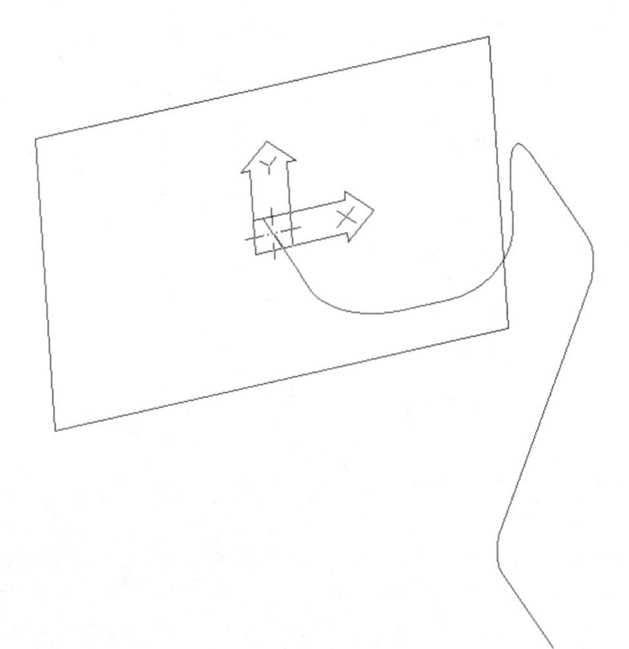

Figure 5.52

13. Edit the 3D sweep and change all the radii to ".5" through the 3D Pipe Path dialog box and when complete, update the part if needed.
14. Save the file as \Md4book\Chapter5\TU5-16.

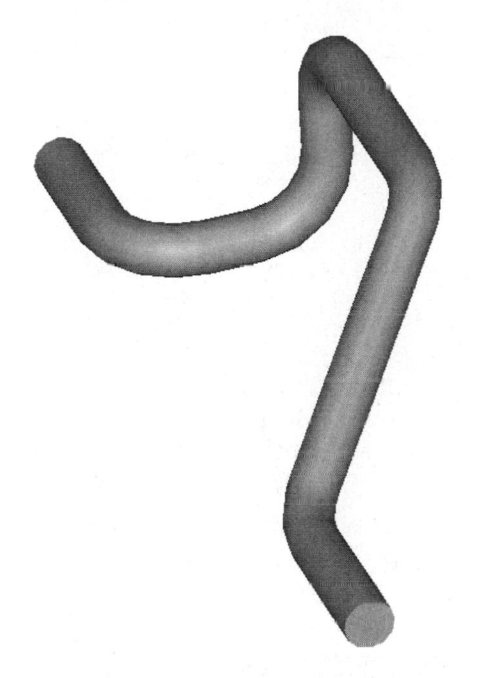

Figure 5.53

TUTORIAL 5.17—3D PATH PIPE CONSTRAINED

For this tutorial, work points have been created and a few have been dimensioned. All the work planes are turned off. A 3D polyline will be sketched near to each work point. If an end point of a line falls within the pick box size of a work point, it will automatically be constrained to it when the path is selected for the pipe path. If the end point is not automatically constrained to the work point, you can right click in the 3D pipe Path dialog box on the constrain icon for the specific point and then select Constrain to Work Point and pick the work point.

1. Open the file as \Md4book\Chapter5\TU5-17.
2. Draw a 3D polyline using the AutoCAD NEAR object snap, and snap to each work point in sequence: number 1 through 7, as shown in Figure 5.54.
3. Issue the 3D Path command **(AM3DPATH)** with the Pipe option, then select the 3D polyline for the polyline path source and then press ENTER.

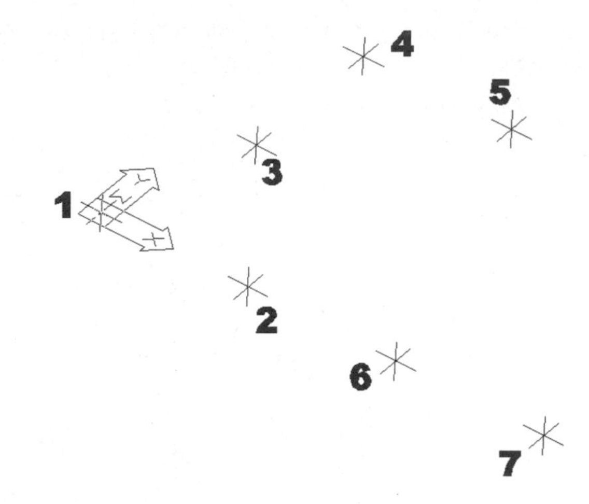

Figure 5.54

4. For the start point, select near the end of the path at the 0,0 coordinate (UCS icon).
5. In the 3D Pipe Path dialog box change the Radius for points 2 through 6 to ".5" as shown in Figure 5.55. Then select the OK button to exit the 3D Pipe Path dialog box.

3D Pipe Path

No.	C	From	Delta X	Delta Y	Delta Z	Length	Angle XY	Angle Z	Radius
1									
2									0.5000
3									0.5000
4									0.5000
5									0.5000
6									0.5000
7									
8									

Create Work Plane | Closed | OK | Cancel | Help

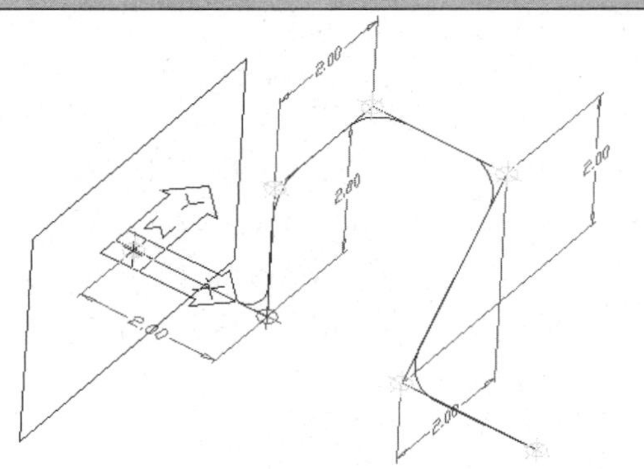

Figure 5.55

6. Press ENTER to accept the default orientation of the *XY* axis. When complete, your screen should resemble Figure 5.56.

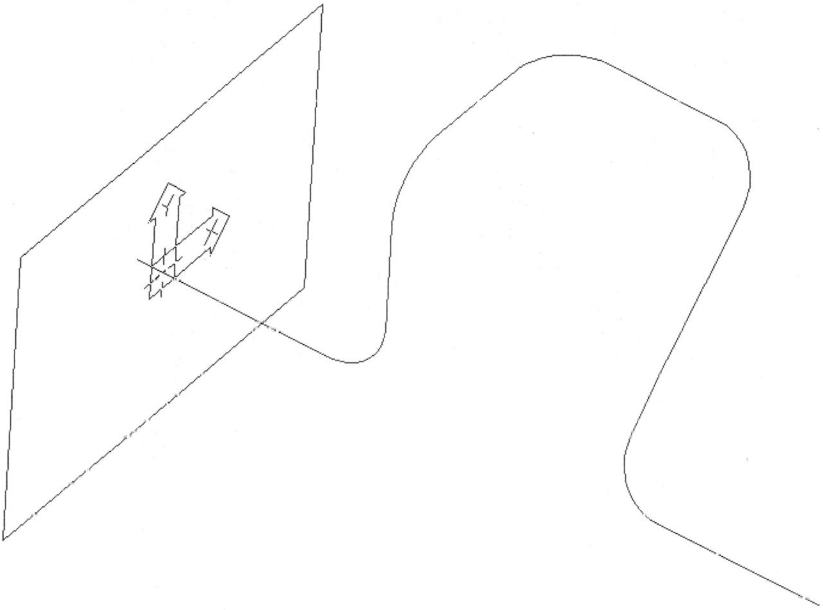

Figure 5.56

TUTORIAL

7. Sketch a circle near the start of the path.
8. Profile the circle then apply a concentric constraint to the circle and the work point, and add a ".5" diameter dimension to the circle. If you have problems constraining to the work point, turn off all the work points and then unhide just the last work point shown in the browser. Then constrain to that work point.
9. Sweep the circle along the path using the Sweep **(AMSWEEP)** command. Accept the defaults to follow the path.
10. Turn off the visibility of the work plane.
11. Shade the part, and when complete, your screen should resemble Figure 5.57.
12. Turn on the visibility of all the work points.
13. Edit the sweep and change a few of the "2.00" dimensions to "3.00" dimensions. When finished, update the part if needed. The part will take on the shape of the locating work points.
14. Save the file.

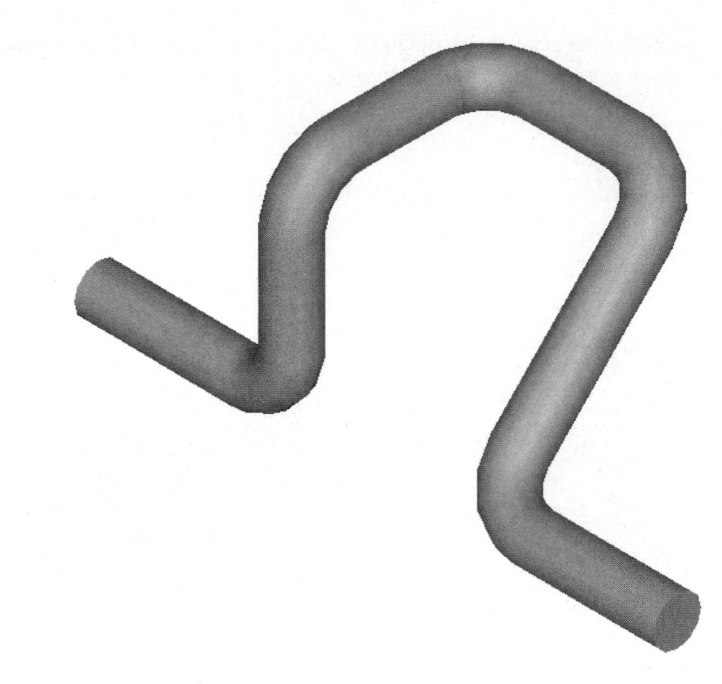

Figure 5.57

Spline:

The last option of the 3-D path **(AM3DPATH)** command is the Spline option which will use a spline as a path. The spline option is similar to the Pipe option except that it gives you a few more options. After issuing the 3-D path **(AM3DPATH)** command with the Spline option you will be prompted to "Select 3D spline path source: You then have two options to use the command.

1. Select a spline. If the spline is not closed, you will be prompted to select a start point. T 3D Spline Path dialog box will then appear as shown in Figure 5.58.

3D Spline Path

No.	C	From	Delta X	Delta Y	Delta Z	C	i	j	k	Weight
1			3.0836	3.8092	0.0000		0.7072	-0.7070	0.0000	0.8618
2			5.3722	3.8984	0.0000		0.4924	0.8647	-0.0994	0.7243
3			5.2236	5.6513	0.0000		0.0298	0.9605	0.2766	0.9837
4			7.2236	7.6513	2.0000		0.8532	0.1473	0.5003	0.1698

☑ Create Work Plane ☐ Closed

OK Cancel Help

Figure 5.58

2. Or, at the prompt, press ENTER and you can input values that represent the spline into the 3D Spline Path dialog box.

The 3D Spline Path dialog box has eleven columns. Each is described below.

No.

Specifies the control point number from the start point, as well as the number of vertices contained within the spline.

C

Specifies whether or not the point is constrained. If constrained to a work point, the icon will change to a small rectangle with a work point in it. If constrained to a point on the spline, the icon will change to magnet that is shaded red. If not constrained, the icon will look like an magnet shaded gray.

From

Specifies the number the current control point is Constrained From. If the point is unconstrained, the column will be blank.

Delta X

Specifies the change in the *X* coordinate from the last control point in the path, or from 0,0,0 of the current UCS for the first point in the path.

Delta Y

Specifies the change in the *Y* coordinate from the last control point in the path, or from 0,0,0 of the current UCS for the first point in the path.

Delta Z

Specifies the change in the *Z* coordinate from the last control point in the path, or from 0,0,0 of the current UCS for the first point in the path.

C

Specifies a lock on the tangency of the control point as defined by the i, j, and k vector controls.

i

Specifies the "*X*" value of the tangency of the control point.

j

Specifies the "*Y*" value of the tangency of the control point.

k

Specifies the "*Z*" value of the tangency of the control point.

TUTORIAL

Weight

Specifies the amount of influence the Tangency value will have. Only the first and last Weights are available for editing.

To change a value in the 3D Spline Path dialog box, click in a field and type in a new value or use the spinner controls in the field. Entire columns can be changed to the same value. To highlight a column: put the cursor in the first (or last) field in the column, hold the mouse button down, and drag down (or up) the column to highlight the the portion you wish to change. Depress the mouse button. When the fields are highlighted, type in a new value and press ENTER. When values in the 3D Spline Path dialog box change the path on screen will update to reflect the change. To constrain a point in the path, a work point needs to exist. Right click on the constrain icon for the point to constrain and then select Constrain to Work Point from the pop-up menu. Then select the work point to constrain the point to. When finished making changes in the dialog box, select OK and a work plane will be created normal to the start point. Create a sketch, profile and constrain it as needed and then issue the Sweep command **(AMSWEEP)** on the Part Modeling toolbar as shown in Figure 5-32. Or you may right-click in the drawing area and from the pop-up menu, select Sweep from the Sketched & Work Features menu. The same Sweep dialog box and options are used for both a 2D and 3D sweep. The sweep can be edited with the edit feature command **(AMEDITFEAT)** and the originating dialog box will appear. Make changes as you would to any other feature.

TUTORIAL 5.18—3D SPLINE PATH PIPE UNCONSTRAINED

1. Start a new drawing.
2. Draw a 3D spline with the following coordinates using the AutoCAD SPLINE command.
   0,0,0
   @2,0
   @0,2
   @0,0,2
   @3,0
   @2,-2,-2
   @2,0
   Then press ENTER to stop issuing points.
3. Press ENTER twice to accept the starting and ending tangencies.
4. Change to an isometric view using the **8** key.
5. Issue the 3D Path command **(AM3DPATH)** with the Spline option, then select the 3D spline for the 3D Spline path source.
6. For the start point select near the end of the line at the 0,0 coordinate (by the UCS icon) and the 3D Spline Path dialog box will appear.

7. Select OK to accept the defaults.
8. Press ENTER to accept the default orientation of the *XY* axis. When complete, your screen should resemble Figure 5.59.

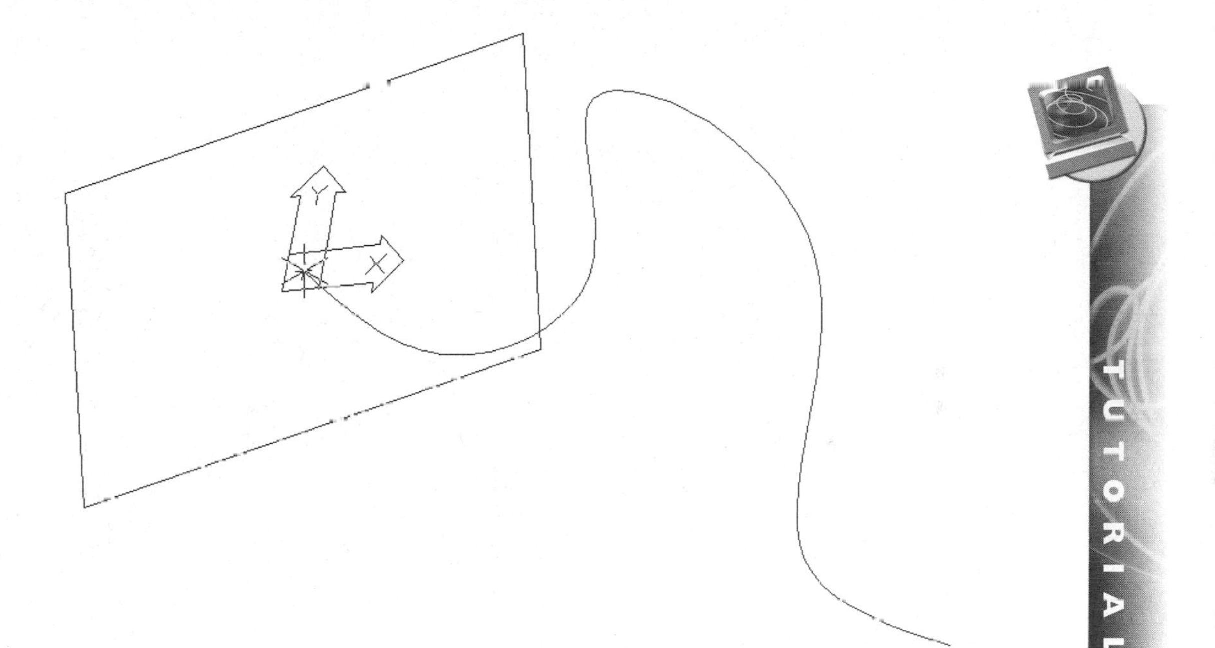

Figure 5.59

9. Sketch a circle near the start of the path.
10. Profile the circle.
11. Delete the fix constraint from the circle.
12. Then apply a concentric constraint to the circle and the work point and add a ".5" diameter dimension to the circle.
13. Sweep the circle along the path using the Sweep **(AMSWEEP)** command. Accept the defaults to follow the path.
14. Turn off the visibility of the work plane.
15. Shade the parts and when complete, your screen should resemble Figure 5.60.
16. Save the file as \Md4book\Chapter5\TU5-18.

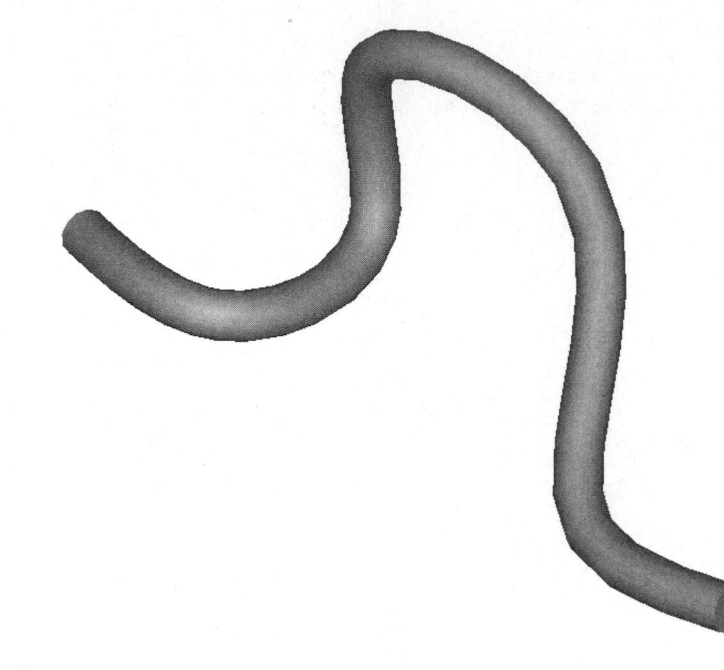

Figure 5.60

LOFTED PARTS

If you have a part that blends from one shape to another and you cannot create it by extruding or revolving, you can create a lofted solid. A lofted solid blends two or more distinct shapes. Before creating a lofted solid, you must have first defined profiles or edges on a part that you want to blend. Issue the Loft command **(AMLOFT)** from the Part Modeling toolbar, as shown in Figure 5.61. Or right-click in the drawing area, and from the pop-up menu, select Loft from the Sketched & Work Features menu. A Loft dialog box will appear, as shown in Figure 5.62.

The dialog box is divided into seven sections. If there is an existing part or loft, some sections may be grayed out.

Operation:

Base: This is the default if it is the base feature, and it cannot be changed.

Join: Add material to the active part.

Cut: Removes material from the active part.

Intersect: Keeps what is common to the active part and the loft.

Split: Splits the part into two parts.

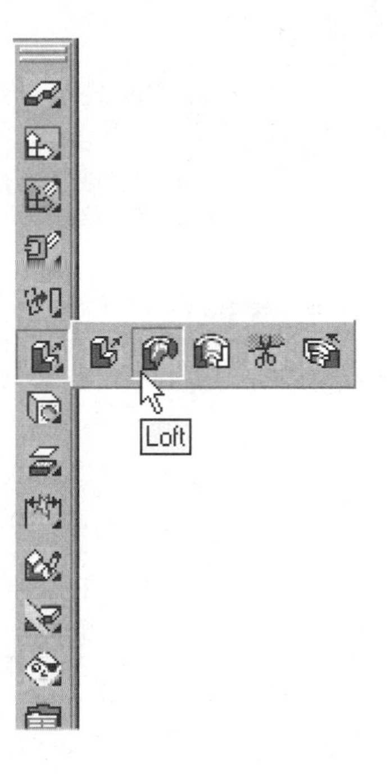

Figure 5.61

Loft

Operation: Base

Termination: Sections

Type: Cubic

Minimize Twist

Sections to loft

Redefine

Start Points

Reorder

Delete

Start Section

Tangent to Adjacent Face

Angle: 90

Weight: 1.0000

End Section

Tangent to Adjacent Face

Angle: 90

Weight: 1.0000

End Section

Weight

End Angle

Start Angle

Weight

Start Section

Preview

OK

Cancel

Help

<<

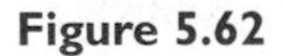

Figure 5.62

Termination:

Sections: The loft will be created from the sections (profiles) you select.

To Face: The loft will start at the first profile and stop at a selected face.

From To: The loft will start and then stop at the selected faces. The selected faces need to be different planes.

Type:

Linear: Between two profiles, straight edges will close the profiles. Linear only works with two profiles.

Cubic: Cubic requires at least two profiles, but more can be selected. The profiles will be blended together in the order in which they were selected.

Closed Cubic: At least three different profiles must be selected to create a loft that is closed. Do not select the first profile a second time for the last profile; Mechanical Desktop understands that the first profile will also be used for the last profile and it does not need to be selected twice.

Minimize Twist:

If checked, Mechanical Desktop will calculate the straightest way to blend the profiles together. When checked, this will override any selected start points. If unchecked, the profiles will possibly have more of a twist as the profiles blend.

Sections to Loft:

Redefine: Select this button to pick the profiles in the order in which the loft will be created.

Start Points: Select this button to define the endpoint in which the loft will start. Circles will always be at the quadrant at 0°.

Reorder: When a loft is edited, this button allows you to reselect the profiles in a different order.

Delete: When a loft is edited, this button allows you to select a profile to remove from the loft.

Start Section:

Tangent to Adjacent Face: Check here if you want the loft to be tangent to the face from which the loft will start. If this is not checked, the loft will start perpendicular to the selected face.

Angle: Controls how the loft will travel. The default is 90, meaning the profile will travel straight upward, or perpendicular to the profile when blending to the next profile. A value smaller than 90 means the profile will lean inward and a value larger than 90 means the profile will lean outward.

Weight: The higher the value, the longer the beginning shape will maintain its original shape before blending into the next profile or face. The values can range from 0 to 10.

End Section:

Tangent to Adjacent Face: Check here if you want the loft to be tangent to the face where the loft will end. If this is not checked, the loft will end perpendicular to the selected face.

Angle: Controls how the loft will travel. The default is 90, meaning the profile will travel straight upward, or perpendicular, when blending to the next profile. A value smaller than 90 means the profile will lean inward and a value larger than 90 means the profile will lean outward.

Weight: The higher the value, the longer the beginning shape will maintain its original shape before blending into the next profile or face. The values can range from 0 to 10.

Preview:

This button, when selected, will allow you to see how the loft will appear by showing lines on the part representing how the loft will be created.

The loft feature is a parametric feature and can be edited. When you are editing a loft feature, the same Loft dialog box that was used to create it will appear on the screen. Each profile should be placed on its own work plane and each can be dimensioned and constrained appropriately. The profiles that will be used for a loft cannot be constrained to each other. However, you can place a work point before profiling the first sketch and then dimension all the profiles to the one work point. When you create a closed loft, the starting profile also needs to be used for the last profile. The first and last profile cannot fall on the same plane. When a loft is edited, the same rules apply as for any other feature, with the addition of adding more profiles.

Note: The more profiles used, the greater the complexity of the part.

Use a work point in the first sketch to constrain all the sketches.

The work points to which the object will be constrained must be placed before profiling the object

After completing a loft, shade the part to see it better.

TUTORIAL 5.19—LOFTED PARTS, LINEAR AND CUBIC

1. Start a new file.
2. Place a work point in the middle of the screen.
3. Draw a circle near the middle of the work point.
4. Profile the circle.
5. Remove the fix constraint on the circle.
6. Dimension the circle with a "2" diameter dimension and two "0" dimensions to the lock the center of the circle to the work point.
7. Switch to an isometric view using the **8** key.
8. Create a work plane and make it the active sketch plane, that is, Planar Parallel and Offset a distance of "2" from the world *XY*. Accept the default offset direction and the orientation of the *X*, *Y* and *Z* axes.

9. Turn off the display of the work plane.
10. Place a work point above the first work point.
11. Draw a square that is centered about the second work point.
12. Profile the square.
13. Remove the fix constraint on the square.
14. Dimension the square as shown in Figure 5.63. The "1.25 dimensions are dimensioned to the work point that is on the first sketch plane.

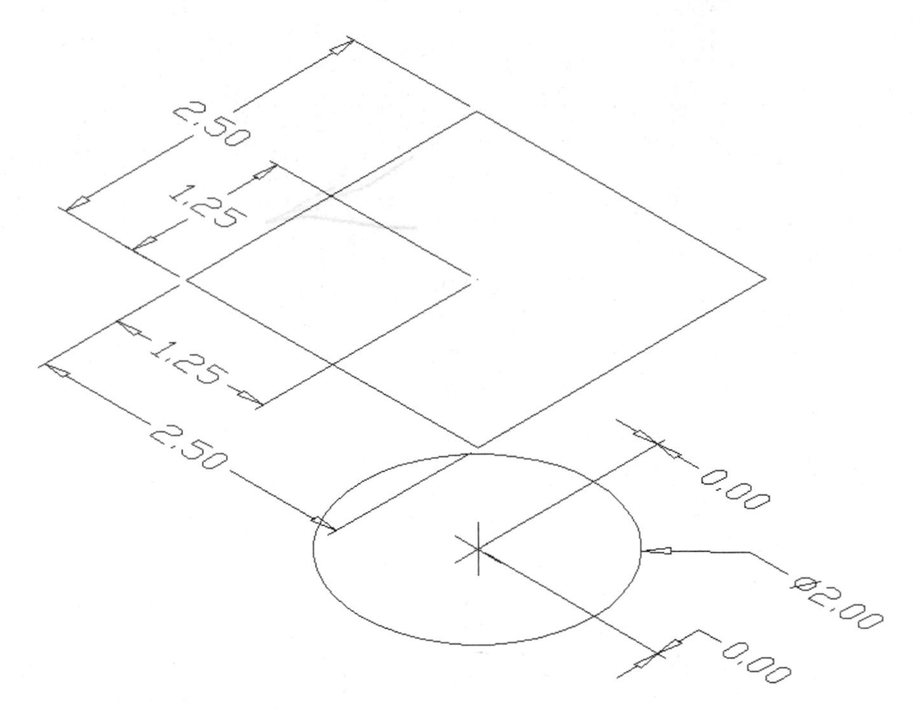

Figure 5.63

15. Issue the Loft command **(AMLOFT)** and select the circle and then the square and press ENTER. The Loft dialog box will appear.
16. Change the type to Linear and select the Preview button and then ENTER to return to the Loft dialog box and then select OK to accept the other defaults.
17. Shade the part, and when complete, your screen should resemble Figure 5.64
18. Edit the loft and change the type to Cubic, select the Preview button and then ENTER to return to the Loft dialog box and then select OK. Update the part if needed, and when complete, your screen should resemble Figure 5.65.

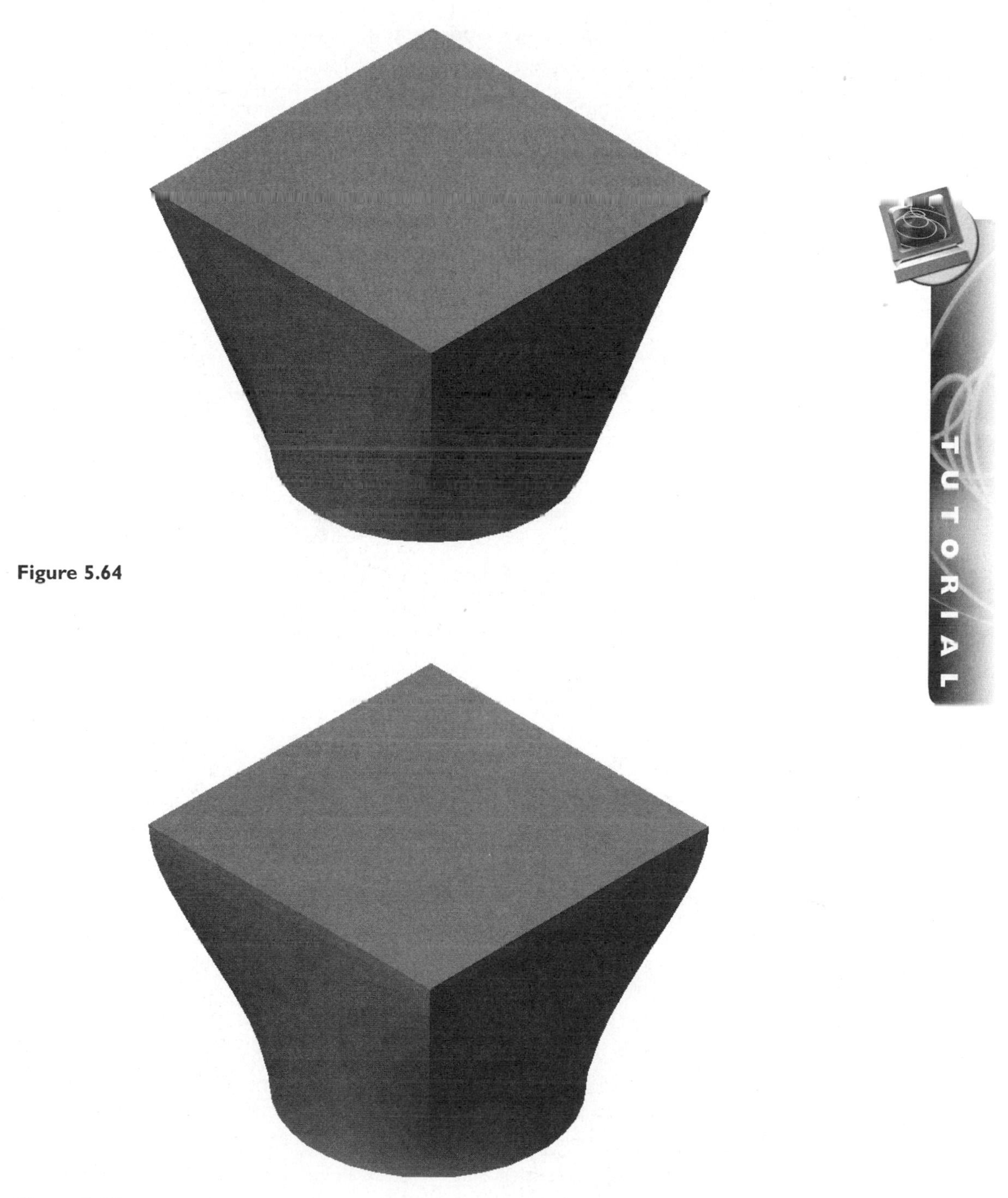

Figure 5.64

Figure 5.65

19. Edit the loft and change the Weight of the start section to "5", select the Preview button and then ENTER to return to the Loft dialog box and then select OK. Update the part if needed, and when complete, your screen should resemble Figure 5.66.

Figure 5.66

20. Edit the loft and change the Weight of the start section back to "1" and the end section to "3", select the Preview button and then ENTER to return to the Loft dialog box and select OK. Update the part if needed, and when complete, your screen should resemble Figure 5.67.
21. Edit the loft, change the Weight of the end section back to "1" and change the Angle on the start section to "120", select the Preview button and then ENTER to return to the Loft dialog box and select OK. Update the part if needed, and when complete, your screen should resemble Figure 5.68.
22. Edit the loft, change the Angle on the start section back to "90" and uncheck Minimize Twist. Select the Preview button and then ENTER to return to the Loft dialog box and select OK. Update the part if needed, and when complete, your screen should resemble Figure 5.69.

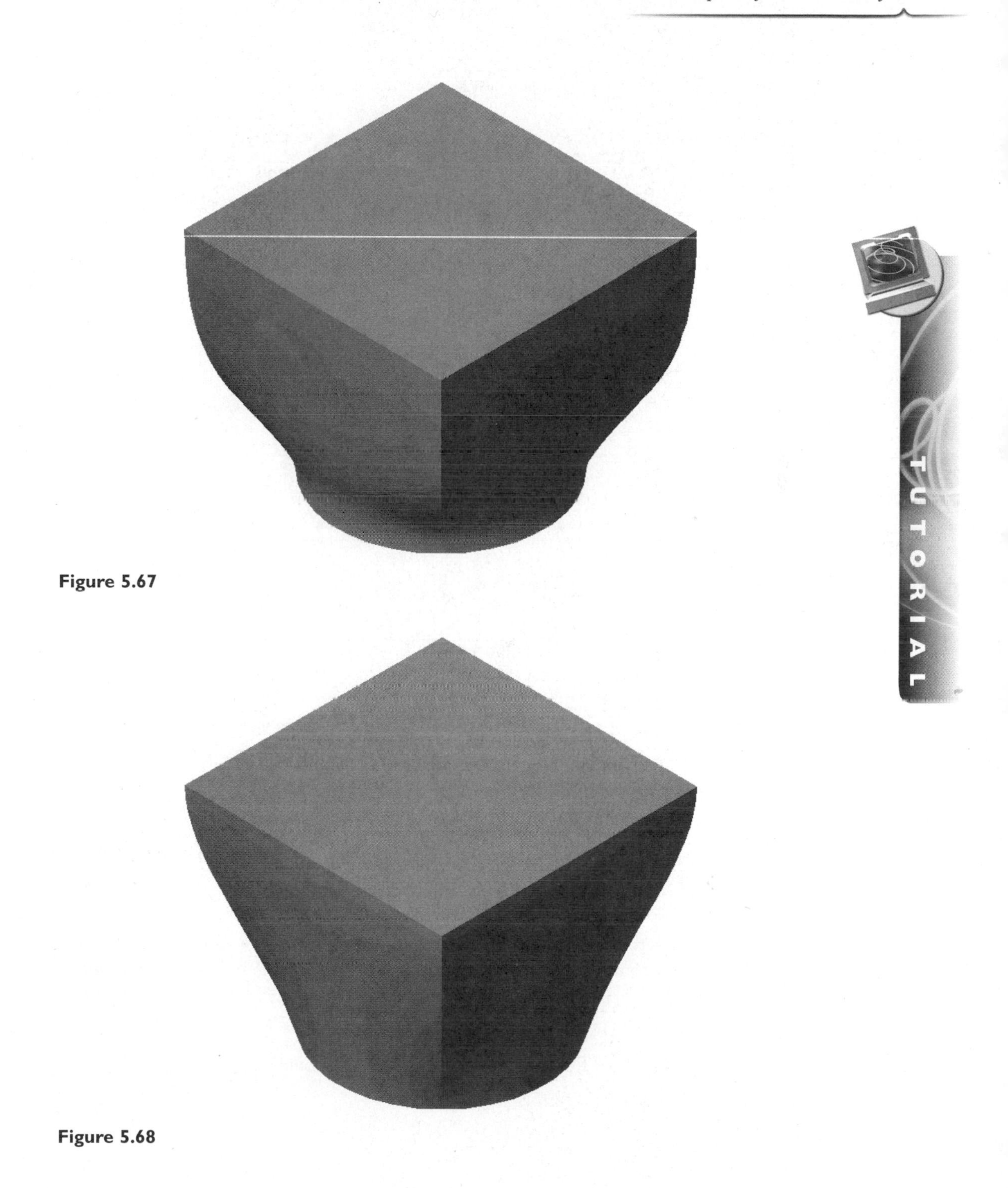

Figure 5.67

Figure 5.68

Figure 5.69

23. Edit the loft and select different start points on the square and update the part if needed to see the results.
24. Save the file as \Md4book\Chapter5\TU5-19.

TUTORIAL 5.20—LOFTED PARTS, CLOSED CUBIC

In this tutorial, four profiles have been created on two work planes.

1. Open the file \Md4book\Chapter5\TU5-20.
2. Issue the Loft command **(AMLOFT)** and in order, select the 3.00", 1.00", 2.00" and then the 1.50" diameter circle. Press ENTER. The Loft dialog box will appear.
3. Change the type to Closed Cubic and select the Preview button and then ENTER to return to the Loft dialog box and select OK to accept the other defaults.
4. Turn off the visibility of the work planes.
5. Shade the part. When complete, your screen should resemble Figure 5.70.
6. Save the file.

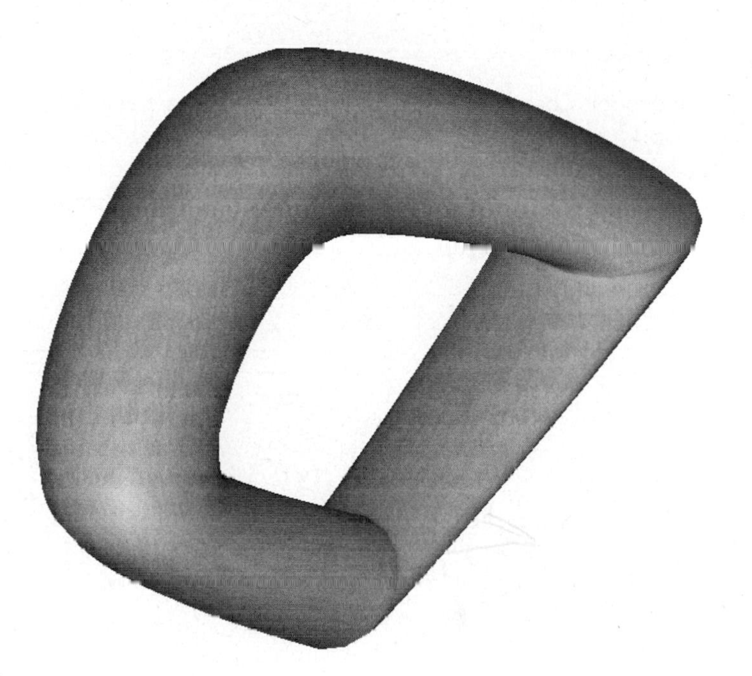

Figure 5.70

EXERCISE

For the following exercise, follow the instructions.

Exercise 5.1—Cylinder

Open the file \Md4book\Chapter5\Cylinder.dwg. Create the part as shown in Figure 5.71. Use work axis, work points and work planes as needed. Then create the views as shown in Figure 5.71. A drawing border already exists in the file. The base, ortho and auxiliary drawing views will be a scale of 1=2 and the isometric view at 1 relative to the parent; reposition dimensions and add hole notes as needed.

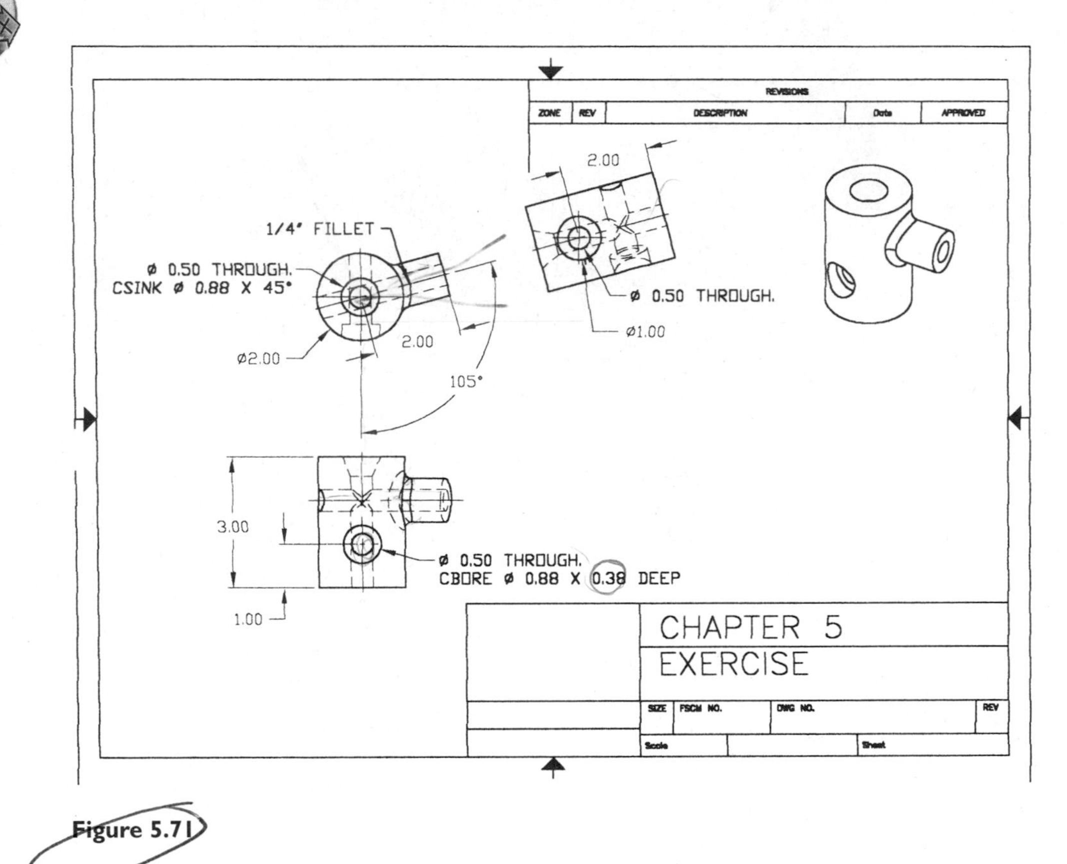

Figure 5.71

REVIEW QUESTIONS

1. A work axis can only be placed in the center of an arc or circular edge of a Mechanical Desktop part. T or F?
2. Name two reasons for creating a work axis.
3. Why should the New Sketch Plane **(AMSKPLN)** command be used instead of the UCS command?
4. Describe a situation in which a work plane needs to be created.
5. How do you make a work plane your active sketch plane?
6. After an angled work plane has been created, how can you change the angle?
7. In a drawing with multiple work planes, a single work plane cannot be turned off. T or F?
8. What condition must exist on the part edges so the edges can be used for a path?
9. A 3D spline or 3D polyline can be used for a 3D path. T or F?
10. A minimum of three profiles are required to create an open cubic loft. T of F?

CHAPTER 6

Advanced Dimensioning, Constraining and Sketching Techniques

In this chapter you will learn how to get more out of Mechanical Desktop by using advanced techniques. You will learn how to use construction geometry to better control the sketch, how to set up relationships between geometries, use power dimension and edit commands and use existing edges to close a sketch.

AFTER COMPLETING THIS CHAPTER, YOU WILL BE ABLE TO:

- Use construction geometry to help constrain profiles.
- Use multiple profiles.
- Use existing edges of a part to close a profile.
- Copy sketches.
- Power dimension profiles.
- Power edit dimensions.
- Use automatic dimensioning.
- Create relationships between dimensions.
- Create both local and global variables.
- Create a table driven part by using an Excel spreadsheet.

CONSTRUCTION GEOMETRY

Construction geometry can help you create part that, without the geometry would be difficult to create. In Chapter 1 you learned that sketches are created in a continuous linetype will be made into the part; any other linetype will be considered construction geometry. Create the sketch with construction geometry, then profile the sketch along with the construction geometry. Construction geometry can be constrained and dimensioned the same way continuous lines can, but they will not seen in the part.

Construction geometry can reduce the number of constraints and dimensions that are required to fully constrain a sketch. They can also be used to dimension the quadrants of arcs and circles. Construction geometry can be used to help define the sketch. For example, a construction circle inside a hexagon could drive the size of the hexagon. Without construction geometry, the hexagon would require six constraints and dimensions; it would require only three constraints and dimensions with construction geometry—the circle will have tangent constraints applied to it. When you edit a feature that was created with construction geometry, the construction geometry will reappear for the editing and disappear again when the part is updated. To aid in the creation of construction geometry you can issue the Construction Line command **(TB_CONSTR_LINE)** or the Construction Circle command **(TB_CONSTR_CIRCLE)** command as shown in Figure 6.1 from the Part Modeling toolbar. A construction line or circle will be placed on the AM_CON layer whose linetype is set to layer and by default is hidden. The current layer does not change. Construction geometry does not need to be created in this manner—this is just another method that is available. Construction geometry can also be added or deleted to sketch sets using the Append to Sketch command **(AMRSOLVESK)**.

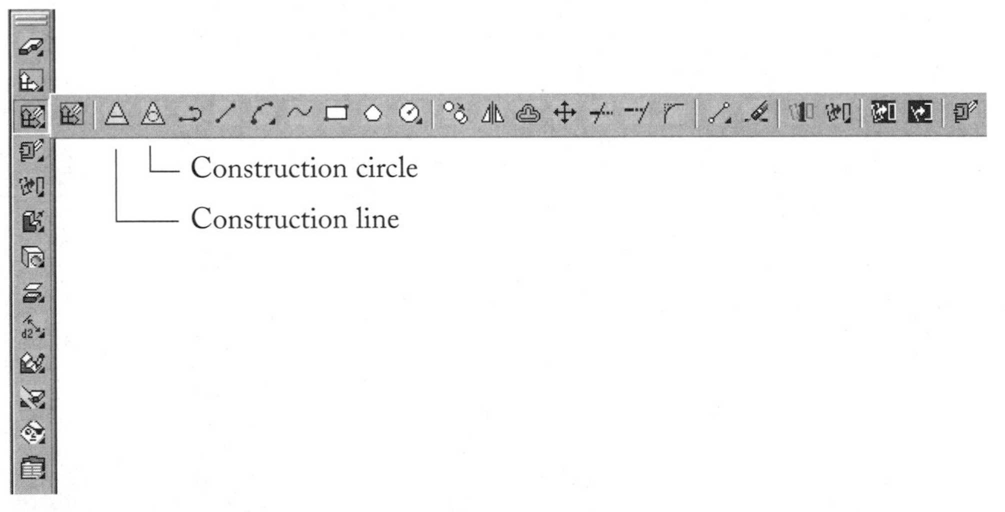

Figure 6.1

TUTORIAL 6.1—USING CONSTRUCTION GEOMETRY IN CREATING A HEXAGON

1. Start a new drawing.
2. Draw a construction circle and a hexagon that is concentric to the circle as shown in Figure 6.2. Approximate size of the circle should be 2" in diameter.

3. Profile both the hexagon and the construction circle.
4. Show all your constraints; there should be six tangent and two sets of parallel constraints. If all six tangent constraints are not there, add them as needed.
5. Add a Y Value constraint to the end point of the line near the left quadrant of the circle and to the construction circle.
6. Add a 60° angle dimension and a 2" diameter dimension as shown in Figure 6.2 (all the constraints are also shown).

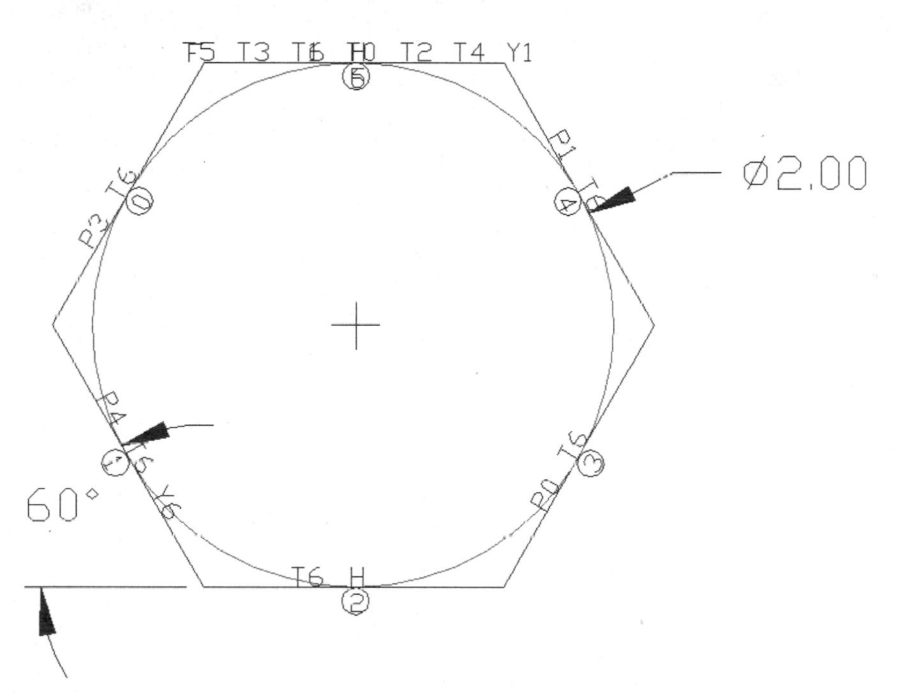

Figure 6.2

7. Use the Change Dimension command **(AMMODDIM)** to change the diameter of the circle to 1.
8. Change to an isometric view (the **8** key).
9. Extrude the profile .5" in the positive Z direction.
10. Edit the part, change the diameter to 3 and update the part.
11. Save the file as \Md4book\Chapter6\TU6-1.dwg.
12. Start a new drawing, draw a hexagon and try to constrain it without using construction geometry. Then try to change the size of the hexagon.

TUTORIAL 6.2—USING CONSTRUCTION GEOMETRY TO DIMENSION A QUADRANT

The goal of this tutorial is to control the overall height of the profile. By using a construction line you will be able to dimension a quadrant. Place the start point of the polyline near the quadrant so when it is dimensioned, it will appear that the dimension is for the quadrant of the arc.

1. Start a new file from scratch.
2. Sketch, profile, constrain and dimension the objects as shown in Figure 6-3. The top horizontal line is a construction line. Apply a Y Value constraint to the top of the two vertical lines and a tangent constraint to the arc and the construction line. The "1" dimension is to the construction line. If drawing the sketch as a polyline, use the second point option of the polyline command to create the non tangent arc.

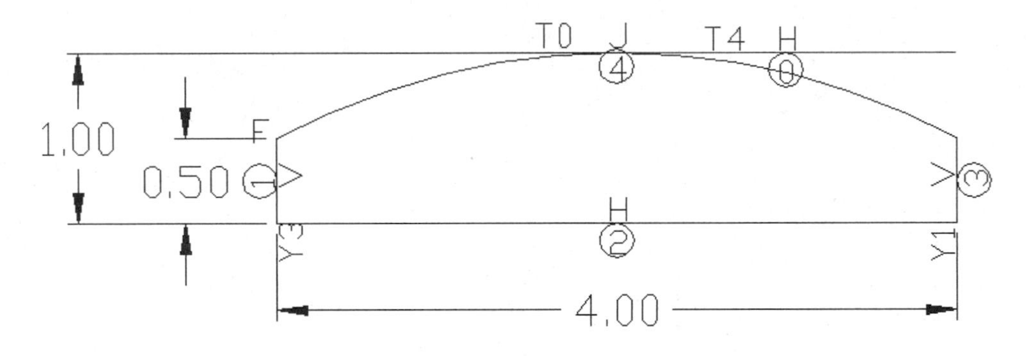

Figure 6.3

3. Change the "1" dimension to ".75".
4. Erase the ".5" dimension and create a "10" radial dimension in its place.
5 Change the "10" radial dimension to "15".
6. Extrude the sketch "4" .
7. Save the file as \Md4book\Chapter6\TU6-2.dwg.
8. Start a new drawing and try to create the same part without construction geometry. Try to control the overall height of the part.

MULTIPLE PROFILES

In Mechanical Desktop 4 you can have multiple sketches that have been profiled individually or profile multiple sketches in a single operation. The result is called a Multiple Loop Profile. If the sketches have been profiled individually and you then execute extrude, revolve or sweep, you will be asked to select the profile or profiles to work with. As long as there is more than one profile you will be asked which profile to use. If you select more than one sketch for the profile the islands will remove material from the part when extruded, revolved or swept. Individual profiles cannot be dimensioned, while sketches in the same profile can be dimensioned to one another. While editing a sketch set, profiles can be added or deleted as necessary using the Append to Sketch command **(AMRSOLVESK)**.

TUTORIAL 6.3— MULTIPLE INDIVIDUAL AND LOOPED PROFILES

1. Start a new file from scratch.
2. Draw a 4" square profile, dimension and extrude it 1" in the default direction.
3. Change to an isometric view using the **8** key.
4. Draw a 1" diameter circle in the upper left corner of the square and then profile it.
5. Draw a 1" diameter circle in the upper right corner of the square and then profile it.
6. Place a 1" diameter dimension on each of the circles.
7. Try to place a dimension between the two profiles. You will receive the prompt:
   ```
   Cannot create dimensions between objects in
   different sketches.
   ```
8. Issue the Mechanical Desktop Extrude **(AMEXTRUDE)** command and select the circle in the upper left corner of the square to extrude. In the Extrusion dialog box set the Operation to Cut and the Termination to Through and select OK to complete the command.
9. Repeat the Mechanical Desktop Extrude **(AMEXTRUDE)** command. You will not have to select a profile since there is only one profile. In the Extrusion dialog box set the Operation to Join, the Termination to Blind, Distance to "2" and select OK to complete the command.
10. Shade the part, and when complete, your screen should resemble Figure 6.4 (shown in shaded mode).
11. Save the file as \Md4book\Chapter6\TU6-3A.dwg.
12. Start another new drawing.
13. Draw a 4" square and create a 1" diameter circle at all four inside corners.
14. Profile the square and circles in one operation.
15. Dimension and constrain the profile as shown in Figure 6.5.

TUTORIAL

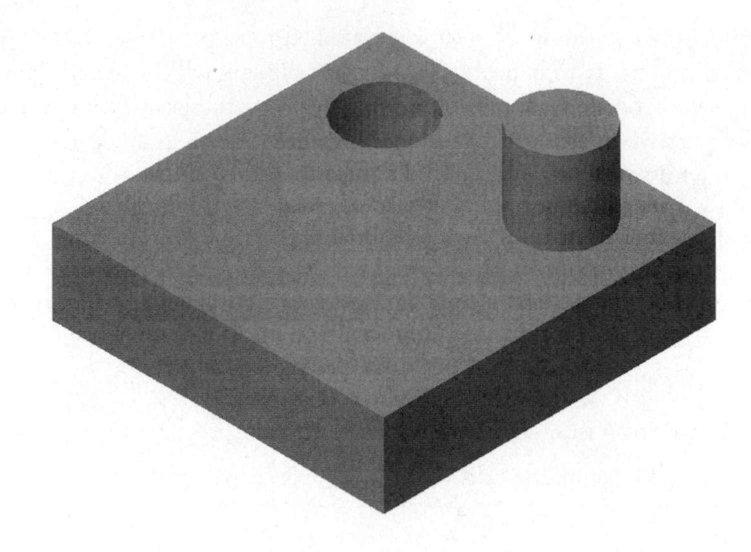

Figure 6.4

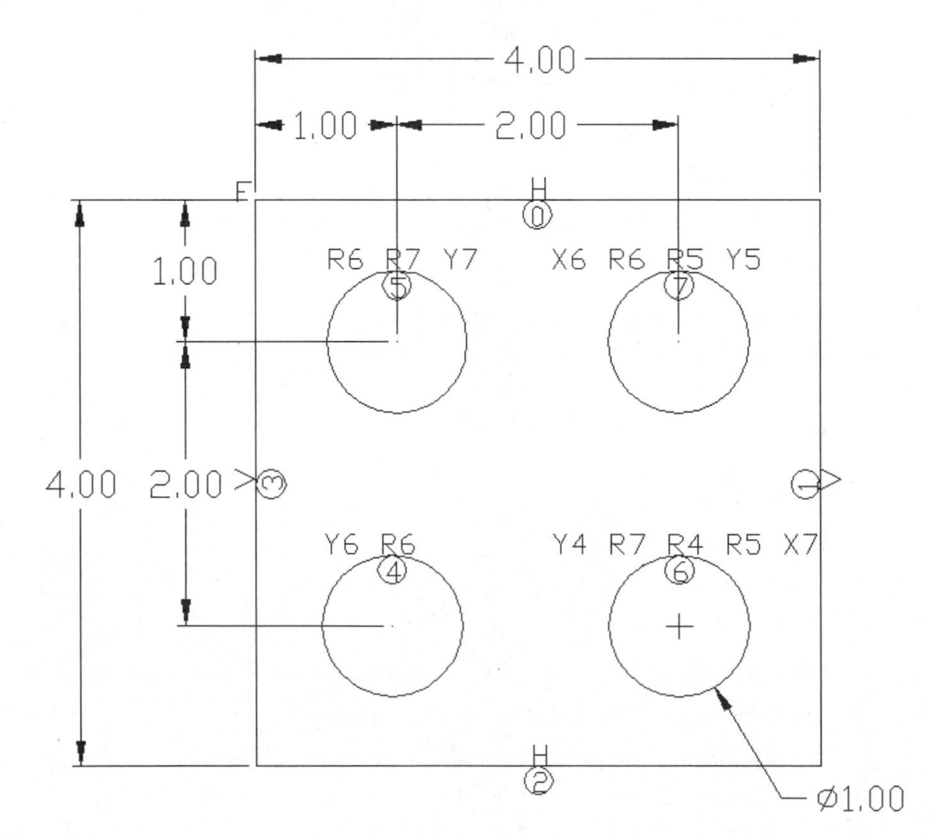

Figure 6.5

16. Change to an isometric view using the **8** key.
17. Issue the Mechanical Desktop Extrude **(AMEXTRUDE)** command. Since this will be the base feature, the Operation will be grayed out. In the Extrusion dialog box set the Termination to Blind, distance to "1" and select OK to complete the command.
18. Shade the part and when complete your screen should resemble Figure 6.5 (shown in shaded mode).

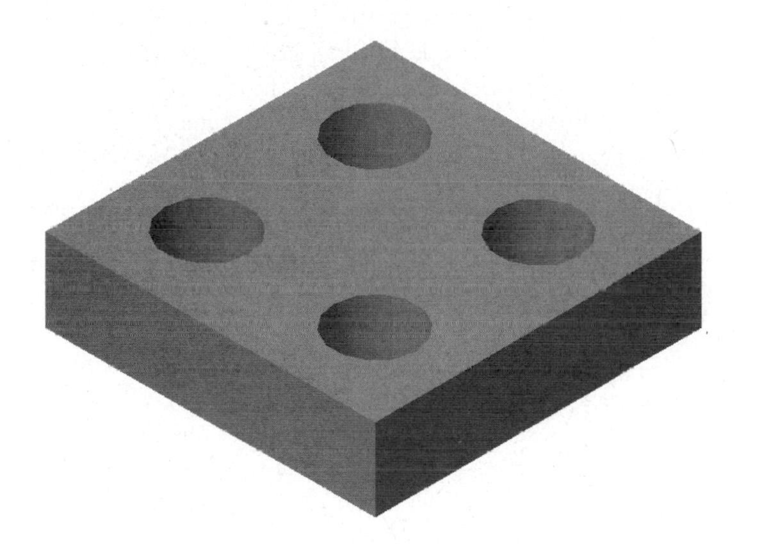

Figure 6.6

19. Edit the sketch of the extrusion. Delete the circle in the lower right corner. Issue the Append to Profile command and press ENTER to tell Mechanical Desktop that the sketch has changed. You didn't need to select new objects since nothing new was added.
20. Update the part if needed. The hole should have been deleted.
21. Edit the sketch of the extrusion. Add a circle in the lower right corner. Issue the Append to Profile command and select the new circle; press ENTER.
22. Update the part if needed, and the hole should appear.
23. Save the file as \Md4book\Chapter6\TU6-3B.dwg.

CLOSE EDGE

In Chapter 1, you learned that the first sketch needed to be closed. In this section, you will learn how to create sketches on existing parts that are opened but use existing edges of the part to close it. If necessary, an open sketch can use more than one edge to close the sketch during profiling. When using the close edge option, you will follow the same steps as previously learned. Make a plane the active sketch plane and draw a sketch. If the sketch can use an existing edge to close the profile, do not draw those side(s). The open geometry cannot have a gap larger than the pickbox size. After you profile the sketch of an open profile, a message will appear:

```
Select part edge to close the profile:
```

Select the edge(s) to close the profile and press ENTER. Silhouette edges cannot be used for closing the sketch unless the system variable DISPSILH is set to 1. It is highly recommended to keep the system variable DISPSILH set to 0 when not needed for silhouette edges. The number of constraints and dimensions that the profile needs will appear on the command line. Constrain and dimension the sketch as needed. Sketches that use a parts edge to close them cannot be arrayed or copied.

TUTORIAL

TUTORIAL 6.4—CLOSING A PROFILE BY SELECTING EDGES

1. Open the file \Md4book\Chapter6\TU6-4.dwg.
2. Make the top horizontal plane the active sketch plane and accept the default *XY* orientation.
3. Draw in the two lines as shown in Figure 6.7.

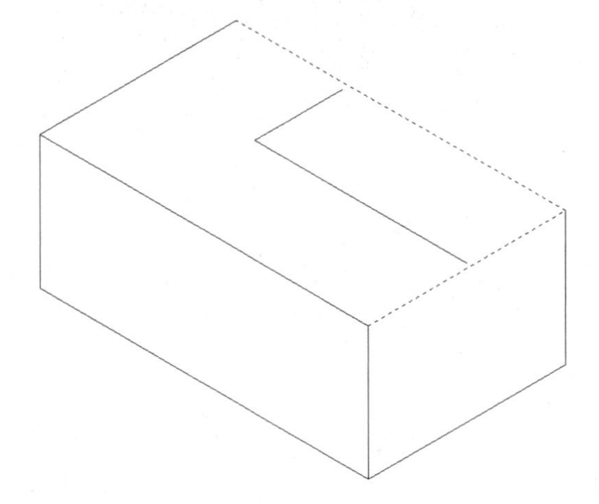

Figure 6.7

4. Profile the two lines and press ENTER and when prompted:
   ```
   Select part edge to close the profile:
   ```
   Select the top back horizontal and the top right vertical edges as highlighted in Figure 6.7 and press ENTER. If you get the same message, stretch the sketch closer to the edges, zoom out from the part or increase the pickbox size and try it again.

5. Dimension the profile with a "1" and "1.5" value as shown in Figure 6.8, shown with lines hidden.

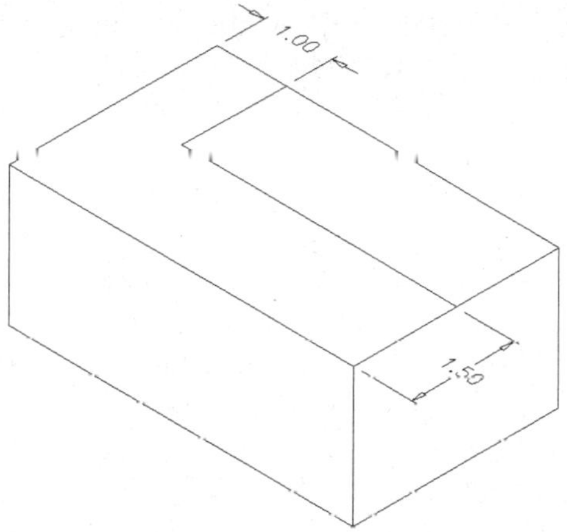

Figure 6.8

6. Extrude the profile ".75" blind with the Join operation. Accept the default extrusion direction and when complete, your screen should resemble Figure 6.9 shown with lines hidden.

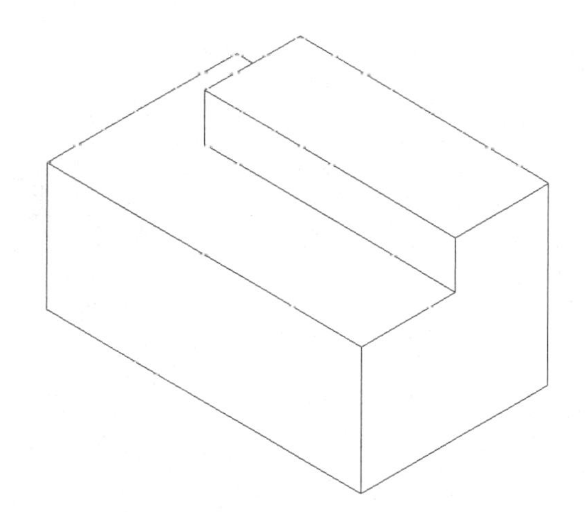

Figure 6.9

7. Make one of the vertical planar faces the active sketch plane. Sketch an open profile and use part edge(s) to close the sketch. Extrude the profile with the cut operation.
8. Save the file.

COPY SKETCH

With the Copy Sketch command **(AMCOPYSKETCH)** from the Part Modeling toolbar, as shown in Figure 6.10, right-click in the drawing area and from the pop-up menu, select Copy Sketch from the Sketch Solving menu. Or you may right-click on the sketch name in the browser and then select Copy from the pop-up menu. You can copy a sketch or a sketch of an existing feature to the active sketch plane of the active part. Sketches can be copied from one part to another. The sketch will be copied to the current sketch plane, along with the dimensions that define its shape. The dimensions that determine the sketch's *X* and *Y* placement will not be copied. If local design variables were used on the sketch, they will be copied only to the same part, otherwise they will be copied as their numeric value. If global design variables were used on the sketch they will be copied regardless of whether or not the same part. Once on this plane you can rotate or move the sketch with AutoCAD commands and add dimensions to lock the sketch down to the current sketch plane. You can have multiple profiles on the part at the same time. When multiple profiles exist and you begin to create a feature, you will be prompted to select the sketch that you want to work with. A profile that used a part's edge(s) cannot be copied. The copied objects should then be profiled and constrained as normal.

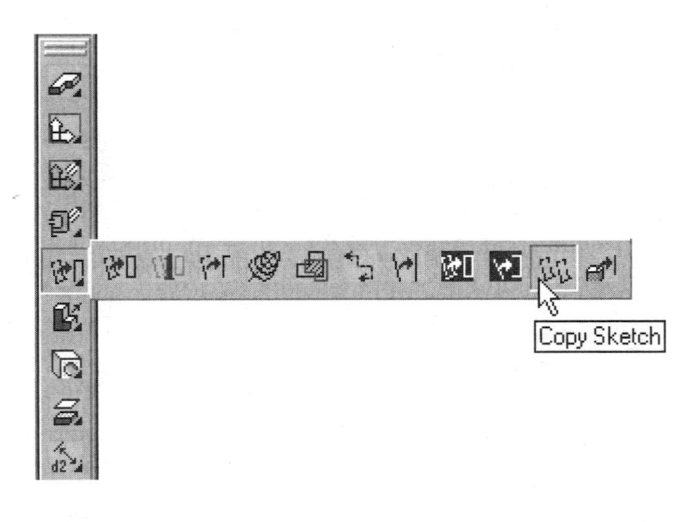

Figure 6.10

SHOW OBJECTS ON THE ACTIVE SKETCH PLANE

With multiple sketches on the screen on different sketch or work planes, you may lose track of what objects are on the active sketch plane. To highlight the objects that lie on the active sketch plane, issue the Show Active Sketch **(AMSHOWSKETCH)** command as shown in Figure 6.11 from the Part Modeling toolbar. When prompted to "Select objects to consider", press ENTER or use any AutoCAD selection technique to create a selection set for Mechanical Desktop to analyze.

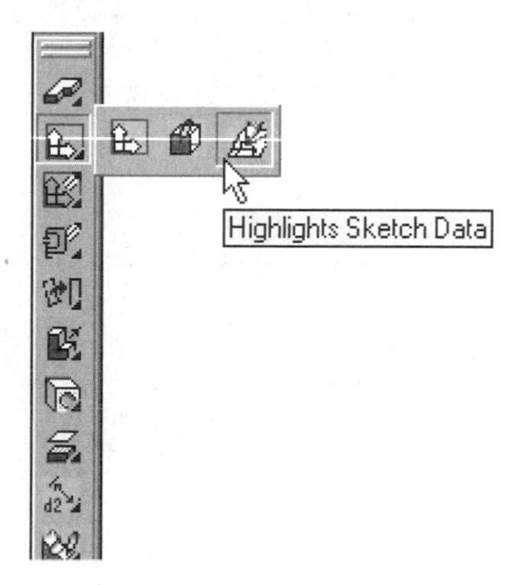

Figure 6.11

PROJECT OBJECTS/FACE

Another method to copy objects or 3D faces is to project them. To project an object or a 3D face, issue the Project to Plane command **(AMPROJECT2PLN)** as shown in Figure 6.12 from the Part Modeling toolbar. Or you may right-click in the drawing area and from the pop-up menu, select Project Objects to Plane from the Sketch

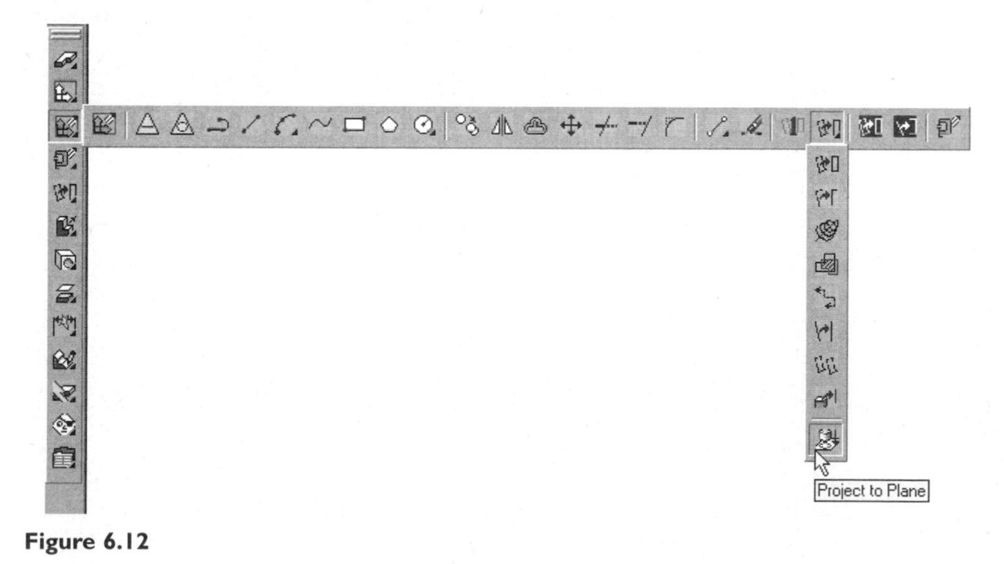

Figure 6.12

Projection options

Projection of:

2D-Objects

3D-Face

Projection to:

Workplane

Sketchplane

3D-Face

OK Cancel

Figure 6.13

Solving menu. The Projection options dialog box will appear as shown in Figure 6.13. In the dialog box select either 2D objects or 3D faces to project onto a workplane, sketchplane or a 3D face. If profiled objects and dimensions are projected, only the objects will be projected and not the dimensions. The projected objects should then be profiled and constrained as normal.

COPY EDGE

While working, you may need to create another sketch that consists of part edges. Instead of resketching the objects you can copy them with the Copy Edge command **(AMPARTEDGE)** as shown in Figure 6.14 from the Part Modeling toolbar or right-click in the drawing area and from the pop-up menu, select Copy Edge from the

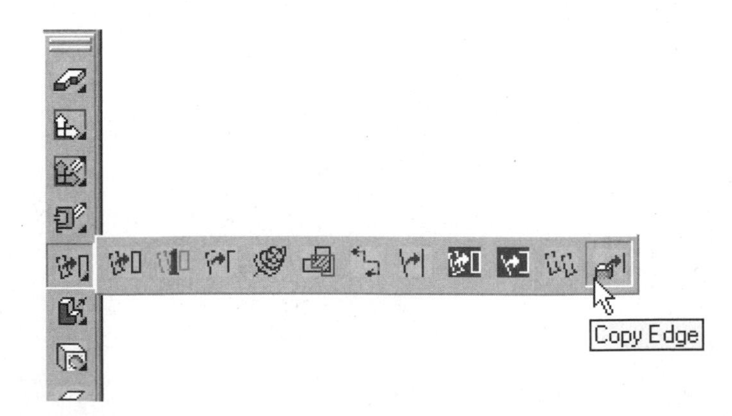

Figure 6.14

Sketch Solving menu. After issuing the command, individually pick the edges to copy or to copy all the edges on a planar face. Type F and ENTER and then select an edge that helps define the planar face. You can either accept by pressing ENTER or cycle to the next face by typing N and ENTER. When the correct face is highlighted, press ENTER. The copied edges can be deleted as needed. The copied edges should then be profiled and constrained as normal.

TUTORIAL 6.5—COPYING AND PROJECTING SKETCHES AND COPYING EDGES.

In this tutorial a rectangle has been extruded and a profile has been revolved.

1. Open the file \Md4book\Chapter6\TU6-5.dwg.
2. Make the top horizontal face the active sketch plane. Accept the default *X Y Z* orientation.
3. Issue the Copy Sketch command **(AMCOPYSKETCH)**
4. Press F and ENTER to copy a sketch of a feature. Select the revolution on the side of the part and press ENTER to accept the sketch of the revolution to copy.
5. Locate the sketch by selecting a point near the lower left corner of the top face as shown in Figure 6.15, and then press ENTER to accept this location. At this point, you could dimension the profile to give *X* and *Y* placement.

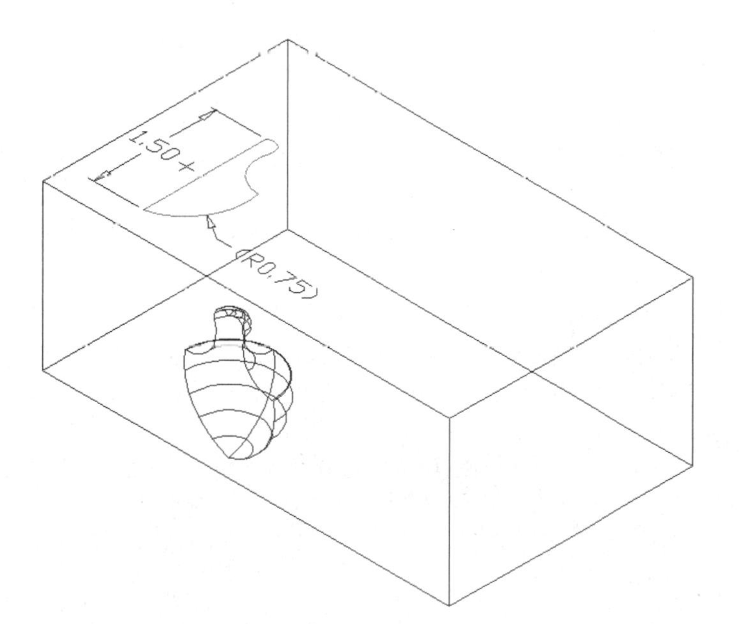

Figure 6.15

6. Make the right side vertical face the active sketch plane. Accept the default *X Y Z* orientation.
7. Repeat the Copy Sketch command **(AMCOPYSKETCH)**.
8. Press ENTER to copy a sketch. Select a point near the left side of the face and press ENTER to accept this location. When complete, your screen should resemble Figure 6.16. At this point, you could dimension the profile to give *X* and *Y* placement.

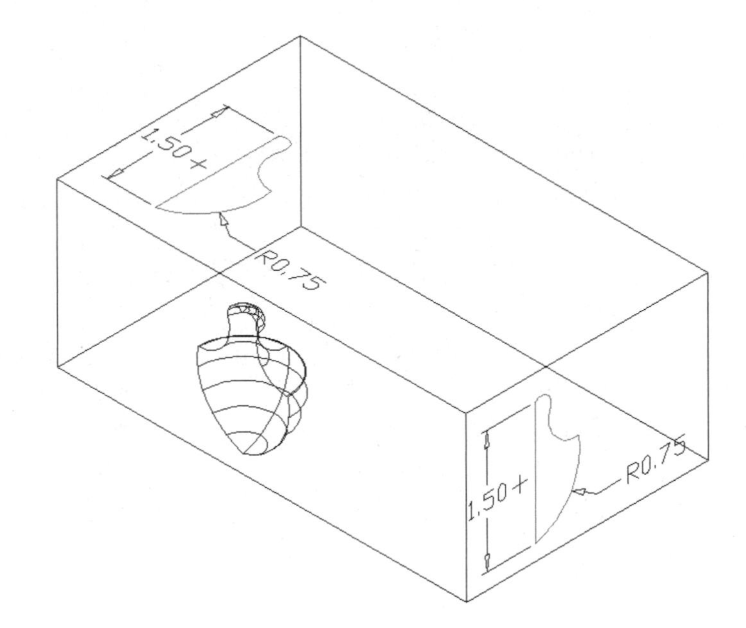

Figure 6.16

9. Make the long back vertical face the active sketch plane and accept the default *X Y Z* orientation.
10. Issue the Project to Plane command **(AMPROJECT2PLN)**.
11. In the Project options dialog box, Projection of = 3D face and Projection to = Sketchplane and then select OK.
12. Select the outside edge of the revolution. The cross hairs in Figure 6.17 show the area. Press N and ENTER to highlight the outside face as highlighted in Figure 6.17 and then press ENTER to complete the command. When complete, your screen should resemble Figure 6.18.

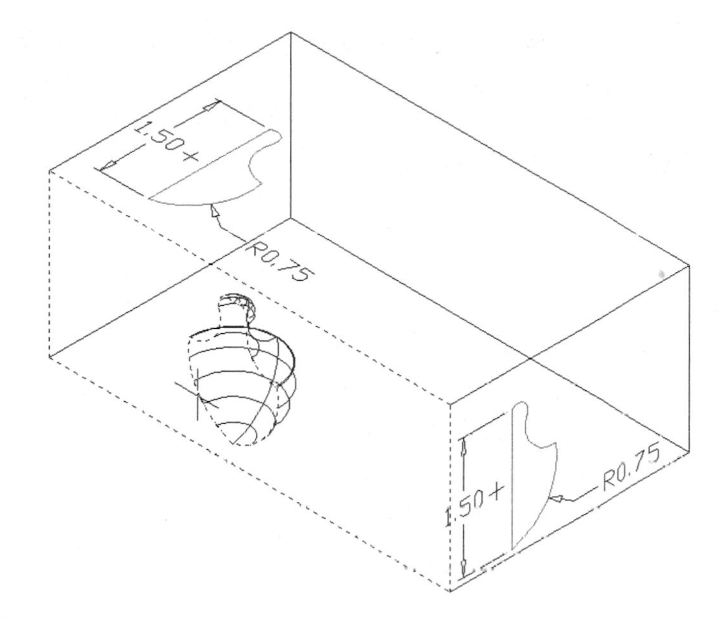

Figure 6.17

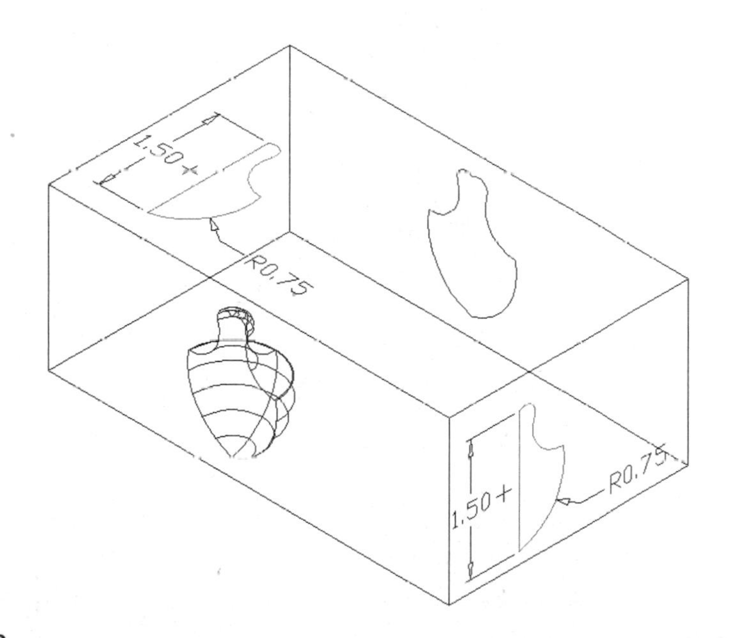

Figure 6.18

13. Practice copying edges using the Copy Edge command **(AMPARTEDGE)**. Select individual edges and all the edges on a face. After copying edges you can turn off the visibility of the part to see the copied edges.

14. Practice extruding and revolving the sketches and profiles.
15. When finished save the file.

POWER DIMENSIONING

Mechanical Desktop gives you another command to dimension profiles called Power Dimensioning **(AMPOWERDIM)** as shown in Figure 6.19 from the Part Modeling toolbar. Or you may also right-click in the drawing area and from the pop-up menu, select Power Dimensioning from the Dimensioning menu. The power Dimenioning command allows you to alter how a dimension will appear. When a line is selected, a Power Dimensioning dialog box will appear as shown in Figure 6.20. The dialog box has the following sections:

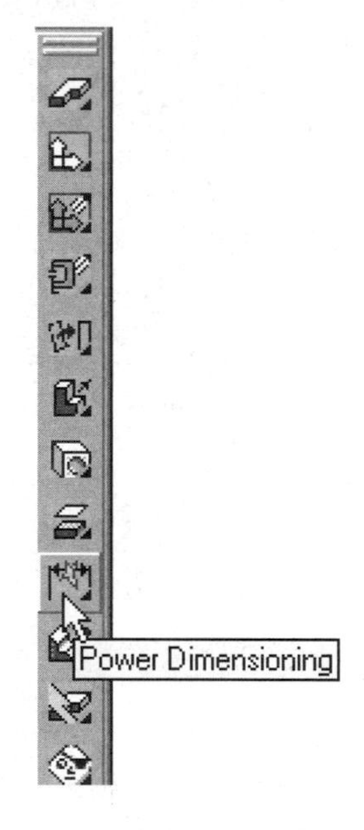

Figure 6.19

Dimension Text: Has three options:

1. Type text in the blank area to appear in the dimension.
2. Select the button with three dots to allow you to pick special characters to insert in the dimension.
3. Selecting the arrow will give you options for designating the type of dimension.

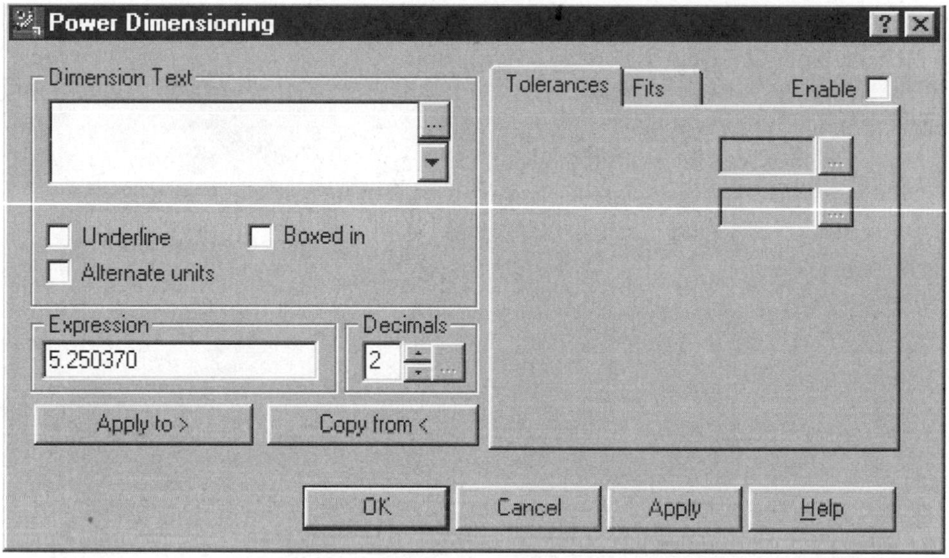

Figure 6.20

Underline: When checked, the dimension will be underlined.

Alternate units: When checked, the alternate units specified in the current dimension style will appear next to the primary unit.

Boxed in: When checked, a box will appear around the entire dimension.

Expression: Type in the value to be used in the dimension.

Decimals: Specify the number of decimal places to appear for the dimension.

Apply to>: When selected, you can choose the dimensions to apply the settings, as well as other characteristics to apply.

Copy from<: When checked you will be prompted to select a dimension from which to copy selected characteristics.

Enabled: When checked, the Tolerance tab will be enabled.

Tolerances: Type in or select both upper and lower deviations from the option dialog box. These values will appear on the dimension. They will also be used for tolerance modeling which will be covered in Chapter 9.

Fits: Select the type of fit that you want represented on the dimension.

If an arc or circle is selected, a dialog box similar to the one shown in Figure 6.21 will appear. Select how you want the dimension to appear from the dialog box. After selecting OK, the Power Dimensioning dialog box will appear as shown back in Figure 6.20.

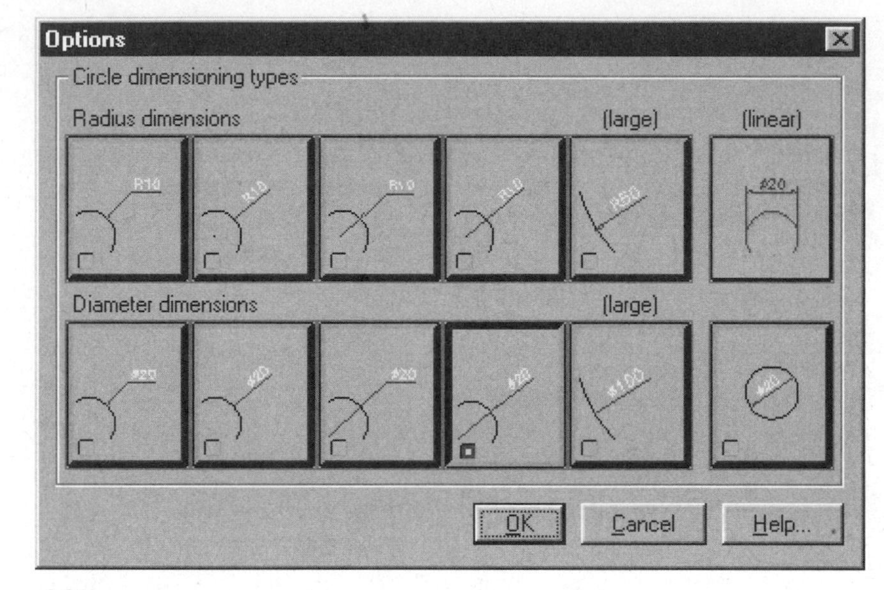

Figure 6.21

POWER EDIT

Once a new dimension or a power dimension has been placed, it can be power edited using the Power Edit command **(AMPOWEREDIT)** as shown in Figure 6.22 from the Part Modeling toolbar. You may right-click in the drawing area and from the pop-up menu, select Power Edit from the Dimensioning menu or double click on a

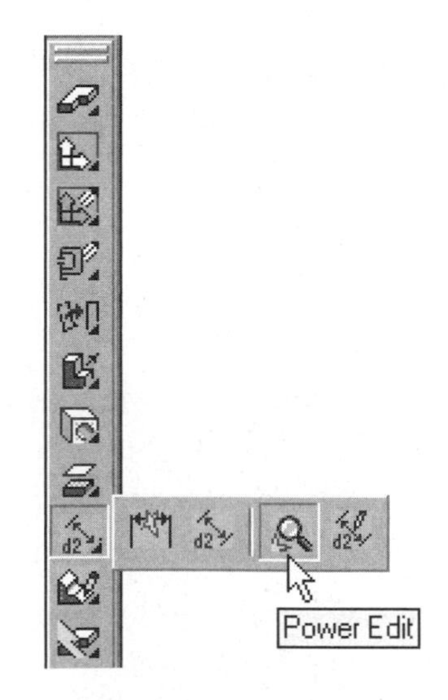

Figure 6.22

dimension. After issuing the Power Edit command, the same Power Dimensioning dialog box that is shown in Figure 6.20 will appear. Edit the dimension in the same manner that you created a power dimension.

AUTOMATIC DIMENSION

To speed the process of dimensioning sketches you can use the Automatic Dimensioning command **(AMAUTODIM)** that can be accessed under the Annotate pull down menu. Or you may right-click in the drawing area and from the pop-up menu, select Automatic Dimensioning from the Dimensioning menu. Automatic dimensioning can place multiple dimensions with only a few picks. First sketch and profile the object that you want to dimension. Then issue the Automatic Dimensioning command **(AMAUTODIM)**. In the Automatic Dimensioning dialog box are three tabs: Parallel, Ordinate and Shaft / Symmetric. Each tab is described below. After selecting the setting you want, select the objects that you want to dimension using any AutoCAD selection technique (window, crossing etc.), then press ENTER. You will be prompted to select "First extension line origin:" Select an origin point for the first extension and then select an origin point for the second extension line or press ENTER to use the previous point. Then select a point to place the dimensions. If you checked Both Axis in the Automatic Dimensioning dialog box, you will follow the same prompts to place the dimensions in the other axis. These automatic dimensions are parametric and by default will take on the value of the object they dimension. Edit the dimension's values as needed with the **(AMMODDIM)** command and proceed as usual.

PARALLEL

As changes are made in the dialog box, the image in the dialog box will update reflecting the changes.

Type: Select either Baseline or Chain.

Both Axes: If checked, both horizontal and vertical dimensions will be created from the selection set.

Rearrange into a New Style: When checked, existing dimension type can be changed from either baseline to chain or from chain to baseline.

Select Additional Contour: When checked you can select other circular edges to dimension or rearrange.

Display Power Dimensioning Dialog: When checked, the Power Dimensioning dialog box will appear every time a dimension location is selected.

ORDINATE

Type: Select Ordinate or Equal Leader Length.

Both Axes: If checked, both horizontal and vertical dimensions will be created from the selection set.

Rearrange into a New Style: When checked, existing dimension type can be changed from either baseline or chain to ordinate.

Select Additional Contour: When checked, you can select other circular edges to dimension or rearrange.

Display Power Dimensioning Dialog: When checked, the Power Dimensioning dialog box will appear every time a dimension location is selected.

SHAFT/SYMMETRIC

Type: Select Full Shaft, Half Shaft, or Symmetric.

Place Dimension Inside Contour: When checked, dimensions will be drawn inside the profile. If unchecked the dimensions will be drawn outside the sketch. This option is only available only for Full and Half Shafts.

Rearrange into a New Style: When checked, existing dimension types can be changed into Shaft/Symmetric type.

Select Additional Contour: When checked, you can change existing dimensions into Shaft/Symmetric type.

Display Power Dimensioning Dialog: When checked, the Power Dimensioning dialog box will appear every time a dimension location is selected.

TUTORIAL 6.6—POWER DIMENSION, POWER EDIT AND AUTOMATIC DIMENSIONING

1. Open the file \Md4book\Chapter6\TU6-6.dwg.
2. Profile the exterior objects and the circle.
3. Issue the Power Dimensioning command **(AMPOWERDIM)**.
4. Press ENTER to select an object to dimension.
5. Select the circle. From the Options dialog box, select a style of your choice and then select OK to exit the dialog box. Then select a point to locate the dimension.
6. In the Power Dimensioning dialog box, change the number of decimal places to "3".
7. Select OK and then press ENTER to complete the command.
8. Issue the Power Edit command by double-clicking on the dimension and then in the Power Dimensioning dialog box. Change the number of decimal places to "2" and select OK to complete the command.

9. Issue the Automatic Dimensioning command **(AMAUTODIM)** and under the Parallel tab set the options as follows:
   Type = Baseline.
   Check Both Axes.
   All other options should be left unchecked.
   Select OK to leave the dialog box.
10. Window around all the objects.
11. Select the lower left corner for the first extension line origin.
12. Press ENTER to use the previous point for the second extension line origin.
13. Place the vertical dimensions on the left side of the profile by selecting a point to the left of the profile.
14. Press ENTER to use the previous starting point for the starting point for next extension line.
15. Place the horizontal dimensions on the bottom of the profile by selecting a point below the profile.
16. Press ENTER to complete the command and then your screen should resemble Figure 6.23. Note that the profile should be fully constrained.

TUTORIAL

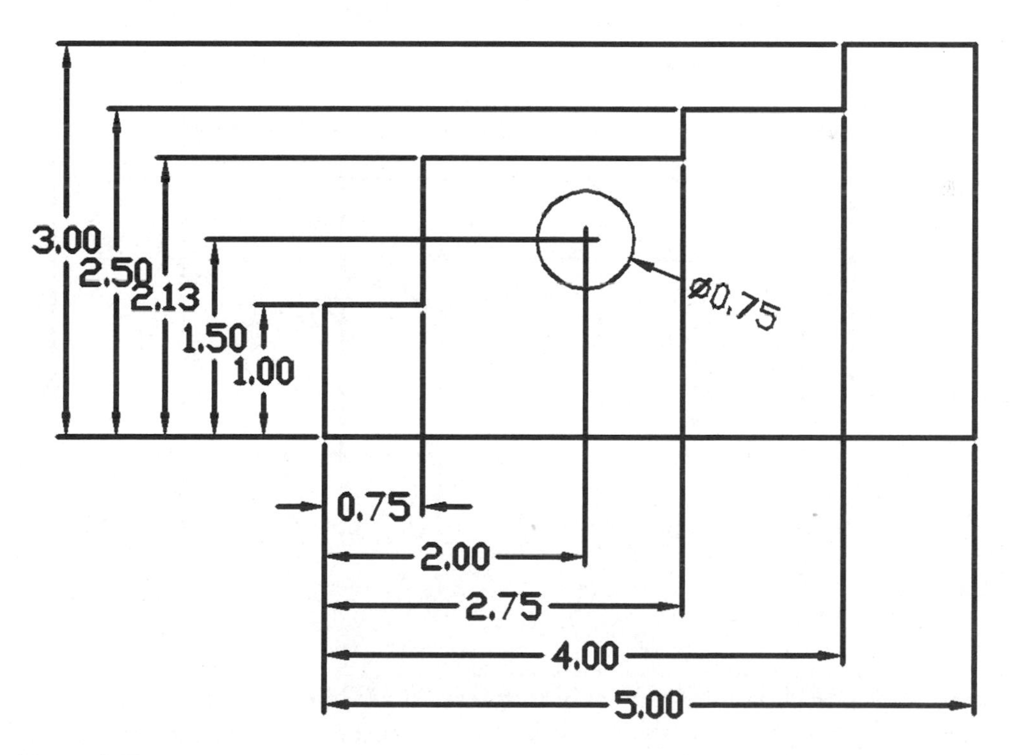

Figure 6.23

17. Issue the Power Edit command **(AMPOWEREDIT)** by double-clicking on one of the extension lines of one of the horizontal dimensions. Under the Parallel tab set the options as follows:
    Type = Chain.
    Uncheck Both Axes if it is checked.
    Check Rearrange into a New Style if it is not already checked.
    All other option should be left unchecked.
    Select OK to leave the dialog box.
18. Select all the horizontal dimensions with a crossing window and then press ENTER.
19. Select the lower left corner for the first extension line origin.
20. Place the horizontal dimensions on the bottom of the profile by selecting a point below the profile.
21. Press ENTER to complete the command and then your screen should resemble Figure 6.24.
22. Practice power editing dimensions.
23. Extrude the profile ".5" and edit a few of the dimensions of the feature.
24. Save the file.

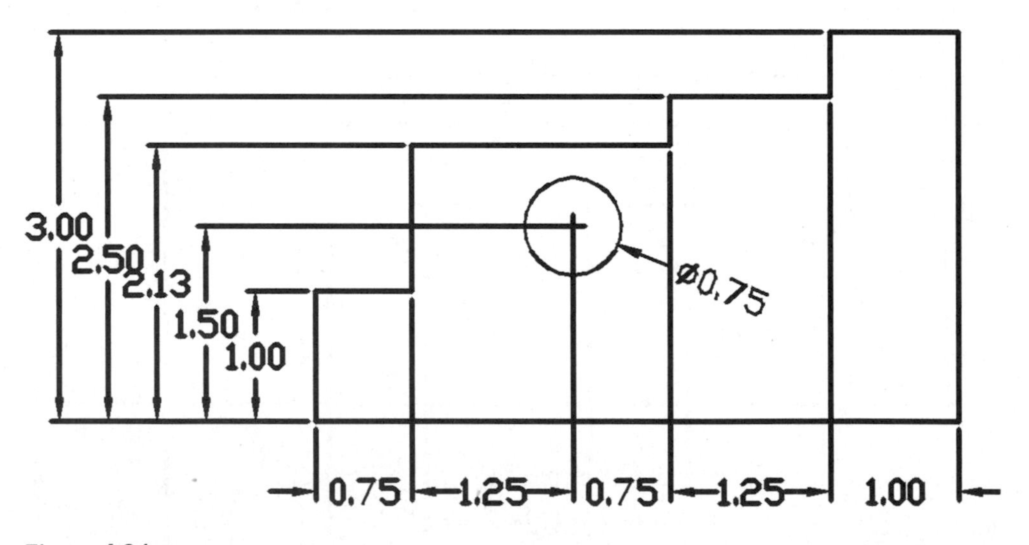

Figure 6.24

DIMENSION DISPLAY AND EQUATIONS

When creating parts, you may want to set up relationships between dimensions. For example, the length of a part may need to be twice that of its width or a hole may always need to be in middle of the part. In Mechanical Desktop, there are a few different methods to set up relationships between dimensions. The first method uses a label that is automatically given to each dimension. Each label starts with the letter "d" and is given a number, for example "d0" or "d27". The first dimension created is given the label "d0" and each dimension that follows sequences up one number at a time. If a dimension is erased, the next dimension does **not** go back and reuse the erased value. Instead, it keeps sequencing from the last value on the last dimension created. To see the "d" values for each, issue the Dimension Display command **(AMDIMDSP)**, as shown in Figure 6.25 or right-click in the drawing area and from the pop-up menu, select either Dimensions As Parameters, Dimensions As Numbers or Dimensions As Equastions from the Dimensioning menu.

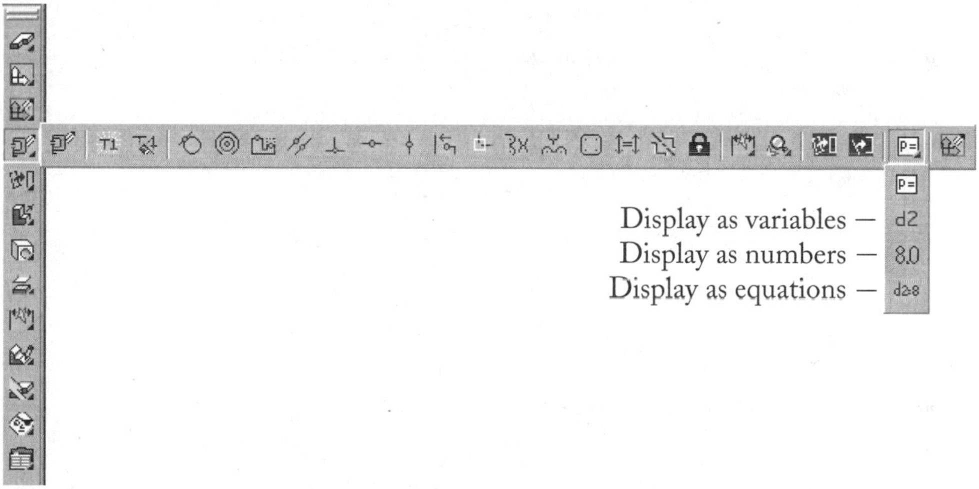

Figure 6.25

There are three different dimensional display modes: parameters, equation and numeric. After you select a dimension display mode, all the dimensions on the screen will change to that mode.

Parameters mode: Displays as the dimension number or "d#" only.

Equation mode: All the dimensions on the screen will change to "d# = #", showing each actual value, for example, d7=4.50.

Numeric mode: The default. Displays the dimensions as actual numbers.

As dimensions are created, they will reflect the current dimension display mode, which has no effect on the dimension's value. To create a dimension using another dimension's "d#", type in the "d#" when prompted, to verify a dimension's value. You can also use mathematical operators like "(d9/4)*2" or "65/9". Figure 6.26 shows the mathematical operators that Mechanical Desktop supports. Use the Edit Dimension command **(AMMODDIM)** from the Part Modeling toolbar to add or remove a relationship between dimensions.

Note: After a dimension with a parameter is created, it can be changed with the Change Dimension command.

When creating relationships between dimensions, use the equation display mode to see the "d#" and the actual value of the dimension.

Operator	**Description**
^	Exponent
+	Add
-	Subtract
*	Multiply
/	Divide
%	Modulus (remainder)
sqrt	Square Root
log	Logarithm
ln	Natural Logarithm
floor	Rounds down to nearest whole number
ceil	Rounds up to nearest whole number
sin	Sine
cos	Cosine
tan	Tangent
asin	Arcsin ($\sin^{-1}$)
acos	Arcos ($\cos^{-1}$)
atan	Arctang ($\tan^{-1}$)
sinh	Hyperbolic Sine
pi	Pi

Figure 6–26

TUTORIAL 6.7—USING DIMENSION DISPLAY AND EQUATIONS

1. Open the file \Md4book\Chapter6\TU6-7.dwg.
2. Profile the sketch.
3. Change the Dimension Display **(AMDIMDSP)** mode to Parameters (tooltip refers to the mode as variables).
4. Create the right vertical dimension with the value of "1". When complete, your drawing should look like Figure 6.27.

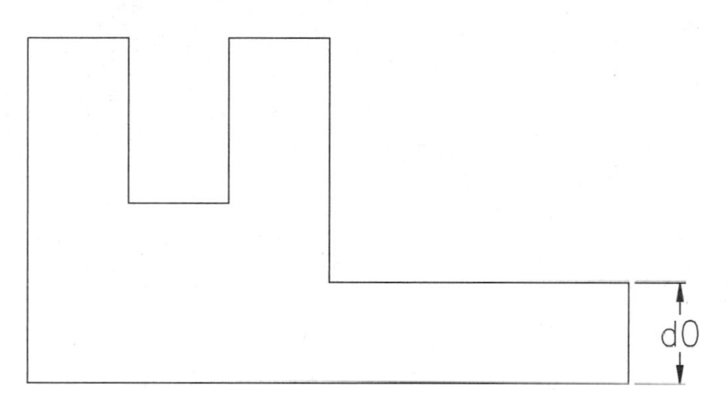

Figure 6.27

5. Create the three horizontal dimensions; dimension them left to right and type in "d0" for each value. When complete, your drawing should look like Figure 6.28. If they were dimensioned in a different order, the "d#" would be different; the order in which the dimensions are placed will have no effect on the actual part.

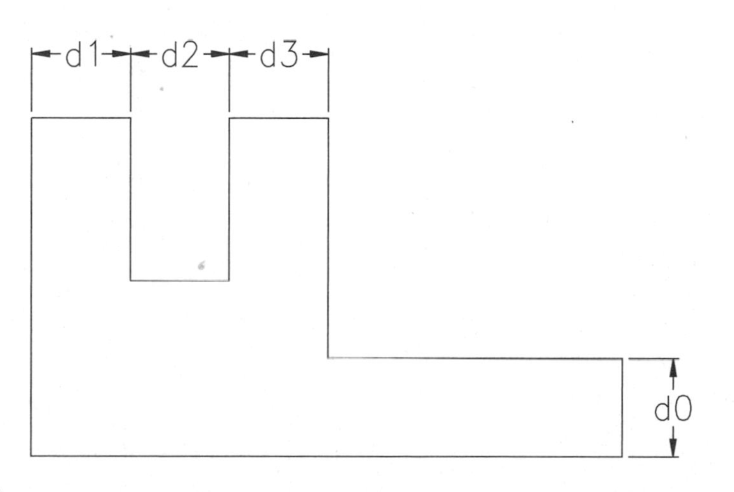

Figure 6.28

6. Change the Dimension Display **(AMDIMDSP)** mode to Equations.
7. Add two vertical dimensions on the left side of the part. Place the inside vertical line as "d0*2" and the outside vertical line as "d0*3". When complete, your drawing should look like Figure 6.29. In the Figure, the three horizontal dimensions have been repositioned for clarity.

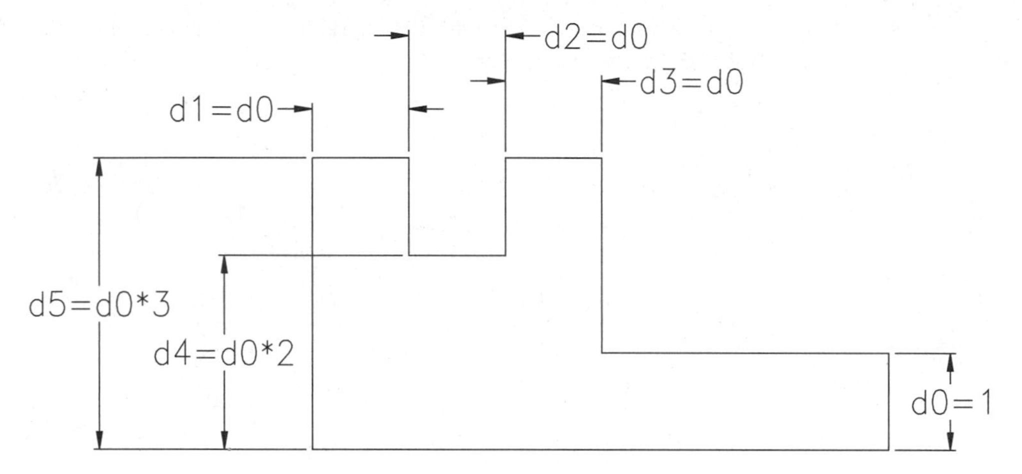

Figure 6.29

8. Dimension the bottom horizontal line with the equation "(d5*2)-d0". When complete, your drawing should look like Figure 6.30.

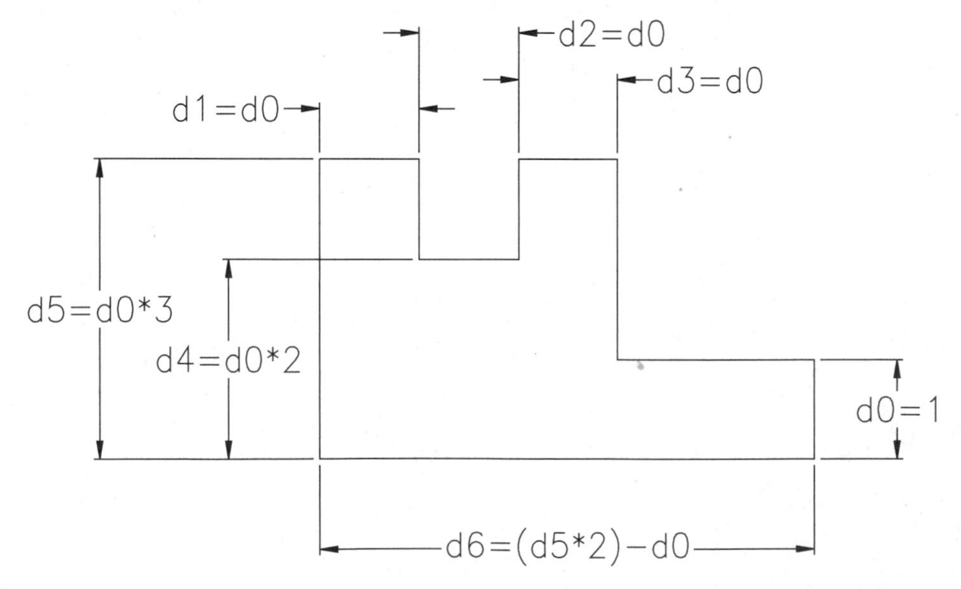

Figure 6.30

TUTORIAL

9. Change the Dimension Display **(AMDIMDSP)** mode to Numeric. When complete, your drawing should look like Figure 6.31

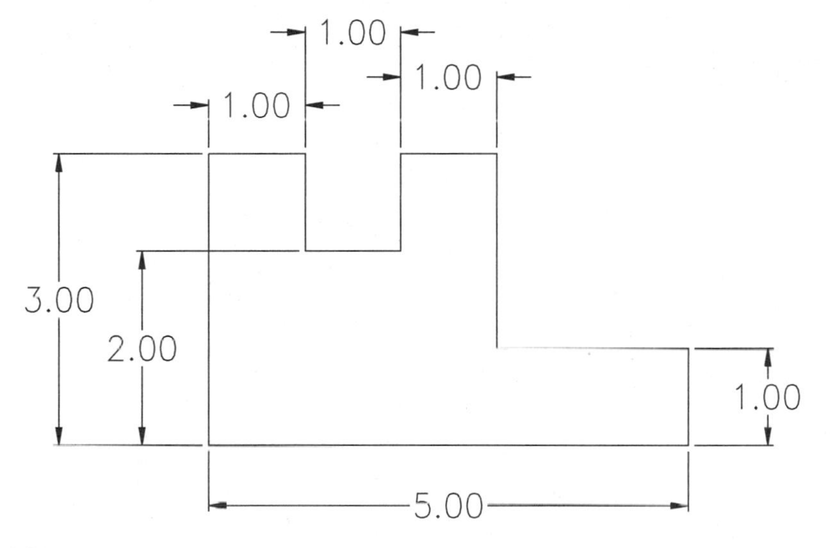

Figure 6.31

10. Change to an isometric view with the **8** key.
11. Extrude the profile; in the extrusion distance, type in "d0". When complete, your part should resemble Figure 6.32, shown with lines hidden.

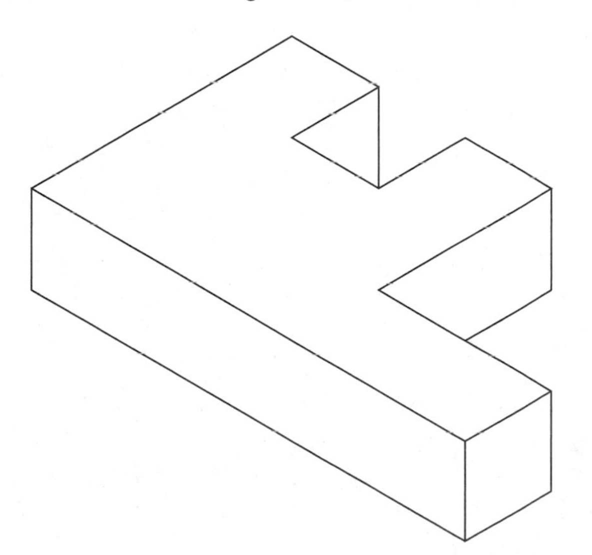

Figure 6.32

12. Edit the part and change the "5" bottom horizontal dimension "(d5*2)-d0" to "4", and change the "d0" (the vertical dimension created first) to "2". Update your part and your drawing should look like Figure 6.33, shown with lines hidden.
13. Save the file.

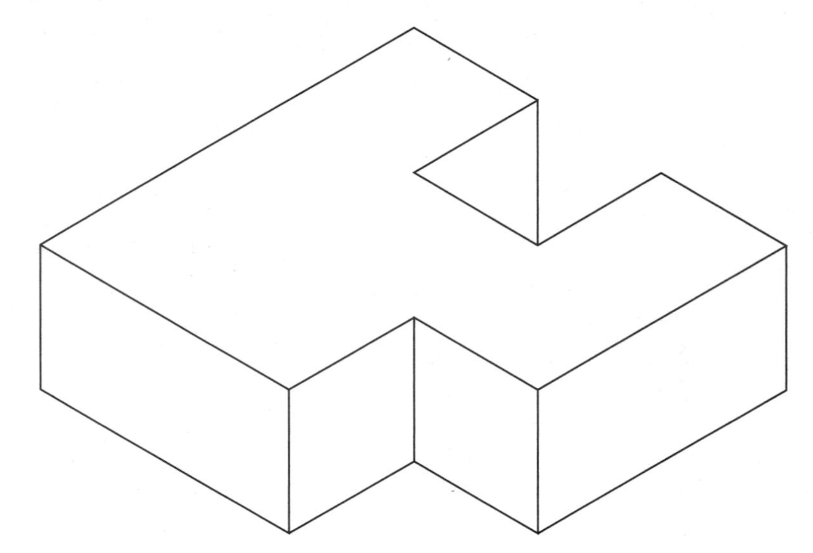

Figure 6.33

DESIGN VARIABLES

The second method of setting up relationships between dimensions is to use design variables. A design variable is a user-defined name that is assigned numeric value either explicitly or through equations. A design variable can be used anywhere a dimension is required. A design variable can be defined by any name except "c# or d#" where # is any number (those letter-number combinations are used for arrays and dimensions), and the single letter command line options of the AMPARDIM command:

```
Command: _ampardim
Select second object or place dimension:
Enter dimension value or
[Undo/Hor/Ver/Align/Par/aNgle/Ord/Diameter/pLace]
```

There are two types of design variables: active part and global. Active part design variables can be used in the active part only. Global design variables can be used in multiple parts in the same file as well as in assembly constraints. If an active and a global design variable exist with the same name in the same file, the active variable takes precedence over the global variable.

ACTIVE PART VARIABLES

To create an active part design variable, you must first have a sketch that has been profiled. After a sketch has been profiled, it becomes the active part. Then an active part design variable can be created. Issue the Design Variables command **(AMVARS)** from the Part Modeling toolbar, as shown in Figure 6.34, or right-click on the part name in the browser and select Design Variables from the pop-up menu. After the command is issued, the Design Variables dialog box will appear as shown in Figure 6.35. There are two tabs: Active Part and Global. To create an Active Part variable, select New and a New part Variable dialog box will appear. Type in information for the name, equation and comment fields.

Figure 6.34

Figure 6.35

Name: Can contain no spaces, can be up to 72 characters in length and must start with a letter. However, for ease of use it is recommended that you keep the names short.

Equation: Any mathematical operator can be used, as described in Figure 6.26. Other design variables can also be used in the equation.

Comment: Place a description of up to 250 characters with spaces. The comment field is used only for the operator's reference and is not required.

After filling in the information, select OK and you will be returned to the Design Variables dialog box. The information that was filled in will appear and the value field will reflect the value of the equation. Along the top row of the dialog box are the letters T D U. They represent:

T = Variable is Table-driven.

D = Duplicated variable in the active part, global variable or table-driven. For duplicate variables, the active part variable takes precedence over global variables.

U = The variable is Unreferenced in the part(s).

The letter T, D, or U will appear on the line of the variable that matches. To edit a design variable, double-click on the area that you want to edit. Type in the new information, press ENTER and then select OK. If you edit the design variables through the browser, the part will update automatically. Otherwise, use the Update Part command **(AMUPDATE)** to update the part. To remove a single design variable that is not being referenced, use the Delete button on the right side of the dialog box. To remove all design variables that are not being referenced, use the Purge button on the right side of the dialog box. If a design variable is used in the part it cannot be deleted.

Global Variables

Global design variables can be created in the same manner as Active part variables except that you select the Global tab in the Design Variables dialog box and fill in the name, equation and comment. Global variables can be used in any part in the file and can also be used for values in assembly constraints. Global variables are often used to set up relationships between parts. For example, a global variable called "Hole_Dia" could be created to represent a hole diameter. Both a plate to replace the hole and a shaft to go through the hole would use the global variable "Hole_Dia". If the global variable changes, both parts would update to reflect the change.

In the Global tab section named **Global Variable File** you have the option to import, link, export and unlink global variables. When variables are exported, they are saved in an ASCII file with the extension "prm". This ASCII file can be edited with any word processor. When saving the file, make sure that it is in ASCII format.

Import...: Brings design variables into the current drawing from a "prm" file. The variables become global variables in the current file.

Link...: Creates a link between a "prm" file and the current drawing. The drawing gets the variables from the file. If the "prm" file changes, you can update the drawing with those changes either by reopening the drawing file or by reestablishing the link (this will bring in the new variables).

Export ...: Take the global variable in the current drawing and create a "prm" file.

Unlink Remove the reference between the "prm" file and the current file. The variables will become global variables in the file.

In the Global tab section named **Copy to Active part** you can copy global variables into active part variables. The options are Selected, Referenced and All.

Selected: The highlighted variable will be copied to an active part variable.

Referenced: All global variables that are referenced in the active part will be copied to active part variables.

All: All global variables will be copied to active part variables.

Another method to create a global variable is to move active part variables to global variables. To move an active part design variable to a global variable, select one of the three options from the Move to Global section in the Design Variables dialog box.

Selected: Moves only the design variables that you choose.

Referenced: Moves only the design variables that are used in the active part.

All: Moves all the design variables.

Note: In order for an active part variable to be created, there must be at least one sketch that has been profiled.

Active part variables take precedence over global variables!

EQUATION ASSISTANT

When creating dimensions you can use the equation assistant to create mathematical equations or use it to select design variable names. To use the equation assistant, issue the New Dimension command **(AMPARDIM)** and when you are prompted to type in a value, right click in the drawing area and select Equation Assistant from the pop up menu as shown in Figure 6.36 or when a value is required in a dialog box right click in the box for the value and select Equation Assistant from the pop-up menu. The Equation Assistant dialog box will appear as shown in Figure 6.37. Create a mathematical equation selecting the numbers, operations or from the design variable list.

A combination of the three can be used. When finished creating the equation, select the Equate button to generate the result and then select OK to pass the result to the New Dimension command. If you want to use the design variable name only DO NOT select Equate as the value will only be passed to the dimension. Instead, select OK to pass the design variable name to the dimension.

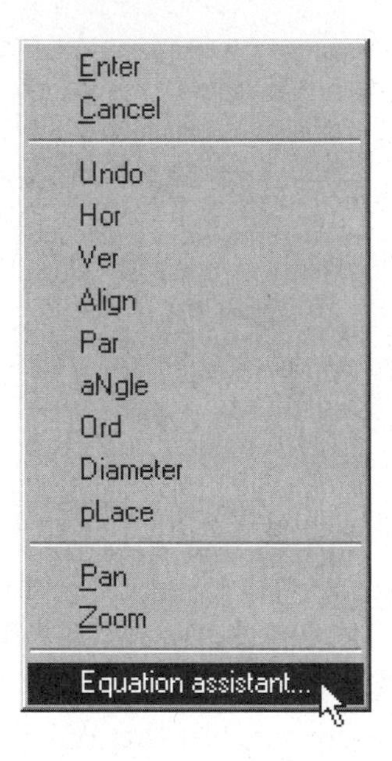

Figure 6.36

Figure 6.37

TUTORIAL 6.8—USING ACTIVE PART DESIGN VARIABLES

1. Open the file \Md4book\Chapter6\TU6-8.dwg.
2. Profile the sketch.
3. Create two active part design variables. Issue the Design Variables command **(AMVARS)**.
4. Select New and give the values Name = LENGTH, Equation = 2, Comment = length of part. Select OK in the new Part Variable dialog box.
5. Create the second design variable, select New and give the values Name = WIDTH, Equation = LENGTH/2, Comment = width of part. Select OK in the new Part Variable dialog box.
6. Select OK in the Design Variables dialog box.
7. Dimension the sketch as shown in Figure 6.38. For clarity, the dimensions are shown in equation mode. Type in the corresponding variables and values or use the equation assistant.

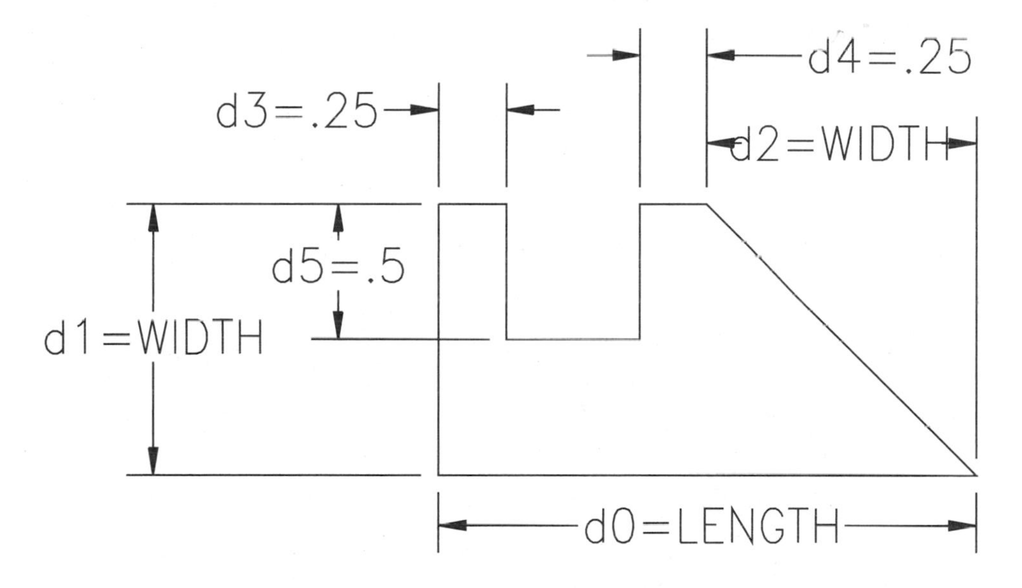

Figure 6.38

8. Change to an isometric view (the **8** key) and extrude the profile with the value "LENGTH/4" and then press ENTER. When complete, your screen should resemble Figure 6.39, shown with lines hidden.

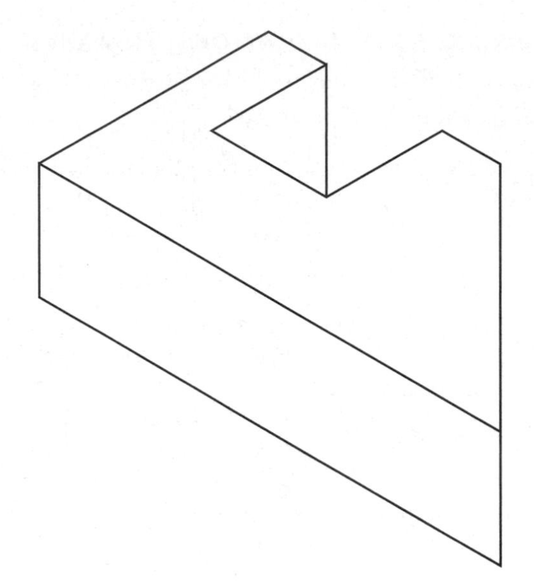

Figure 6.39

9. Edit the design variable "LENGTH so that its Equation = "3".
10. Edit the design variable "WIDTH so that its Equation = "1".
11. Update the part if needed. When complete, your screen should resemble Figure 6.40, shown with lines hidden.
12. Save the file.

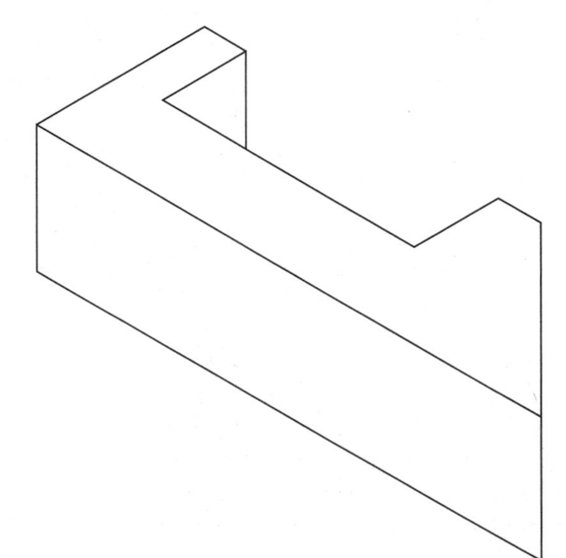

Figure 6.40

TUTORIAL 6.9—USING GLOBAL DESIGN VARIABLES

In this tutorial, two parts have been created and assembled using assembly constraints. A global design variable named "Hole_Dia" with a value of ".5" has also been created and used for the diameter of the shaft. Multiple parts and assembly information will be covered in Chapters 7 and 8.

1. Open the file \Md4book\Chapter6\TU6-9.dwg.
2. Issue the Design Variables command **(AMVARS)**.
3. Change the equation of the global variable "Hole_Dia" to ".2" and then select OK to complete the command. When complete, your screen should resemble Figure 6.41, shown with lines hidden. The shaft updated but the hole in the plate did not, because the hole is not referencing the design variable.

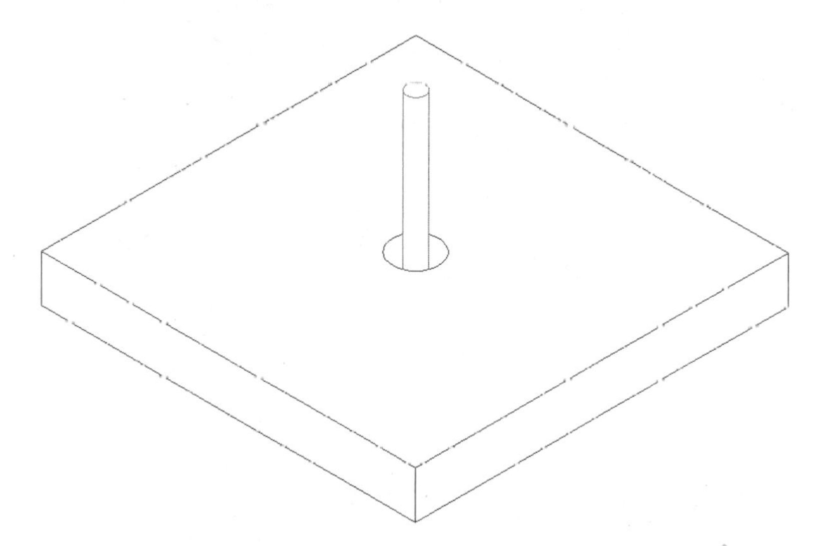

Figure 6.41

4. Edit the value of the diameter of the hole in the plate to "Hole_Dia". Update the part, if needed, and when complete, your screen should resemble Figure 6.42, shown with lines hidden.
5. Issue the Design Variables command **(AMVARS)**. Select the Global tab and change the equation of the design variable "Hole_Dia" to "1". Select OK to complete the command. Both the hole and the shaft should have the same 1" diameter.
6. Create a global design variable by issuing the Design Variables command **(AMVARS)**. Select the Global tab, select New and give the values Name = Clearance, Equation = .1, Comment = Hole Clearance. Select OK in the new Part Variable dialog box.

TUTORIAL

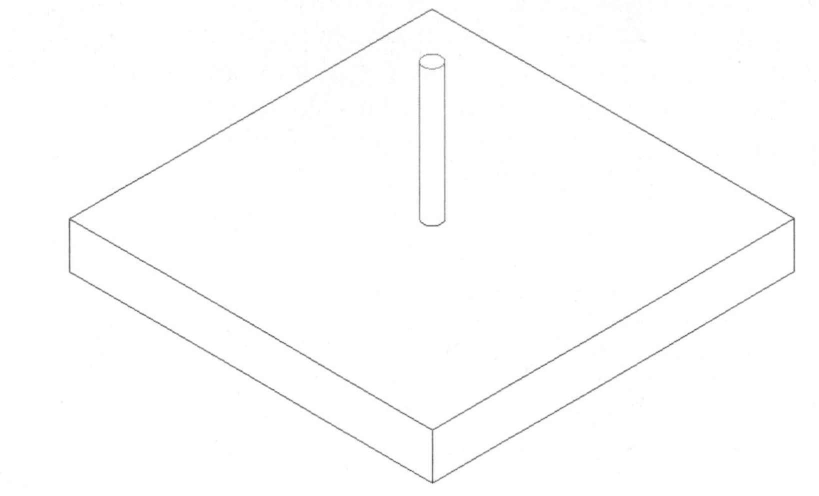

Figure 6.42

7. Edit the value of the diameter of the hole in the plate to "Hole_Dia + Clearance". Update the part if needed, and when complete your screen should resemble Figure 6.43, shown with lines hidden.
8. Save the file.

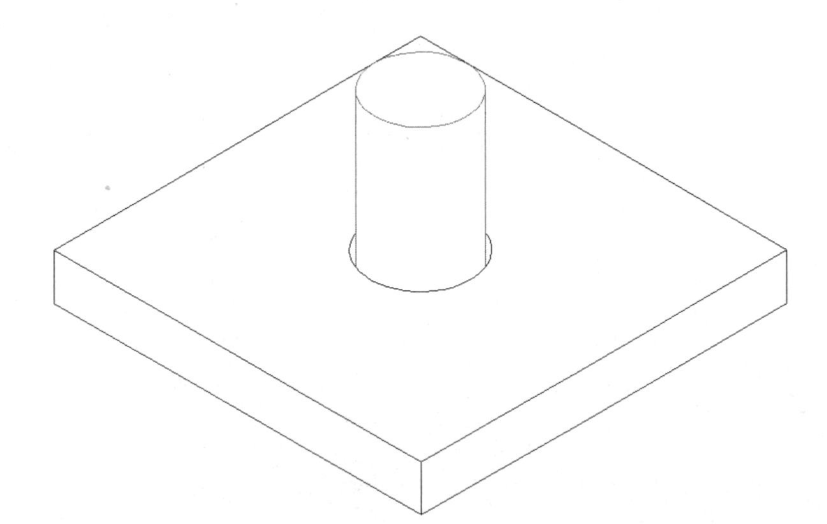

Figure 6.43

TABLE-DRIVEN PARTS

Table-driven variables are similar to design variables except that the name and equation are stored in an Excel Spreadsheet. The spreadsheet can have different values of the same variable if multiple rows are created under the variable name, as shown in Figure 6.44. The table-driven variables will be used as active or global design variables. Table-driven variables are used in the modeling process just like design variables; they can be placed in dimensions of a sketch or existing part. If the part is using an active design variable with the same name as an Active Part table-driven variable, the active design variable will be overridden by the table-driven-design variable. If a part has global design variables that are table-driven and a duplicate name for an active part design variable the active part design variable will be used. The spreadsheet is linked to the part, and changes in Excel can be updated inside Mechanical Desktop to change the part.

To create a table driven variable, you must first have a sketch that has been profiled. From the Part Modeling toolbar, issue the Design Variables command **(AMVARS)** or right-click on the part name in the browser and select Design Variable from the pop-up menu. The tab that is current Active Part or Global will determine if the table driven design variables will be for the active part or global.

	A	B	C	D
1		Length	Width	Depth
2	Part A	5	4	0.25
3	Part B	7	3	0.5
4	Part C	8	7	0.75
5	Part D	12	10	1

Figure 6.44

There are eight basic steps to create and use active part or global table-driven design variables:

1. To create an active part table-driven design variable create a sketch and profile it, then right-click on the part name in the browser and select Design Variables from the pop-up menu or select the Design Variables command **(AMVARS)** from the Part Modeling toolbar. To create a global table-driven design variable, select the Design Variables command **(AMVARS)** from the Part Modeling toolbar if a part does not exist or right-click on the part name

in the browser and select Design Variables from the pop-up menu and make the Global tab active.

2. Select Setup in the lower left corner of the Design Variables dialog box and a Table Driven Setup dialog box will appear like the one shown in Figure 6.45.

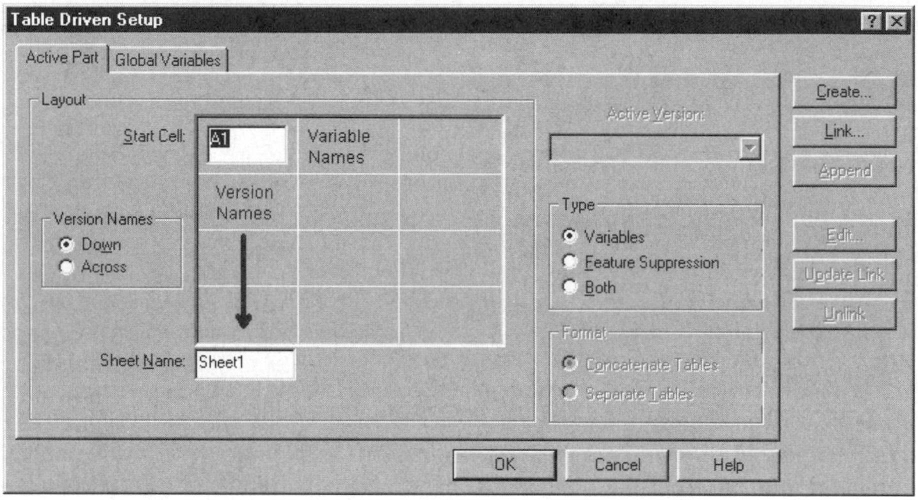

Figure 6.45

3. Fill in the information in the Table Driven Setup dialog box. The dialog box has the following options:
   Version Names: Choose whether the version (design variable) names should go down or across the spreadsheet.
   Start Cell: Type in the start cell location
   Sheet name: Type in the sheet name in which the information will be located.
   Variables: If selected, only design variables will be shown in the table.
   Feature Suppression: If selected, only features will be shown in the table.
   Both: If selected, both design variables and features will be shown in the table.
   Concatenate Tables: If selected, both design variables and features will appear in the same sheet. A blank column or row must exist between the design variable names and the feature names.
   Separate Tables: If selected, the design variables and features will appear on different sheets.
   Create: Selecting this button will allow you to type in a new file name and directory for the Excel spreadsheet.
   Link: Select this button if you want to create a link to an existing Excel spreadsheet.

Append: Select this button to append new design variables to the linked spreadsheet.
Update Link: Select this button to read new values into Mechanical Desktop in the linked Excel spreadsheet.
Unlink: Select this button if you want to remove the link to an existing Excel spreadsheet. The active version of variables will become active part or global design variable depending upon the mode you are working in.

4. Select Create Table and a Create Table dialog box will appear. Give the spreadsheet a name and location, select Save and Excel will load, if it is not already running. If there are active design variables, they will be written out to this Excel file with a name of Generic. Generic can be renamed to any name without spaces. Or if an Excel spreadsheet already exists, select Link and then select the Excel file.
5. In Excel, add the data to the rows and columns of the spreadsheet as needed and then save the file. The A1 cell is reserved and cannot be used.
6. Return to Mechanical Desktop and select the Update Link button. The variables will be added to the Design Variables dialog box and the versions will appear in the browser.
7. Dimension the sketch or edit the part to add the table driven design variables. Finish the part as required.
8. To change between versions, double-click on the version in the browser or, in the Design Variables dialog box, select the version from the Active Version drop-down list.

The data in the Excel spreadsheet can be linked to a Mechanical Desktop drawing by highlighting the cells in Excel that you want linked to Mechanical Desktop. Copy them to the clipboard using CTRL+C or right-click on the highlighted area and then select Copy from the pop-up menu. Switch back to Mechanical Desktop and choose Paste Special from the Edit pull-down menu and select a point to locate the chart. Now that the chart is linked to the spreadsheet, when the variable(s) change inside Excel, they will automatically be updated in the chart inside Mechanical Desktop.

To update the variables in Mechanical Desktop, either right-click on the top level of the table in the browser and select Update from the pop-up menu or issue the Design Variables command **(AMVARS)** and select Update Link from the dialog box. The new values will be current. If you open a drawing and the table is red in the browser, this is a warning that there is a problem in locating the spreadsheet. Right-click on the table in the browser and select Resolve Conflict from the pop-up menu and choose the new location of the table.

Note: Table-driven parts require Excel '95 or later.

TURN IN

TUTORIAL 6.10—USING TABLE DRIVEN PARTS AS DESIGN VARIABLES

In this tutorial, the part is complete with active part design variables. You will create a table and then link the Excel spreadsheet to the part.

1. Open the file \Md4book\Chapter6\TU6-10.dwg.
2. Issue the Design Variables command **(AMVARS)** or right-click on Part1_1 and select Design Variables from the pop-up menu.
3. Select Setup from the Table Setup area of the Design Variables dialog box and then select the Create button in the Table Driven Setup dialog box.
4. Save the spreadsheet with the following name and location: \Md4book\Chapter6\TU6-10.xls.
5. In Excel, fill in the spreadsheet as shown in Figure 6.46 and then save the file.

	A	B	C	D	E	F	G
1		Head_dia	Head_thick	Shaft_length	Shaft_dia	Hole_dia	Hole_depth
2	Part A	1	.125	2	.5	.0625	.5
3	Part B	1	0.125	3	0.5	0.0625	.5
4	Part C	1.25	0.125	4	0.625	0.125	.5
5	Part D	1.25	0.1875	5	0.625	0.125	.5
6	Part E	1.5	0.1875	6	0.75	0.125	.5
7	Part F	1.5	0.25	7	0.75	0.1875	0.625

Figure 6.46

6. Make Mechanical Desktop the active application.
7. Select Update Link from the Table Driven Setup dialog box.
8. The active version table driven design variables will fill in the Design Variable dialog box and a T will appear in the T column for each variable. Select OK to complete the command.
9. Double-click on the different versions in the browser to change the part.
10. Make the version Part B the active version.
11. Make the drawing mode current. Drawing views should update.

TUTORIAL

12. Make Excel the active application and highlight all the cells that contain information. Highlight the cells by selecting a cell in one of the corners and, while the mouse button is depressed, drag it to the opposite corner.
13. Copy this to the clipboard: either select Copy from the Edit pull-down menu, hold down CTRL + C at the same time or right-click and select Copy.
14. Make Mechanical Desktop the active application and from the Edit pull-down menu select Paste Special and then select Paste Link from the dialog box. The Microsoft Excel worksheet should be highlighted. Then select OK and the chart will appear in the drawing. Locate the pasted data until your screen resembles Figure 6.47.

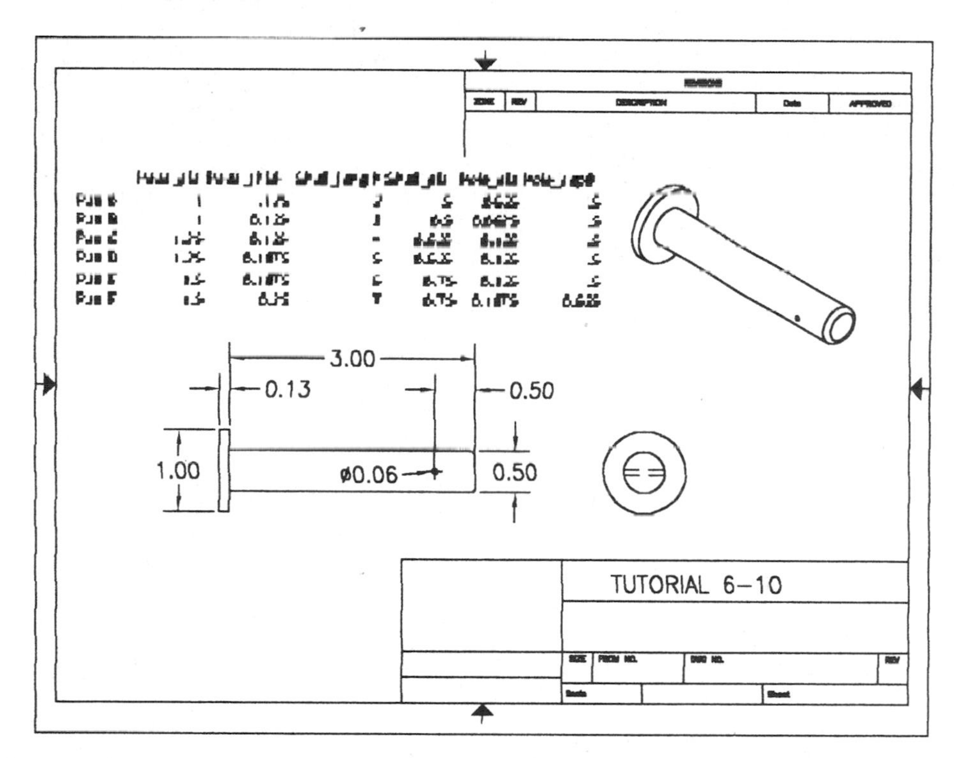

Figure 6.47

15. Make Excel the active application by double-clicking in the linked chart and change a few of the cells in the spreadsheet. Then save the file.
16. Make Mechanical Desktop active; the linked chart should automatically be updated. To update the variables, right-click on the name "Table (TU6-10.xls)" in the browser and select Update.
17. Double-click on the different versions in the browser to change the part update the part if needed.

18. To change the appearance of all the dimensions to show the design variable name do the following: under the Drawing pull down menu select Parametric Dim Display and then select Dimensions As Parameters. Type in ALL and press ENTER to update all displayed dimensions.
19. Save the file.

EXERCISES

For the following three exercises, follow the instructions at the beginning of each exercise.

Exercise 6.1—Vice Base

From scratch, create the part as shown in Figure 6.48. When creating this part, try using close edge, copy sketch, and extrude To Plane and From Hole. For clarity, the part is shown with no fillets. The _" ACME thread has a diameter of .58 and a tap diameter of .77. When the part is finished, try to create different types of fillets. Create drawing views as time permits. Use the \Md4book\Chapter6\ansi_a.dwg as the border and the drawing views should be at a scale of 1=2. When finished, save the file as \Md4book\Chapter6\ViceBase.dwg.

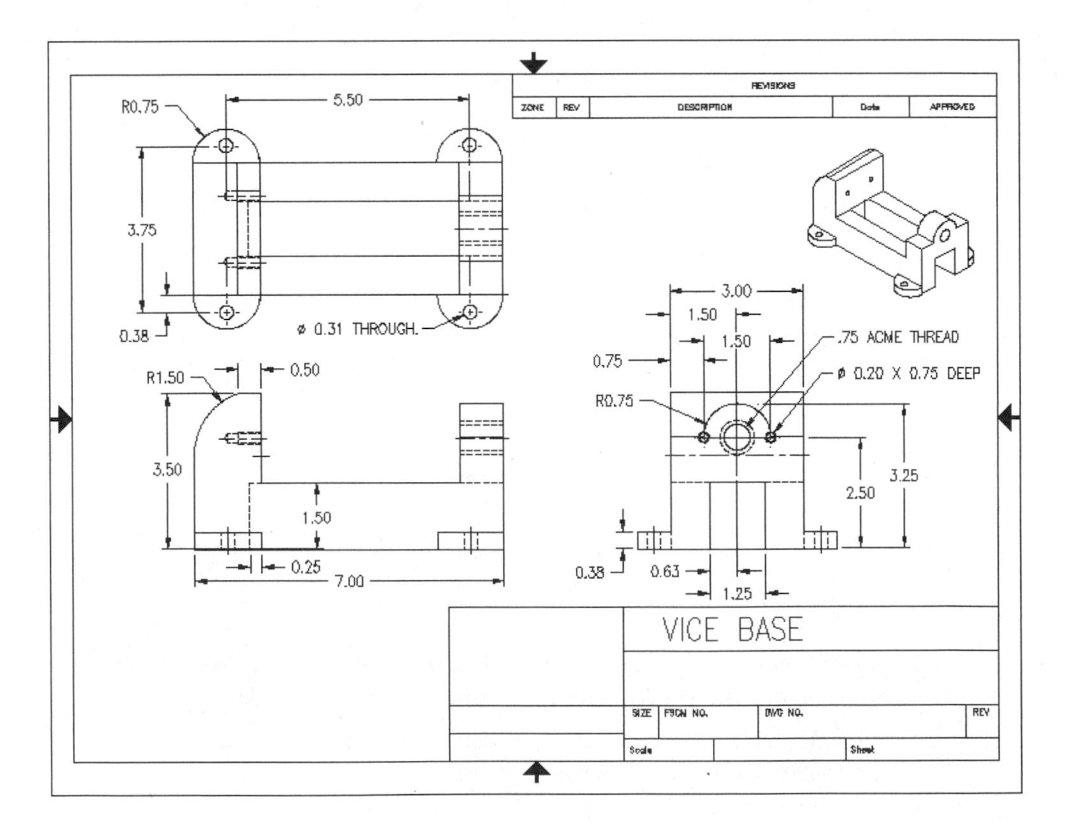

Figure 6.48

Exercise 6.2—Table Driven Angle

In this exercise, draw an angle, create a table that will drive the profile. Use the names and values as shown in the chart in Figure 6.49. Dimension the profile as shown in Figure 6.49 and extrude the profile 1". Insert the file \Md4book\Chapter6\ansi_a.dwg as the border in Layout1 and then create the full scale drawing views as shown in Figure 6.49 and link the spreadsheet data into the drawing layout. When complete, save the file as \Md4book\Chapter6\Angle.dwg.

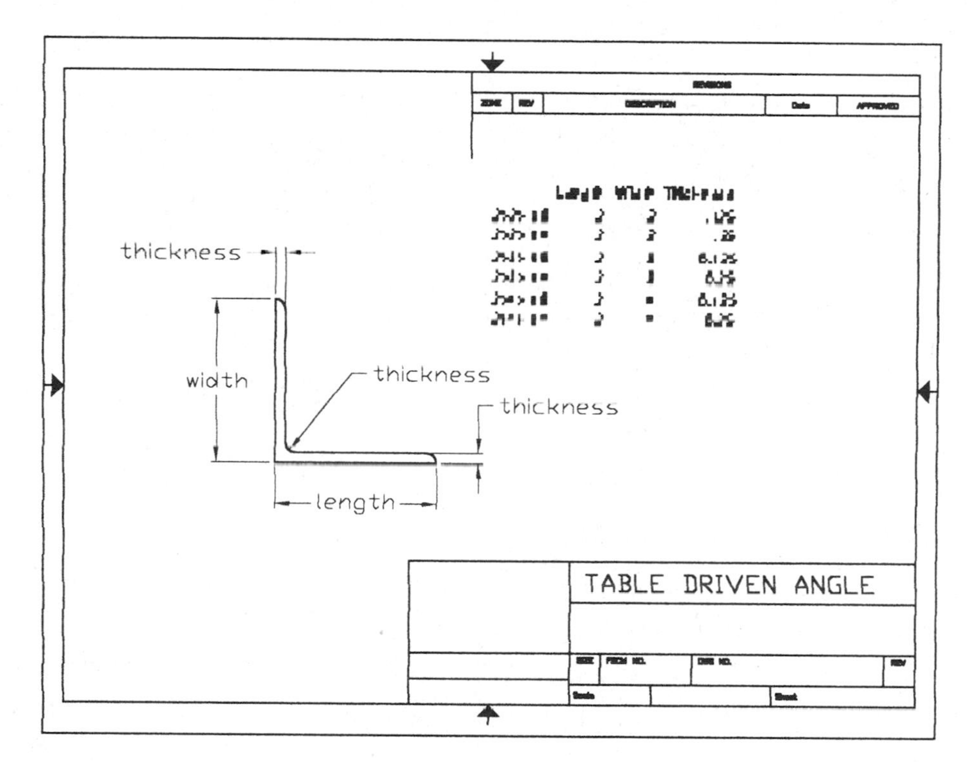

Figure 6.49

EXERCISE

Exercise 6.3—Piston

From scratch, create the part as shown in Figure 6.50. Use diameter dimensions where possible. Create drawing views as time permits. Use the file \Md4book\Chapter6\ansi_a.dwg as the border and the drawing views should be at a scale of .375. The iso views should be .5 scale relative to the parent. When finished, save the file as \Md4book\Chapter6\Piston.dwg.

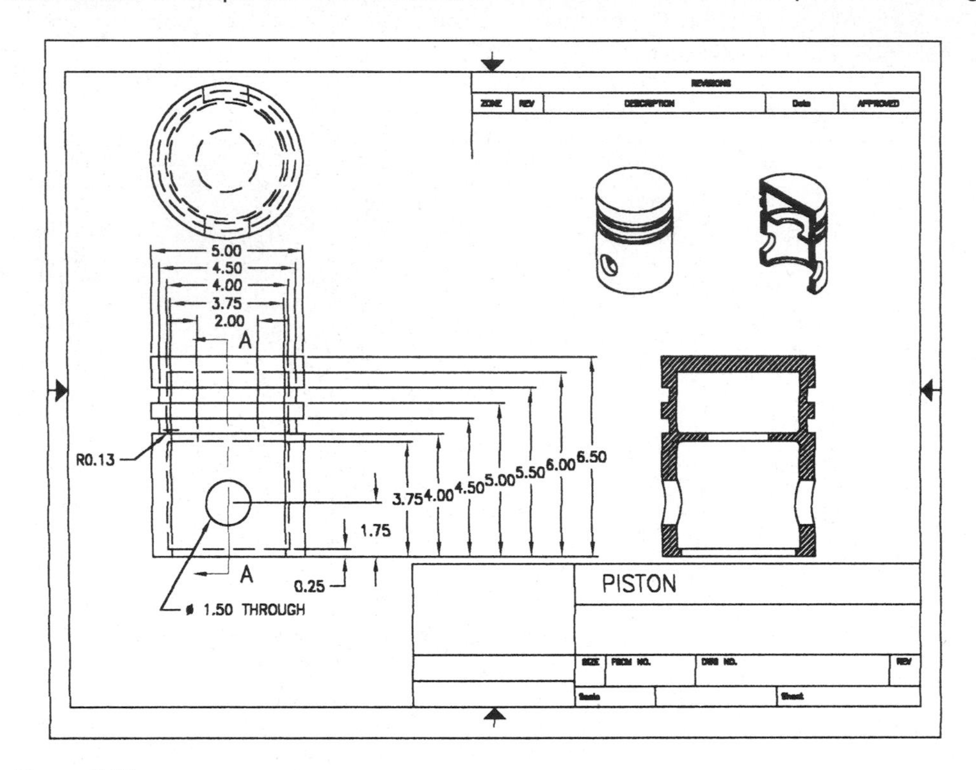

Figure 6.50

REVIEW QUESTIONS

1. When would you use construction geometry?
2. Only hidden lines can be used as construction geometry. T or F?
3. There can only be one sketch in a profile. T or F?
4. After a parametric dimension is erased, the next parametric dimension created will use the "d#" of the erased dimension. T or F?
5. Dimensions that are placed with the Automatic Dimensioning command **(AMAUTODIM)** are not parametric and need to be profiled. T or F?
6. When creating design variables, you can use any alpha numeric or numeric alpha combination you want. T or F?
7. Table-driven variables can be created with Lotus. T or F?
8. What is the difference between an active part variable and a global design variable?
9. An active part variable takes precedence over a global variable. T or F?
10. Explain what the Equation Assistant could be used for.

CHAPTER 7

Advanced Modeling Techniques

Up to now, you have learned basic modeling techniques. In this chapter you will learn advanced modeling techniques for suppressing features, scaling parts, shelling parts, adding face draft, creating multiple parts in the same file, mirroring a part, scaling a part, creating a thin-walled part using a shelling technique, copying features, creating multiple parts in the same file and combining two parts in a single part (using a technique called parametric booleans). You will also learn how to replay the steps in the creation of a part and get mass property information from a part and multiple parts.

AFTER COMPLETING THIS CHAPTER, YOU WILL BE ABLE TO:

- Suppress features.
- Scale a part.
- Create a shelled part.
- Create and understand an "instance".
- Mirror a part.
- Split a part into two parts.
- Add face draft.
- Array features.
- Copy features.
- Reorder features.
- Create multiple parts in a single file.
- Combine two parts in one part.
- Edit combined parts.
- Replay the steps in the creation of a part.
- Obtain mass property information from a part.
- Obtain mass property information from multiple parts.

FEATURE SUPPRESSION

Feature suppression adds functionality to parts creation by allowing features to be hidden in a part without deleting them. Feature suppression is like freezing the objects on a layer. Feature suppression can be used to simplify parts, which also increases system performance. Feature suppression also shows the part in different states through the manufacturing process and creates different parts and reaches faces and edges that you otherwise would not be able to reach. For example, if you wanted to dimension to the theoretical intersection of an edge that was filleted, you could suppress the fillet and add the dimension and then unsuppress the fillet. There are three ways to activate the feature suppression and unsuppression command **(AMSUPPRESSFEAT)**.

1. Through the browser, right-click on the feature that you want to suppress or unsuppress and select Suppress, Unsuppress or Unsuppress + from the pop-up menu. Figure 7.1 shows the suppress options in the pop-up menu.

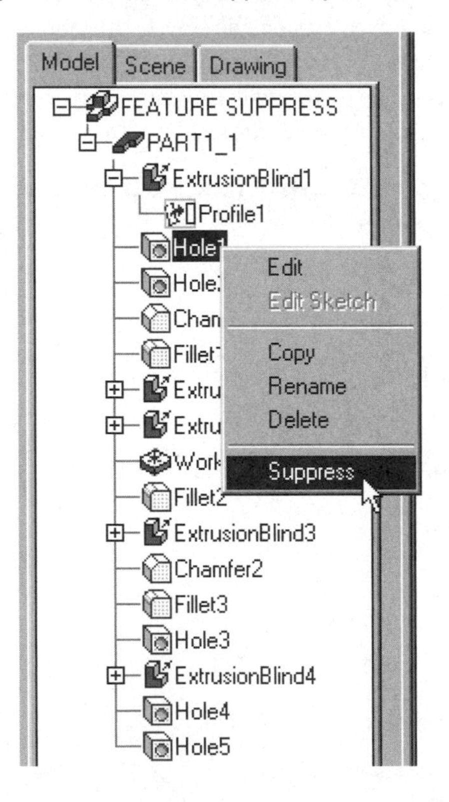

Figure 7.1

2. Through the browser, right-click on the name of the part containing features you want to suppress or unsuppress and select Suppress by Type, Unsuppress by Type or Unsuppress All, as shown in Figure 7.2. The Suppress By Type dialog

box will appear, as shown in Figure 7.3, or the Unsuppress By Type dialog box will appear. Check the type of feature(s) you want to suppress or unsuppress and then select OK. The feature types that are grayed out in the dialog box do not exist in the file.

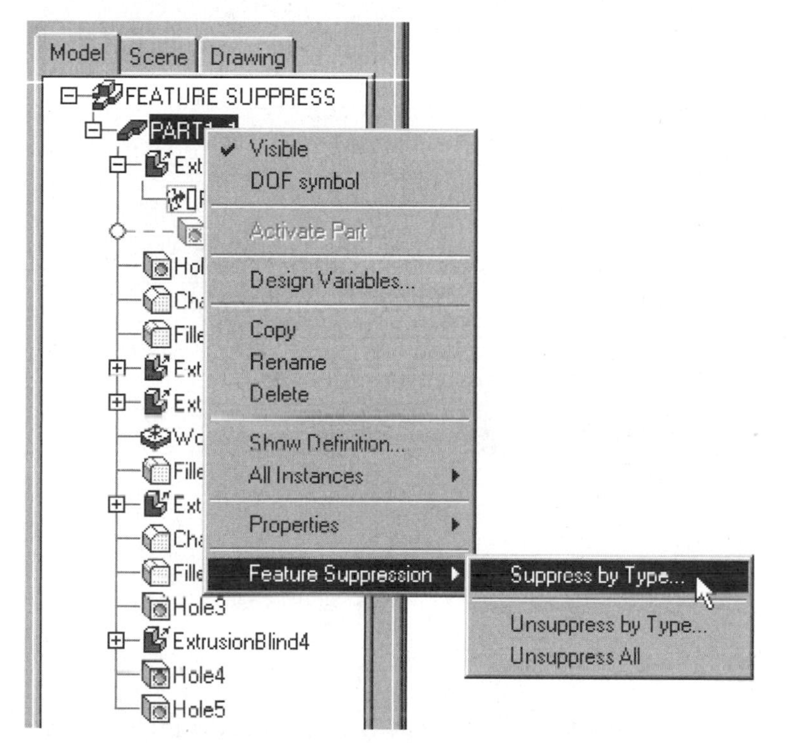

Figure 7.2

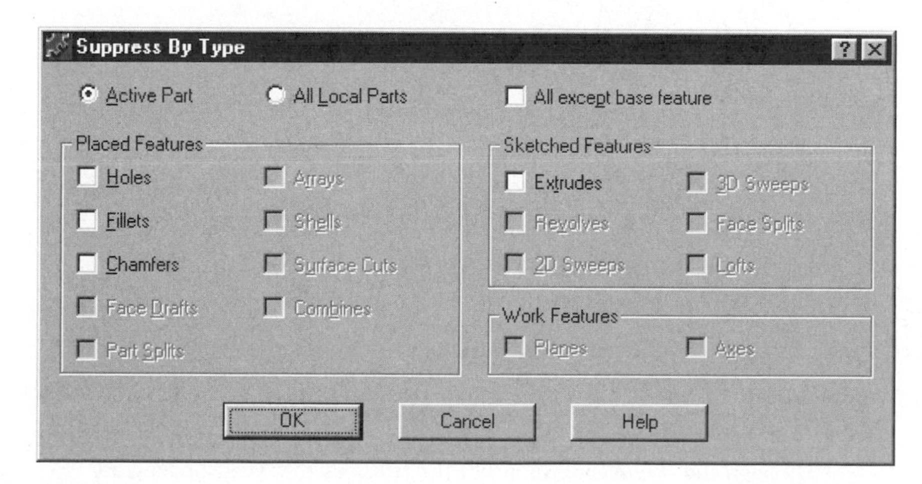

Figure 7.3

3. Select one of the icons from the Part Modeling toolbar, as shown in Figure 7.4.

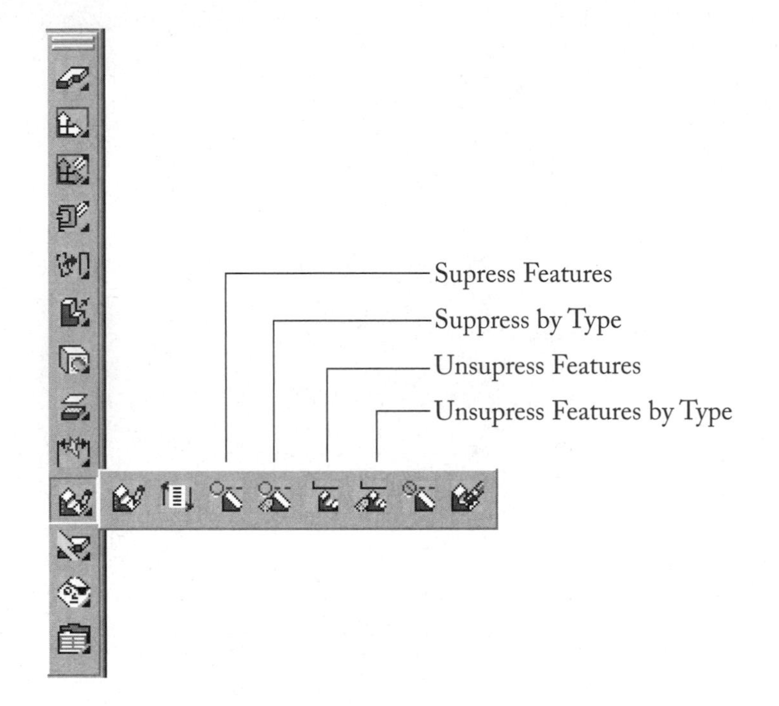

Figure 7.4

If a feature that is suppressed has a feature that is dependent upon it and you right-click on the feature in the browser, you will have two options: Unsuppress and Unsuppress +.

Unsuppress will unsuppress the selected feature.

Unsuppress + will unsuppress the selected feature and any feature that is dependent to it. For example, a hole that was created concentric to a fillet would be dependent on the fillet. When a feature is suppressed, its icon in the browser will be grayed out and a circle and hidden line will go from the browser to the feature name.

TUTORIAL 7.1—FEATURE SUPPRESSION

1. Open the file \Md4book\Chapter7\TU7-1.dwg.
2. Expand the Mechanical Desktop browser tree to see all the features.
3. Suppress a few features through the browser. Watch the icon of the selected feature change appearance.
4. Through the browser, unsuppress the features that you suppressed.

5. Select the Suppress By Type icon from the Part Modeling toolbar and suppress all the holes and fillets.
6. Select the Unsuppress by Type icon to unsuppress only the Holes by unchecking Fillets in the dialog box. This will leave the fillets suppressed.
7. Through the browser, suppress the feature "ExtrusionBlind3".
8. Through the browser, unsuppress the feature "ExtrusionBlind3" with the Unsuppress + option. The extrusion and the holes that were dependent on it reappear.
9. Select the Unsuppress Features icon from the Part Modeling toolbar and press ENTER to unsuppress all features.
10. Save the file.

SUPPRESSING FEATURES THROUGH A TABLE

In the last section, you learned how to suppress features within a file. In this section, you will learn how to suppress features through a table using Excel. Before you suppress features through a table, a part must exist that contains features. It may also have design variables, but it is not required. Feature suppression through a table can only be done for the active part. Issue the Table Driven Suppression Access command **(MNU_TABLESUPP)** from the Part Modeling toolbar, as shown in Figure 7.5, or

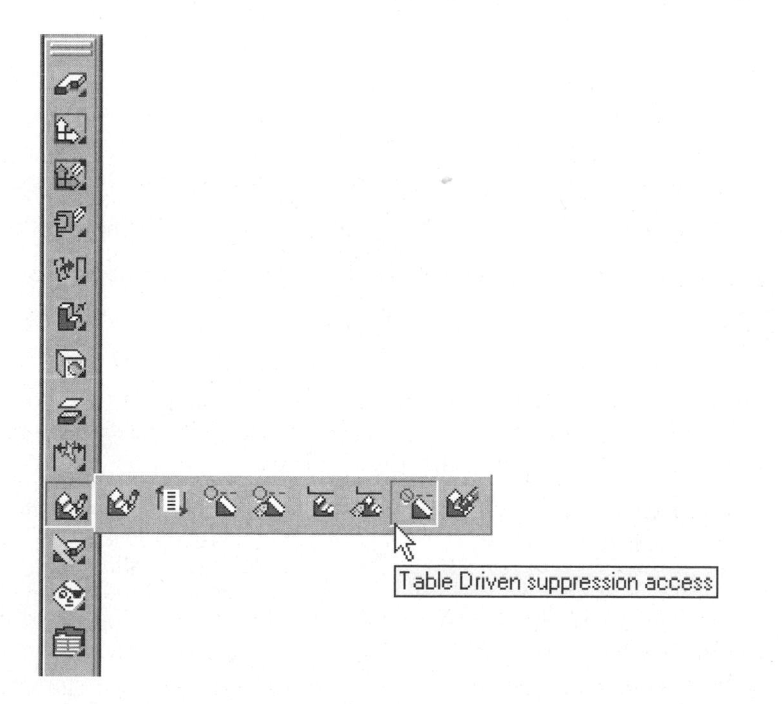

Figure 7.5

TUTORIAL

right-click in the browser and select the Design Variables command **(AMVARS)**. Make the Active Part tab current and click Setup from the Table Driven section of the dialog box. A Table Driven Setup dialog box will appear, as shown in Figure 7.6. There are two areas in the Table Driven Setup dialog box for feature suppression: Type and Format.

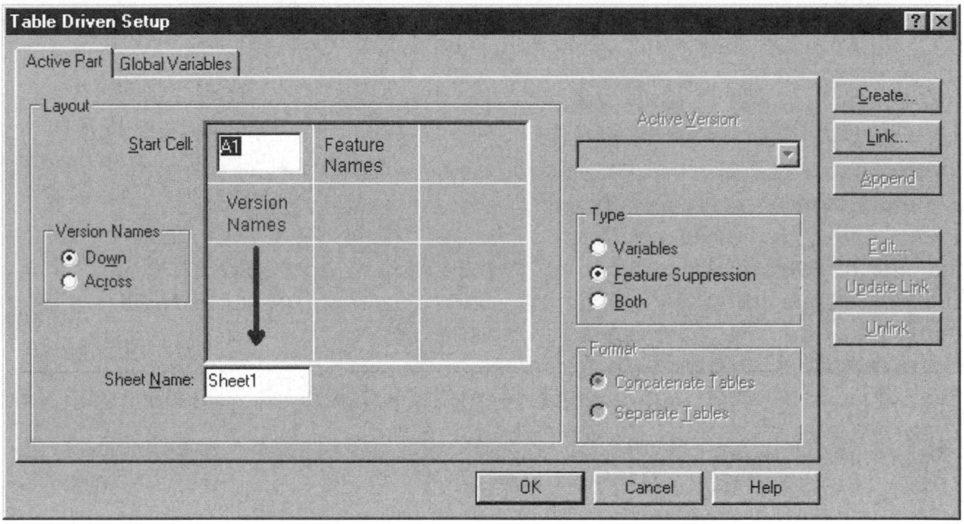

Figure 7.6

Type:

Variables: If selected, only design variables will be shown in the table.

Feature Suppression: If selected, only features will be shown in the table.

Both: If selected, both design variables and features will be shown in the table.

Format: This section will only be used if the Type is set to Both.

Concatenate Tables: If selected, both design variables and features will appear in the same sheet. A blank column or row must exist between the design variable names and the feature names.

Separate Tables: If selected, the design variables and features will appear on different sheets.

After making your choices, select Create from the upper right side of the dialog box and give the Excel file a location and name. Then to suppress a feature in the table, place any character or number in the cell that represents the feature. To unsuppress the feature through a table, remove the character or number in the cell of the feature you want to unsuppress.

Note: Mechanical Desktop looks for a blank cell for unsuppressed features. If the cell contains a character, it will be suppressed. It would be a good practice to always use "S" or "X" or a specific character of your choice for suppression via the Excel spreadsheet.

In the Mechanical Desktop browser, rename the features to better describe them. Then the feature names will be easier to identify in Excel.

Try to keep the feature names short; otherwise your columns in the spreadsheet could become very wide.

The active version name is in parentheses next to the part name.

TUTORIAL 7.2—FEATURE SUPPRESSION USING A TABLE

In this tutorial, you will suppress all the features and then unsuppress them, showing the different steps for making this part.

1. Open the file \Md4book\Chapter7\TU7-2.dwg.
2. Issue the Table Driven Suppression Access command **(MNU_TABLESUPP)**.
3. Select Feature Suppression for the Type and, if it is not already checked, check Down for the Version Names.
4. Select Create from the upper right corner of the dialog box and give the name and location of the file as \Md4book\Chapter7\TU7-2.xls.
5. In Excel, type in the values as shown in Figure 7.7.

	A	B	C	D	E	F	G	H	I	J	K
1		Cylinder	CylinderWeld	OutsideFillets	CutOut1	CutOut2	BaseHole1	BaseHole2	BaseHole3	BaseHole4	CBoreHole
2	Step1	S	S	S	S	S	S	S	S	S	S
3	Step2		S	S	S	S	S	S	S	S	S
4	Step3			S	S	S	S	S	S	S	S
5	Step4				S	S	S	S	S	S	S
6	Step5					S	S	S	S	S	S
7	Step6						S	S	S	S	S
8	Step7							S	S	S	S
9	Step8								S	S	S
10	Step9									S	S
11	Step10										S
12	Step11										

Figure 7.7

6. Save the Excel file by selecting the *X* in the upper right corner of the spreadsheet.
7. Make Mechanical Desktop the current application.
8. Select Update Link from the right side of the Table Driven Setup dialog box.
9. To complete the command, select OK twice in the dialog boxes.
10. In the browser under the Table feature, there should be Step1 through Step11,

as shown in Figure 7.8. Double-click on Step1 to make it the active part. Continue down through Step11 to see the completed part. Notice that the active version name is in parentheses next to the part name.

11. Save the file.

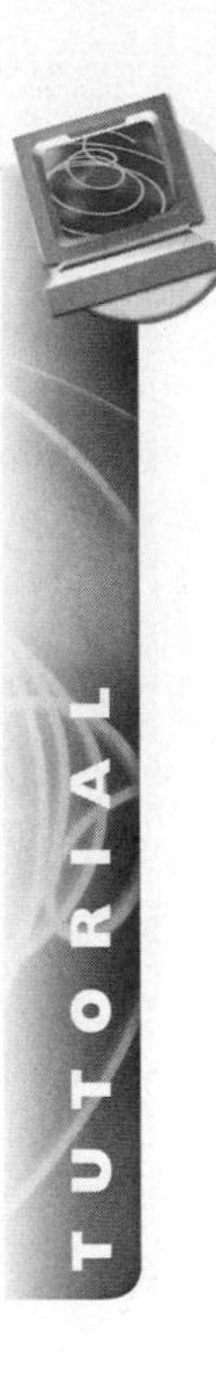

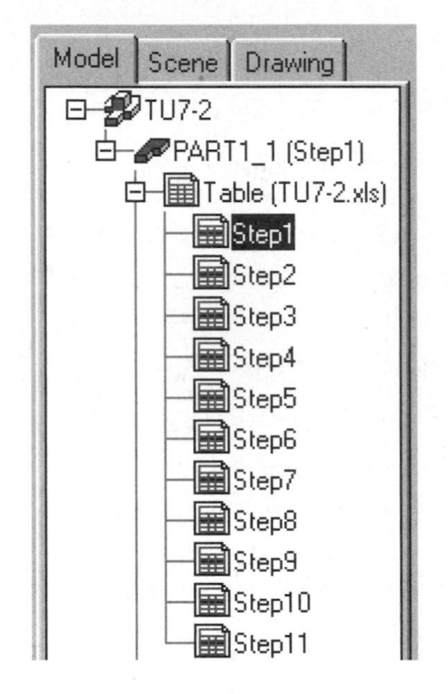

Figure 7.8

SCALING A PART

When creating parts, you may want to scale a part up or down or convert it from inches to mm or vice versa. The Scale Part command **(SCALE)** will scale a Mechanical Desktop part and maintain full parametrics. Issue the Scale Part command from the Part Modeling toolbar as shown in Figure 7.9 or right-click in the drawing area and from the pop-up menu, select Scale Part from the Part menu. After issuing the command select the part or parts that you want to scale; use any AutoCAD selection technique (pick, window, crossing etc.). When finished, press ENTER, then select a base point—you can use object snaps. The base point will be the point from which the part is scaled up or down. Then type in the scale factor to scale the part. A value greater than 1 will increase the size and any number less than 1 will decrease the part size. Each dimension will be scaled according to the scale factor. For example, a factor of 2 will double the size of the part, while .25 will decrease the part to one quarter of the original size. Angle dimensions will not be scaled. For example, a part with a 45° angle will still be 45° after scaling. Design variables will be converted to actual numbers. For example, if a design variable of length = 5 is scaled by the factor of .5, the result would be a "2.5" dimension. Under-constrained sketches can also be scaled. Table-driven parts cannot be scaled.

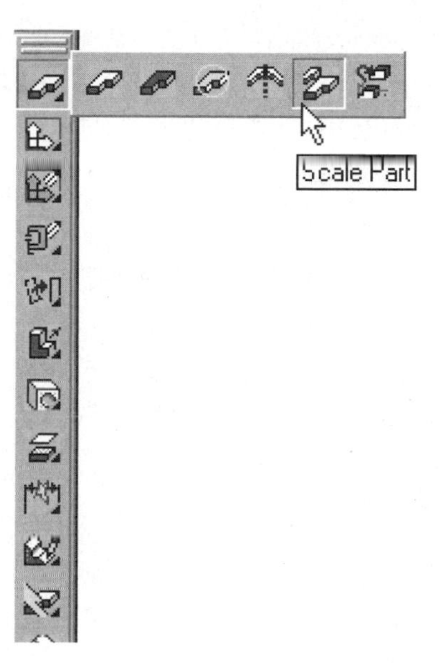

Figure 7.9

Tip: If you convert many drawings between inches and millimeters, you can set up a global variable as a multiplier and use it everywhere you need a dimension. For example, set a global variable as mm=1. Then when you create the part, use mm as a multiplier for every number in the part. For example, for a "4" inch dimension that will also be converted to millimeters, type in "4*mm". Then when you want to convert the file to millimeters, change the global variable mm from 1 to 25.4 and the entire model will be converted to millimeters.

TUTORIAL 7.3—SCALING

1. Open the file \Md4book\Chapter7\TU7-3.dwg.
2. Issue the Edit Feature command **(AMEDITFEAT)** to verify the length of the part is "6.50".
3. UNDO to return the part to its original state.
4. Issue the Scale Part command **(SCALE)**.
5. Select the part and press ENTER.
6. Select the center of the front circular edge.
7. Type in a value of ".5"and press ENTER.
8. Issue the Edit Feature command to verify the length of the part is "3.25".
9. Save the file.

TUTORIAL 7.4—SCALING A PART WITH DESIGN VARIABLES

1. Open the file \Md4book\Chapter7\TU7-4.dwg.
2. Verify that the design variables in this drawing are Length=4 and Width = 3.
3. Issue the Edit Feature command **(AMEDITFEAT)** to verify that the dimensions are driven by design variables.
4. Issue the Scale Part command **(SCALE)**.
5. Select the part and press ENTER.
6. Select a point near the center of the part.
7. Type in a value of ".5"and press ENTER.
8. Issue the Edit Feature command to verify the design variables have been replaced with a "2" and "1.50" dimension.
9. Save the file.

SHELLING

While creating parts, you may need to create a thin walled part. The easiest way to create a thin walled part is to create the main shape and then use the shell command to remove material from the part. The term shell refers to giving the outside shape of a part a thickness (wall thickness) and removing the remaining material on the inside, like scooping out the inside of a part, leaving the walls a specified thickness. A part can only be shelled once, but individual faces of the part can have different thicknesses. If a wall has a different thickness than the shell thickness, this is referred to as a thickness override. If a face that you select for a thickness override has faces that are tangent to it, those faces will also have the same thickness override. Faces can be excluded from being shelled and these faces will be left opened. Once a part has been created that you want to shell, issue the Shell command **(AMSHELL)** from the Part Modeling toolbar as shown in Figure 7.10. Or you may right-click in the drawing area and from the pop-up menu, select Shell from the Placed Features menu and the shell feature dialog box will appear as shown in Figure 7.11. The Shell Feature dialog box contains the following sections: Default Thickness, Excluded Faces and Multiple Thickness Overrides.

Default Thickness:

Inside: Offsets the wall thickness by the given value into the part.

Outside: Offsets the wall thickness by the given value out of the part.

Mid-plane: Offsets the wall thickness evenly into and out of the part by the given value. A mid-plane offset cannot have overrides.

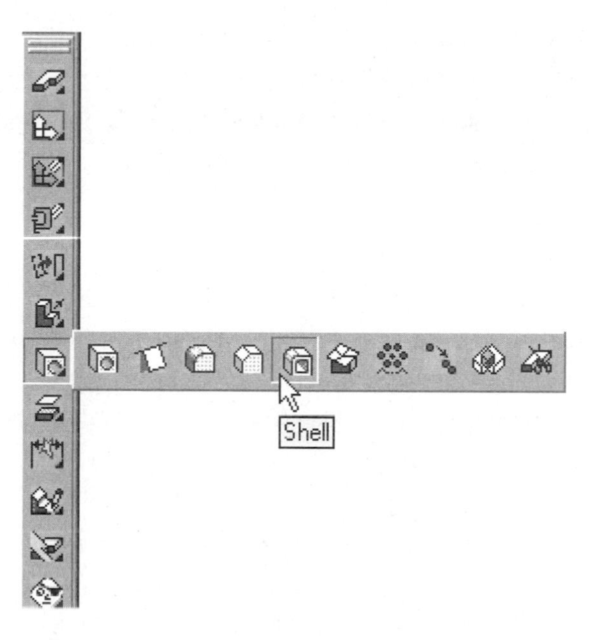

Figure 7.10

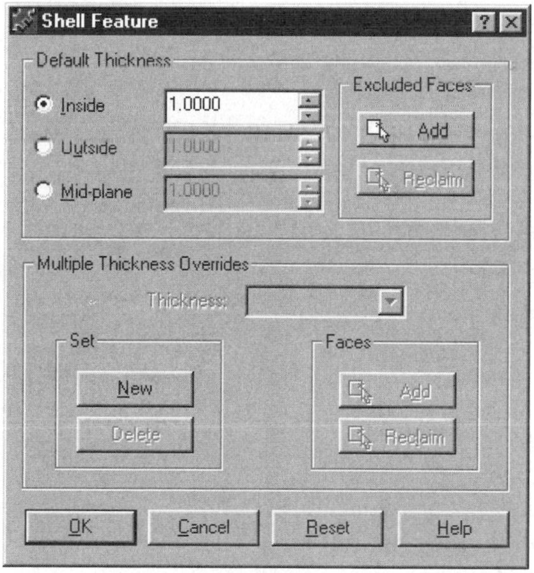

Shell Feature
Default Thickness
Inside 1.0000
Outside 1.0000
Mid-plane 1.0000
Excluded Faces
Add
Reclaim
Multiple Thickness Overrides
Thickness:
Set
New
Delete
Faces
Add
Reclaim
OK Cancel Reset Help

Figure 7.11

Excluded Faces:

Add: The selected face will be left open.

Reclaim: The selected open face will be shelled or remained closed.

Multiple Thickness Overrides:

Thickness: The override value of the offset. You can have multiple overrides for a single part applied to different faces.

Set:

New: This is where you set the thickness override, then use Add to apply it to the selected face.

Delete: The selected overridden face will be set back to the default thickness.

Faces:

Add: The selected face(s) will be given the override value for the thickness.

Reclaim: The faces that were overridden with the current override will be highlighted. Then selecting any of the highlighted faces will return them to the default thickness.

After filling in the dialog box, select OK and the part will be shelled. To edit a shell, issue the Edit Feature command **(AMEDITFEAT)** or double-click on the shell in the browser. When you edit a shell, the same dialog box will appear as when the shell was created. You can change the default thickness here and add or reclaim faces, as well as add or delete any of the thickness overrides that have been set. When editing a shell, follow the same steps that you would to create the shell.

TUTORIAL 7.5—SHELLING AND EDITING A SHELL

1. Open the file \Md4book\Chapter7\TU7-5.dwg.
2. Issue the Shell command **(AMSHELL)**.
3. Type in a value of ".03" for the Inside value and select OK. The inside of the box should be totally shelled out and resemble Figure 7.12.
4. Edit the shell with the Edit Feature command **(AMEDITFEAT)** or from the browser, double-click on Shell1. From the dialog box, select Add from the Excluded Faces section and select the top face, as shown in Figure 7.13. Press ENTER to return to the Shell Feature dialog box.
5. Change the default thickness to ".1" and then select OK to complete the command.
6. If the part did not automatically update, update it now. When complete, your part should resemble Figure 7.14, shown with lines hidden.

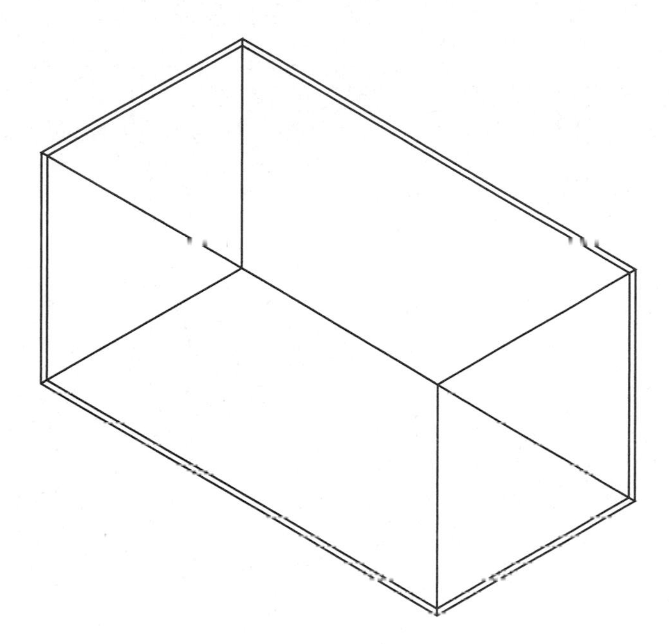

Figure 7.12

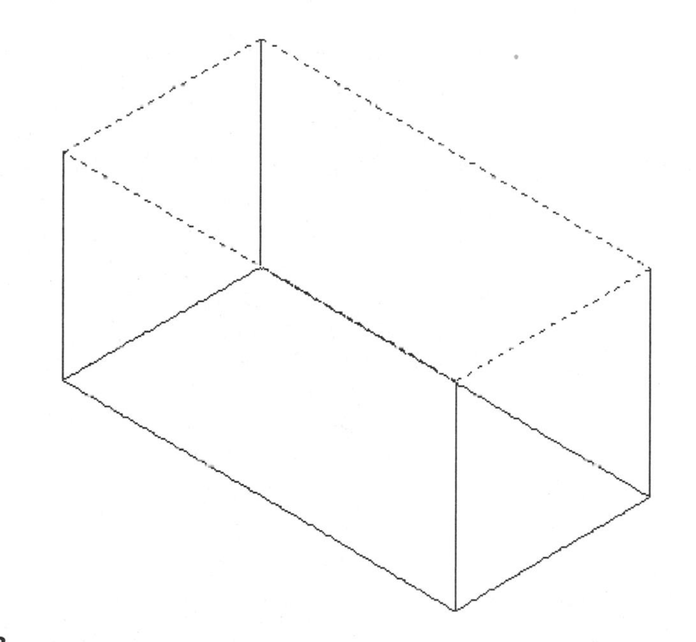

Figure 7.13

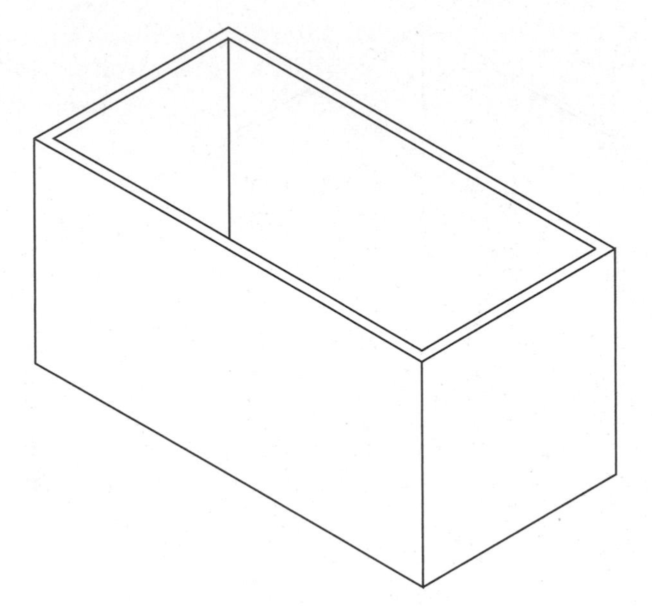

Figure 7.14

7. Edit the shell, select New from the Multiple Thickness Overrides section and type in a thickness of ".2".
8. Select the Add button under Faces, select the front left face, as shown in Figure 7.15, and press ENTER twice to return to the Shell Feature dialog box.

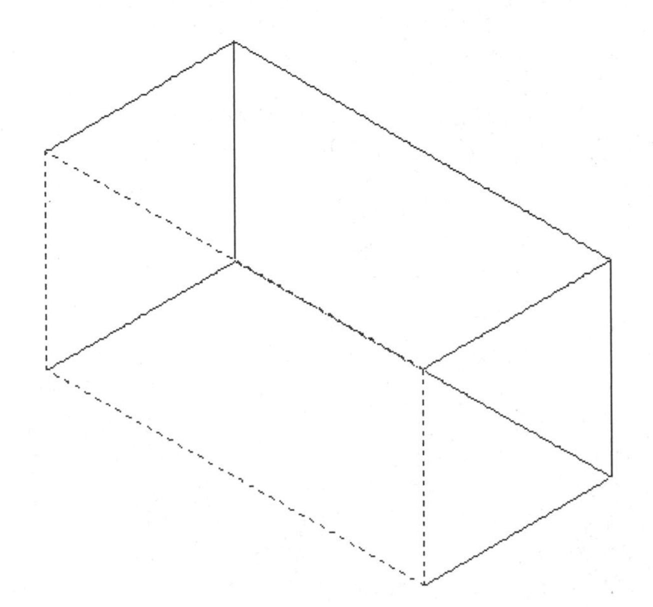

Figure 7.15

9. Select OK to complete the command. If the part did not automatically update, update it now. When complete, your part should resemble Figure 7.16, shown with lines hidden.

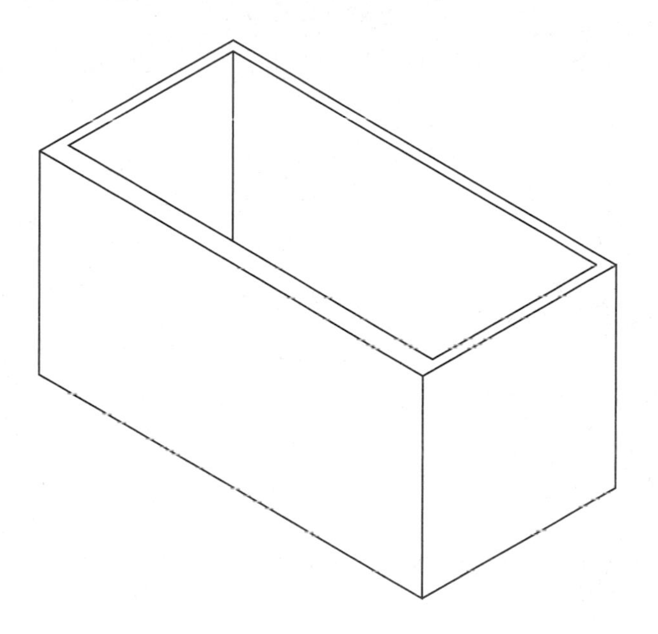

Figure 7.16

10. Edit the shell and Select Reclaim from the Faces section near the bottom right of the dialog box.
11. Select the same face that you did in step 8 and the highlight will disappear. Press ENTER to return to the Shell Feature dialog box.
12. Select OK to complete the command and update the part if necessary. When complete, your part should again resemble Figure 7.14.
13. Save the file.

TUTORIAL 7.6—SHELLING

1. Open the file \Md4book\Chapter7\TU7-6.dwg.
2. Issue the Shell command **(AMSHELL)**.
3. Give a value of ".06" for an Outside value.
4. Select Add from the Excluded Faces section and select the top and bottom faces of the revolution, as shown in Figure 7.17. Select the top face and press ENTER (you may need to cycle to the correct face), then select the bottom face (you may need to cycle to the correct face). Press ENTER twice to return to the Shell Feature dialog box.

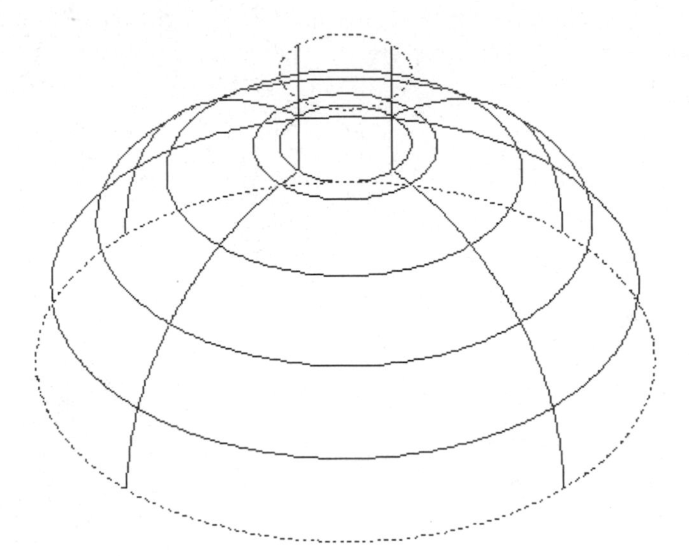

Figure 7.17

5. Select New from the Multiple Thickness Overrides section and type in a value of ".125".
6. Select the Add button under the Faces section and select the top cylinder as shown in Figure 7.18. Press ENTER to return to the Shell Feature dialog box and then select OK to complete the command.

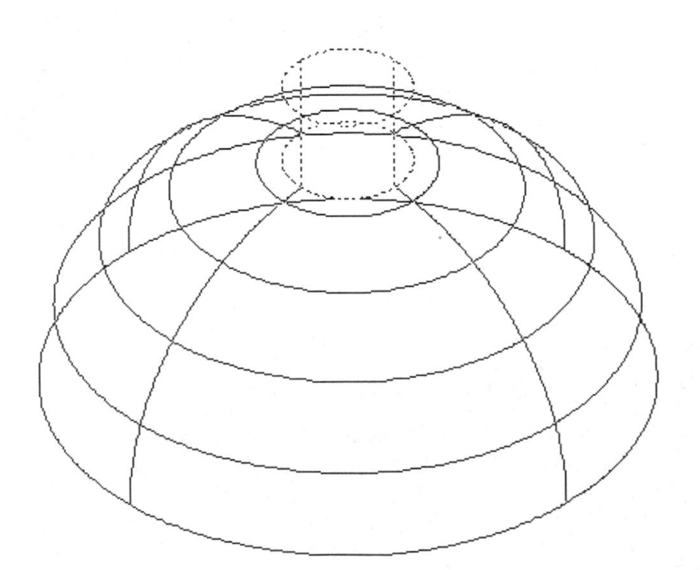

Figure 7.18

7. Update the part if necessary. When finished, your part should resemble Figure 7.19.
8. To better see the part, shade and rotate it.
9. Save the file.

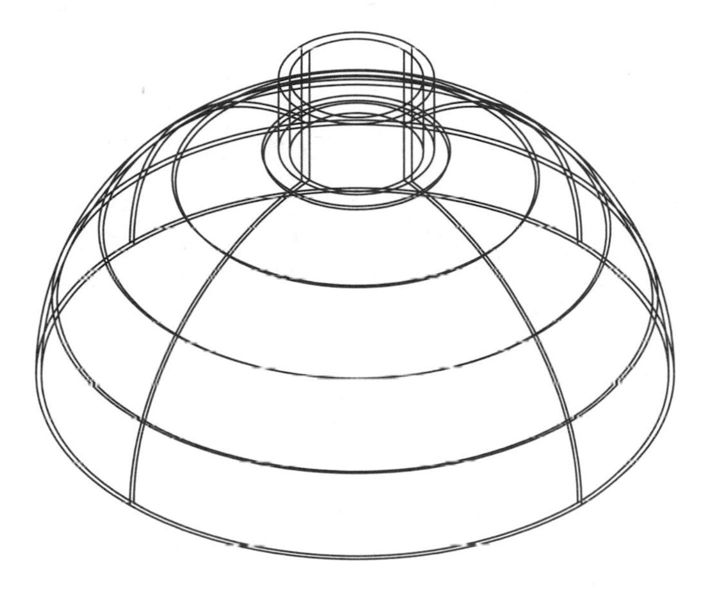

Figure 7.19

SPLIT LINE

If you want to split a part into two or split a face into two faces, one method to use is the split line. A split line is a sketched feature. A split line can be projected through a part, cutting it in two. A split line can also be projected onto a part, splitting a set of faces. An edge is created at the point the split faces touch. This edge can be used to create a face draft. Face draft is used frequently in manufacturing to remove a part from its cast or mold.

To create a split line, first create a part and then draw a sketch a line on the current sketch plane. The sketched split line does not need to be closed; you can use construction geometry to constrain it, but it does not need to be fully constrained. Once the split line has been sketched, issue the Split Line command **(AMSPLITLINE)** from the Part Modeling toolbar, as shown in Figure 7.20 or right-click in the drawing area and from the pop-up menu, select Split Line from the Sketch Solving menu. Next, select the split line and constrain it the same way you would any other profile. The split line will then be used to split a part or split a set of faces. Splitting faces will be covered next and splitting of parts will be covered later in the chapter.

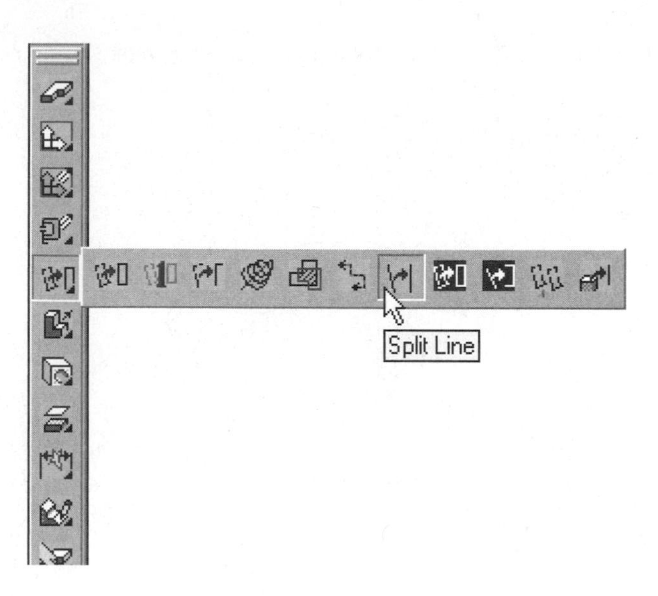

Figure 7.20

FACE SPLIT

Face split refers to splitting a set of faces into different faces. After the faces are split, the part will be maintained as a single part but the selected faces will be split. The face split is a feature and can be edited or deleted like any feature. There are two ways to split faces on a part: intersect the active part with a work plane or project a sketch onto a selected set of faces, creating a split line. Issue the Face Split command **(AMFACE-SPLIT)** from the Part Modeling toolbar, as shown in Figure 7.21 or right-click in the drawing area and from the pop-up menu, select Face Split from the Sketched & Work Features menu. After issuing the command, select from one of two options at the command line: Planar and Project.

Planar: Used to create a face split at the intersecting location of a work plane and the active part.

Project: A split line must exist before this option is used. The split line will be projected onto a set of selected faces, splitting them.

After making the selection press ENTER and you will then be prompted:

```
Select faces to split or [All]:
```

Press A and ENTER to project the split line on all of the faces, or select individual faces. After selecting the faces to split, press ENTER to complete the command or you may also remove faces that you do not wish to be included by type R and press ENTER and then select the faces to remove from the selection set. This new edge can be used to create a face draft.

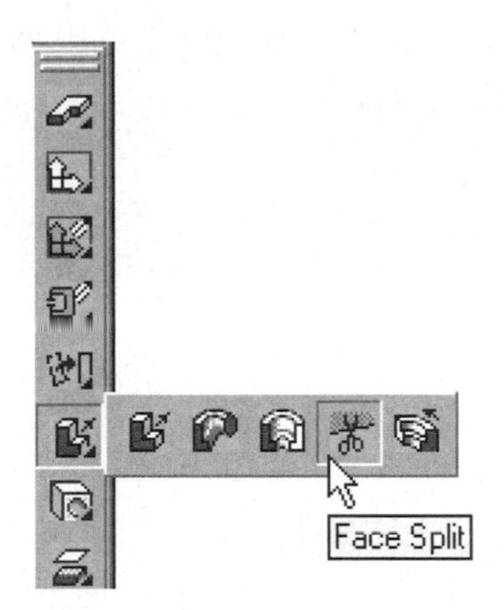

Figure 7.21

TUTORIAL 7.7—SPLIT LINE AND FACE SPLIT

1. Open the file \Md4book\Chapter7\TU7-7.dwg.
2. Draw in an angled line as shown in Figure 7.22. The front vertical face is already the current sketch plane. The line is shown with thickness and lines hidden for clarity.
3. Issue the Split Line command **(AMSPLITLINE)** and select the angled line. Press ENTER twice to complete the command.
4. Dimension the angled line, as shown in Figure 7.22. Add Project constraint to the ends of the split line and the outside edges of the part. Note: If the Split Line was sketched within 4° of horizontal you will need to delete the horizontal constraint before adding both dimensions.

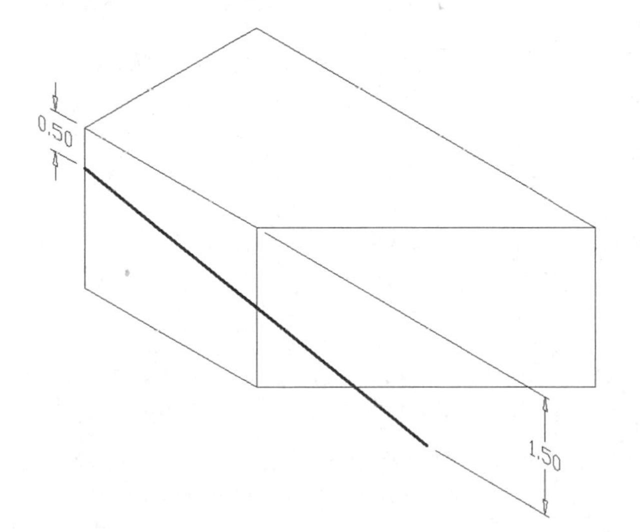

Figure 7.22

5. Issue the Face Split command **(AMFACESPLIT)**.
6. Press ENTER to accept the Project option and project the angled line onto the part.
7. Press A and ENTER twice to select all of the faces and complete the command. When finished, your part should resemble Figure 7.23.
8. Save the file. The file will used to apply face draft in the next section.

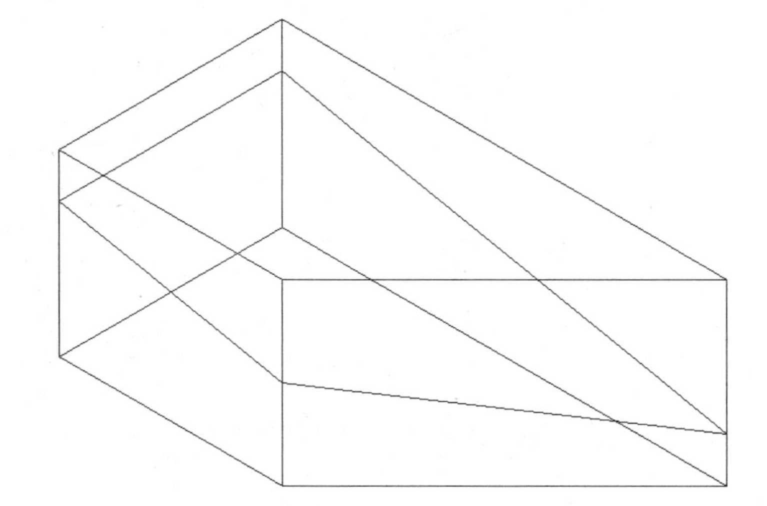

Figure 7.23

FACE DRAFT

Face draft is a feature that allows you to angle any face. Face draft can be applied to shelled parts or any specified face. A part edge or a split line can be used for the edge from which to base the face draft. To apply face draft to an existing face on a part, issue the Face Draft command **(AMFACEDRAFT)** from the Part Modeling toolbar, as shown in Figure 7.24 or right-click in the drawing area and from the pop-up menu, select Face Draft from the Placed Features menu. A Face Draft dialog box will appear as shown in Figure 7.25. There are four sections in the Face Draft dialog box.

Type:

From Plane: The selected plane will be where the draft angle starts from or it will apply the draft measuring the angle from the height of the draft plane.

From Edge: The selected edge or split line will be where the draft angle starts from. It allows the selected edge to remain fixed as the faces are defined for the draft.

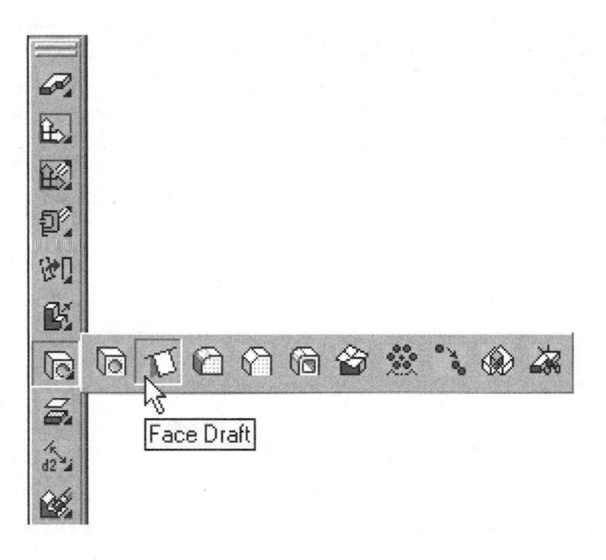

Figure 7.24

Face Draft
Type From Plane
Angle 1
Draft Plane
Faces to Draft
Add
Reclaim
Include Tangencies
OK Cancel Help <<
Return to Dialog

Figure 7.25

Shadow: Allows draft to be applied where a tangency condition needs to be maintained. For example, you might select a fillet to apply shadow draft to a face that is tangent to the fillet and the face will move and the fillet will remain tangent. The face being selected for shadow draft must be a ruled surface, that is, a fillet.

Angle: The amount of draft must stay within the range of 0 and 80. No negative values are allowed. The direction of the draft will be determined by the direction of the arrow when the draft plane is selected.

Draft Plane: This is the plane that determines the starting point of the draft angle and it should be perpendicular to the faces that need to be drafted. The edges that define the draft plane will not move. The opposite side(s) of the faces where the draft is being applied will move to form the draft. When a face is selected, you can left-click to cycle through the faces and then press ENTER when the correct face is highlighted. Then a cone with an arrow icon will appear; the direction of the arrow determines if the draft will go in or out. The cone can also help show the direction of the draft, because it is sitting in the middle of the selected plane. The draft will follow the direction of the cone and will follow the angle of the cone from that plane. You can left-click to flip the cone and arrow and then press ENTER when the direction is correct. If the arrow is going out from the part, the draft will go outward, adding material. If the arrow is pointing into the part, the draft will go inward, removing material.

Faces to Draft:

Add: Select the face or faces to apply the draft. Left-click to cycle through the faces until the correct face is highlighted. Then select another face or press ENTER to return to the dialog box. If multiple faces are selected in this manner, they will appear as a FaceDraft feature in the browser.

Reclaim: Allows you to remove a face that was selected for a draft angle to be applied or to remove a face that has a draft angle applied to it.

Include Tangencies: When checked, any faces that are tangent to the selected face will also be drafted, such as fillets.

TUTORIAL 7.8—DRAFT ANGLE, FROM PLANE

1. Start a new drawing from scratch.
2. Sketch, profile and dimension a 4" square. Extrude it 2" in the default direction.
3. Change to an isometric view using the **8** key.
4. Issue the Face Draft command **(AMFACEDRAFT)**.
5. In the Face Draft dialog box, if From Plane is not the current Type, select it. Change the Angle to "5" and then select Draft Plane.
6. Select the top plane of the rectangle and then press ENTER. Press ENTER again to accept the default direction of the arrow, as shown in Figure 7.26 with lines hidden for clarity.
7. Select Add from the Face Draft dialog box and then select the right vertical face, as shown in Figure 7.27 with lines hidden for clarity. Press ENTER to return to the Face Draft dialog box and then select OK to complete the command.
8. Repeat the Face Draft command **(AMFACEDRAFT)**. In the Face Draft dialog box change the Angle to "20" and then select Draft Plane.

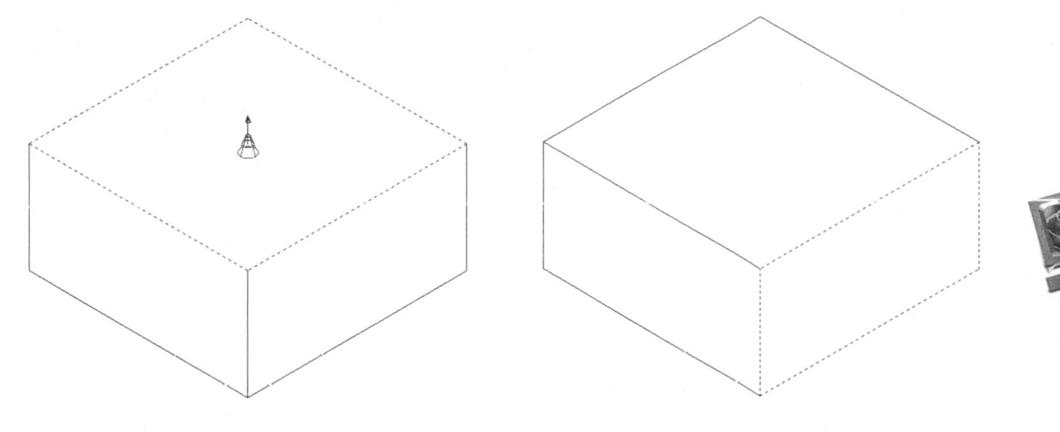

Figure 7.26  **Figure 7.27**

9. Select the top plane of the rectangle and then press ENTER. Flip the direction of the arrow so that it is pointing into the part, as shown in Figure 7.28 (with lines hidden for clarity) and then press ENTER to return to the Face Draft dialog box.

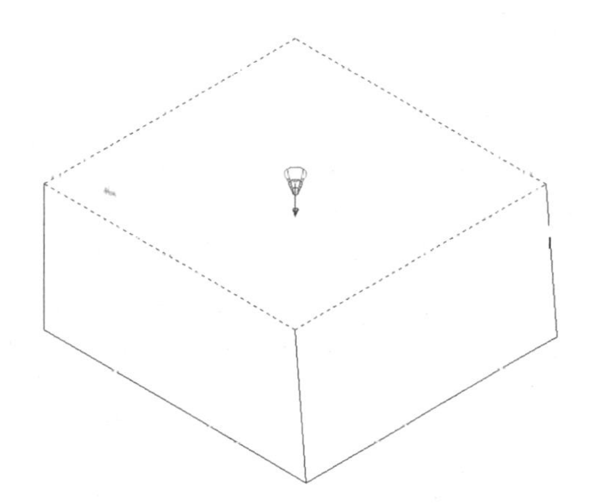

Figure 7.28

10. Select Add from the face Draft dialog box and then select the three vertical faces as shown in Figure 7.29. Press ENTER to return to the dialog box and then select OK to complete the command. When complete, your part should resemble Figure 7.30.

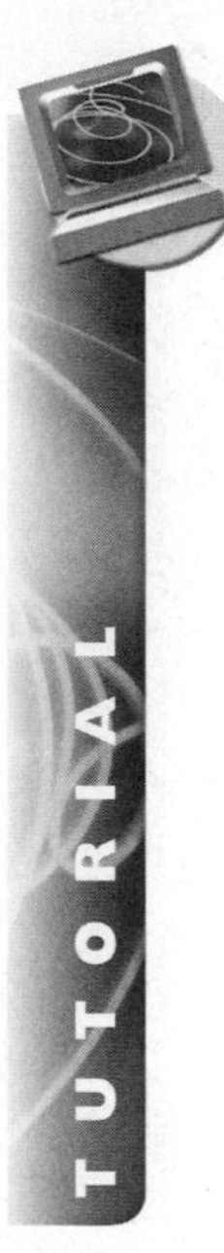

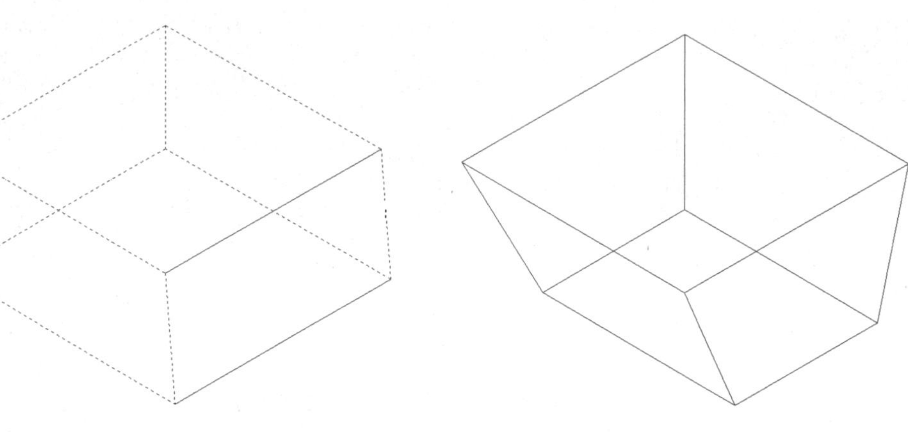

Figure 7.29 Figure 7.30

11. Shade and rotate the part to better see the first face drafting outward and the other three faces drafting inward.
12. Edit the face drafts with different angles and try changing the arrow direction.
13. When finished, save the file as \Md4book\Chapter7\TU7-8.dwg.

TUTORIAL 7.9—DRAFT ANGLE ON A SHELLED PART USING FROM PLANE

1. Open the file \Md4book\Chapter7\TU7-9.dwg.
2. Issue the Face Draft command **(AMFACEDRAFT)**.
3. In the Face Draft dialog box, if From Plane is not the current Type, select it. Change the Angle to "5" and then select Draft Plane.
4. Select the top plane of the shell, as shown in Figure 7.31. Press ENTER twice to accept this plane and the direction of the arrow out away from the part.

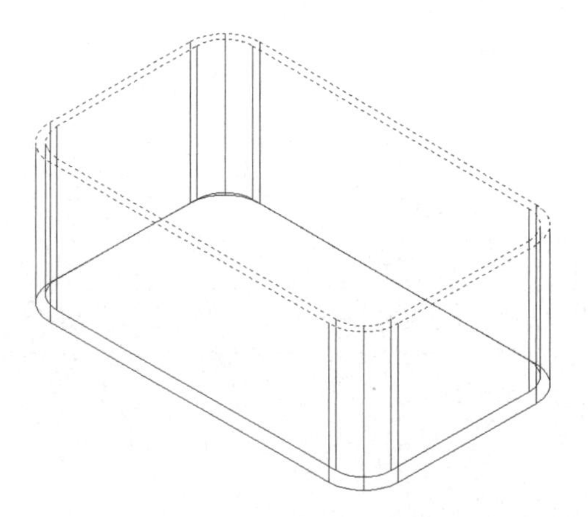

Figure 7.31

5. In the Face Draft dialog box, check Include Tangencies, if it is not already checked, and then select Add. Select one of the inside vertical faces; they will all highlight, as shown in Figure 7.32. Press ENTER twice to accept these faces. Back in the Face Draft dialog box, select OK to complete the command. When finished, your part should resemble Figure 7.33.

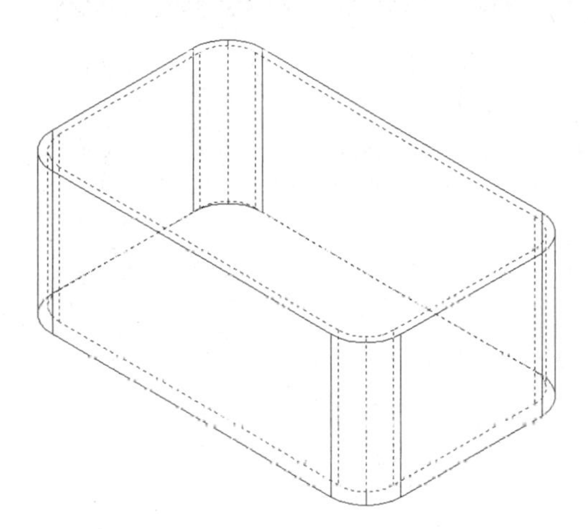

Figure 7.32

6. Shade and rotate the part to better see the inward draft of the part.
7. Save the file.

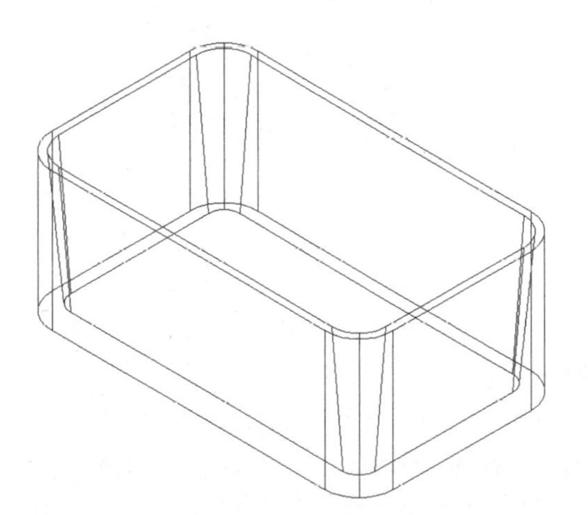

Figure 7.33

TUTORIAL

TUTORIAL 7.10—DRAFT ANGLE USING A SPLIT LINE

For this tutorial, you will use the file from Tutorial 7.7. If you did not save the file go back and complete it now.

1. Open the file \Md4book\Chapter7\TU7-7.dwg.
2. Issue the Face Draft command **(AMFACEDRAFT)**.
3. In the Face Draft dialog box, make From Edge the current Type, change the Angle to "5" and then select Draft Plane.
4. Select the top horizontal plane of the part, as shown in Figure 7.34. Press ENTER twice to accept this plane and the direction of the arrow out away from the part.

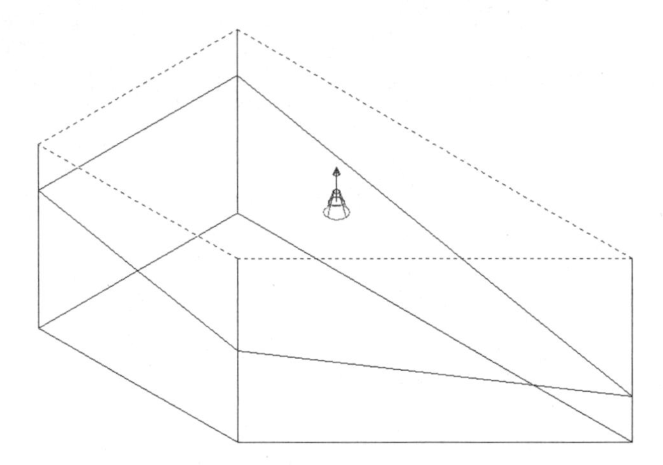

Figure 7.34

5. Back in the Face Draft dialog box, select Add and then select all of the top vertical faces, as shown in Figure 7.35. Press ENTER to accept these faces.
6. For the prompt:
   `Select fixed edge:`
   Select the split line that was projected onto the part in Tutorial 7.7 and then press ENTER.
7. In the Face Draft dialog box, select OK to complete the command.
8. Repeat the Face Draft command (AMFACEDRAFT).
9. In the Face Draft dialog box, make From Edge the current Type, change the Angle to "5" and then select Draft Plane.
10. Select the bottom horizontal plane of the part, as shown in Figure 7.36. Press ENTER twice to accept this plane and the direction of the arrow down and away from the part.

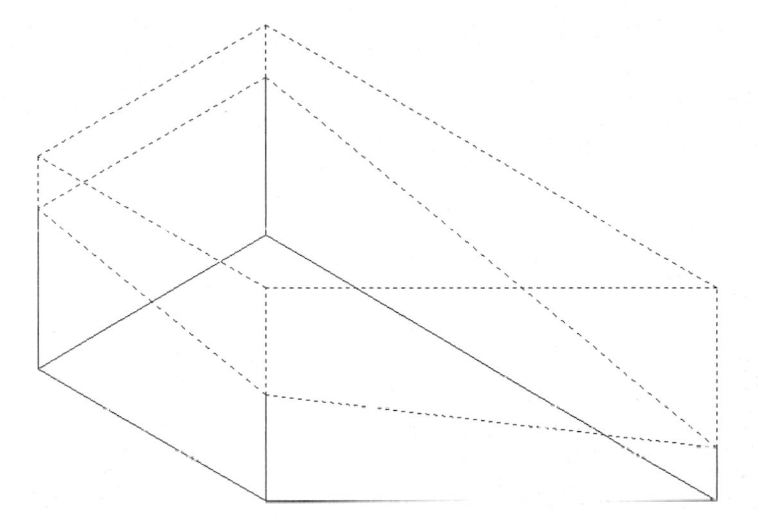

Figure 7.35

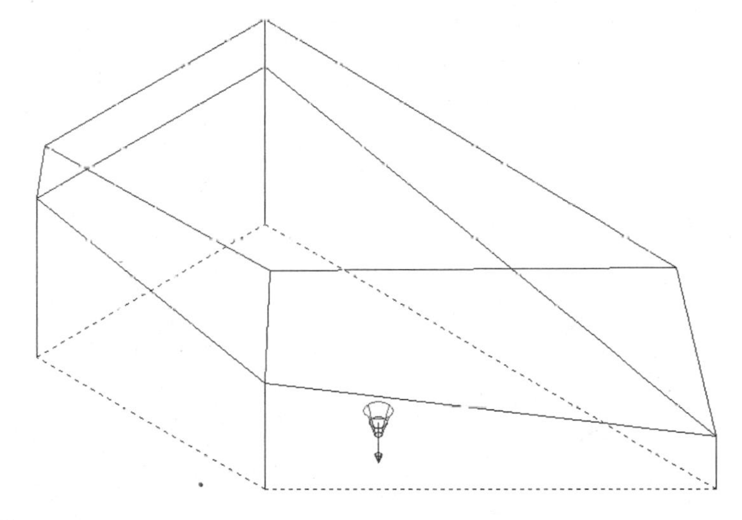

Figure 7.36

11. Back in the Face Draft dialog box, select Add and then select all of the bottom vertical faces, as shown in Figure 7.37. Press ENTER to accept these faces.
12. At the prompt:
    `Select fixed edge:`
    Select the split line that was projected onto the part in Tutorial 7.7 and then press ENTER.
13. In the Face Draft dialog box, select OK to complete the command. When done your part should resemble Figure 7.38.
14. Save the file.

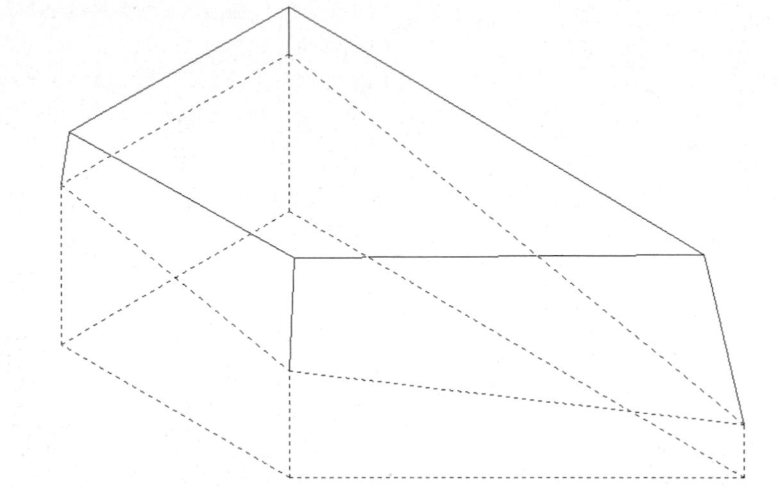

Figure 7.37

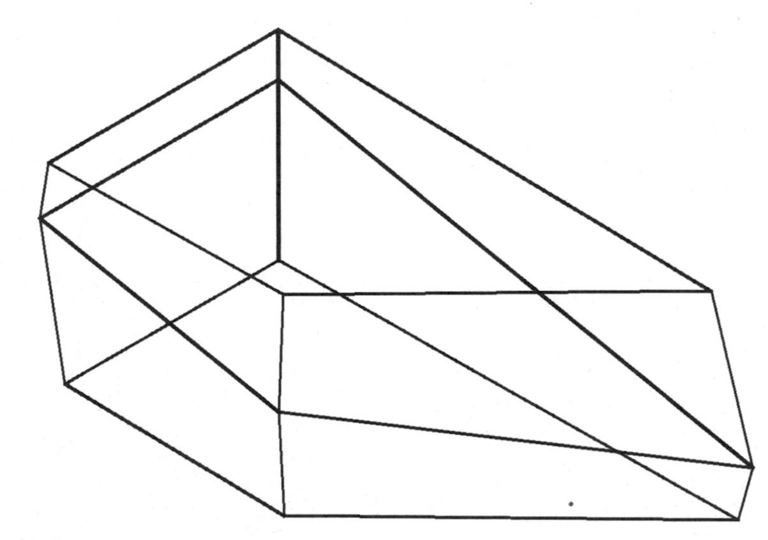

Figure 7.38

ARRAYS

In Mechanical Desktop, there are two type of arrays: rectangular and polar. Before creating an array make the plane the active sketch plane upon which the array will be based. Before creating a polar array, you must have a work point, work axis or cylindrical edge about which the feature will be rotated. Both the rectangular and polar arrays are accessed from the same command, Feature Array **(AMARRAY)**, found on the Part Modeling toolbar as shown in Figure 7.39. Or you can right-click in the drawing area and from the pop-up menu, select Array from the Placed Features menu. After issuing the command, you are prompted to select a feature to array. After you select a feature, an Array dialog box will appear, as shown in Figure 7.40. If you move the dialog box away from the part, you will see an image that shows the direction for rows, columns and the positive angle for the selected feature. In the Array dialog box, select either the rectangular or polar tab.

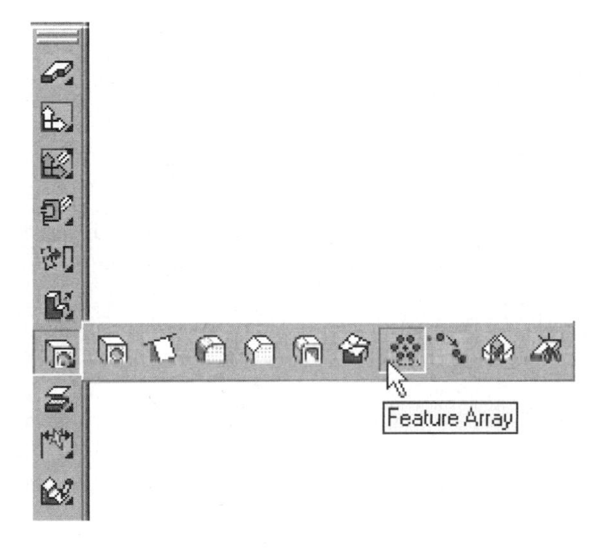

Figure 7.39

Figure 7.40

RECTANGULAR ARRAYS

Under the rectangular tab, you will input the number of rows, columns and the spacing between each. The spacing can be a positive or a negative value. If negative, the features to be arrayed in the negative direction.

POLAR ARRAYS

Under the Polar tab, you have the following options:

Number of Instances: Type in the number of features that the result of the array will equal. The selected feature counts as one feature.

Full Circle: Type in a value for the Number of Instances and that number will be evenly spaced around 360°.

Included Angle: Type in a value for the Number of Instances and the Angle. The number of instances will be evenly spaced within the angle specified.

Incremental Angle: Type in a value for the Number of Instances and the Angle. Each instance will be separated by the number of degrees specified by the value of the angle.

Rotate as copied: When checked, the Rotate as Copied option allows the features to be rotated as they are copied—otherwise they will maintain the orientation of the selected feature. After filling in the information, select OK and then select a work point or work axis as the point of rotation, and the command will be complete.

EDITING ARRAYS

For both rectangular and polar arrays, after the feature is arrayed, the arrayed features or instances have a child relationship to the feature that was arrayed—the parent. If the size of the parent feature changes, all the child features will also change. If a hole is arrayed, and the parent hole type changes, the child holes will also change. You can edit the arrayed set by using the Edit Feature command **(AMEDITFEAT)** or through the browser. After issuing the command, select any of the arrayed holes. Then you can change the number of rows, columns, or arrayed features, the spacing of the rows and columns and the angle for polar arrays. After the arrayed set is selected, the dimensions or an angle will appear on the selected feature, as well as "C#=". These "C#=" represent the number of rows, columns or number of polar arrayed features. These "C#=" will be incremented by one as each new array is created. To edit or delete an arrayed feature by itself, issue the Edit Feature command **(AMEDITFEAT)** or select the array from the browser and press I to edit the selected feature independently of the set. Then select the feature. Once a feature is independent of the array, it has no relationship to the arrayed set and it can be edited or deleted independently. If the number of rows or columns of the arrayed set changes, the location of the independent feature will be maintained. This feature will not be constrained to the part. To constrain the feature, use the Edit Feature command with the Sketch option, as described in Chapter 2.

Note: A base part or feature cannot be arrayed.

A feature that used another edge to close it can be arrayed.

For polar arrays, a work point, work axis or cylindrical edge must exist before you issue the Feature Array command (**AMARRAY**); any one can be used as the axis of rotation.

TUTORIAL 7.11—CREATING AND EDITING A RECTANGULAR ARRAY

1. Open the file \Md4book\Chapter7\TU7-11.dwg.
2. Issue the Feature Array command **(AMARRAY)**.
3. Select the hole in the plate.
4. Type "5" for the Number of Columns, "1" for the column Spacing, "2" for the Number of Rows and "2" for the row Spacing and then select OK. When complete, your part should resemble Figure 7.41.

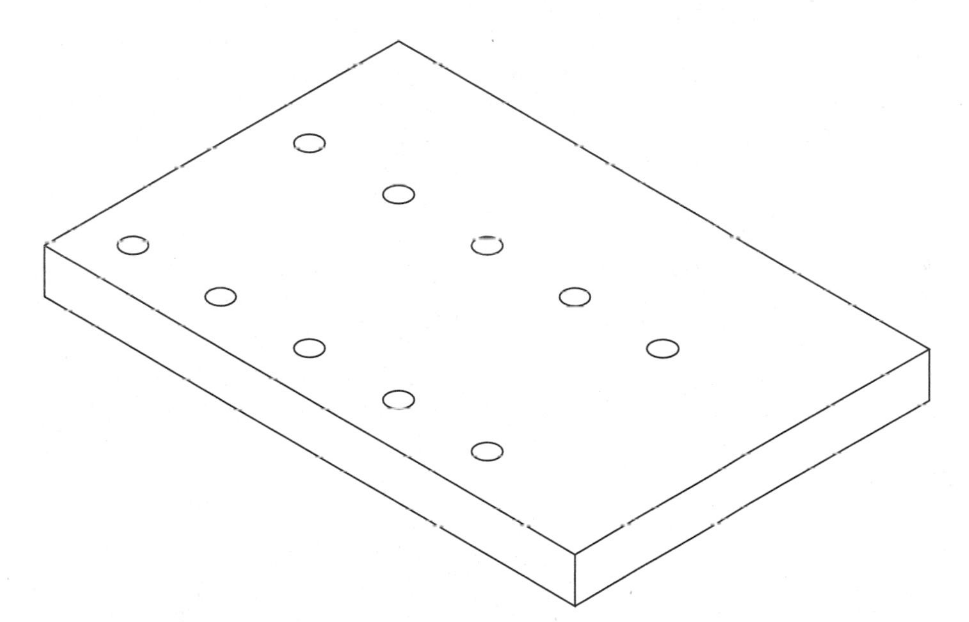

Figure 7.41

5. Change to the plan view with the **9** key.
6. Issue the Edit Feature command **(AMEDITFEAT)** or from the browser, double-click on RectArray1. If you used the icon to issue the command, select any

one of the arrayed features. Then the dimensions for the spacing will appear on the selected feature along with two C#=#, as shown in Figure 4.42. The C0 represents the number of rows and C1 represents the number of columns.

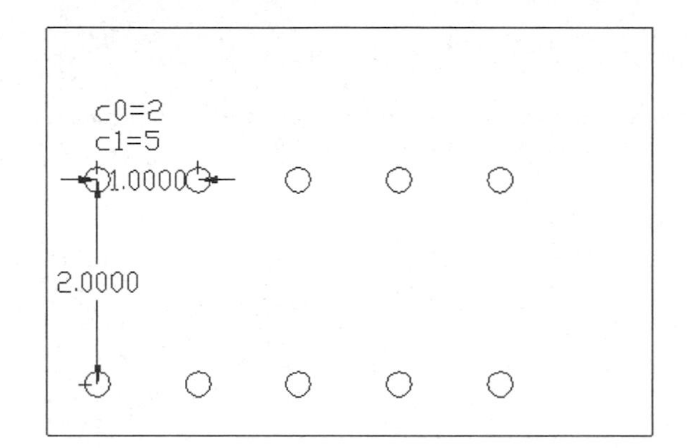

Figure 7.42

7. Change the number of rows by selecting C0=2 and typing in "4" and then press ENTER. Change the number of columns by selecting C1=5 and typing in "6" and then press ENTER. Change the row spacing from "2" to "1" and press ENTER. Update the part if necessary. When complete, your part should resemble Figure 7.43.

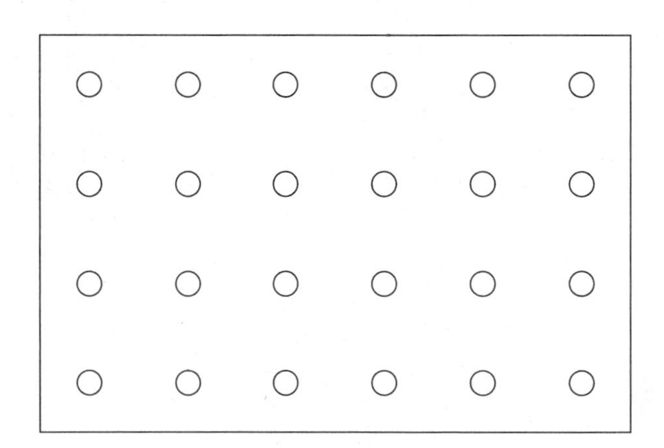

Figure 7.43

8. Repeat the Edit Feature command.
9. Press I and ENTER to edit the Independent array instance.

TUTORIAL

10. Select the hole in the upper right corner and you will be returned to the command prompt.
11. Press ENTER to repeat the Edit Feature command.
12. Select the hole in the upper right corner and change its diameter to .375; update the part.
13. Delete the hole in the upper right corner. When finished, your part should resemble Figure 7.44.
14. Save the file.

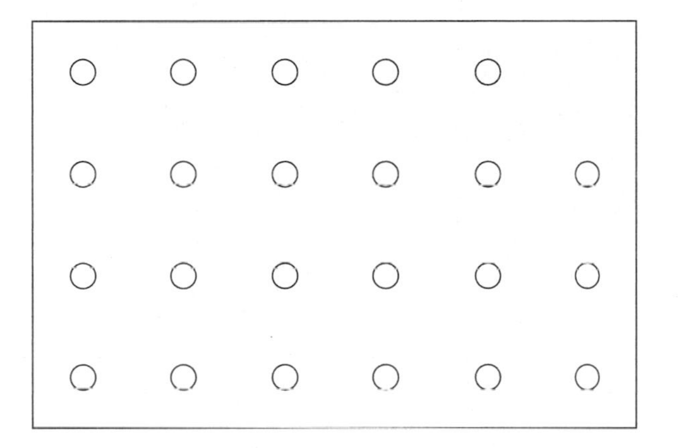

Figure 7.44

TUTORIAL

TUTORIAL 7.12—CREATING A POLAR ARRAY: FULL CIRCLE

1. Open the file \Md4book\Chapter7\TU7-12.dwg.
2. Issue the Feature Array command **(AMARRAY)** and select the counter bore hole on the bottom of the part "Hole1".
3. Select the Polar tab and type "3" for the Number of instances, select Full Circle, check Rotate as Copied and then pick OK.
4. Select the work axis and, when complete, your part should resemble Figure 7.45.
5. Repeat the Feature Array command **(AMARRAY)** and select the counter bore hole on the top of the part "Hole2".
6. Select the Polar tab and type "3" for the Number of instances, select Included Angle, change the Angle to "–120", check Rotate as Copied and then pick OK.
7. Select the work axis and, when complete, your part should resemble Figure 7.46.

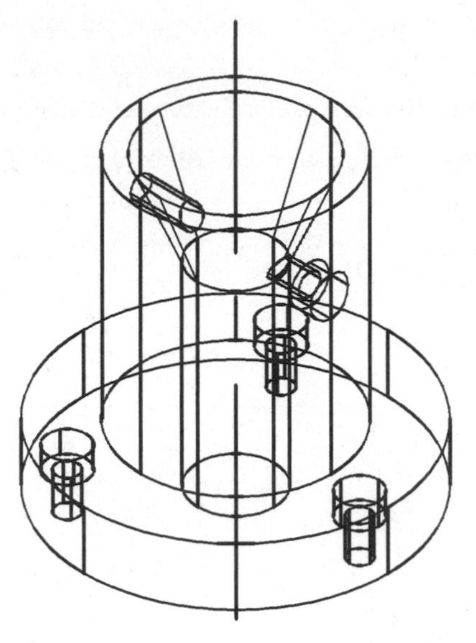

Figure 7.45

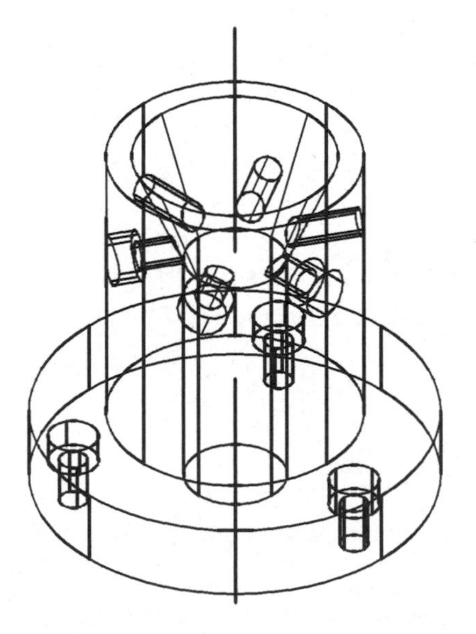

Figure 7.46

8. Delete the last polar array.
9. Issue the Feature Array command **(AMARRAY)** and select the counter bore hole on the top of the part "Hole2".

10. Select the Polar tab and type "3" for the Number of instances, select Incremental Angle, change the Angle to "–50", check Rotate as Copied and then pick OK.
11. Select the work axis and, when complete, your part should resemble Figure 7.47.
12. Save the file.

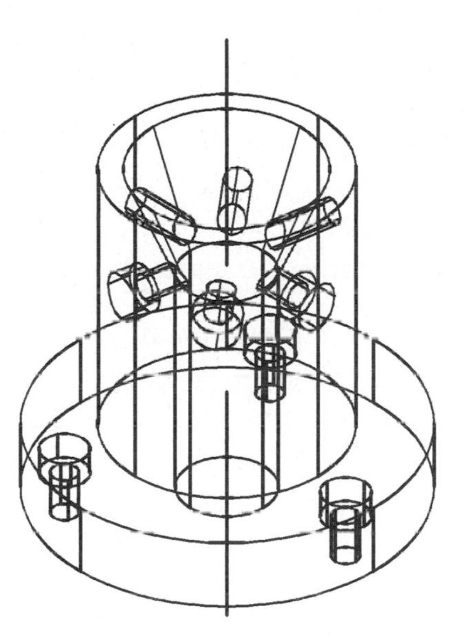

Figure 7.47

COPYING A FEATURE

In Chapter 6, you learned that you can copy a sketch and then create a feature based on that sketch. In this section, you will learn how to copy a feature to the current sketch plane. Features can be copied to the same part or to a different part. However, the parts need to be in the same file and the part that the feature is being copied to must be the active part. The following features cannot be copied: the base feature, fillets, chamfers, shells, features within a toolbody (parametric Boolean), features where the sketch was closed by an edge and features that have a termination To Plane, To Face and From/To.

Before copying a feature, make the target part active. Then make the plane where the feature will be placed the active sketch plane. Issue the Copy Feature command **(AMCOPYFEAT)** by selecting the command from the Part Modeling toolbar as shown in Figure 7.48. Or you can right-click in the drawing area and from the pop-up menu, select Copy from the Edit Features menu or right-click on the feature's name in the browser and choose Copy from the pop-up menu. If you issue the command from the toolbar or pop-up menu from the drawing area select the feature to be copied in the drawing screen. Depending on the feature that was selected, you may be prompted to accept or cycle to the next feature. If the command was selected from the browser, the feature will already have been selected. Select a point with the left mouse button where you want the feature to be placed on the active sketch plane. The copied feature will appear in blue. To change the position, keep selecting points with the left mouse button until it is in the correct position and press ENTER. Before pressing ENTER you will see the following prompt:

```
Specify location on the active part or
[Parameters/Rotate/Flip]:
```

Depending on the geometry that is being copied, not all of the options may appear.

Parameters: Allows you to set the copied feature as a dependent or independent feature of the original feature. Dependent means that if the parent dimension changes, so will the copied features. A dependent feature will have its dimensions d# equal to the d# of the original dimension. You can break this relationship by editing the feature and typing in a new value for any dimension. Now it will be independent, with no relationship to the original feature.

Rotate: Rotates the copied feature in 90° increments until it is in the correct orientation.

Flip: Mirrors the copied feature about the *ZX* plane of the current UCS.

After the feature is copied, it will not be constrained to the plane or part. If the feature needs to be constrained to the plane, use the Edit Feature command **(AMEDITFEAT)** with the Sketch option and add dimensions or constraints as needed.

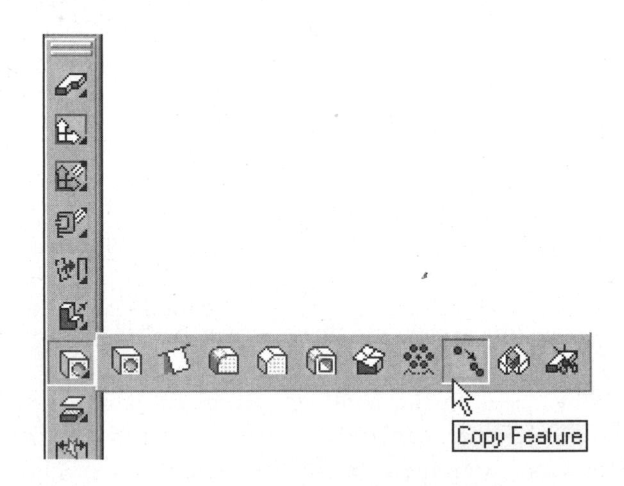

Figure 7.48

Then update the part. For holes that are copied, a work point is used to locate them, hence a work point will appear when you return to the sketch. Add dimension or constraints to the work point and update the part.

TUTORIAL 7.13—COPYING A FEATURE

1. Open the file \Md4book\Chapter7\TU7-13.dwg.
2. Make the angled face the active sketch plane and orient the positive X so it points into the screen.
3. Issue the Copy Feature command **(AMCOPYFEAT)** and select the extruded feature or right-click on the name ExtrusionBlind2 in the browser, choose Copy from the pop-up menu and then press ENTER. Select a point near the middle of the angled plane but do not press ENTER.
4. Press P and ENTER; press D and ENTER to set up a dependency between the two features.
5. Press R and ENTER to rotate the feature so that it resembles Figure 7.49, shown with lines hidden for clarity. Press ENTER to exit the command.

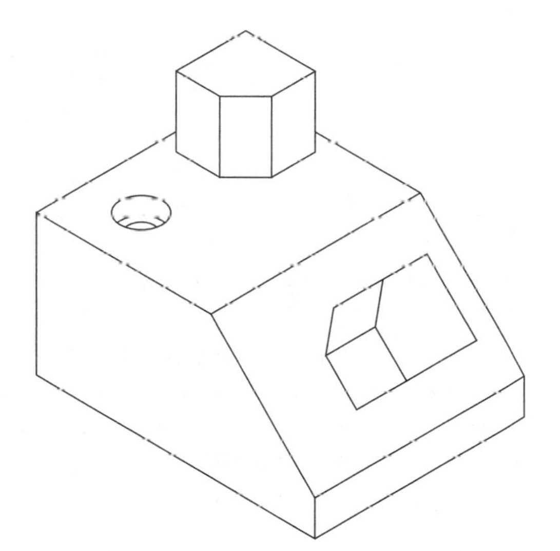

Figure 7.49

6. Edit the sketch of the copied feature and add both a ".5" and a "1" dimension, as shown in Figure 7.50, shown with lines hidden for clarity.

TUTORIAL

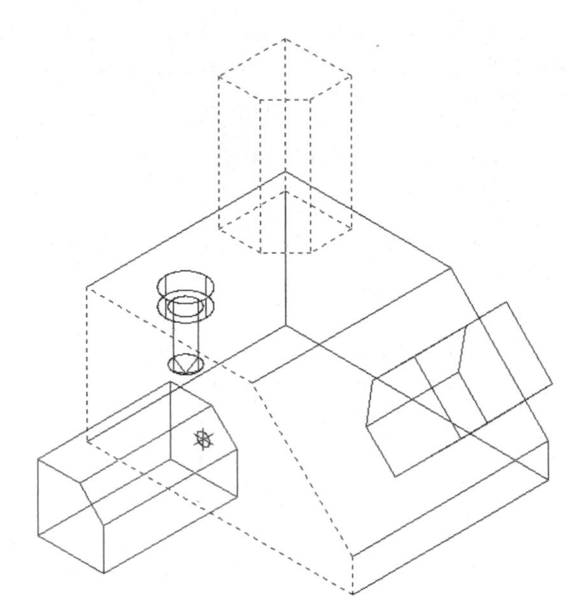

Figure 7.50

7. Update the part if needed.
8. Edit the feature "ExtrusionBlind2" and change the extrusion distance to 2. If needed, update the part. Both extrusions should look the same, since a dependency was set when the feature was copied.
9. Make the front left vertical face the active sketch plane and accept the default orientation of the *XYZ* axis.
10. Issue the Copy Feature command **(AMCOPYFEAT)** and select the top extruded feature or right-click on the name ExtrusionBlind2 in the browser, choose Copy from the pop-up menu and then press ENTER. Select a point near the middle of the front left vertical plane but do not press ENTER.
11. Press P and ENTER; press I and ENTER to have the copied feature be independent.
12. Press R and ENTER to rotate the feature so that it resembles Figure 7.51. Press ENTER to exit the command.
13. Edit the feature "ExtrusionBlind2" and change the extrusion distance to .5. If needed, update the part. The first and second extrusion should look the same, since a dependency was set when the feature was copied and the third extrusion should still have an extruded distance of 2.
14. Save the file.

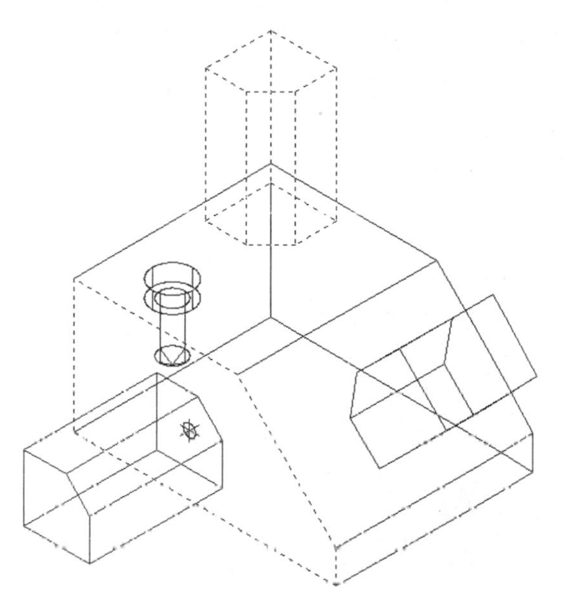

Figure 7.51

REORDERING

As you build your parts, you may get an unexpected result after a feature is created. After looking at the part, you may see that if a feature had been created in a different order, you might have achieved a different result. Instead of recreating the part, you can reorder the history of the features in the browser. Reordering features can be done in the browser or issue the Reorder Feature command **(AMREORDFEAT)** by selecting the command from the Part Modeling toolbar as shown in Figure 7.52 or right-click in the drawing area and from the pop-up menu, select Reorder from the Edit Features menu. To use the browser, with the left mouse button select a feature's name in the browser tree and, while keeping the mouse button depressed, move the feature up or down in the tree structure. As you move the feature through the tree structure, you will see a circle with a diagonal line through it. This means that the feature cannot be moved to this location. Keep moving the feature until a horizontal line appears in the browser. This tells you that the feature may be placed in this location. If this is the correct location, release the left mouse button. Otherwise, keep moving the feature until it is in the correct location. If you prefer not to use the browser, you can issue the Reorder Feature command **(AMREORDFEAT)** and for the prompt:

```
Select feature to reorder
```

Select the feature that you want to reorder: Then for the prompt:

```
Select destination feature:
```

Then select the feature that you want the first feature to come after.

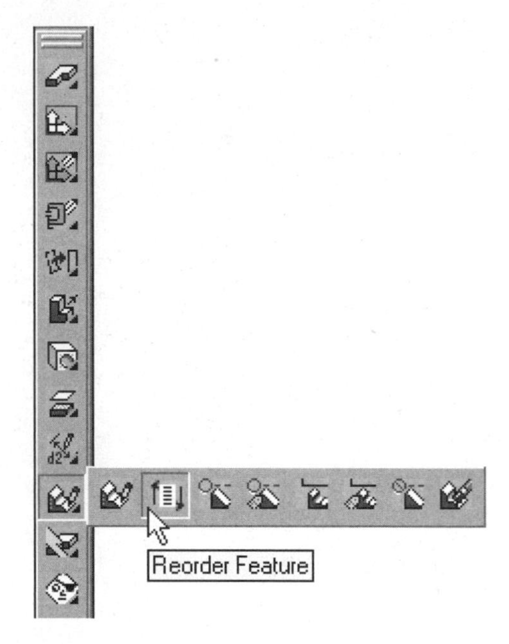

Figure 7.52

TUTORIAL 7.14—REORDERING

1. Start a new drawing from scratch.
2. Create a 2" square and extrude it 1".
3. Change to an isometric view using the **8** key.
4. Create a .5" drill through hole in the middle of the extrusion. Use the Through for the termination and 2 Edges for the placement option. Locate the hole 1" in from each edge from the bottom of the extrusion.
5. Shell the cube with a thickness of .1 and exclude the top face (so that it is open). Your part should look like Figure 7.53. The drilled hole appears as a boss because the shell adds thickness to the hole.
6. If the result intended is a drilled through hole instead of a boss, it should have been created after the shell. In the browser, select the hole and, keeping the left mouse button depressed, move it so that it is after the shell, as shown in Figure 7.54. Then release the left mouse button. It did not matter if you reordered the hole after the shell or the shell before the hole. When complete your part should resemble Figure 7.55.

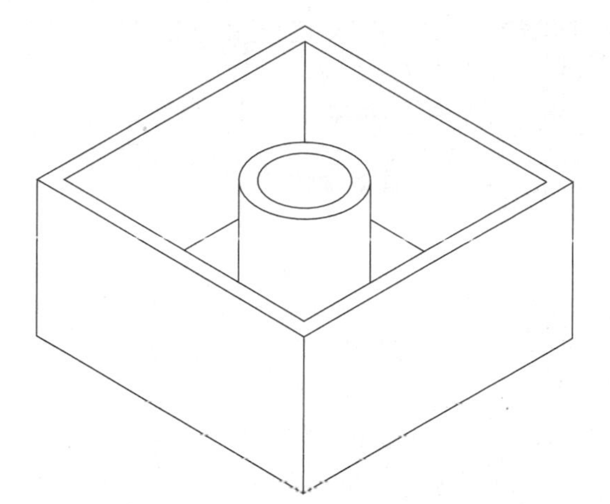

Figure 7.53

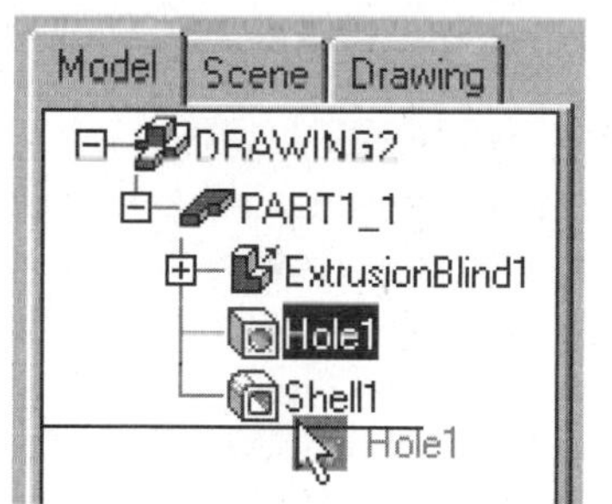

Figure 7.54

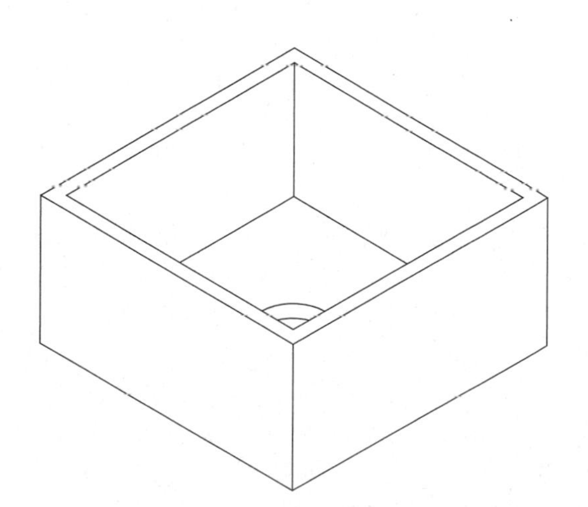

Figure 7.55

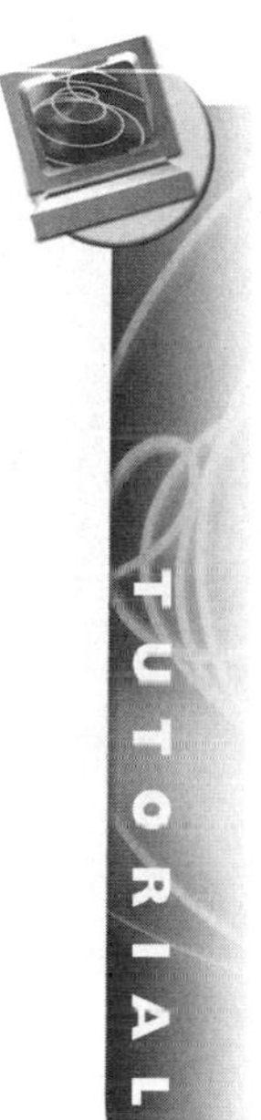

7. Move the shell so that it is after the hole and the part should again look like Figure 7.53.
8. Save the file as \Md4book\Chapter7\TU7-14.dwg.

ADDING A SECOND PART TO A FILE

As you work on a file, you may want to create multiple parts in the same file. These parts do not need to be assemble together but they may be. If you want to create multiple parts in the same file do not use the New Part File command, as this will only allow a single part per file. After a part exists in a file there are three methods that can be used to create a new part in a file.

The first method is by using the browser. Click the right mouse button in a blank area in the browser. A pop-up menu will appear for selecting New Part. At the command line, you will see a prompt:

```
Select an object or enter new part name <PART2>:
```

You can then type in a name for the part or press ENTER to accept the default name. The default naming prefix can be set through Desktop options. By default, the naming prefix is set to Part. The first part created will have a name Part1_1, and each part created afterwards will increment by one, Part2_1, Part3_1. After each part has been created, its name will appear in the browser. The first number in the suffix represents the order in which the part was created. The second number refers to the instance number of a part. An instance is like an AutoCAD block. When a part is copied in the same file, it is created as an instance of the original part. The AutoCAD copy command will create an instance. If one of the instances changes, so will the other instances. (Instances will be covered in more detail in the next section.)

After the part has been given a name, create a sketch and profile it, as you have learned to do in the previous chapters. The Select option from the pop-up menu is used to convert a 3D solid (AutoCAD native solid) to a part that Mechanical Desktop can use. The part will not be parametric. However, features created to this part after the conversion will be parametric. Select a 3D solid and give it a name or press ENTER to accept the default name.

The second method is to issue the New Part command **(AMNEW)** from the Part Modeling toolbar as shown in Figure 7.56. After issuing the command, you will be prompted for the same information as in the browser method.

The third method is to make a new part based on an existing part, but this new part will not be an instance of the original part; instead it will be a new part definition. To create another part that is based on an existing part, use the Assembly Catalog command **(AMCATALOG)** found on the bottom of the browser as shown in Figure 7.57. The Assembly Catalog dialog box will appear; select the All tab and all the part names will appear. Right click on the part name to base a new part on and select Copy

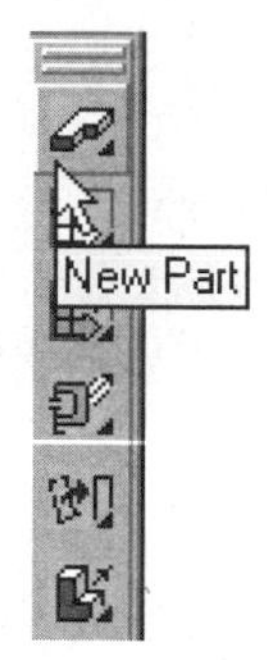

Figure 7.56

Definition as shown in Figure 7.58. Type in a new name in the Copy Definition dialog box and you will be returned to the drawing where you can select a new point to insert the new part into the file. This new part has no relationship to the original part unless global design variables were used. The Assembly Catalog command is discussed in greater detail in Chapter 8.

Figure 7.57

Figure 7.58

INSTANCES

An instance is a copy of an existing part. An instance has the same name as the original part, but the number after the underscore will be sequenced. For example, if the original part has a name Part4_1, the copy will be Part4_2. An instance acts like a block in AutoCAD. If the original part or an instance of the part changes, all the parts will reflect the change. To create an instance, you can use the AutoCAD copy command, right-click on the part name in the browser and select Copy, or use the Part Catalog command **(AMCATALOG)** just as you would to create a copied definition except select the Instance option. If you want a copy of the original part to have no relationship to the original part, use the Copy Definition option in the Assembly Catalog command.

Active Part

If there are multiple parts in the same file, only the active part can be edited. If you try to perform an operation on a part that is not active, an error message will appear at the command line:

```
Invalid selection. Keep trying. Bad selection.
Please try again.
```

Only one part can be active at a time. Looking at the Desktop browser, you can tell what part is active by the icon next to the name—if it is gray and the background for the part name is blue it is the active part. To make a part active, you can double-click on the part in the drawing area, you can double-click on the file name in the browser, select the part name in the browser with the right mouse button and select Activate Part, or issue the Activate Part command **(AMACTIVATE)** from the Part Modeling toolbar as shown in Figure 7.59. After issuing the command select a part to make active.

Figure 7.59

TUTORIAL 7.15—CREATING A NEW PART AND ACTIVATING A PART

1. Open the file \Md4book\Chapter7\TU7-15.dwg.
2. Select in the white area of the browser with the right mouse button.
3. Choose the New part option from the pop-up menu.
4. Press ENTER to accept the default name.
5. Draw a circle.
6. Profile the circle; it should require one constraint or dimension. This is because Part2_1 has no relationship to Part1_1.
7. Add a "2" diameter dimension to the circle.
8. Extrude the circle 2 units in the default direction.
9. Activate Part1_1 (the plate with the hole) by double-clicking on it
10. Try to make the top plane of the circle that you just extruded the active sketch plane. You will see the message "Active part must be selected".
11. Activate Part2_1 by right-clicking on the name Part2_1 and choose Activate Part.
12. Make the top plane of the circle the active sketch plane.
13. Issue the Part Catalog by selecting its icon from the bottom of the browser, go to the All tab, and copy the definition of part2 and give it a new name: Test2. Insert Test2 into the drawing by selecting an insertion point in the middle of the screen and press ENTER to complete the command.
14. Edit the diameter of the part Test2_1 to ".5" and update the part. The Part2_1 should not have changed.
15. Use the AutoCAD copy command and copy part Test2_1 to a blank area on the screen. Then make this copy, Test2_2, the active part by double-clicking on the name Test2_2 in the browser.
16. Edit the diameter of the part Test2_2 to ".2" and update the part. The copy of the Test2_1 part should have also changed.
17. Save the file.

TUTORIAL

MIRRORING A PART

After creating a part, you may need to create a second part that is a mirrored image of the first. A part can be mirrored about a planar face on the part, a work plane or a line. A line can be a regular AutoCAD line or two digitized points. When using the line method, you can use object snaps. Issue the Mirror Part command **(AMMIRROR)** from the Part Modeling toolbar, as shown in Figure 7.60 or right-click in the drawing area and from the pop-up menu, select Mirror Part from the Part menu. Then at the command line you will see the prompt:

```
Select part to mirror:
```

Select the part to mirror (do not press ENTER after selecting the part, this would exit the command) and then you will be prompted:

```
Select planar face to mirror about or [Line]:
```

If you want to mirror about a planar face, select in the middle of the planar face, select an edge that defines the planar face or a work plane. The plane must be part of the part that is being mirrored. You will have an option either to accept the highlighted plane or cycle to the next plane. When the correct planar face is highlighted, press ENTER. To mirror about a line press L and ENTER. Then you will be prompted to select two points. You can use object snaps to select the two points or pick two points with ortho on or off. The points can be selected from part edges, an AutoCAD line, a work axis or two arbitrary points. The next prompt will ask you either to:

```
Enter an option [Create new part/Replace instances]
<Create new part>:
```

If you press ENTER, you will be prompted to give the new part a name. You can either press ENTER to accept the default name or type in a different name. The part that was mirrored will be left alone and a new part will be created. This new part will have all the same dimensions and constraints of the original part, but it will not have a relationship to the first part (unless it was created with global design variables). If the variables are active part variables, the variables will be copied to the new part and have no relationship to the original part. If a global variable is used, the new part will maintain the relationship to the global variable. If you select the Replace option, the original part will be deleted and the mirrored part will take its place. There are no assembly constraints automatically applied to the parts. They could have assembly constraints applied if desired. Assembly constraints will be covered in Chapter 8.

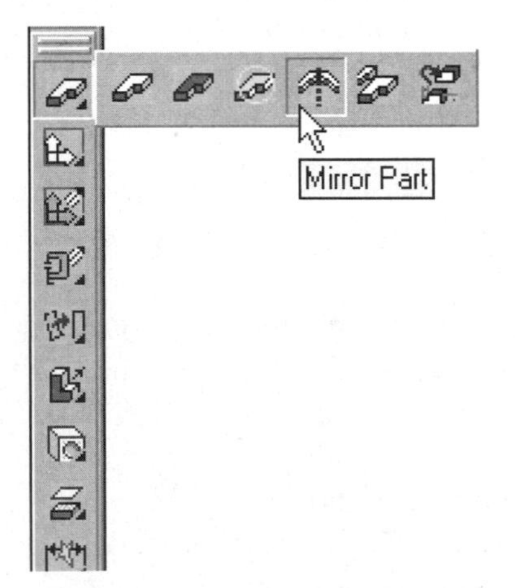

Figure 7.60

TUTORIAL 7.16—MIRRORING ABOUT A PLANE

1. Open the file \Md4book\Chapter7\TU7-16.dwg.
2. Issue the Mirror Part command **(AMMIRROR)**.
3. Select the part on an edge but do not press ENTER, because this will exit the command.
4. Select the face as shown in Figure 7.61. Once it is highlighted, press ENTER to accept this planar face.

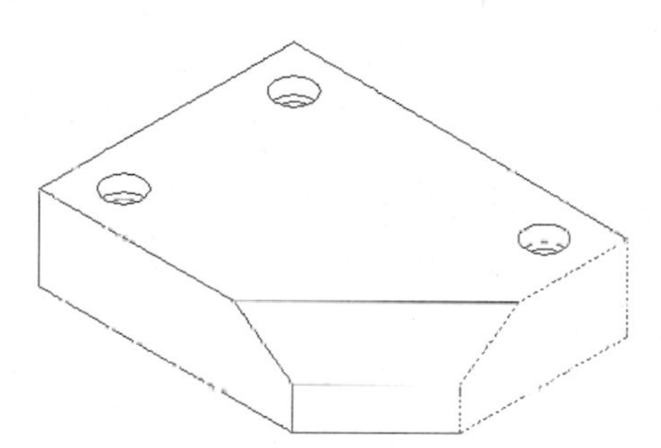

Figure 7.61

5. Press ENTER to create a new part.
6. Press ENTER to accept the default part name. When finished, your screen should resemble Figure 7.62.
7. Save the file.

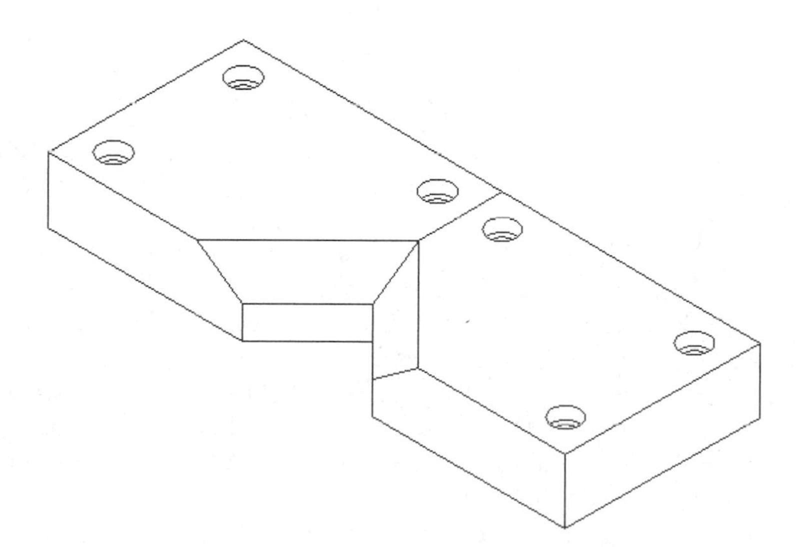

Figure 7.62

TUTORIAL 7.17—MIRRORING ABOUT A LINE AND REPLACING THE PART

1. Open the file \Md4book\Chapter7\TU7-17.dwg.
2. Issue the Mirror Part command **(AMMIRROR)**.
3. Select the part on an edge but do not press ENTER, because this will exit the command.
4. Press L and ENTER to mirror about a Line.
5. Use object snaps to select the endpoints of the line.
6. Press R and ENTER to Replace instances. When complete, your screen should resemble Figure 7.63.
7. Save the file.

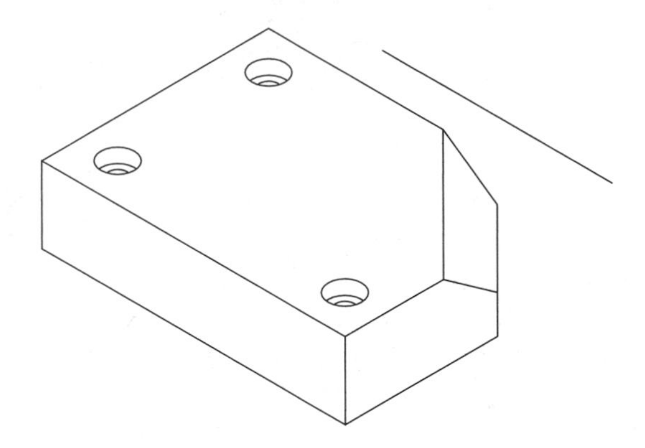

Figure 7.63

SPLITTING A PART INTO TWO PARTS

In Chapter 3 you learned about creating a new part by using the split operation when a profile was extruded, revolved, lofted or swept and in the last few sections you learned a few different methods for creating new parts. Another option to create a new part is to use the Part Split command **(AMPARTSPLIT)**. The Part Split command will divide the active part into two parts along a planar face, work plane or split line. Each part will maintain all the features of the part before it was split but the feature(s) that make up the other part will be suppressed. Issuing the Part Split command from the Part Modeling toolbar, as shown in Figure 7.64 or right-click in the drawing area and from the pop-up menu select Part Split from the Placed Features menu. Then select a planar face, work plane or a split line and then press ENTER. An arrow will appear and the command line will prompt:

```
Define side for new part [Flip/Accept] <Accept>:
```

The direction the arrow is pointing will be the side that becomes the new part. When the arrow is pointing in the correct direction, press ENTER and then you will be prompted to accept the default part name or type in a name.

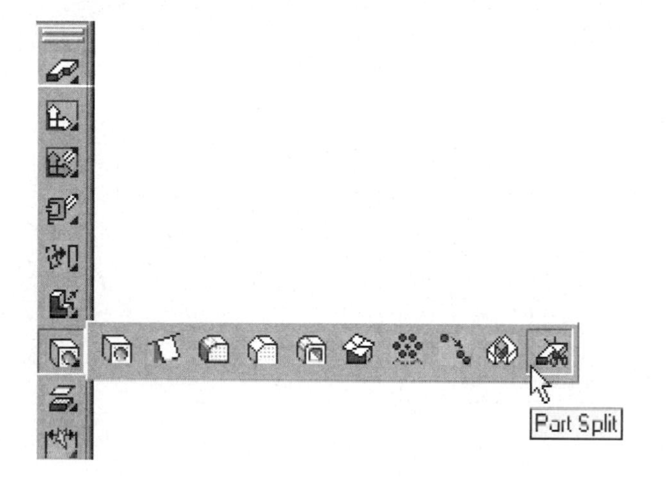

Figure 7.64

TUTORIAL

TUTORIAL 7.18—PART SPLIT BY PLANAR FACE AND A WORK PLANE

1. Open the file \Md4book\Chapter7\TU7-18.dwg.
2. Issue the Part Split command (AMPARTSPLIT), select the middle horizontal plane, as shown in Figure 7.65, and press ENTER.

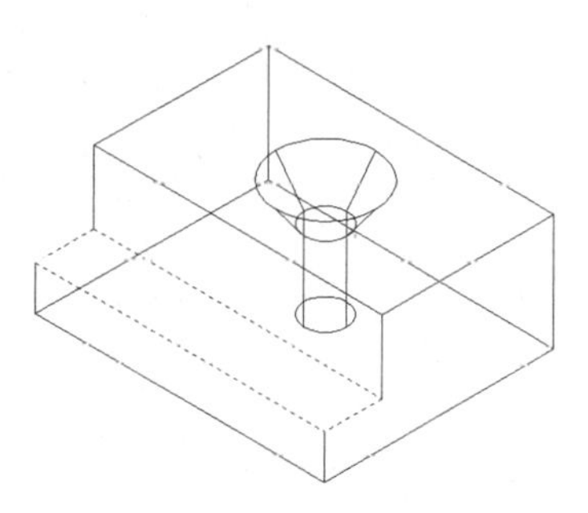

Figure 7.65

3. An arrow will appear going up and away from the part. Press ENTER to accept this direction.
4. Press ENTER to accept the default name of Part2.

5. Move PART2 (the top half) to the right of the screen. When complete, your screen should resemble Figure 7.66.

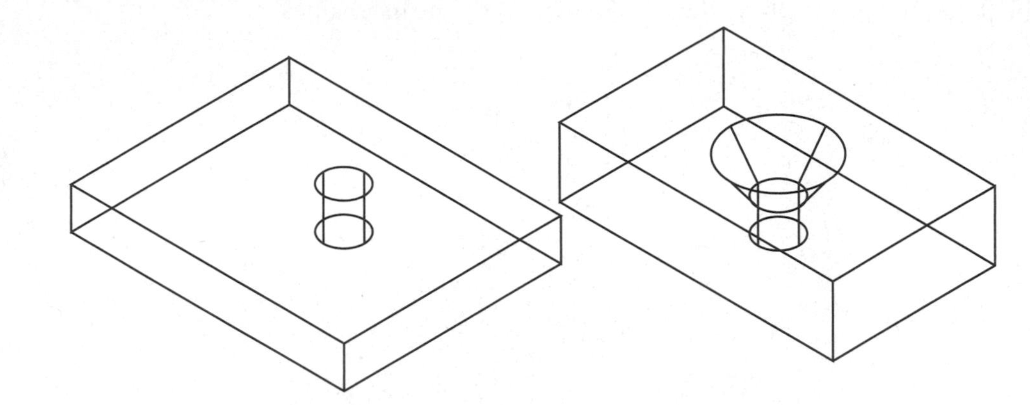

Figure 7.66

6. Make Part2 the active part.
7. Zoom in on Part2, the top half of the part.
8. Create a work plane that is parallel and offset a distance of "2.5" in from the rightmost vertical face.
9. Issue the Part Split command, select the work plane and flip the direction of the arrow so that it is pointing out of the screen, as shown in Figure 7.67.

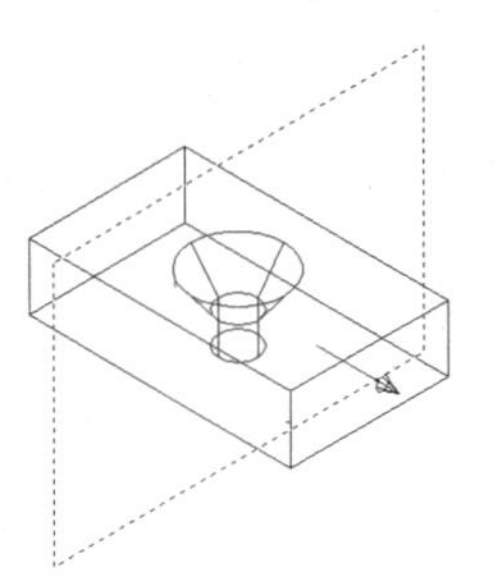

Figure 7.67

10. Press ENTER twice to accept the arrow direction and the default part name of Part3.
11. Turn off the visibility of the work plane in Part2.

12. Make Part3 the active part.
13. Turn off the visibility of the work plane in Part3.
14. Move Part3 to the right. When complete, your drawing should resemble Figure 7.68.
15. Save the file.

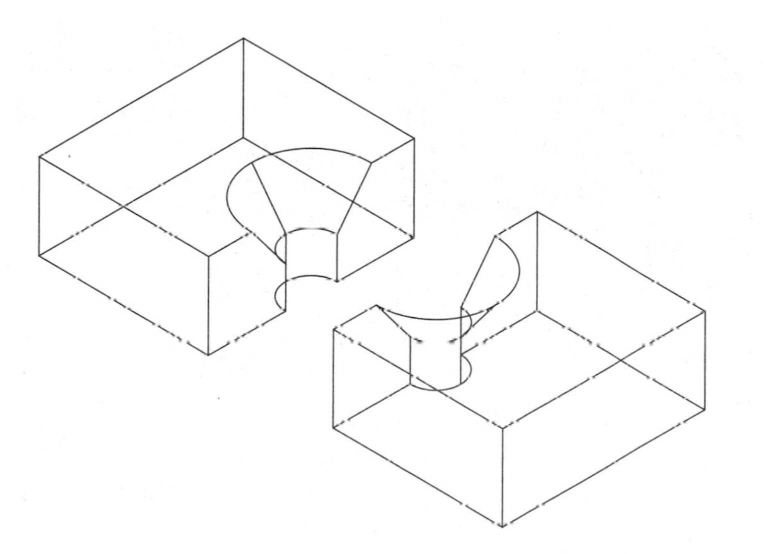

Figure 7.68

TUTORIAL 7.19—PART SPLIT BY A SPLIT LINE

For this tutorial, a split line has already been created.

1. Open the file \Md4book\Chapter7\TU7-19.dwg.
2. Issue the Part Split command **(AMPARTSPLIT)** and select the split line.
3. Flip the direction of the arrow so that it is pointing up out of the part and then press ENTER.
4. For the name, type in "Mouse Top" and press ENTER.
5. Move the top of the mouse off to the right of the bottom of the mouse.
6. Shade the parts. When complete, your screen should resemble Figure 7.69.
7. Save the file.

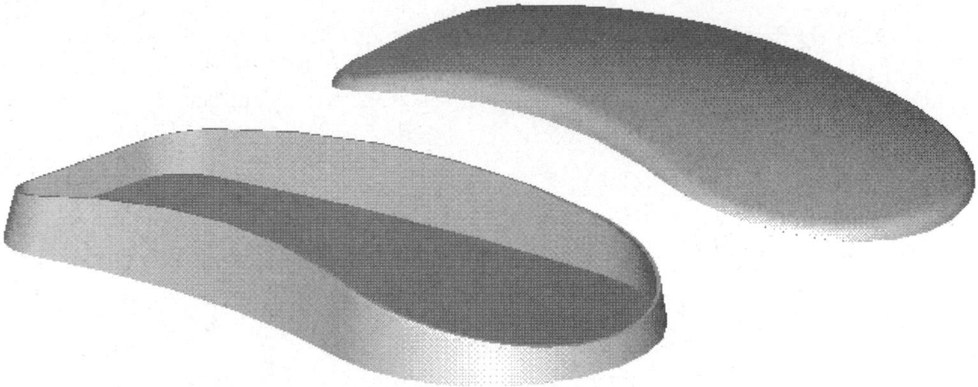

TUTORIAL

Figure 7.69

CREATING COMBINED PARTS

As you start creating parts, you may want to create a part that is made up of two individual parts. Instead of recreating this part, you can issue the Combine command **(AMCOMBINE)** from the Part Modeling toolbar, as shown in Figure 7.70 or right-click in the drawing area and from the pop-up menu, select Combine from the Placed Features menu. You have the option to cut, add or keep what is common between the two parts. For example, if you need to create a mold for a part that already exists, you could remove the part from a rectangular part. To create a part that consists of two different parts, start by creating two parts. The base part must be the active part in the active assembly. The part that will be cut or joined to the base part may be external and needs to have been created with the New Part File command. Make the part that will be the base part active (the part that will have material added to it or removed from it). Then you can use AutoCAD commands to align the parts in the correct position or use assembly constraints (which will be covered in Chapter 8) to position the two parts in the correct location. If assembly constraints are used to locate the two parts, there will be a relationship between the parts. If one part changes shape or location, the second part will maintain the relationship that was set with assembly constraints. The two parts do not need to be touching; there can be space between them and they still can be joined together. Of course, there will be a gap even after they are joined.

After issuing the Combine command **(AMCOMBINE)** you will be prompted at the command line:

```
Enter parametric boolean operation
[Cut/Intersect/Join] <Cut>:
```

After selecting an operation, you will be asked to select a toolbody. A toolbody refers to the part that will be consumed by the base part. After you select the toolbody, the command will be complete. In the browser, the toolbody will appear as a part nested under the name Combine#, which itself is nested under the active part.

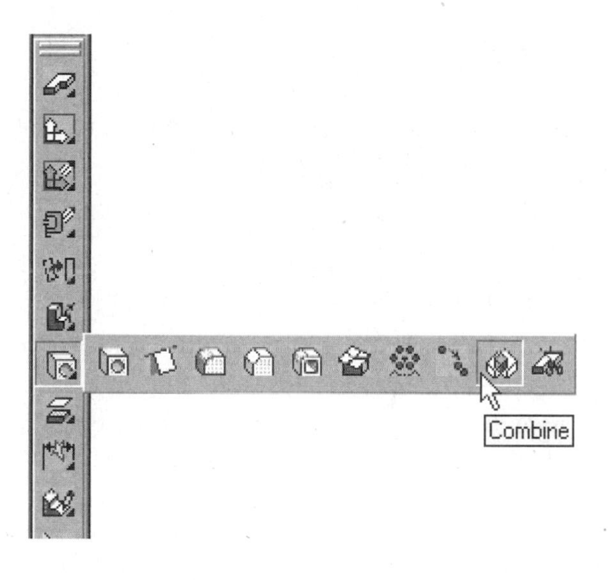

Figure 7.70

EDITING COMBINED PARTS

After the combine operation, the parametric dimensions for both parts are still intact. Once a combined part has been created, there are three methods for editing both parts, as explained below.

The first method is to use the Edit Feature command **(AMEDITFEAT)** and select any of the features. Once the feature is selected, its dimensions will appear on the screen. Select the specific dimension to edit, give it a new value and then update the part as you would any other part.

The second method also starts with the Edit Feature command, but instead of selecting the feature to edit, press T to edit the toolbody. This will place the part in a rolled back state, which means that the parts will appear as they were before the Combine command. Repeat the Edit Feature command to edit the toolbody features. The part that was the parent part cannot be edited in a rolled back state. After editing the toolbody, update the part. You will have the following options at the command prompt:

```
"Enter an option [active Part/Full] <active Part>:
```

Full: Updates the edits for the toolbody, any assembly constraints and then returns the two parts into a combined state.

Active Part: Updates the edits for toolbody only.

If you update the Active Part, another update will be required to return the two parts into a combined state. After issuing the update the command you will have see the following prompt:

```
Enter an option [Full/posiTioning] <Full>:
```

Full: Will update the assembly constraints and return the two parts into a combined state.

Positioning: Will update the assembly constraints. Another update will be required to return the two parts into a combined state.

The third method for editing combined parts is to select the part features to edit from the browser.

There are six main steps to follow for combining two parts:

1. Create two parts. The main part needs to be local. The second part can be local or referenced in.
2. Make the main part the active part (the main part is the part to which geometry will be added to or from which geometry will be removed).
3. Position the parts with the AutoCAD commands like move, rotate etc. or use assembly constraints. Assembly constraints are recommended since they will hold the parts in the desired position. Assembly constraints will be covered in Chapter 8.
4. Issue the Combine command **(AMCOMBINE)**.
5. Select the Boolean operation Cut, Join or Intersect.
6. Select the part to be used as the toolbody.

TUTORIAL 7.20—COMBINING WITH CUT

For this tutorial, two parts have been created and positioned.

1. Open the file \Md4book\Chapter7\TU7-20.dwg.
2. Verify that the parts are individual parts by selecting each of the part names individually in the browser; they will highlight independently.
3. Rotate the parts in shaded mode to show that there is no cavity.
4. Make the part "MoldBase" the active part.
5. Issue the Combine command **(AMCOMBINE)**, use the Cut option and select the Mouse Part.
6. Rotate the parts in shaded mode to show the cavity.

7. Use the Edit Feature command **(AMEDITFEAT)** to edit the cavity of the mouse part, change the "1" depth dimension to "1.25" and change the draft angle to –20.
8. Update the part. When complete, your screen should resemble Figure 7.71.
9. Save the file.

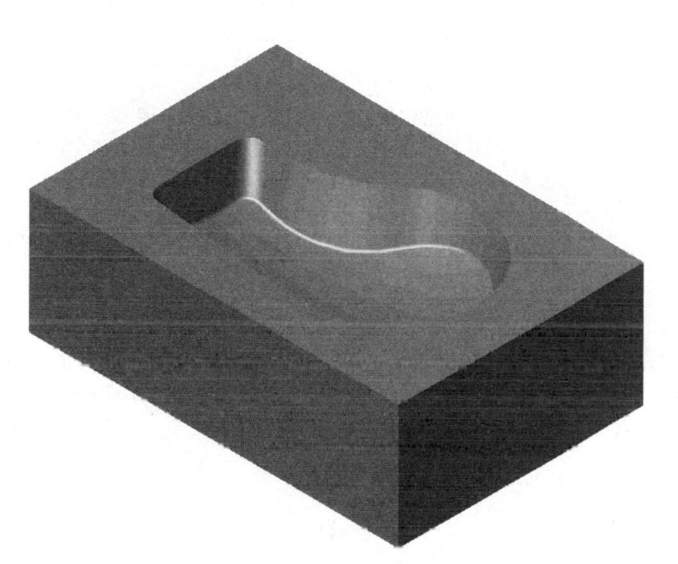

Figure 7.71

TUTORIAL 7.21—COMBINING WITH JOIN

1. Start a new drawing from scratch.
2. Create a new part named Shaft.
3. Draw, profile and dimension the geometry as shown in Figure 7.72.

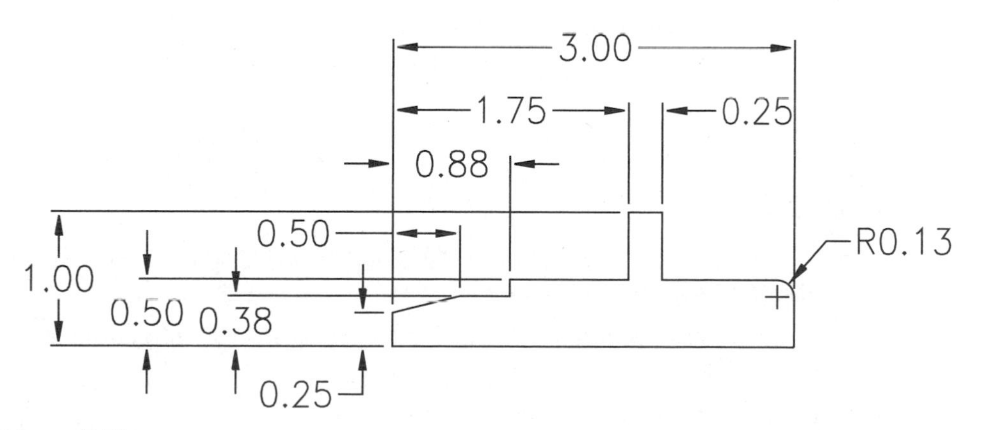

Figure 7.72

4. Do a full revolution along the bottom horizontal line.
5. Create a new part named Base.
6. Draw a 4" square and extrude it 2" in the down direction. This will keep the top of the base and the middle of the revolution in the same plane.
7. Move the center of the Shaft to the approximate middle of the Base. Assembly constraints can be used to position the parts. Assembly constraints will be covered in Chapter 8.
8. Issue the Combine command **(AMCOMBINE)**, use the Join option and select the Shaft. When complete, your screen should resemble Figure 7.73.

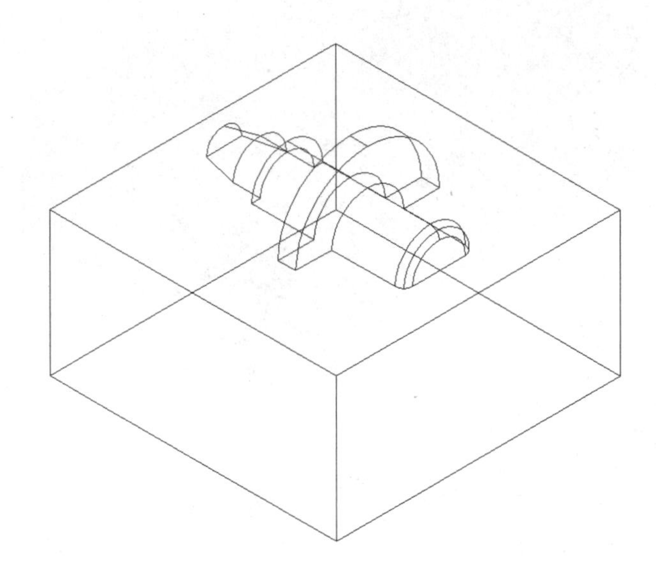

Figure 7.73

9. Use the Edit Feature command **(AMEDITFEAT)** and use the Toolbody option. This will take the Shaft back to its condition before the Combine command.
10. Issue the Edit Feature command again and select the shaft, change the overall length from "3" to"4".
11. Perform an update for the active part.
12. Move the revolution so the center point of the back circular edge is in the middle of the top edge of the part.
13. Perform an update with the Full option. When complete, your screen should resemble Figure 7.74
14. Save the file as \Md4book\Chapter7\TU7-21.dwg.

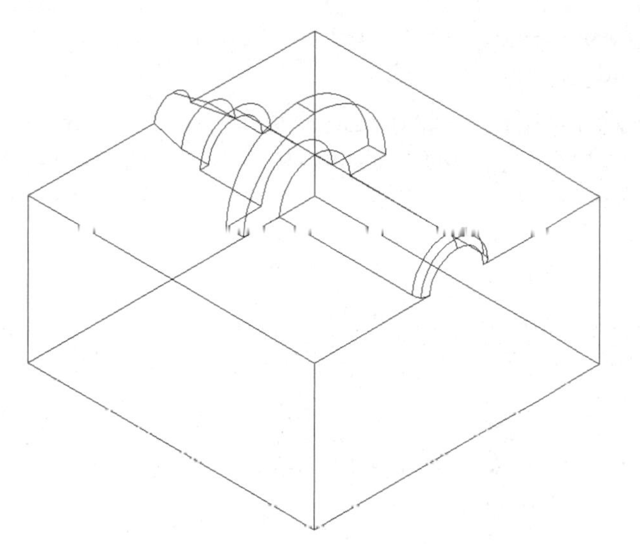

Figure 7.74

REPLAYING

In AutoCAD, there is the UNDO command that steps back one command at a time. Since Mechanical Desktop is running on top of AutoCAD, the UNDO command is still a valid command. In AutoCAD there is also a REDO command that undoes an undo, but this will only undo the last undo. In Mechanical Desktop, there is a command called Feature Replay **(AMREPLAY)** that is similar to redo because it steps for ward through the command sequence in which the part was created starting at the first step. The Feature Replay command can be used to see how a part was built, to step through a part and stop at a specific point and take a different approach. If you receive an error message while modeling, it can be used to replay through the part and truncate the part at the point before the error message appears. While stepping through the part, at the command line you will see the commands that were used. Also, as you step through the replay, the feature names will highlight in the browser in the order in which they were created.

Issuing the Feature Replay command from the Part Modeling toolbar as shown in Figure 7.75 or right-click in the drawing area and from the pop-up menu, select Replay from the Part menu,. You will be prompted:

```
Enter an option [Display/Exit/Next/Size/Truncate]
<Next>:
```

Display: Turns on all the constraint symbols. Once they are turned on, they cannot be turned off for the duration of the Feature Replay command.

TUTORIAL

Exit: Exits the command. The part will be returned to its state before the Feature Replay command was issued.

Next: This option, which is the default, will advance the part through the command sequence in which the part was created, one step at a time. After stepping all the way through the building process, you will be returned to the command prompt.

Size: Changes the height of the constraints. This can be adjusted throughout the command.

Truncate: Stops the replay command at the current step and the part will be left in that state. All features that were created after this point will be erased.

Note: While in the Feature Replay command, you can transparently pan and zoom.

The Feature Replay command is not used to reorder the part. If you want to reorder the part, use the reorder command.

After a part is truncated, the features that were created after this point will be discarded from the part.

Tip: After receiving complex parts, use the Feature Replay command to gain insight into how the part was created. While running through the sequence, think about other methods that could have been used to create the part.

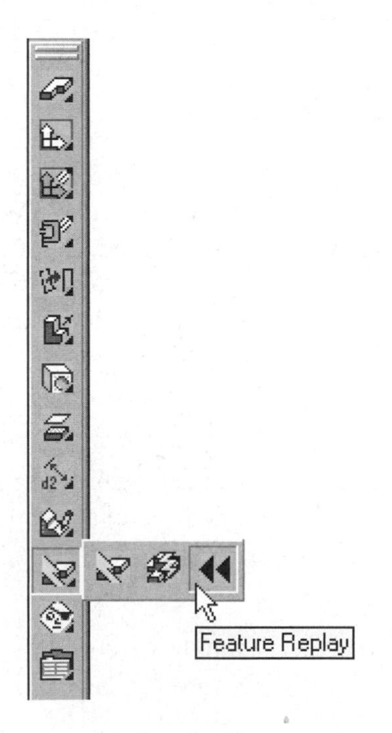

Figure 7.75

TUTORIAL 7.22—REPLAYING THE COMMAND SEQUENCE

1. Open the file \Md4book\Chapter7\TU7-22.dwg.
2. Issue the Feature Replay command **(AMREPLAY)**.
3. Press D and ENTER to turn on all constraint symbols.
4. Press S and ENTER to adjust the constraint symbols to a larger or smaller size so they can easily be seen and then select OK. If the size is not legible, go back and readjust the size.
5. Press ENTER to cycle all the way through the part, watching the command line and the browser as you cycle through.
6. Repeat the Feature Replay command **(AMREPLAY)**.
7. Cycle through the part and then truncate it after Fillet2 is replayed. When complete, your screen should resemble Figure 7.76, shown with lines hidden.

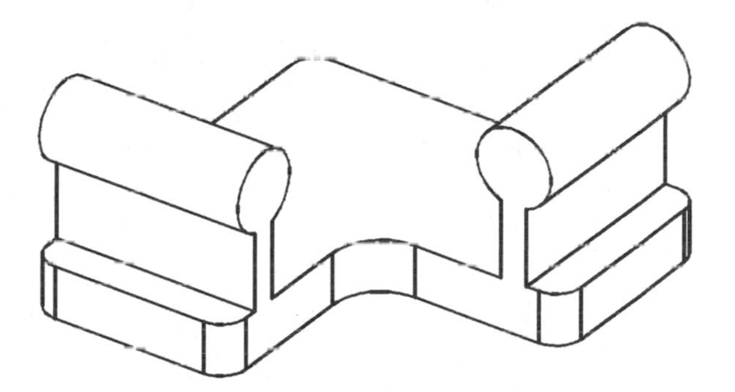

Figure 7.76

8. Add five separate ".125" diameter through holes that are concentric to the filleted edges (Fillet2). When complete, your screen should resemble Figure 7.77, shown with lines hidden.

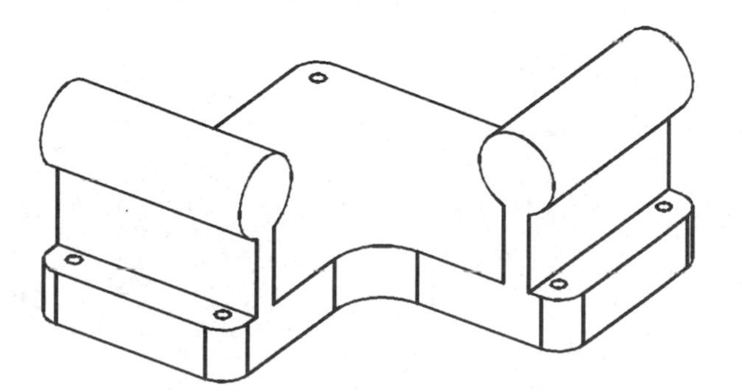

Figure 7.77

9. Issue the Feature Replay command **(AMREPLAY)**.
10. Cycle through the entire part.
11. Save the file.

MASS PROPERTIES INFORMATION FOR THE ACTIVE PART

To obtain mass property information about a part, first make it the active part. Then issue the Acitive Part Mass Properties command **(AMPARTPROP)** from the Part Modeling toolbar as shown in Figure 7.78. Or you can right-click in the drawing area and from the pop-up menu, select Mass Properties from the Part menu and the Part Mass Property dialog box will appear. Type in the density of the material and press ENTER, and the mass property information will be updated to reflect the new density. To create an ASCII file of this information, select the File button and give the file a name and location.

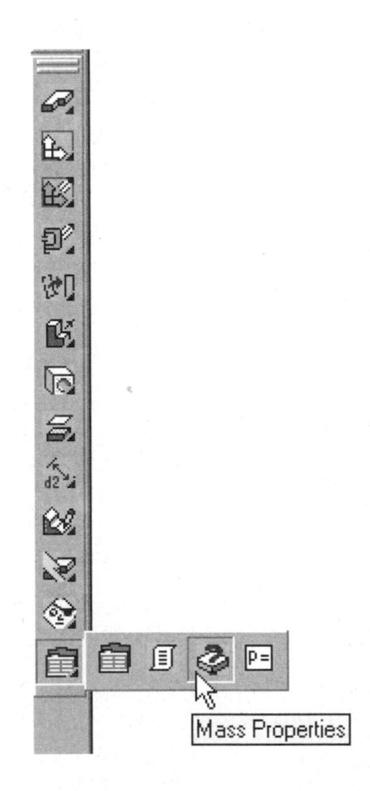

Figure 7.78

MASS PROPERTIES FOR MULTIPLE PARTS

If you want mass property information for multiple parts or want more control of how the values are calculated, issue the Assembly Queries command **(AMASSMPROP)** from the Assembly Modeling toolbar as shown in Figure 7.79 or right click in the drawing area and select Assembly menu from the bottom of the pop-up menu. Then select Mass Properties from the Analysis menu. After issuing the command, you will be prompted to

```
Enter an option (parts/subassemblies) [Name/Select]
<Select>:
```

If you know the subassembly name, you can type it in or press ENTER to select a single part or multiple parts. When selecting parts pick the parts individually. After making the selection, press ENTER and the Assembly Mass properties dialog box will appear containing five areas: % Error, Coordinate System, Material, Assembly Units and Mass Units.

% Error: The default is "1". The lower the number, the more accurate the calculations, but the longer the calculation time.

Coordinate System: The choices are:

Parts CG (center of gravity).This can only be used for a single part.

UCS, the current coordinate system.

WCS, the world coordinate system.

Material: Selecting on Material will bring up the Select Material dialog box. On the right side of the dialog box select the part or parts that you want to assign a specific material. To select multiple parts, hold down the CTRL key while selecting the part names. After selecting the part(s), select in the material area, or select the down arrow, and the available materials will appear. Select a material, then select Assign and the material will be assigned to the part(s). Continue until all parts have a material assigned to them. To add materials to the list, open the ASCII file named MCAD.mat in the \Mechanical Desktop 4\Desktop\Support directory and add the property information.

Assembly Units: Select the units that the parts were drawn in: In, cm or mm.

Mass Units: Select the units that you want the mass in: Lbs, g or Kg.

After filling in the information, select OK and the mass property information will appear in a dialog box. To write the information out to an ACII file, select File or to exit the command, select Done.

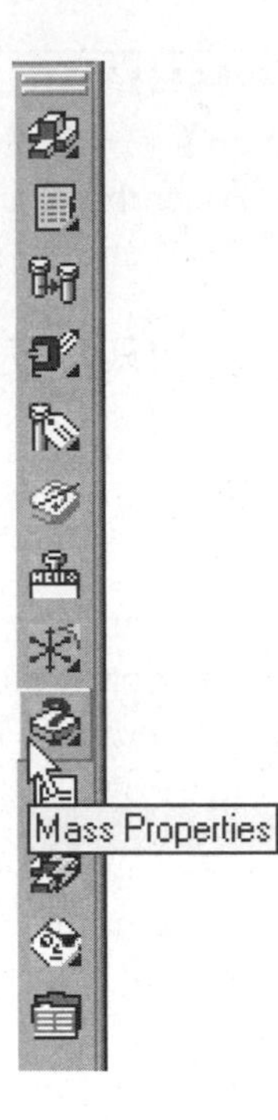

Figure 7.79

Tip: To receive mass property information that is always based upon the same UCS orientation, use the world coordinate system and do not move the part.

TUTORIAL 7.23—OBTAINING MASS PROPERTY INFORMATION

1. Open the file \Md4book\Chapter7\TU7-23.dwg.
2. Issue the Acitive Part Mass Properties command **(AMPARTPROP)**.
3. Type in a density of ".2" and press ENTER.
4. Write the mass property information out to a text file: \Md4book\Chapter7\TU7-23.PPR.
5. Issue the Assembly Queries command **(AMASSMPROP)** and select both parts and press ENTER.
6. Set the Coordinate System to WCS, Assembly units to In, Mass units to Lbs.
7. Assign the Material for Part1 to Mild steel and Part2 to Aluminum.
8. To see the mass property information, select OK in both dialog boxes.
9. Select File and create an ASCII file: \Md4book\Chapter7\TU7-23.MPR.
10. Save the file.

SURFACE STITCHING

There may be times when you get a model from another person that was created using NURBS surfaces. To utilize this part in Mechanical Desktop it will need to be converted to a solid. The surfaces need to form a closed or nearly closed part. Issue the Stitching command **(AMSTITCH)** as shown in Figure 7.80 from the Surface pull down, select all the surfaces that make up the part. The Surface Stitching dialog box will appear. Following are the options.

Stitching Type:

Optimal Stitching: When selected, Mechanical Desktop will calculate the best settings based on the model.

Custom Stitching: When selected, the Operation area of the dialog box will become available.

Operation:

Stitch Surfaces: When checked, surfaces will be combined together.

Heal Gaps: When checke,d gaps will be closed.

SimplifyObjects: When checked, complex surfaces may be simplified.

Enclosed Surfaces Output:

Part: The stitched surfaces will create a Mechanical Desktop part. The features that form this part are not parametric. However, all features added to this part will be parametric.

Quilt: The stitched surfaces will be joined together to form a single NURBS surface.

Preview:

When selected, you will see a preview image of the result.

Settings:

When selected, there will be two tabs that will be available in the Custom Settings dialog box.

Display: Under this tab, you can specify the colors of objects will appear in the Preview button.

Stitch/Heal/Simplify: When Custom Stitching is selected as the Stitching Type, you can set gap and stitching options.

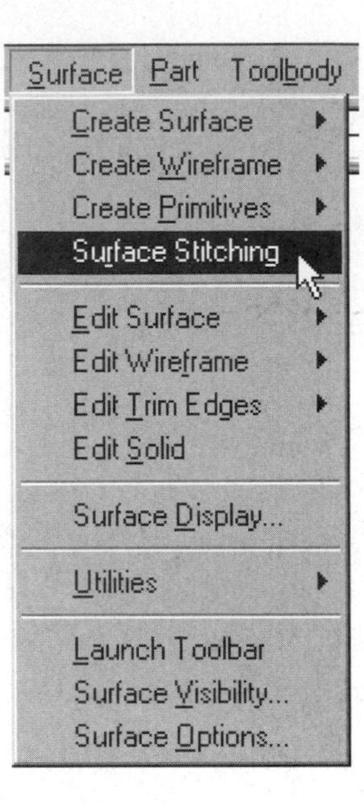

Figure 7.80

TUTORIAL 7.24—SURFACE STITCHING

1. Open the file \Md4book\Chapter7\TU7-24.dwg.
2. Issue the Stitching command **(AMSTITCH)**.
3. Select all the surfaces.
4. Select OK to accept all the defaults.
5. Place a few holes and fillets.
6. Edit the holes and fillets.
7. Save the file.

EXERCISES

Follow the instructions before each exercise.

Exercise 7.1—Bottom Half of Utility Knife

Create the bottom half of a utility knife, as shown in Figure 7.81. When finished, save the file as: \Md4book\Chapter7\\Uknife Bottom.dwg.

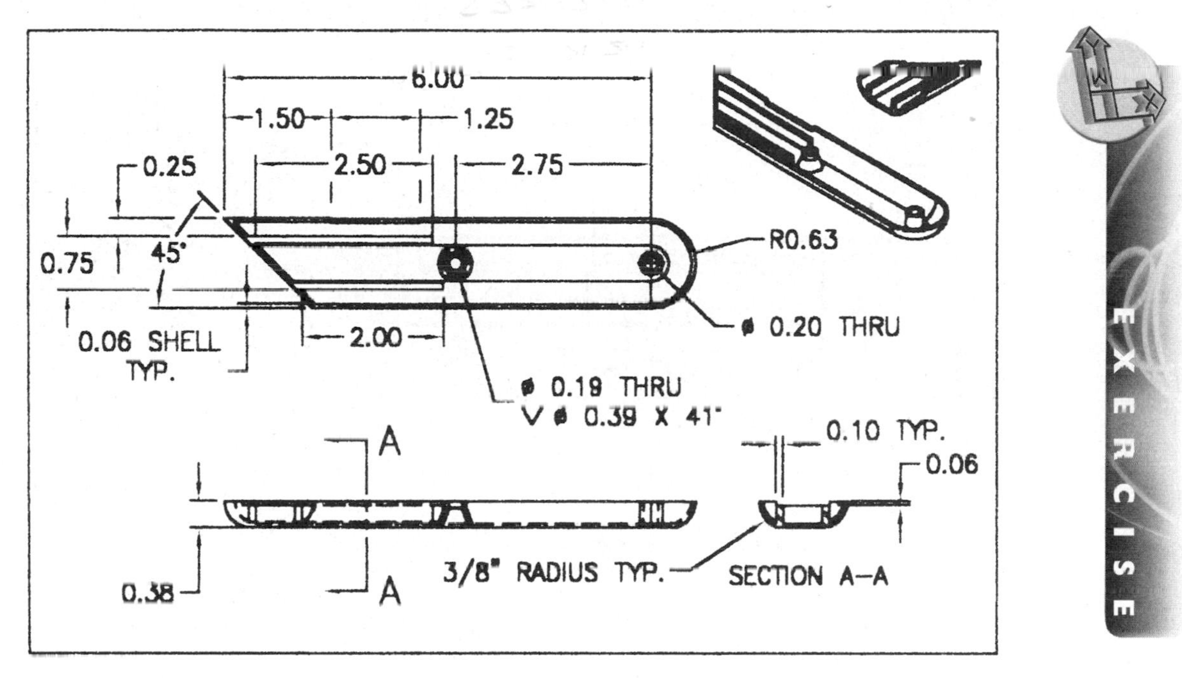

Figure 7.81

Exercise 7.2—Top Half of Utility Knife

Create the top half of a utility knife, as shown in Figure 7.82. Mirror the part created in Exercise 7.1 to help create the top half of the utility knife. When finished, save the file as: \Md4book\Chapter7\Uknife Top.dwg.

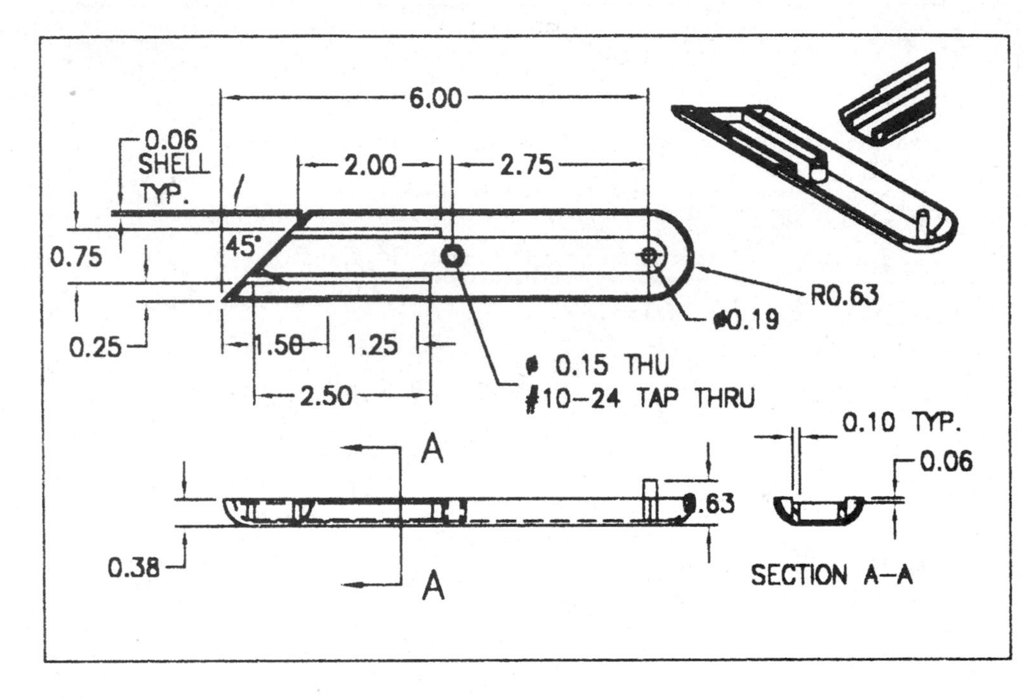

Figure 7.82

REVIEW QUESTIONS

1. After a part is mirrored with the Create new part option, the new part has no relationship to the original, unless if it was created with global variables. T or F?
2. What happens to the design variables in a part that is scaled?
3. When a face is split, the new face becomes a new part. T or F?
4. A feature cannot be copied to a different part. T or F?
5. A feature can only be edited from the active part. T or F?
6. A copied feature is automatically fully constrained by X Value and Y value constraints. T or F?
7. What is the difference between copying a part and copying a part's definition?
8. A part's edge can be used as the axis of rotation for a polar array. T or F?
9. When a part is shelled that has no tangent faces, the part can have multiple wall thicknesses. T or F?
10. In shelling, what does an "excluded face" refer to?
11. After two parts are combined, parametric dimensions are lost for both parts. T or F?
12. When editing a combined part, you can get into a rolled back state. What is a rolled back state? What is the option to get there?
13. Mass property information is always based on the World Coordinate System. T of F?
14. What is a toolbody?
15. The replay command can reorder features. T or F?

PART
↳ WORK FEATURES
↳ WORK PLANES

CHAPTER 8

Assemblies

In Chapter 7 you learned how to create multiple parts in the same file. In this chapter, you will learn other techniques for creating assemblies. By definition an assembly file is a file that contains more than a single part. The parts may be contained within the assembly file referred to as local parts, may be external to the assembly file referred to as referenced parts or the assembly may contain both local and referenced parts. The two methods of assembly design are referred to as "top down" and "bottom up".

In this chapter, you will learn how to create "top down" and "bottom up" assemblies, constrain the parts to one another using intelligent (intelli) constraints, edit the constraints, check for interference and create exploded views. Mechanical Desktop uses a variational solver for constraining parts; this type of assembler is needed to assemble linkage types of mechanisms.

AFTER COMPLETING THIS CHAPTER, YOU WILL BE ABLE TO:

- Create top down assemblies.
- Create sub assemblies.
- Create bottom up assemblies.
- Constrain parts together using assembly constraints.
- Edit assembly constraints.
- Check for interference.
- Create scenes and exploded assembly views.
- Edit scenes and add trails.

AMCATALOG
AMMATE

MECHANICAL DESKTOP OPTIONS – ASSEMBLY TAB

AMFLUSH
AMANGLE
AMINSERT

Before creating an assembly, we will take a look at the Assembly tab of the Edit Options command. Figure 8-1 shows the Assembly tab; below, each option is introduced.

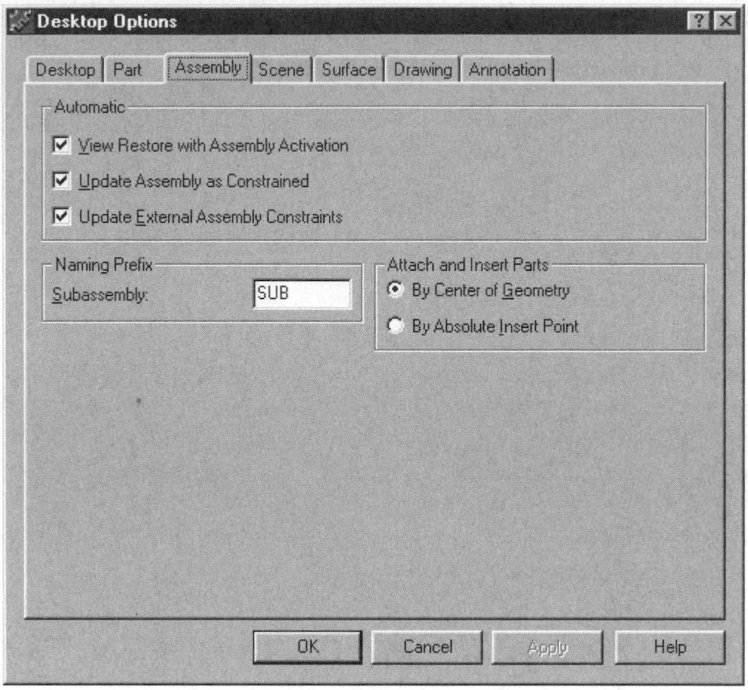

Figure 8.1

View Restore with Assembly Activation:

When checked, the parts will moved to their constrained positions when the assembly is activated.

Update Assembly as Constrained:

When checked, parts will move as assembly constraints are applied. If unchecked, the assembly would need to be updated to reflect the new assembly constraints.

Update External Assembly Constraints:

When checked, edited constraints in external files will be updated.

Naming Prefix – Subassembly:

Type in a default name for sub-assembles. Each subsequent sub assembly will be appended a number. A sub-assembly can be renamed or given a different name when it is created.

Attach and Insert Parts:

Select the radio button to either have the parts inserted at the center of the part (By Center of Geometry) or at its AutoCAD insertion point (By Absolute Insert Point).

CREATING ASSEMBLIES

As discussed earlier by default, Mechanical Desktop starts up in assembly mode when you start a new file. The Model tab in the browser will be the active tab. In Model mode, you can create a single part or multiple parts.

If you want to create just a single part, start a new file under the file pull-down menu with the New Part File option. With that option, you cannot create more than a single part in that file and assembly-related commands will be disabled.

There are three types of assemblies that Mechanical Desktop can create: top down, bottom up and assemblies that are a combination of both top down and bottom up techniques.

A top down approach refers to an assembly in which all the parts are located in the same file. In other words, the user creates each part from within the top level assembly. These parts can later be externalized and referenced to the assembly. The benefit to top down assemblies is that all the parts are stored in an easy to manage single file. The down side is that the file size can grow substantially, slowing down the modeling process.

Bottom up refers to an assembly where all the parts are external to the assembly file and referenced in. The *user* creates the parts in *their* own files and then builds the assembly by referencing the parts to it. The advantage to bottom up assemblies is that the assembly file is usually much smaller in file size since the files are only referenced in and others can easily use the same parts in other assemblies. The down side to bottom up assemblies is that file management can become an issue. Each company needs to evaluate which method is best for them.

TOP DOWN APPROACH

In Chapter 7 you learned how to create multiple local parts in the same file as well as instances. When there are multiple local files in the same file they can be referred to as a top down assembly. If you want to create multiple parts in the same file do not use the New Part File command, as this will only allow a single part per file. For review, there were three methods covered in Chapter 7 that can be used to create a new part in a file. The first three methods will be reviewed and a fourth method called drag and drop will be covered.

The first method is by using the browser. Click the right mouse button in a blank area in the browser. A pop-up menu will appear for selecting New Part. At the command line, you will see a prompt:

```
Select an object or enter new part name <PART2>:
```

The second method is to issue the New Part command **(AMNEW)** from the Part Modeling toolbar or right click in the drawing area and select New Part from the Part menu. After issuing the command, you will be prompted for the same information as in the browser method.

The third method is to make a new part based on an existing part. To create another part that is based on an existing part, use the Assembly Catalog command **(AMCATALOG)** found on the bottom of the browser. The Assembly Catalog dialog box will appear; select the All tab and all the part names will appear. Right click on the part name to base a new part on and select Copy Definition. Type in a new name in the Copy Definition dialog box and you will be returned to the drawing where you can select a new point to insert the new part into the file. This new part has no relationship to the original part unless global design variables were used. The Assembly Catalog command will be discussed next.

Multiple Documents and Drag and Drop Parts

In Mechanical Desktop 4.0 you can open as many files as needed. This is referred to as multiple document environment. The screen can be split as needed to show all the files that are open by using the options under the Window pull-down menu. You can switch between the open files to model, edit or inquire the files as needed. With the multiple document environment you can also drag a part from one file to another. To do so open the file that the part exists in that will be copied then open the file that the part will be copied into. Each file needs to be in model or assembly mode. The copied part will have no relationship to the original part. Split the screen so both files are visible. With the left mouse button select the part that you want to copy. Then press the right mouse button and keep the button pressed down and drag the cursor to the file that the part will be copied into and then release the mouse button. A popup menu will appear giving you the following options:

Copy Here: The part will be copied to the position of the cursor and will be a Mechanical Desktop part.

Paste as Block: The part will be copied to the position of the cursor and will be an AutoCAD block. The block will need to be exploded before it will be recognized as a Mechanical Desktop part.

Paste to Orig Coords: The part will be copied to the same coordinates in the new file as it is in the original file. The copied part will be a Mechanical Desktop part.

Cancel: This will cancel the command sequence.

Once the part has been copied into the second file, the second file will be the active file. Then proceed modeling or assembling parts as normal.

CATALOG

While working on assemblies you will need to see what parts are in the file, determine if they are local or external, change a parts name, copy a part or insert an instance or reference into the assembly. The Assembly Catalog command will accomplish all of these tasks. Issue the Assembly Catalog command **(AMCATALOG)** by clicking on the Catalog icon on the bottom of the browser as shown in Figure 8.2 or right-click in the drawing area and from the pop-up menu, select Assembly Menu from the bottom of the pop-up menu (if it is not current) then select Catalog. An Assembly Catalog dialog box will appear, as shown in Figure 8.3. In the Assembly Catalog dialog box, you will see what parts are in the file, and you can insert an instance or reference in a file. The Assembly Catalog dialog box has two tabs: External and All. The External tab shows all the external files that are attached to the current assembly file. Details about this tab will be covered in the bottom up assembly section. The All tab shows all the external files that are attached to the file in the left area of the dialog box and all the local files in the assembly on the right side of the dialog box. If there are local parts in the assembly, you can right-click on the part name and choose one of the six options from the pop-up menu as shown in Figure 8.4.

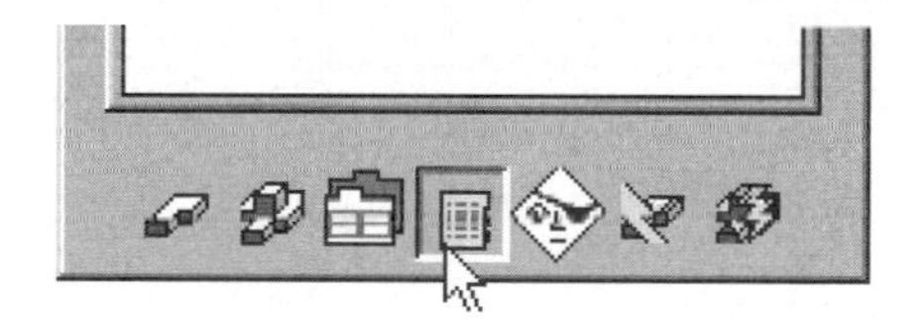

Figure 8.2

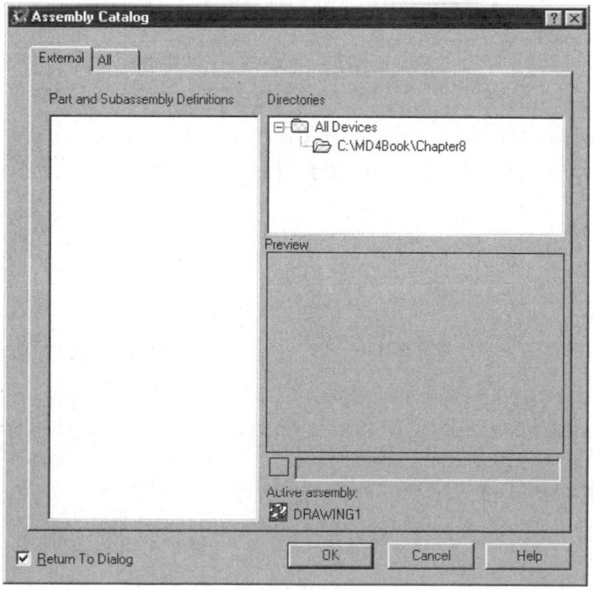

Figure 8.3

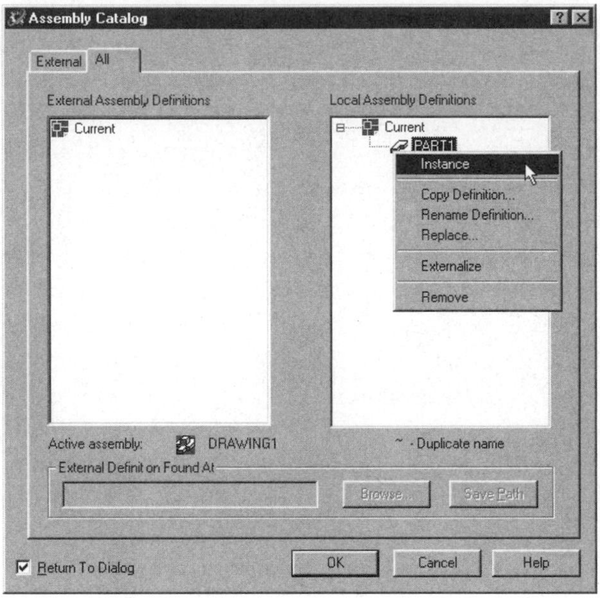

Figure 8.4

Instance: Creates an instance of the part, which is like an AutoCAD block. If one instance changes, so will the other instances.

Copy Definition: Creates a new part based on the selected part. This new part has no relationship to the original part.

Rename Definition: Gives the selected part a new name.

Replace: Replaces the selected part with another part in the current file. The selected part will maintain its current name, but the geometry will be replaced by the new part.

Externalize: Writes the selected part out to a new file. This file is then linked or referenced to the current file.

Remove: Erases the part from the current file.

Another method to move a local part to become an external part or move an external part to a local part is to drag it from one section in the dialog box to the other.

TUTORIAL 8.1—TOP DOWN ASSEMBLY

For this tutorial you will create a top down assembly, and later in the chapter you will apply assembly constraints to position the parts.

1. Start a new file from scratch.

2. Draw, profile and dimension a "4" square and extrude it ".5" in the default direction.
3. Create a new part with the default name Part2.
4. Draw an "L" bracket, profile and dimension it as shown in Figure 8.5.

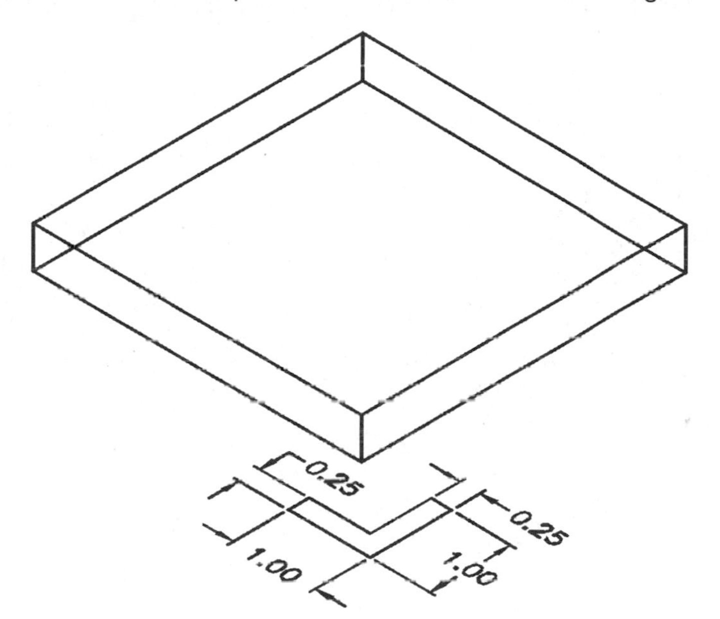

Figure 8.5

5. Extrude the "L" bracket ".25" in the default direction.
6. Create a new part with the default name Part3.
7. Draw, profile and dimension a ".5" diameter circle, extrude it "2" in the default direction. When complete, your screen should resemble Figure 8.6
8. In the browser right-click on the name Part2_1, select the copy option from the pop-up menu and select a point to the right side of the current "L" bracket.
9. Issue the Assembly Catalog command **(AMCATALOG)**.
10. Select the All tab and notice the names of the parts.
11. Right-click on the name Part2, select Copy Definition and give it a name "Lfillet" and then place 2 instances of the new part near the left side of the part.
12. Make the part "Lfillet_2" the active part.
13. Add a ".5" fillet to the inside edge of the bracket and update the part. You may want to switch your viewpoint to see the inside edge better. When complete, your screen should resemble Figure 8.7.

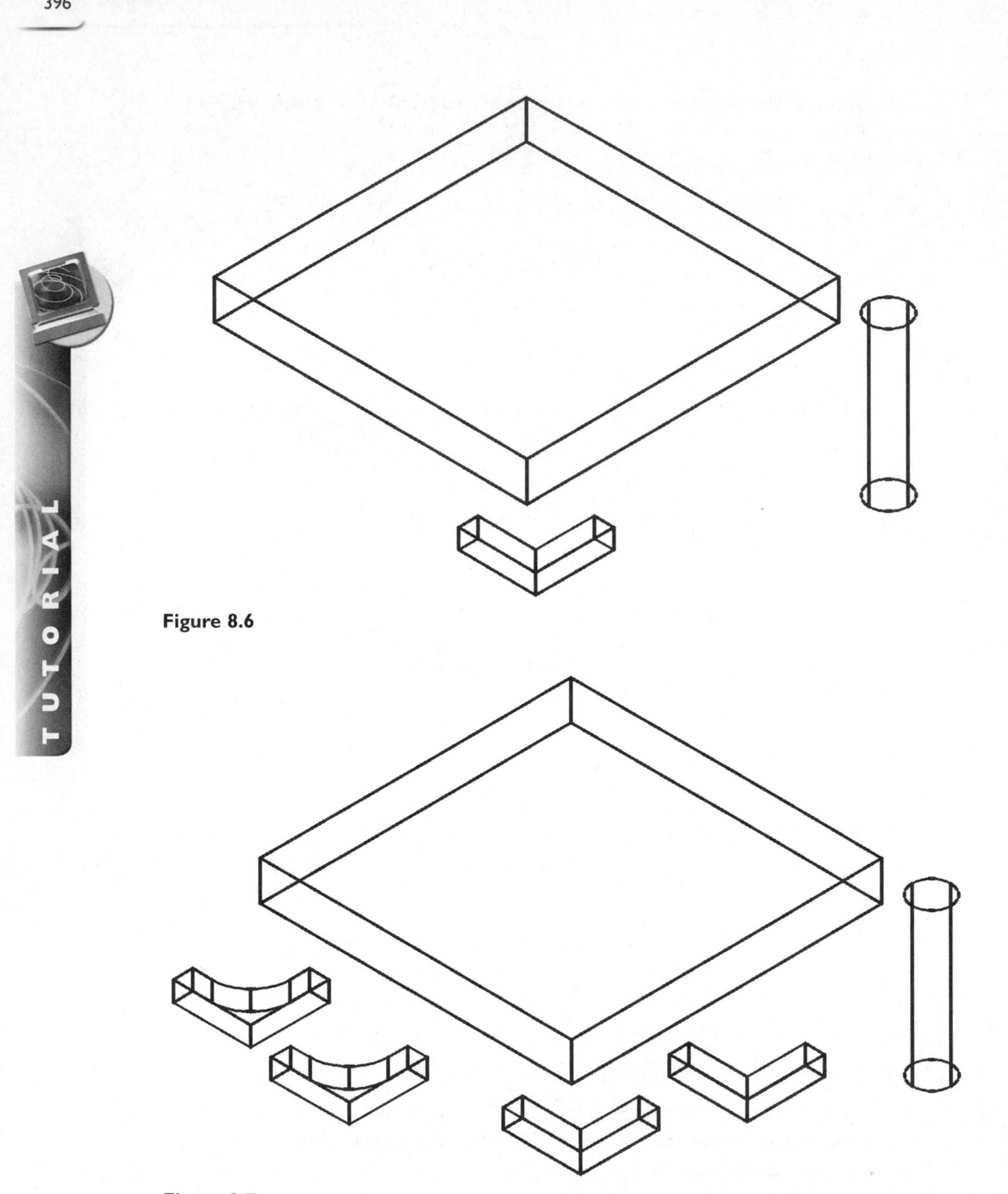

Figure 8.6

Figure 8.7

14. Issue the Assembly Catalog command **(AMCATALOG)**, then right-click on Part3 and select the Remove option. Select OK to confirm that this part definition will be deleted from the file.
15. Repeat the Assembly Catalog command, then right-click on Part2 and select the Replace option. From the drop-down list select "Lfillet" if it is not already selected. Then select OK twice and then select "Yes" to remove the definition of Part2. Part2 will be deleted from the file. When complete, your drawing should resemble Figure 8.8.

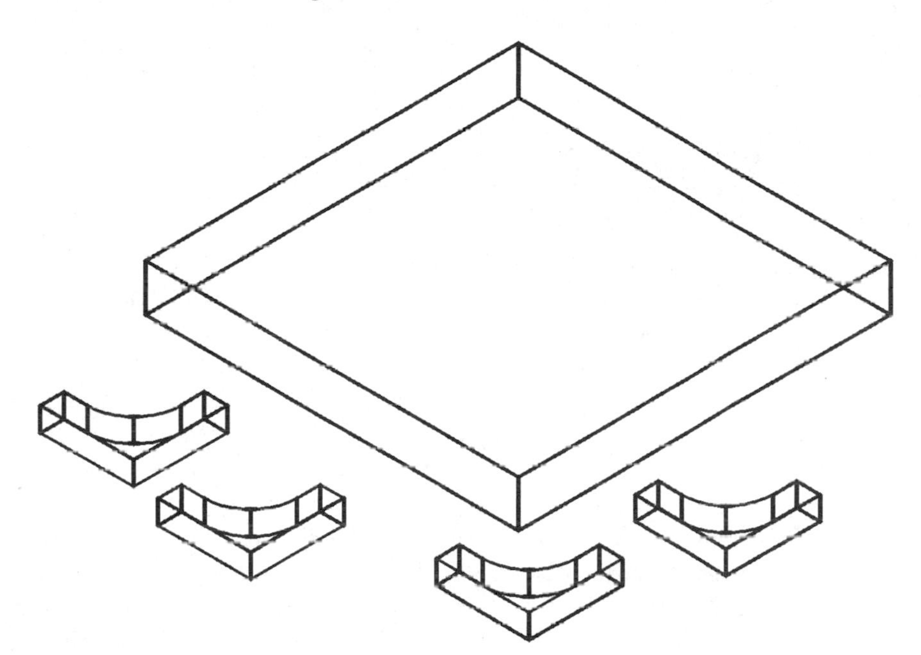

Figure 8.8

16. Save the file as \Md4book\Chapter8\TU8-1.dwg.

TUTORIAL 8.2—MULTIPLE DOCUMENTS AND DRAG AND DROP

1. Start a new file from scratch.
2. Open the file \Md4book\Chapter8\Plate.dwg.
3. Open the file \Md4book\Chapter8\Pin.dwg.
4. From under the Window pull-down select Tile vertically. All three files should be visible on the screen.
5. Select the Plate from the Plate file with the left mouse button, press the right mouse button and keep it pressed.
6. Drag the Plate from into the new file and then release the mouse button.

7. Select Paste to Orig Coords from the pop-up menu.
8. Select the Pin from the Pin file with the left mouse button, press the right mouse button and keep it pressed.
9. Drag the Pin from into the new file and then release the mouse button.
10. Select Paste to Copy Here from the pop-up menu.
11. To show that the copied files are Mechanical Desktop parts, make the Plate the active part in the new file. Edit the diameter of the drilled hole to ".375" and update the part if needed.
12. In the new file make the Pin the active part. Edit the diameter of the drilled hole to ".375" and update the part if needed.
13. Save the new file as \Md4book\Chapter8\TU8-2.dwg.

BOTTOM UP APPROACH

The bottom up assembly approach uses files that are referenced to an assembly file. If you have used xref's before, you will see a resemblance to the way files are referenced to an assembly. Even though the procedure resembles xref's, do not use the AutoCAD XREF command. To create a bottom up assembly, create the parts in their individual files. If there are multiple parts in single file, they will be brought in as a subassembly. Any drawing views that were created in these files will not be brought into the assembly file. After the parts are created, start a new drawing, which will become the assembly file. In the new drawing, issue the Assembly Catalog command **(AMCATALOG).** The Assembly Catalog dialog box will appear. Select the External tab if it is not already visible (see Figure 8.9). This tab is divided into two main sections. The Part and Subassembly Definitions section is on the left side, which shows the files in the current directory. The right side, Directories section, shows the directories where the files will be found. Not all the directories where the external parts are located will necessarily be shown by default. You can add individual files from directories not listed by using the Browse and Attach option, which will be discussed in the next section. To add or remove directories from the list, right-click in the white area of the Directories area of the External tab. A pop-up menu will appear as shown in Figure 8.9 with the following choices.

Add Directory: An explorer-like browser will appear, allowing you to select the directory to add. You can repeat the procedure to add multiple directories, but only one directory can be current at a time.

Release Directory: Right-click on a directory name in the dialog box. It will be removed from the current set of directories.

Browse & Attach: Allows you to select individual files from any directory.

Release All: Removes all directories from the existing set of directories.

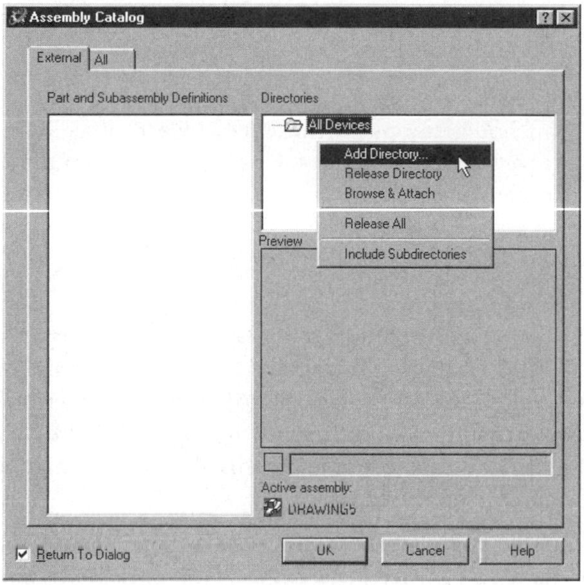

Figure 8.9

Include Subdirectories: Includes all subdirectories in the directory listing. This option needs to be set before you add a directory.

After adding a directory, or directories, make one of the directories current by selecting it with the left mouse button. The files in that directory will appear on the left side of the dialog box.

Viewing:

To see an image of a file, left-click on the file name in the part and Subassembly Definitions area of the external tb and a thumbnail image will appear in the preview window. This preview of the image shows the file as it appeared on the screen when it was last saved.

Sorting:

To sort the files alphabetically or chronologically, right-click in the white area of the Part and Subassembly Definitions section and select alphabetical or chronological.

Inserting:

To insert a file, you can either double-click on the file name or right-click and select Attach. You will then be returned to the drawing area to select a point where you want the part inserted. To insert another instance of the part, select another point, or press ENTER to return to the Assembly Catalog dialog box. The gray background will be removed from the file that was inserted to show that the part is referenced into the file.

Editing:

To edit an external assembly or part, there are three methods.

1. **Open and Edit**
   The first method is to open the part file in Mechanical Desktop, edit the part, update the part and save the part. An easy way to open the file is to right click on the part name in the browser and select Open to edit from the pop-up menu. Then open the assembly file; the changes to the part will be reflected in the assembly or update the external files.

2. **Localize and Edit**
   The second method is to localize the part using the Part Catalog command **(AMCATALOG).** A localized part will have no relationship to the external part file. To localize a part, select the part name in the Assembly Catalog dialog box with the right mouse button and then select Localize or from the All tab drag and drop the part from the External Assembly Definitions area to the Local Assembly Definitions area.

Note: If the external part has drawing views already created, they will be lost when localized. Therefore, it is NOT recommended to use this option when drawing views have been created.

3. **Edit in Place**
   The third method is to edit the part in place, which means that the referenced part will be edited in the assembly and the changes will be saved back to the file. There are three main steps to edit a referenced part in place.

   1. Activate the part to edit: There are two methods. Using the Activate command **(AMACTIVATE)** as shown in Figure 8.10. From the part Modeling toolbar then select the part to edit or in the browser double-click on the part name to activate it. After activating the part to edit all other parts in the file will be grayed outed.

   2. Edit the part as though it was a local part. Make sure that the part is updated after making the edits.

   3. Save the changes back to the referenced file. Issue the Update command **(AMUPDATE)** and use the commit option or right click in the drawing area and select Commit External Edit. You will be prompted to "Save the changes to the external part file? [Yes/No] <Yes>:" selecting Yes will commit the changes; selecting No will not save the changes back to the referenced file. However, if you choose No, the changes will be committed when you save the assembly file.

If, after making a change to a referenced file, you can UNDO back and then commit the changes back to the referenced part. If you have assembly open, another user can open any of the parts, as long as XLOADCTL is set to 0 or 2. To see the updates to referenced parts in the assembly use Refresh command **(AMREFRESH)** as shown in Figure 8.11 from the Assembly toolbar or right-click in the drawing area and from

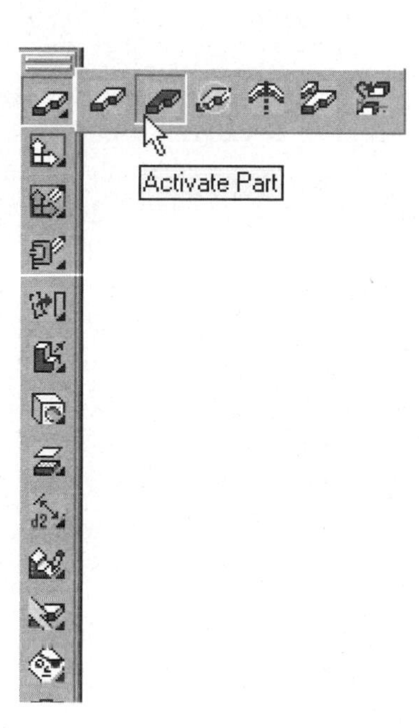

Figure 8.10

the pop-up menu, select Assembly Menu from the bottom of the pop-up menu (if it is not current) then select Refresh from the Assembly menu. To make the main assembly active double-click on the top icon in the Desktop browser.

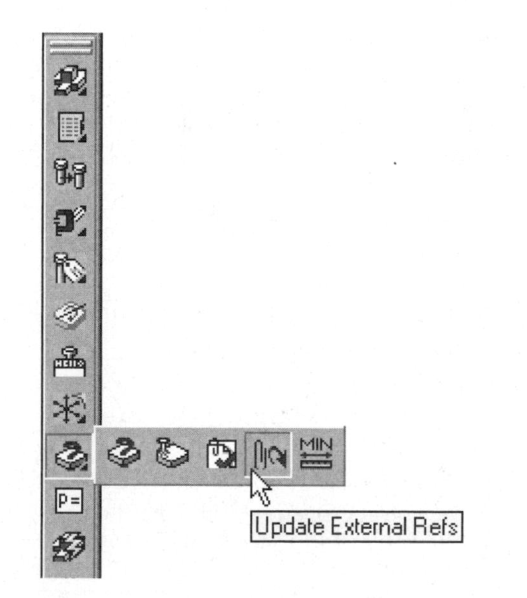

Figure 8.11

EXTERNALIZING

There are times when you would like to take a local part and write it to its own file for detailing etc. This is referred to as externalizing. An externalized part is automatically referenced to the assembly file. To externalize a part, issue the Part Catalog command **(AMCATALOG)** and select the All tab. On the right side of the dialog box the section named Local Assembly Definitions there will be a list of all local parts. Select the part name to externalize with the right mouse button and then choose Externalize from the pop-up menu. Or you can select the name local part name with the left mouse button and keep the button pressed and drag it to the external area of the dialog box and release the mouse button. Then select a directory and name for the file. If there are drawing views associated with the part, they will not go with the externalized part; only the part information is externalized.

Note: The important things to remember are that a file can be attached to an unlimited number of files and that only the part information is inserted in the assembly. No drawing view information will come across.

A single file can be attached to an unlimited number of assemblies.

Changes made to an external part will be reflected in the assembly when it is next opened or refreshed.

An assembly file can have both local and external parts.

TUTORIAL 8.3—BOTTOM UP ASSEMBLY

1. Start a new file from scratch.
2. Issue the Assembly Catalog command **(AMCATALOG)**.
3. Select the External tab.
4. From the Directories area, add the directory where the Chapter 8 files are located: \Md4book\Chapter8.
5. From the files in the Part and Subassembly Definitions section, insert one Base, one Gasket and one Cover in the file. Either double-click on the file name or right-click and select Attach.
6. Insert six bolts from the file HHCS1 file. Again double-click on the file HHCS1 or right-click on the file and select Attach.
7. Switch to an isometric view.
8. Save the file as \Md4book\Chapter8\TU8-3.dwg.
9. Open the file \Md4book\Chapter8\Cover.dwg by right clicking on the part name Cover_1 and select Open to Edit from the list. Create a 1" diameter hole through the center of the part. Save the file.

10. Make the file \Md4book\Chapter8\TU8-3.dwg the active file, then refresh the assembly with the (AMREFRESH) command. There should then be a hole through the center of the cover.
11. Through the Assembly Catalog command, localize the cover.
12. Make the part named Cover active by expanding Cover_1 and double-clicking on Part1_1. Then delete the hole in the center of the part. Update the part if needed.
13. Through the Assembly Catalog command, externalize the cover and save it as \Md4book\Chapter8\Cover2.dwg.
14. Open both files, \Md4book\Chapter8\Cover.dwg and \Md4book\Chapter8\Cover2.dwg, to verify that they are different.
15. Make the file \Md4book\Chapter8\TU8-3.dwg the active file.
16. Activate one of the bolts.
17. Edit the length of the extrusion to "1".
18. Update the part if needed and when finished, your screen should resemble Figure 8.12.

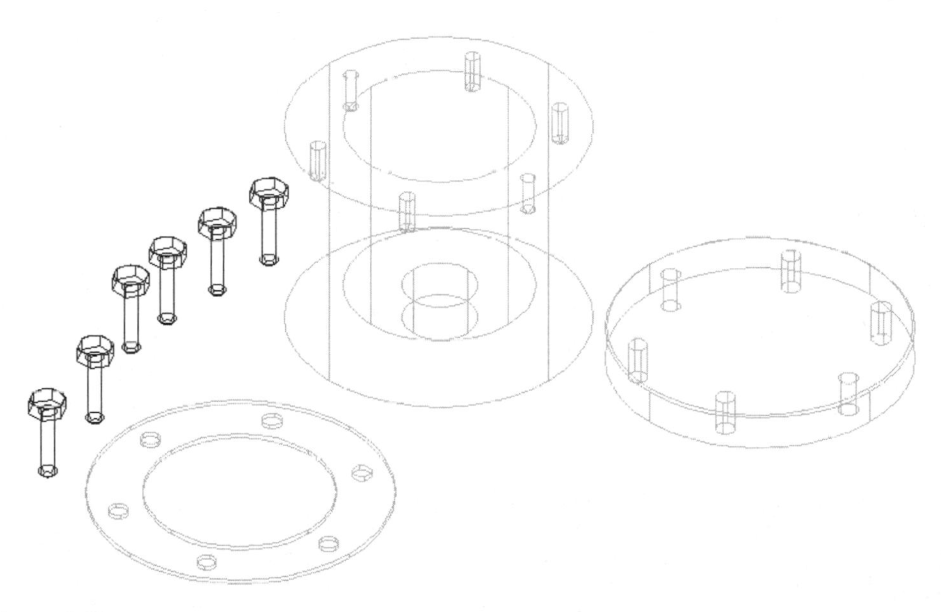

Figure 8.12

19. Commit the changes back to the file \Md4book\Chapter8\HHCS1.dwg using the Update command **(AMUPDATE)** with Commit option.

20. Open the file \Md4book\Chapter8\HHCS1.dwg to verify that the change did occur.

21. Save and close all the files.

SUBASSEMBLIES

While working you may want to group parts together into a subassembly. Each part of the subassembly needs to be created or inserted while the subassembly is active. The parts of a subassembly can be local or external. To create a new subassembly issue the **New** Part/Subassembly command **(AMNEW)** with the subassembly option from the Assembly Modeling toolbar as shown in Figure 8.13, right-click in the drawing area and from the pop-up menu, select Assembly Menu from the bottom of the pop-up menu (if it is not already the current menu) then select New Subassembly from the Assembly menu or right-click in the browser and select New Subassembly from the pop up menu. You will be prompted for a name or press ENTER to accept the default name. Make the new subassembly active by double-clicking on the sub assembly name in the browser or issue the **(AMACTIVATE)** with the assembly option. When a subassembly is active only the parts of the sub assembly will be visible and in the browser the other parts and sub assemblies' names will be grayed out. While in the subassembly create and assemble part as needed. To return to the main file double click on the top icon in the browser. While working at the top level of the file the parts that make up the subassembly will act as one part.

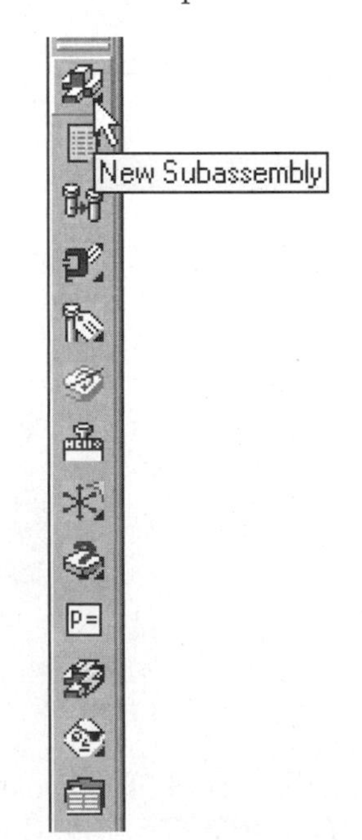

Figure 8.13

TUTORIAL 8.4—SUBASSEMBLIES

1. Start a new file from scratch.
2. Save the file as \Md4book\Chapter8\TU8-4.dwg.
3. Change to an isometric view using the **8** key.
4. Sketch, profile and dimension a 4" square. Extrude it 1" in the default direction.
5. Create a 1" diameter through hole 2" in from each edge.
6. Create a new subassembly by issuing the **AMNEW** command with the Subassembly option. Type in a name of "Wheelassembly" and press ENTER.
7. Make the subassembly "Wheelassembly_1" active by double clicking on its name in the browser. Part1 will disappear from the screen.
8. Create a new part named Wheel.
9. Sketch, profile and dimension a 3" diameter circle. Extrude it 1" in the default direction.
10. Create a 1" diameter through hole that is concentric to the circular extrusion.
11. Create a new part named Shaft.
12. Sketch, profile and dimension a .9" diameter circle. Extrude it 3" in the default direction.
13. Position the parts together using object snaps. (Note: Assembly constraints are the recommended method of positioning parts, assembly constraints will be covered in the next section). Use the Center object snap to move the top of the Shaft to the center of the top of the Wheel. When complete, your screen should resemble Figure 8.14.

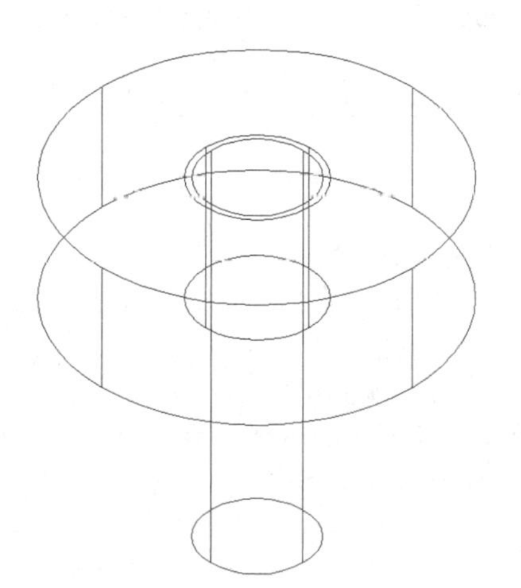

Figure 8.14

TUTORIAL

14. Make the main assembly active by double-clicking on the icon 8.4 in the browser.
15. Position the parts together using object snaps. Use the Center object snap to move the bottom of the Wheel to the center of the top of Part1. When complete your screen should resemble Figure 8.15.

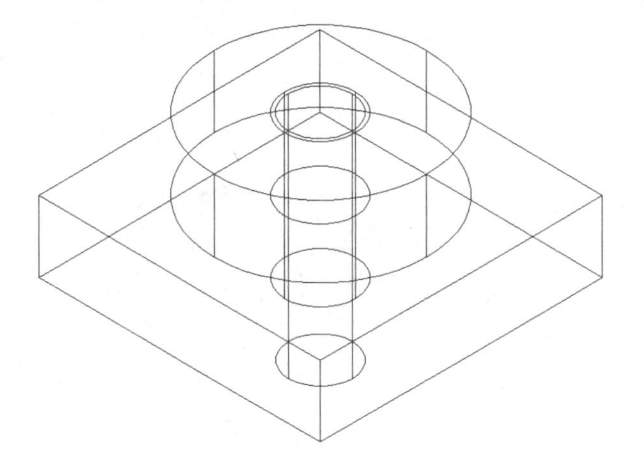

Figure 8.15

TUTORIAL

16. Make the subassembly "Wheelassembly_1" active by double-clicking on its name in the browser. Part1 will disappear from the screen.
17. Make the Shaft the active part if it is not already.
18. Edit the extrusion length to 5", update the part if needed.
19. Reposition the top of the Shaft to the top of the Wheel if needed.
20. Make the main assembly active by double clicking on the icon 8.4 in the browser.
21. Save the file.

ASSEMBLY CONSTRAINTS

In the previous sections you learned how to insert multiple parts in an assembly file, but the parts did not have any relationship to each other. For example, if a bolt was placed in a hole, and the hole moved, the bolt would not move to the new hole position. Assembly constraints are used to create relationships between parts. So, if the hole moves, the bolt will move to the new hole location. In Chapter 1, you learned about geometric constraints. When geometric constraints are applied, they remove the number of dimensions, or constraints, required to fully constrain a profile. When assembly constraints are applied, they reduce the degrees of freedom (or DOF) that the parts can move freely in space. There are six degrees of freedom: three translational and three rotational. Translational means that a part can move along an axis: *X*, *Y* or *Z*. Rotational means that a part can rotate about an axis: *X*, *Y* or *Z*. As assembly constraints are applied, the number of degrees of freedom will be decreased. The first part created or added to the assembly will have zero degrees of freedom. This is the base component, also referred to as grounded. Other parts will move in relation to this part. To see a graphical display of the degrees of freedom remaining on a part, select a part's name in the browser with the right mouse button, then choose the DOF symbol from the pop-up menu or to see all the DOF symbols for all the parts issue the Degree of Freedom command from the Assembly Modeling toolbar as shown in Figure 8.16. An icon and number will appear in the center of the part that shows the degrees of freedom remaining on a part. The number represents the order in which the part appears in the browser. The color of the DOF symbol is controlled by the color of grips and can be altered with the grip Options command **(DDGRIPS).** To see the degrees of freedom of a part in a text format, issue the List Part Data command **(AMLISTASSM)** from the Assembly Modeling toolbar as shown in Figure 8.17, right-click in the drawing area and from the pop-up menu, select Assembly Menu from the bottom of the pop-up menu (if it is not already the current menu) then select Query from the Assembly menu or use the AutoCAD LIST command. After issuing the command, select a part and you will get a listing of the degrees of freedom remaining for the part.

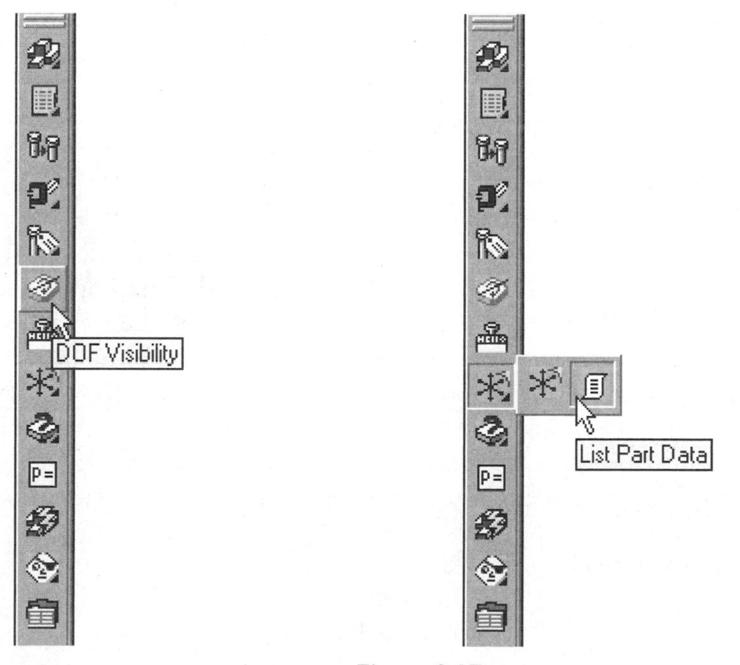

Figure 8.16 **Figure 8.17**

The assembly modeler uses a variational solver, which allows parts to be assembled in an under constrained condition. For example a linkage will have parts that still have degrees have freedom and as the angle changes other parts will adjust accordingly. The first part that appears in the browser is the base part. When inserting or adding parts it is recommended to create them in the order in which they will be assembled. The order will be important when placing assembly constraints and creating scenes. If the parts are not in the correct order they can be reordered using the browser. The process is similar to reordering features. Instead of selecting the feature name to reorder select the part name in the browser with the left mouse button and, keeping the button depressed, move the part up or down until it is in the correct location. Like profiles, assemblies do not need to be fully constrained.

When constraining parts to one another, you will need to understand the terminology used. A list of terminology that is used with assembly constraints follows.

Line: Can be the center of an arc or circular edge, a selected edge or a work axis.

Normal: A vector that is perpendicular to the outside of a plane. With Mechanical Desktop you can flip the normal direction once the plane is selected.

Plane: Can be defined by the selection of a plane or face, two non-collinear lines, three non-linear points, one line and a point that does fall on the line. When edges and points are used to select a plane, this is referred to as a construction plane.

Point: Can be an endpoint or a midpoint of a line or the center of an arc or circular edge.

Offset: The distance between two selected lines, planes, points or any combination of the three.

TYPES OF CONSTRAINTS

Mechanical Desktop has four types of assembly constraints: Mate, Flush, Insert and Angle. The constraints can be accessed through the individual constraint commands Mate **(AMMATE)**, Flush **(AMFLUSH)**, Angle **(AMANGLE)** and Insert **(AMINSERT)** from the Assembly Modeling toolbar as shown in Figure 8.18 or right-click in the drawing area and from the pop-up menu, select Assembly Menu from the bottom of the pop-up menu (if it is not already the current menu) then select the specific constraint from the 3D Constraints menu. No dialog box will appear when applying assembly constraints. Next is a description of the four types of assembly constraints.

Mate:

There are three types of mate constraints: plane, line and point.

> **Mate plane:** assembles two parts so that the normals on the selected planes will be opposite one another when assembled.

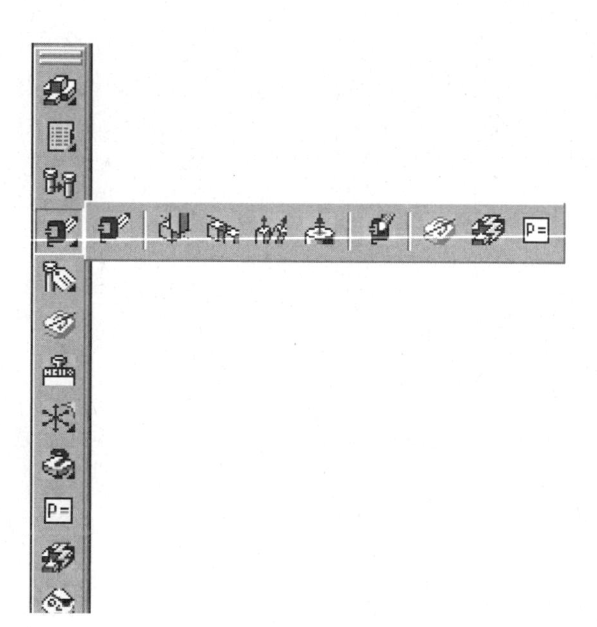

Figure 8.18

Mate line: assembles to line or center lines of arcs and circular edges together.

Mate point: assembles two points (center of arcs and circular edges, endpoints and midpoints) together.

Flush:

The normals on the selected planes will be pointing in the same direction when assembled.

Angle:

You will specify the degrees between the selected planes.

Insert:

Select the circular edges of two different parts and the centerlines of the parts will be aligned and a mate constraint will also be applied to the planes defined by the circular edges. The Insert constraint takes away five degrees of freedom with one constraint. It only works with parts that have circular edges. Circular edges define a centerline/axis and a plane.

The figures below show geometry before and after assembly constraints are applied.

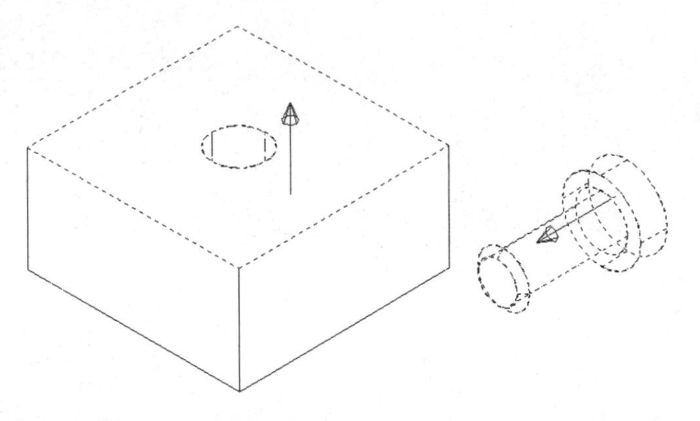

Figure 8.19 *Before applying mate constraint.*

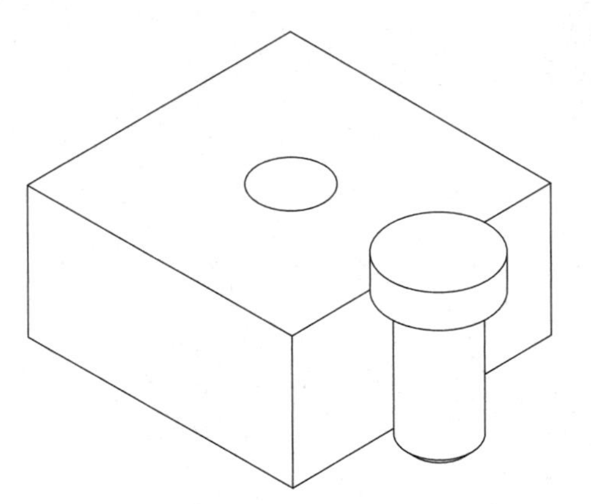

Figure 8.20 *After applying mate constraint.*

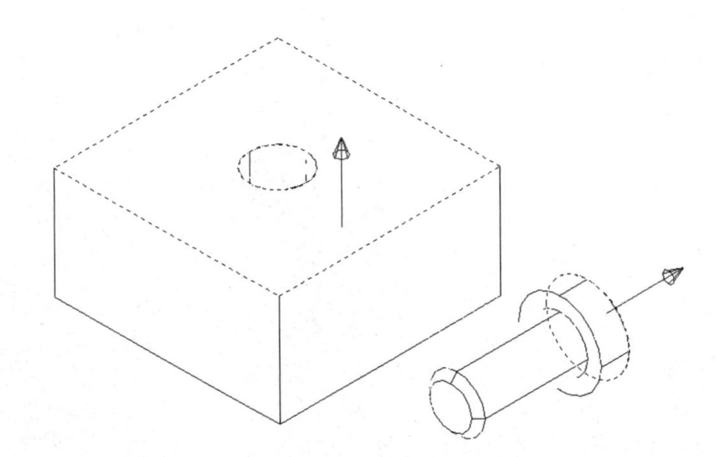

Figure 8.21 *Before applying flush constraint.*

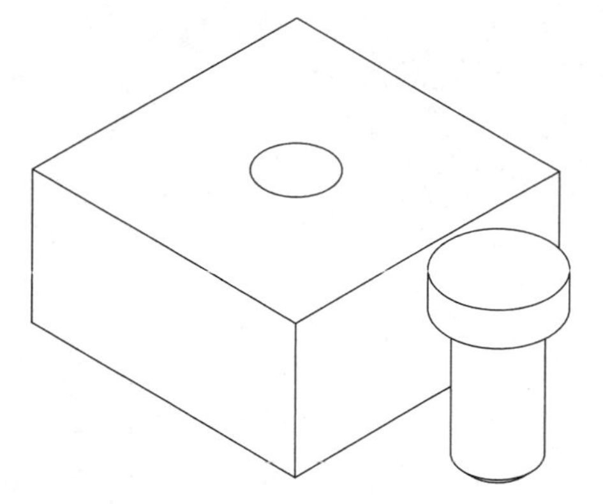

Figure 8.22 *After applying flush constraint.*

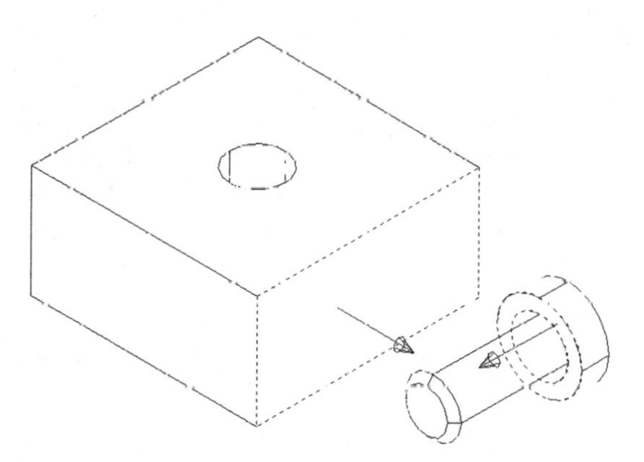

Figure 8.23 *Before applying angle constraint.*

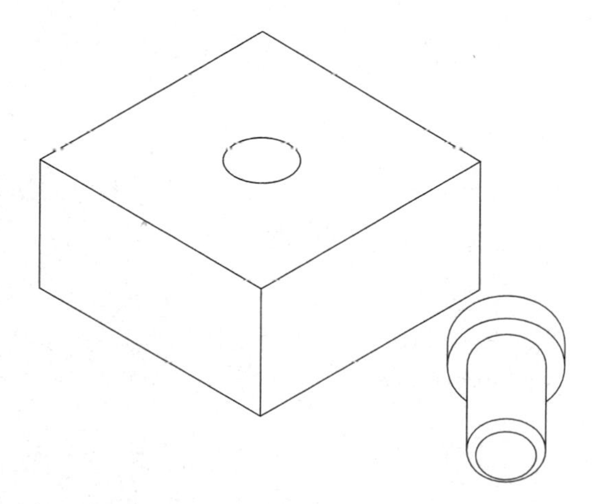

Figure 8.24 *After applying angle constraint.*

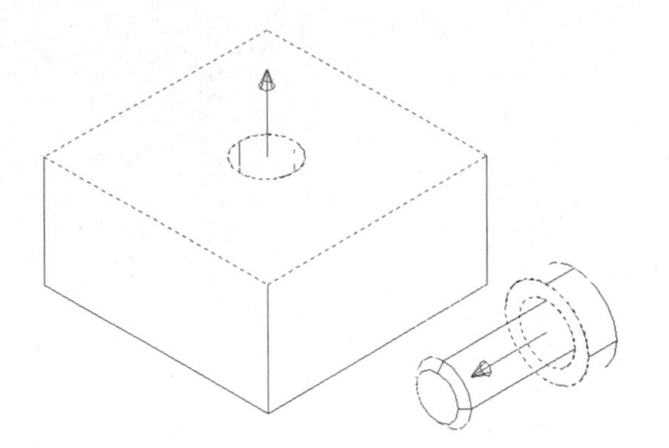

Figure 8.25 *Before applying insert constraint.*

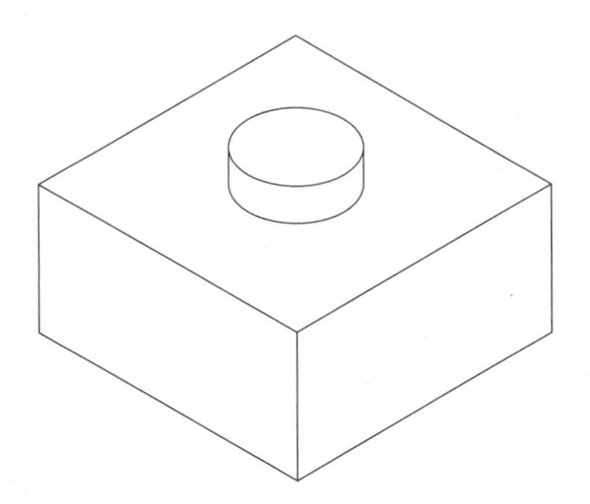

Figure 8.26 *After applying insert constraint.*

After selecting the assembly constraint that you want to apply, you will be prompted to:

```
Select first set of geometry:
```

Depending on the geometry you select, Mechanical Desktop will solve for that selection and allow you to cycle through all the possibilities or to select another edge or point to define the plane. If you select in the middle of a plane, you can cycle through the planes. If you select on an edge, you will cycle through the endpoints, midpoint, the line or edge itself and the two planes that are adjacent to the edge and if a circular edge is selected you will cycle through the center line (axis), center point, plane and then the tangent face.

After you make the first selection, an icon representing a mouse will appear on the screen. On the screen the left mouse button will be flashing red. By pressing the left mouse button you will cycle through the possible geometry options. The different geometry options willl highlight as you cycle through. If the selected location appears over multiple parts, you will cycle through all the geometry options on all the parts. This object cycling is referred to as Intelligent Constraints. Mechanical Desktop analyzes your choices and cycles through them. When the correct geometry is highlighted, press ENTER or click the right mouse button that is green.

You can also define a construction plane by selecting an edge or point and then moving the mouse over another line or point and selecting it. Continue with this technique until a plane is shown and then press ENTER.

To select a plane that is tangent to an arc or circle, select an isometric line or select on the cylindiral edge that is closest to the point you want, to increase accuracy of the selected point object snaps may be used. Cycle through with the left mouse button until the correct object is highlighted, then press ENTER, or select another object that will define a plane. The next prompt will ask you to:

```
Select second set of geometry:
```

Follow the same steps as for the first set. After defining the second set, you will be prompted for an offset distance. Type in a number and press ENTER. If the geometry does not automatically update, select the Update Assembly icon **(AMUPDATE** with the assemble option) as shown in Figure 8.27 or select the icon on the bottom right side of the browser. Under the Assembly options, you can set Mechanical Desktop to

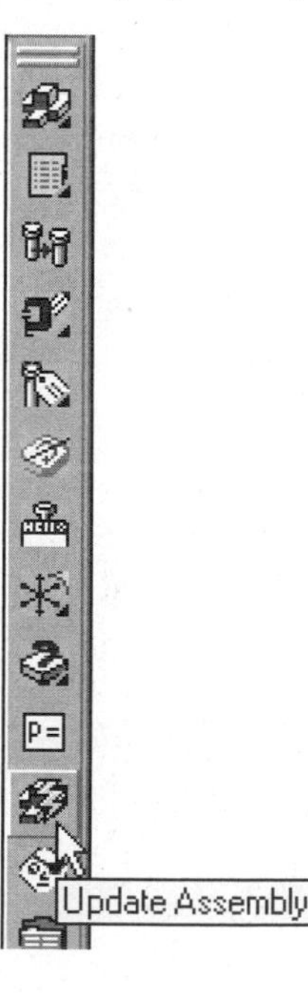

Figure 8.27

automatically Update Assembly as Constrained, which is the default. If this is not checked, it will leave the parts in place until the assembly is updated manually. If the assembled parts make it hard to select new points, use the AutoCAD move command to reposition the parts and then update the assembly. The parts will be moved back to their constrained positions.

Note: You can create theoretical or construction planes or edges by moving the mouse cursor over an edge and cycling through the geometry and then moving it to the next edge and cycling through in the same manner until the plane or edge is defined. Then press ENTER.

A work plane can be used as a plane with assembly constraints.

A work axis can be used to define a line.

TUTORIAL 8.5— CONSTRAINING WITH MATE

1. Open the file \Md4book\Chapter8\TU8-5.dwg.
2. Turn on the degrees of freedom for both side plates by selecting their names in the browser with the right mouse button and selecting the DOF symbol or select Degree of Freedom icon from the Assembly Modeling toolbar. Watch the symbol change as constraints are applied.
3. Issue the Mate Constraint command **(AMMATE)**.
4. Select the bottom plane of the side plate and the back plane of the base, as shown in Figure 8.28 with lines hidden and DOF symbols turned off for clarity. You may need to cycle through with the left mouse button.
5. Press ENTER to accept the default offset distance of "0.00".

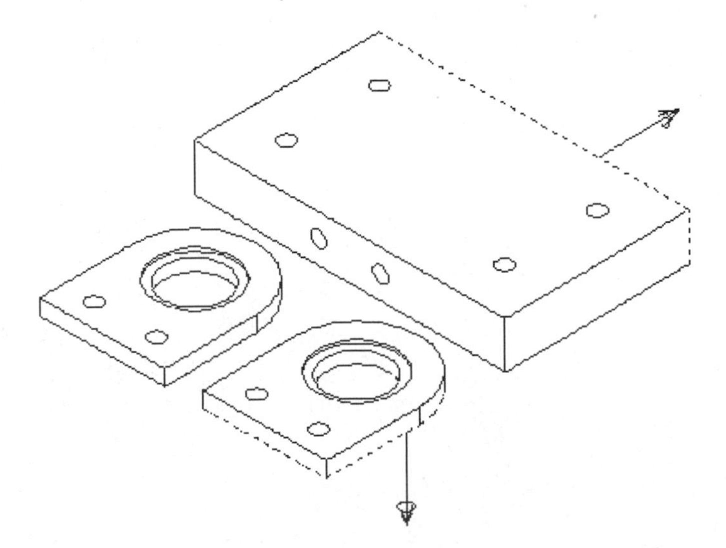

Figure 8.28

6. Repeat the Mate Constraint command.
7. Select the back left hole of the base plate and the upper right hole of the side plate, as shown in Figure 8.29. Your side plate may be in a different position on the screen from that shown. For clarity, the figure is shown with lines hidden and the DOF symbols turned off.

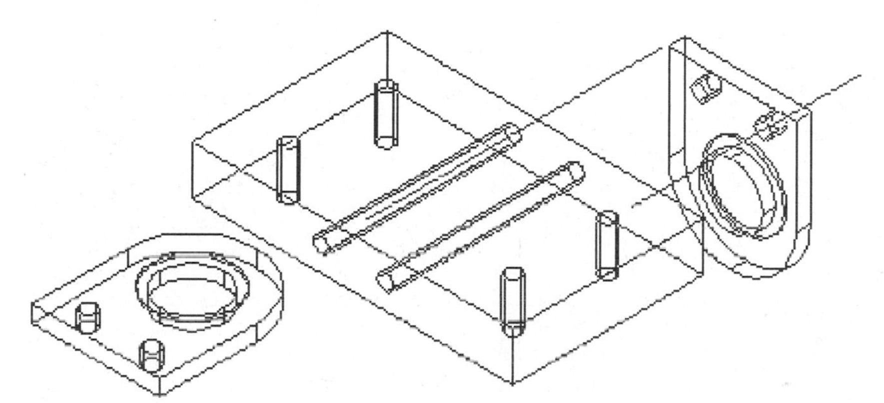

Figure 8.29

8. Press ENTER to accept the default offset distance of "0.00".
9. Repeat the Mate Constraint command.
10. Select the back right hole of the base plate and the left hole of the side plate, as shown in Figure 8.30. Your side plate may be in a different position on the screen from that shown.

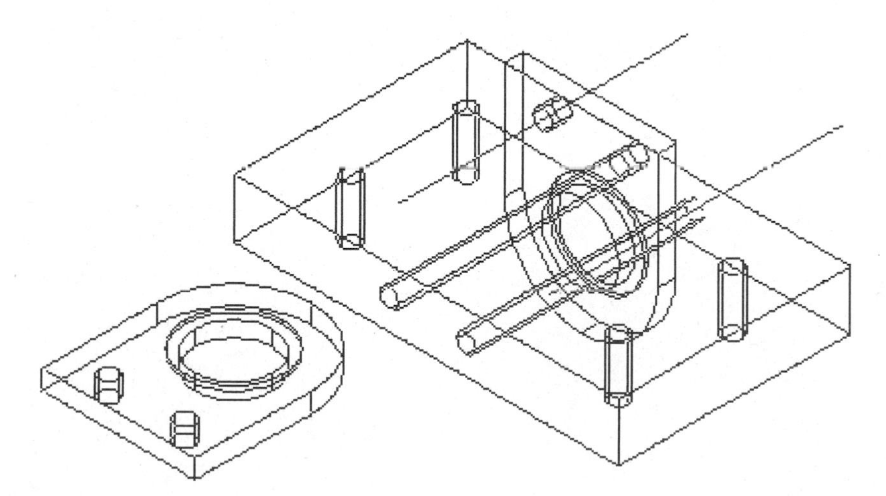

Figure 8.30

11. Repeat the above steps to assemble the other side plate to the Base. When complete, your screen should resemble Figure 8.31.
12. Save the file.

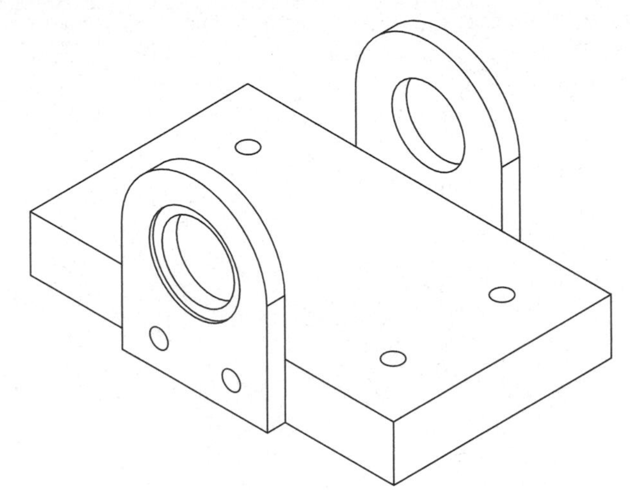

Figure 8.31

TUTORIAL 8.6—CONSTRAINING WITH FLUSH

1. Open the file \Md4book\Chapter8\TU8-6.dwg.
2. Issue the Mate Constraints command **(AMMATE)**.
3. Select the bottom plane of the left guide and the top plane of the base, as shown in Figure 8.32 with lines hidden. You may need to cycle through with the left mouse button.

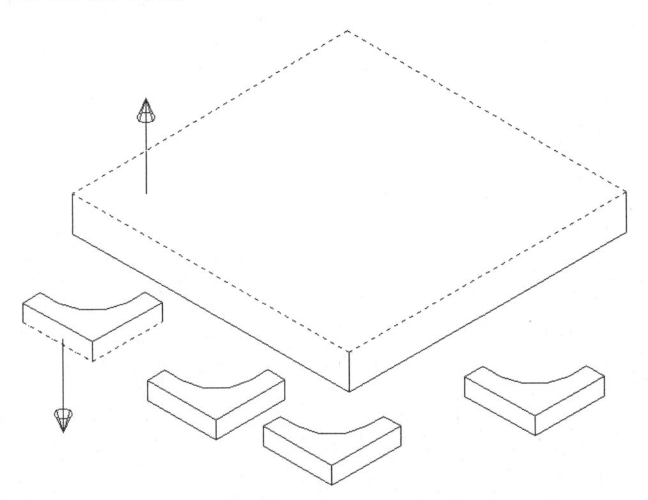

Figure 8.32

4. Press ENTER to accept the default offset distance of "0.00".
5. Issue the Flush Constraint command **(AMFLUSH)**.
6. Select the front left face of the guide and the left back of the base, as shown in Figure 8.33.

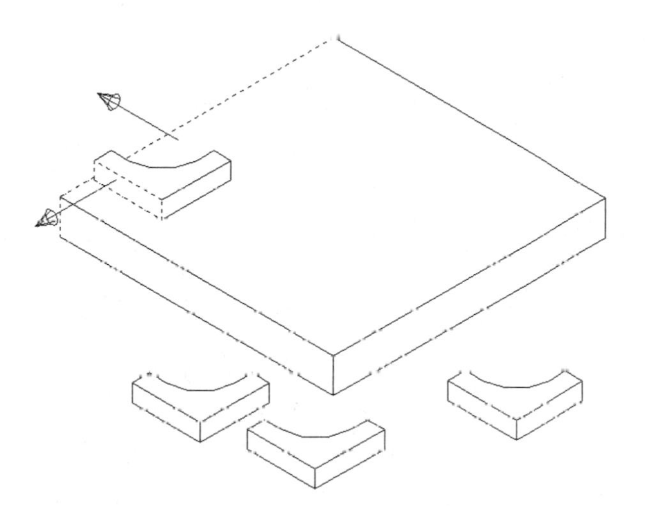

Figure 8.33

7. Repeat the Flush Constraint command.
8. Select the front left face of the guide and the left front of the base, as shown in Figure 8.34.

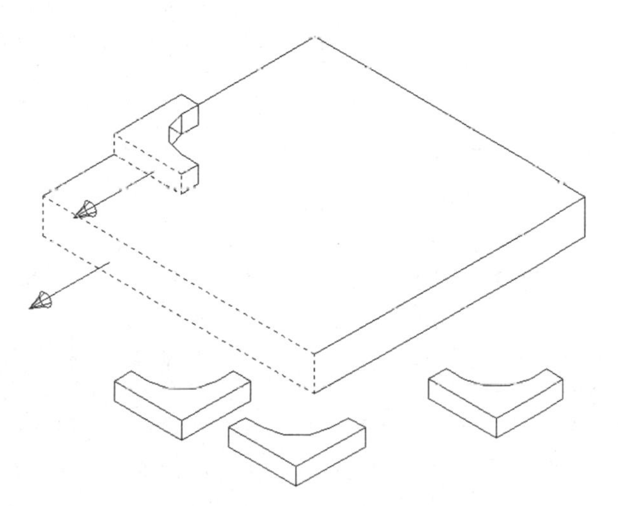

Figure 8.34

9. Press ENTER to accept the default offset distance of "0.00".
10. Constrain the other three guides with the Mate and Flush constraints. When you are finished, your drawing should look like Figure 8.35.
11. Save the file.

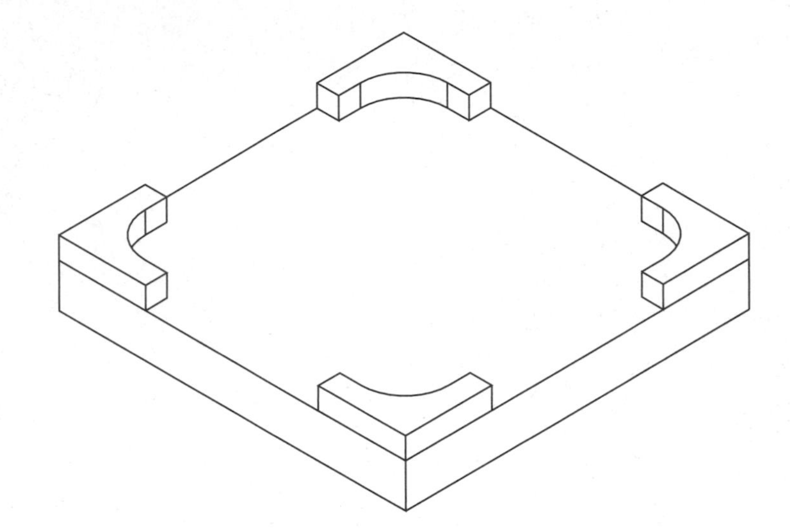

Figure 8.35

TUTORIAL 8.7—CONSTRAINING WITH INSERT

1. Open the file \Md4book\Chapter8\TU8-7.dwg.
2. Issue the Insert Constraint command **(AMINSERT)**.
3. Select one of the top circular edges of the base and a circular edge on the bottom of the gasket, as shown in Figure 8.36. Since the circular edges are concentric, it does not matter which edge in the plane is selected.
4. Press ENTER to accept the default offset distance of "0.00".
5. Issue the Mate Constraint command **(AMMATE)**.
6. To align the holes of the base and gasket; select one of the circles from the top of a hole in the base. The centerline should be highlighted. If it is not, cycle through until it is and press ENTER.
7. Select a hole from the outside perimeter of the gasket and cycle through until the centerline is highlighted and press ENTER.
8. Press ENTER to accept the default offset distance of "0.00".
9. Follow the same steps to insert and align the cover.

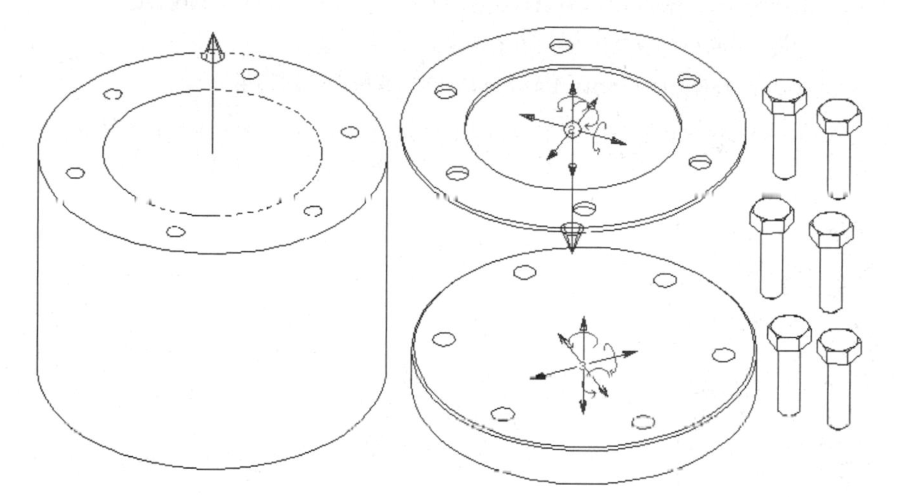

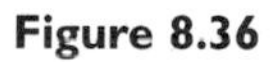

Figure 8.36

10. Then use the Insert constraint to align the six bolts into the six holes of the cover. When complete, your screen should resemble Figure 8.37. Since we do not care how the bolts are rotated in the holes, we will leave them under-constrained.
11. Save the file.

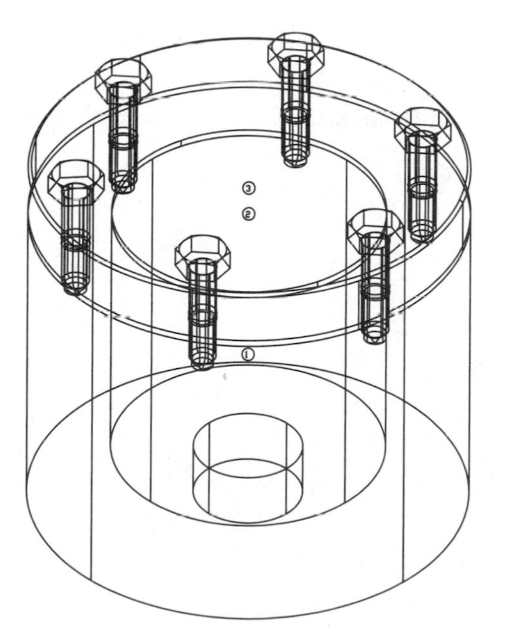

Figure 8.37

TUTORIAL 8.8—CONSTRAINING WITH INSERT AND ANGLE

1. Open the file \Md4book\Chapter8\TU8-8.dwg.
2. Issue the Insert Constraint command **(AMINSERT)**.
3. Select the top left arc of the bottom arm and the lower left arc of the left arm, as shown in Figure 8.38 with lines hidden for clarity.

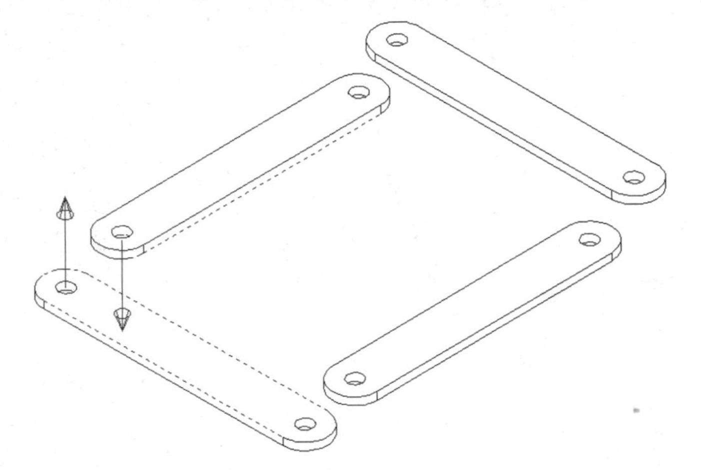

Figure 8.38

4. Press ENTER to accept the default offset distance of "0.00".
5. Press ENTER to repeat the Insert constraint.
6. Select the top right arc of the bottom arm and the lower left arc of the right arm, as shown in Figure 8.39 with lines hidden for clarity.

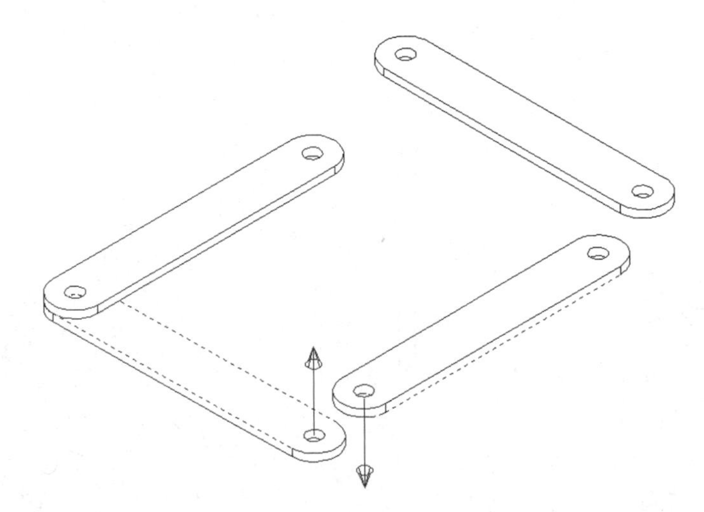

Figure 8.39

7. Press ENTER to accept the default offset distance of "0.00".
8. Press ENTER to repeat the Insert constraint.
9. Select the bottom right arc of the right arm and the top right arc of the top arm, as shown in Figure 8.40 with lines hidden for clarity.

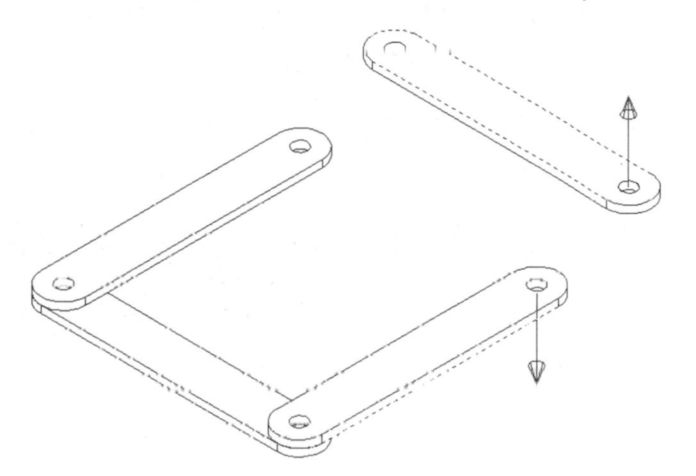

Figure 8.40

10. Press ENTER to accept the default offset distance of "0.00".
11. Turn on the DOF symbol for Arm_4.
12. Use the Mate constraint with the line option to align the center of the left arm and the center of the top arm, as shown in Figure 8.41, lines are hidden for the first three parts for clarity.

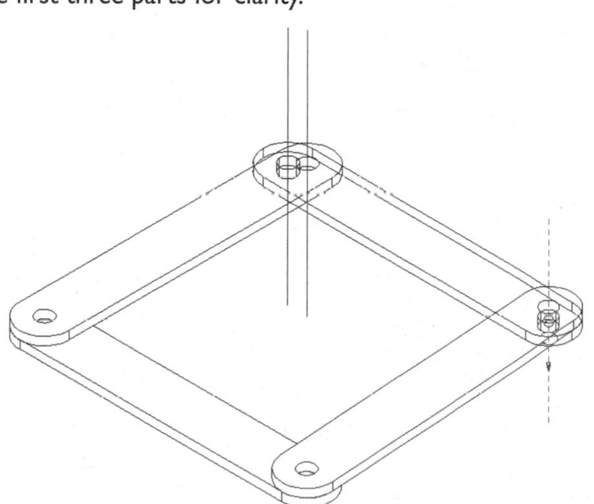

Figure 8.41

13. Issue the Angle Constraint command **(AMANGLE)** to fully constrain the assembly. Select the inside plane of the bottom arm and the inside plane of the left arm, as shown in Figure 8.42.

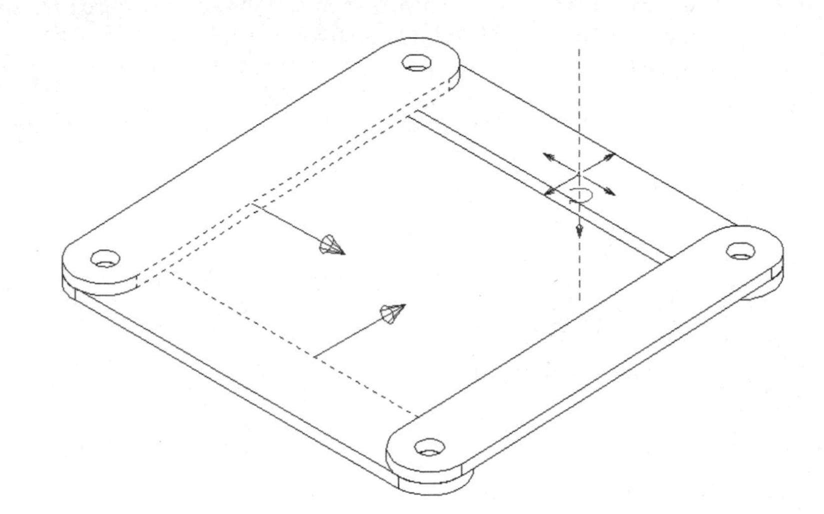

Figure 8.42

14. Type in a value of "-30" and your drawing should look like Figure 8.43.
15. Save the file.

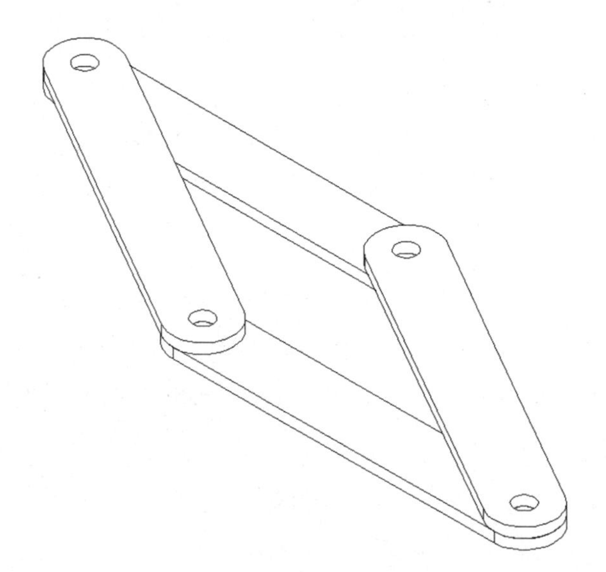

Figure 8.43

TUTORIAL 8.9—CONSTRAINING WITH POINT AND LINE

1. Open the file \Md4book\Chapter8\TU8-9.dwg.
2. Issue the Mate Constraint command **(AMMATE)**.
3. Select the top plane of the base (green) and the bottom plane of Part2 (magenta), as shown in Figure 8.44, with lines hidden for clarity.

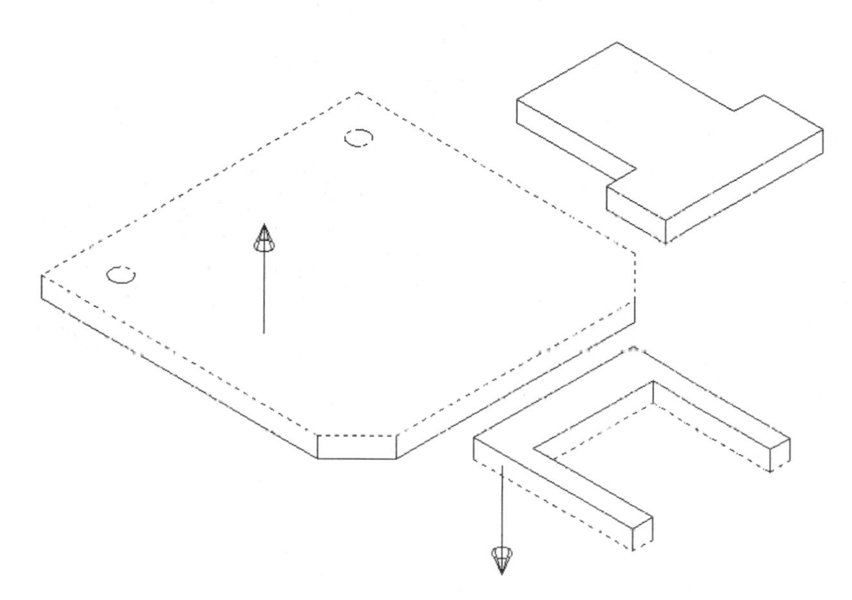

Figure 8.44

4. Press ENTER to accept the default offset distance of "0.00".
5. Press ENTER to repeat the Mate constraint.
6. Select the top back edge of Part1 (green) and cycle through until the middle point is highlighted. Do the same for the back bottom edge of Part2 (magenta), as shown in Figure 8.45 with lines hidden for clarity.
7. Press ENTER to accept the default offset distance of "0.00".
8. Press ENTER to repeat the Mate constraint.
9. Select the top plane of part1 (green) and the bottom plane of Part3 (red), as shown in Figure 8.46 with lines hidden for clarity.

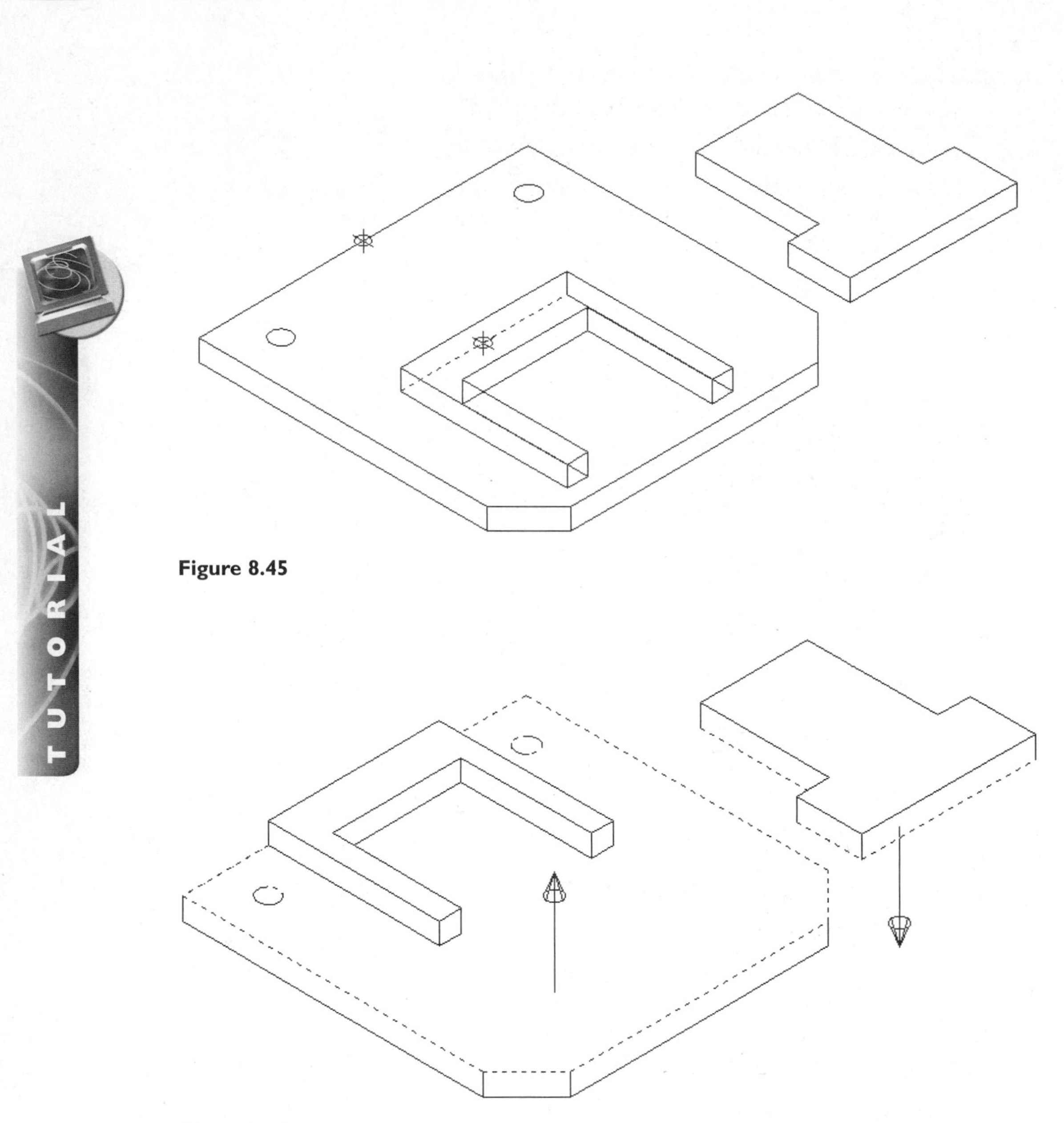

Figure 8.45

Figure 8.46

10. Press ENTER to accept the default offset distance of "0.00".
11. Press ENTER to repeat the Mate constraint.
12. Select the middle plane of Part2 (magenta) and the back plane of Part 3 (red), as shown in Figure 8.47, with lines hidden for clarity.

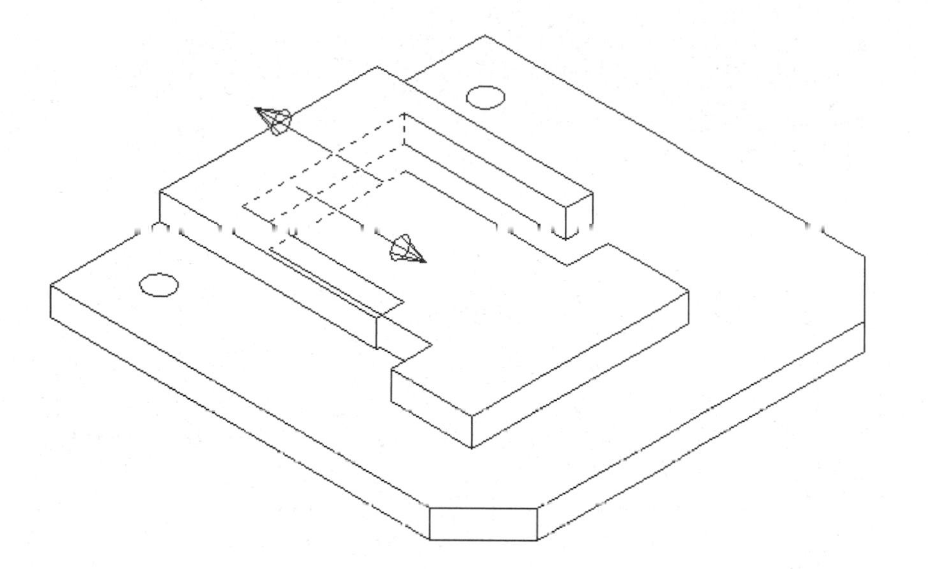

Figure 8.47

13. Type in an offset distance of ".5".
14. Press ENTER to repeat the Mate constraint.
15. Select the front edge of Part2 (magenta) and the front edge of Part 3 (red), as shown in Figure 8.48, with lines hidden for clarity.

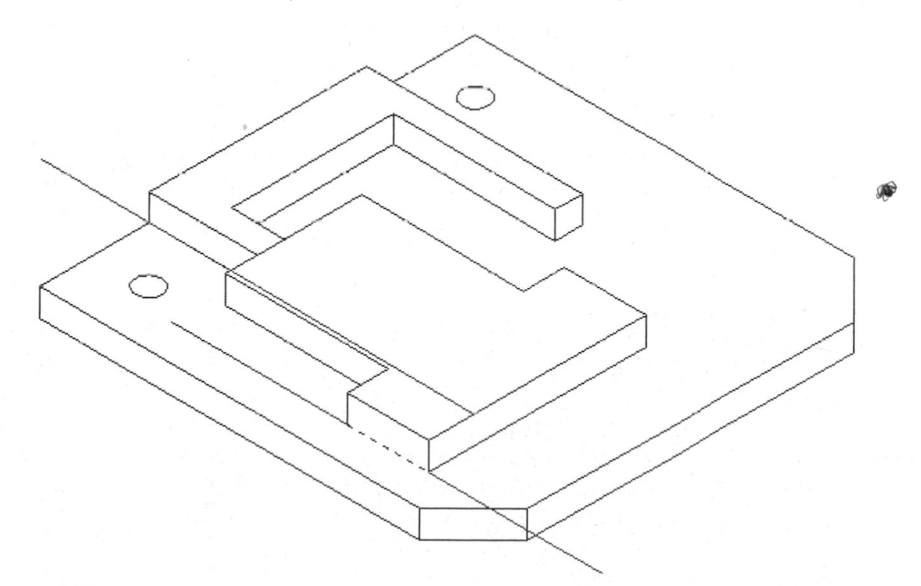

Figure 8.48

16. Press ENTER to accept the default offset distance of "0.00" and when complete your screen should resemble Figure 8.49, with lines hidden for clarity.
17. Save the file.

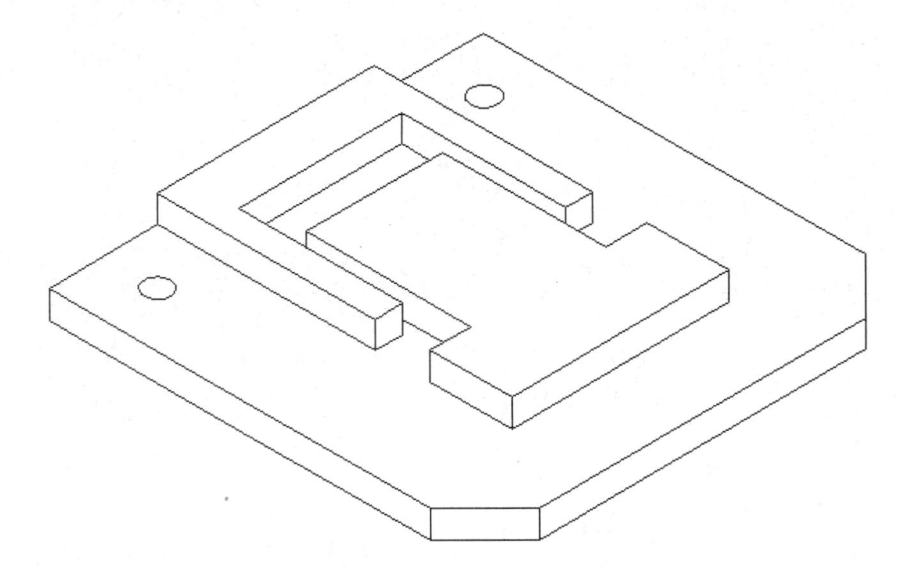

Figure 8.49

TUTORIAL 8.10— CONSTRAINING WITH LINE/PLANE AND POINT

1. Open the file \Md4book\Chapter8\TU8-10.dwg.
2. Issue the Mate Constraint command **(AMMATE)**.
3. Select the top plane of the base (green) and the bottom plane of Part2 (magenta), as shown in Figure 8.50, with lines hidden for clarity.
4. Press ENTER to accept the default offset distance of "0.00".
5. Press ENTER to repeat the Mate constraint.
6. Create a construction plane by selecting the leftmost vertical edge of the bottom plate, cycling through until the edge is highlighted and a line extends beyond the part; do not press ENTER. Then move the mouse to the opposite right vertical edge and select it. Press ENTER to accept this construction plane and the default orientation of the arrow. When complete the construction plane will be positioned through the middle of the part and your screen should resemble Figure 8.51, with lines hidden for clarity.

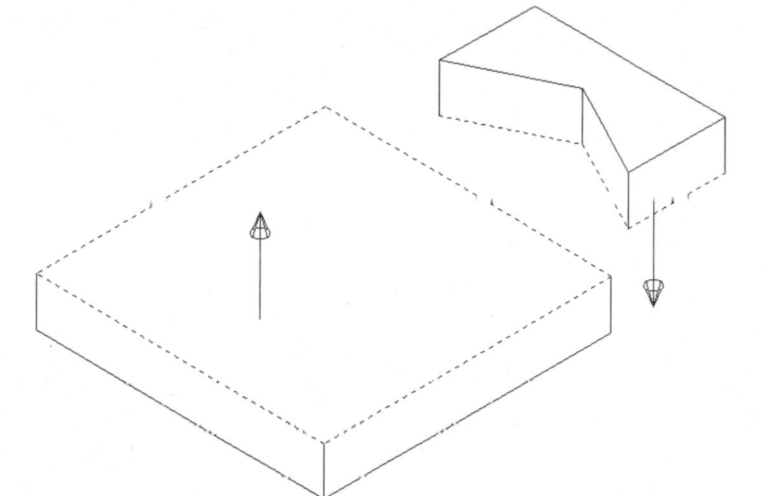

Figure 8.50

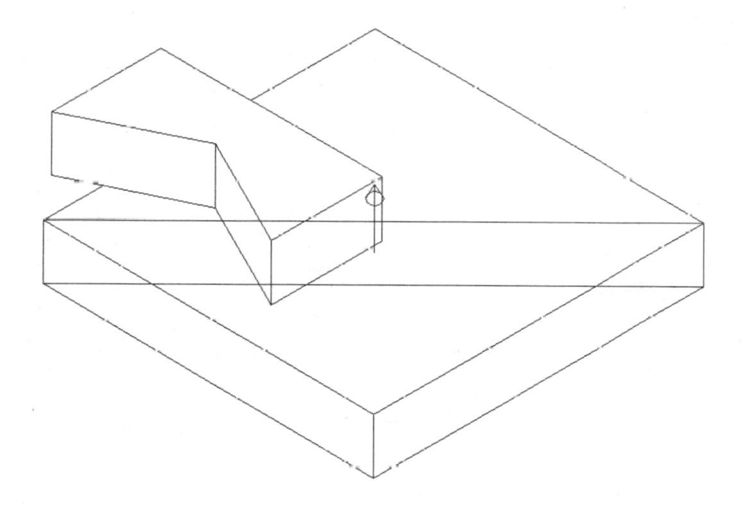

Figure 8.51

7. Create another construction plane by selecting the front left vertical edge in Part2 and then select the front right vertical edge in Part2.
8. Flip the arrow so it is pointing inward as shown in Figure 8.52 (lines hidden for clarity) and then press ENTER.

TUTORIAL

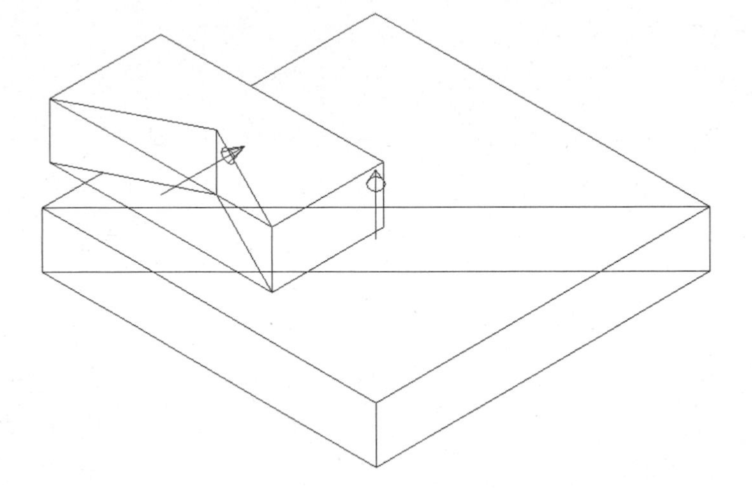

Figure 8.52

9. Press ENTER to accept the default offset distance of "0.00".
10. Issue the Mate Constraint command **(AMMATE)**.
11. Select near the top left edge of Part1 (green), cycle through until the point is highlighted and then press ENTER. Select near the bottom left edge of Part2 (magenta), cycle through until the point is highlighted as shown in Figure 8.53 and then press ENTER.

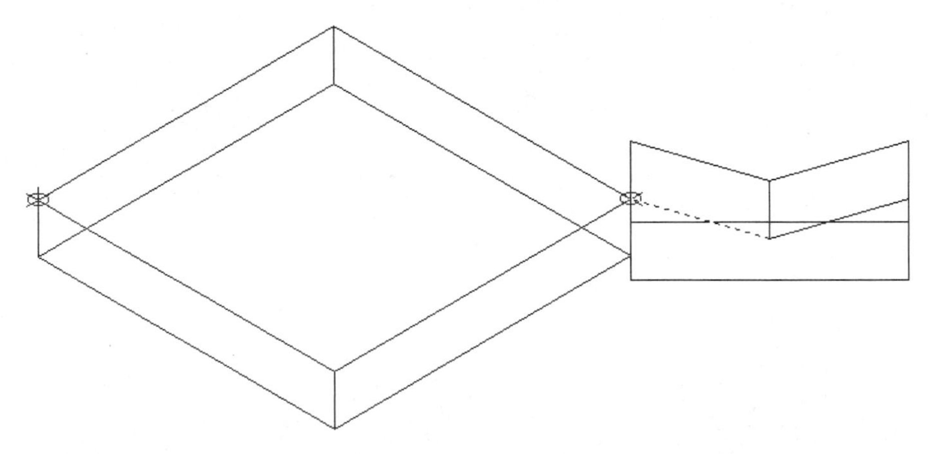

Figure 8.53

12. Type in an offset distance of ".5" and press ENTER. When complete, your screen should resemble Figure 8.54.
13. Save the file.

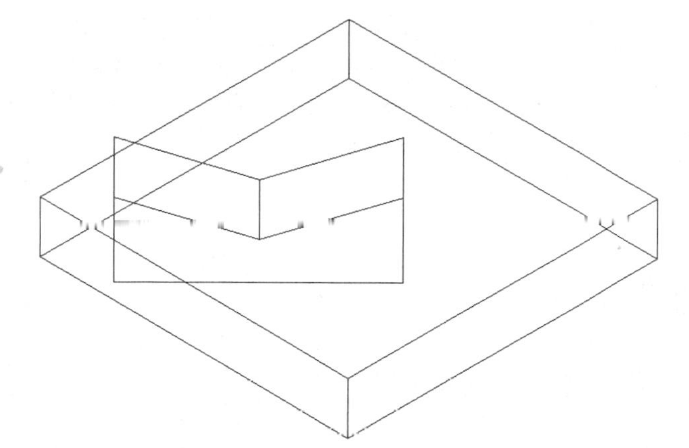

Figure 8.54

EDITING ASSEMBLY CONSTRAINTS

After an assembly constraint has been placed, you may want to edit or delete it to reposition the parts. There are three methods for editing or deleting assembly constraints.

The first two methods are through the browser. In the browser, expand the part name and you will see the assembly constraints. Double-click on the constraint name and an Edit 3D Constraint dialog box will appear, allowing you to edit the constraints. The second method is to right-click on the constraint and a pop-up menu will appear. Choose either Edit or Delete. If you choose Delete, the constraint will be deleted from the part. If you choose Edit, an Edit 3D Constraint dialog box will appear. In the dialog box, type in a new value and update the assembly by selecting the Update icon (lightning bolt) in the same dialog box.

The third method is to select the Edit Constraints command **(AMEDITCONST)** from the Assembly Modeling toolbar as shown in Figure 8.55 or right-click in the drawing area and from the pop-up menu, select Assembly Menu from the bottom of the pop-up menu (if it is not already the current menu) then select Edit from the 3D Constraints menu. After you issue the command, the Edit 3D Constraints dialog box will appear as shown in Figure 8.56. On the left side of the dialog box you will see a listing of the parts that have constraints. When a part name is selected, it will highlight in the drawing window. If you are unsure of a part's name, select the Select button and you will be returned to the drawing area, where you can select a part. You will then be returned to the dialog box. Once a part is highlighted, you can cycle through the constraints by selecting Show; two double arrows will appear. Cycle through the constraints by selecting on the arrows to move forward or backward. Below the arrows you will see the type of constraint, and the constraint will be visible in the draw-

TUTORIAL

ing area. To edit the offset value, type in a new value from the Edit Constraints area. If the assembly does not automatically update, select the lighting bolt icon to re-assemble the parts. To delete the current constraint, select Current from the Delete Constraints area. To delete all the constraints from the current part, select All from the Delete Constraints area.

If you want to change a constraint to a different type, you need to first delete the original constraint and then apply the new constraint. If you do not delete the constraint first and you try to apply a different constraint in its place, you will get the message at the command line "Solve failed" with the constraint name that was being added. For example: "Solve failed. Flush pl/pl constraint not added".

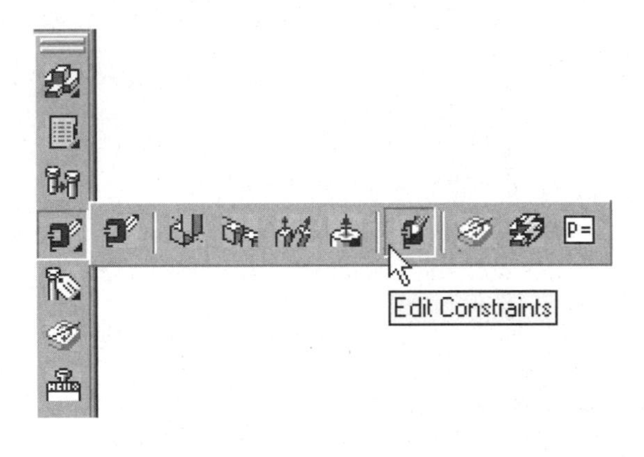

Figure 8.55

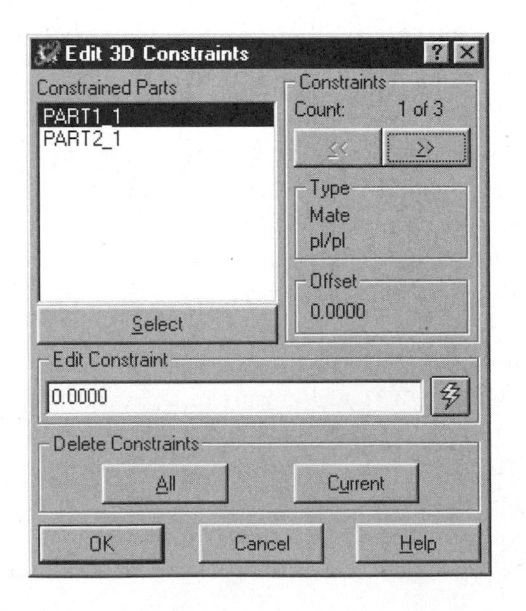

Figure 8.56

TUTORIAL 8.11—EDITING CONSTRAINTS

1. Open the file \Md4book\Chapter8\TU8-11.dwg.
2. Issue the Edit Constraints command **(AMEDITCONST)**.
3. From the Edit 3D Constraints dialog box select U-Chan_1.
4. Select the Show button, then select the >> and cycle to the Third constraint 3 of 3 and change the offset value from ".5" to "0" and then your screen should resemble Figure 8.57.

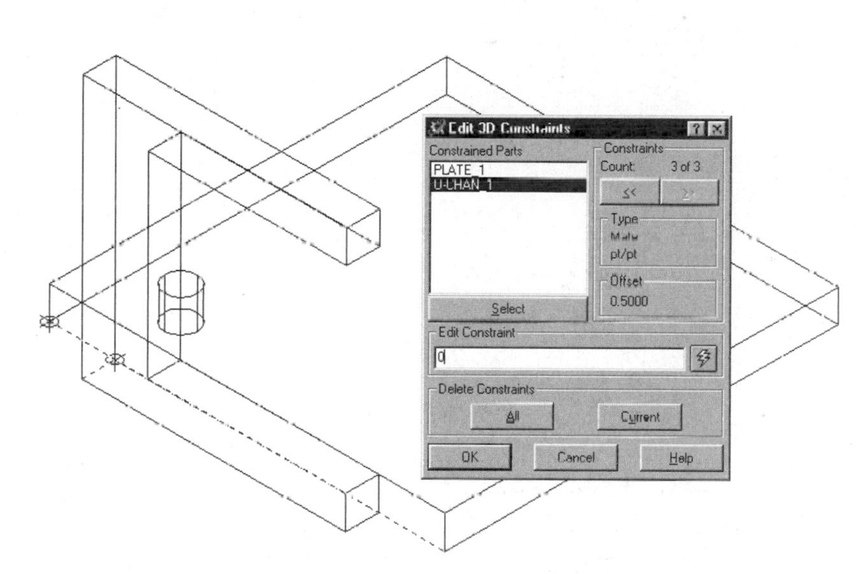

Figure 8.57

5. Select the lightning bolt in the dialog box to update the assembly.
6. While in the Edit 3D Constraints dialog box, select the << button to cycle to the second constraint 2 of 3.
7. Select the Current button to delete the Mate In/In constraint.
8. Select OK to exit the command.
9. Apply a Mate constraint to the bottom of the Plate and to the back side of the U-Chan as shown in Figure 8.58.
10. Press ENTER to accept the default offset distance of "0.00". When complete, your screen should resemble Figure 8.59, shown with lines hidden.
11. Save the file.

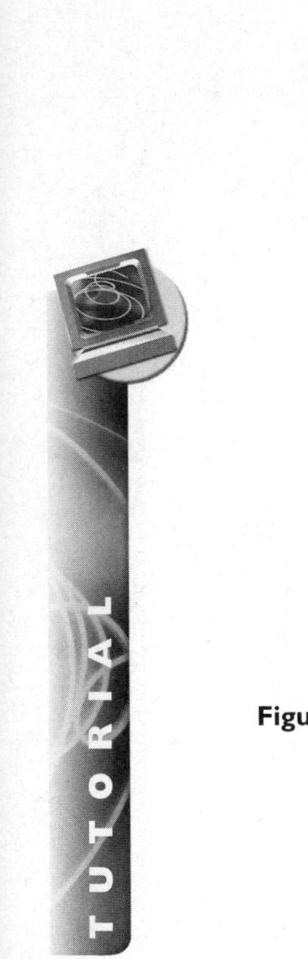

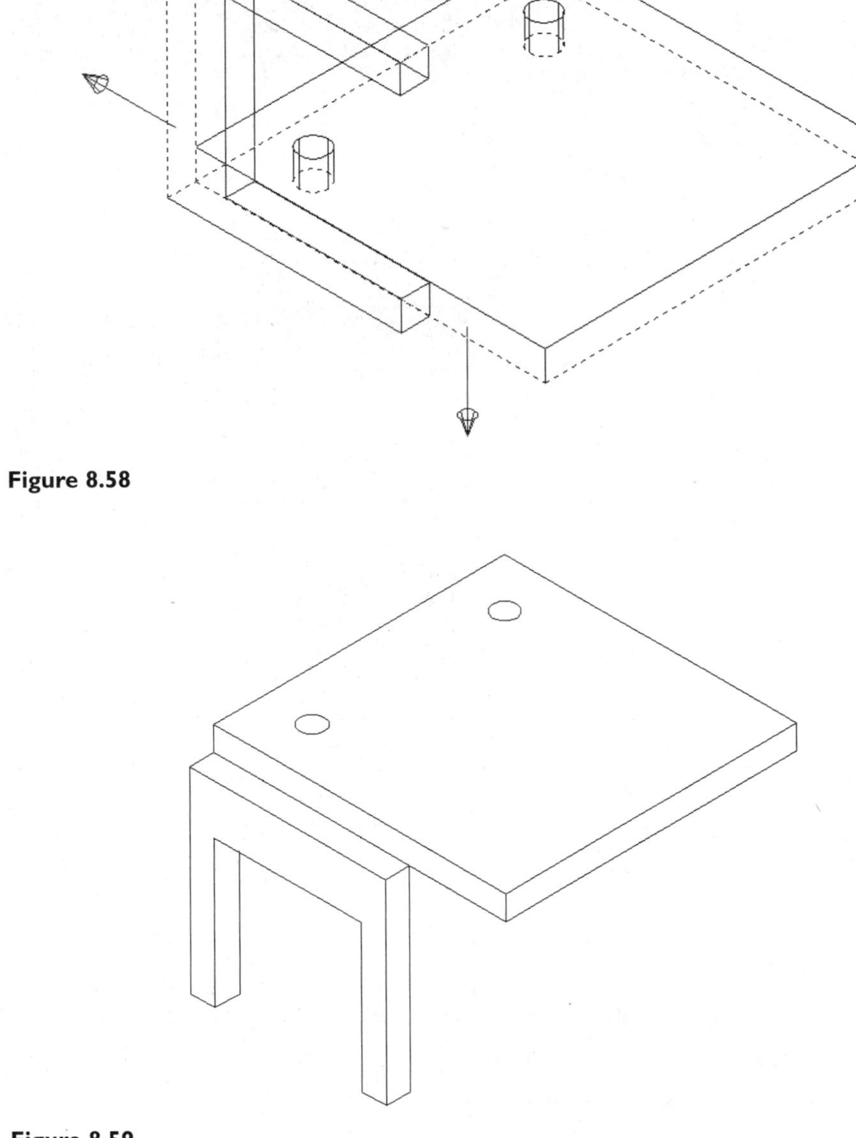

Figure 8.58

Figure 8.59

INTERFERENCE CHECKING

To check the interference between parts, issue the Check Interference command **(AMINTERFERE)** from the Assembly Modeling toolbar as shown in Figure 8-60 or right-click in the drawing area and from the pop-up menu, select Assembly Menu from the bottom of the pop-up menu (if it is not already the current menu) then select Check Interference from the Analysis menu. This command checks interference between two sets of geometry. After issuing the command you will be prompted:

```
Nested part or subassembly selection? [Yes/No] <No>:
```

If you select yes, you will select the subassembly that you want to work with. If you press ENTER, you will be prompted to select the first set. You can select a single part or multiple parts using any AutoCAD selection method; press ENTER. Then select the second set to be checked against the first set and press ENTER to check for interference among the first select set; press ENTER. If no interference is found, you will see a message at the command line:

```
Parts/subassemblies do not interfere.
```

If interference is found, the names of the interfering parts will appear at the command line. You may need to flip screens to see the message. At the command line, you will see a prompt:

```
Create interference solids? [Yes/No] <No>:
```

If you press Y and ENTER, a 3DSolid will be created representing the interference. The 3DSolid will be red and placed on the current layer. Another prompt will appear at the command line:

```
 Highlight pairs of interfering parts/subassemblies?
Yes/<No>:
```

If you press Y, the interfering parts will be highlighted in the drawing area. If there are multiple interferences, press N to advance to the next set. To exit the command, press X.

After exiting the command, you can move the solid or interference and use the AutoCAD distance command along with object snaps to define the amount of interference. The distance reported back is accurate to the number of decimal places set through the **DDUNITS** command. Mass property information can also be checked. After finding out the amount of interference, go back and edit the part(s) to fix the interference. Another option is to use the Combine command **(AMCOMBINE)** to remove or add the 3DSolid that was created after the interference was found. The down side to this method is that the solid or interference is a 3D solid and not a parametric part and it cannot easily be changed.

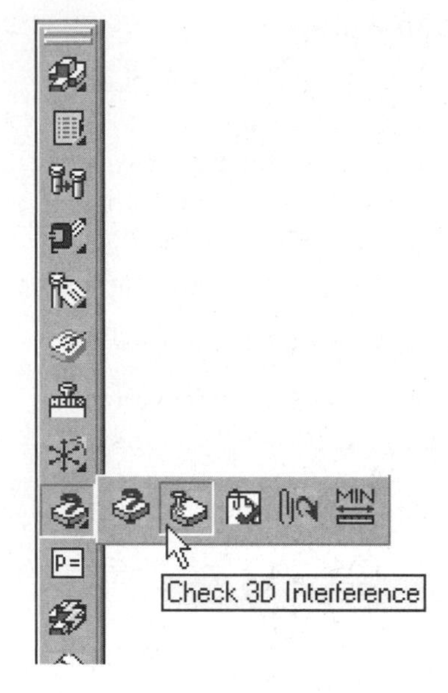

Figure 8.60

TUTORIAL 8.12—INTERFERENCE CHECKING

1. Open the file \Md4book\Chapter8\TU8-12.dwg.
2. Issue the Check Interference command **(AMINTERFERE**).
3. For the prompt:
   Nested part or subassembly selection? [Yes/No] <No>:
   Press ENTER to accept the default.
4. Select both parts and press ENTER.
5. To check for interference between both parts press ENTER.
6. Press the F2 key on your keyboard to see that interference was found in both Part1 and Part2. Press the F2 key to return to the drawing screen.
7. Press Y and ENTER to create an interference solid.
8. Press Y and ENTER to highlight the interfering pairs.
9. Press X and ENTER to exit the command.
10. Use the AutoCAD move command with the "L" option to move the last object created (the interfering solid) to the side.
11. Use the AutoCAD distance command with the quadrant object snap to find how much interference exists.
12. Change one of the part's diameters to remove the interference.

13. Issue the Check Interference command **(AMINTERFERE)** and recheck for interference. If there is still interference, go back through steps 11 through 13 until there is no interference.
14. Save the file.

SCENES

MECHANICAL DESKTOP OPTIONS – SCENE TAB

Before creating a scene we will take a look at the Scene tab of the Edit Options command. Figure 8.61 shows the Scene tab; below, each option is introduced.

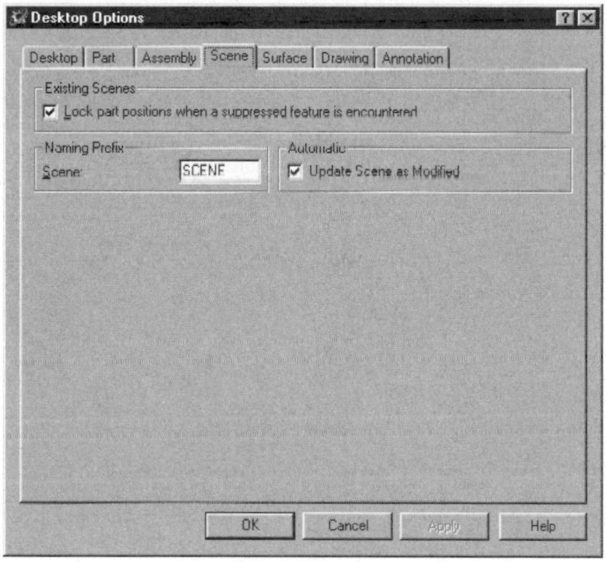

Figure 8.61

Lock part positions when a suppressed feature is encountered:

When checked, any part that has a feature suppressed will be locked in place with regard to the scene(s) and the applied assembly constraints that affect the part. If this is unchecked, a part will be free to move as determined by the applied assembly constraints and a scene's explosion factor, etc.

Scene:

Type in the default name that will be applied when a scene is created. Each scene will be sequenced with an appended number. A different scene name can also be typed in when it is created.

Update Scene as Modified:
When this tab is checked the parts in the scene will move when assembly constraints are applied.

CREATING SCENES

After creating an assembly file, you may want to show all of the parts in different positions, like an exploded view, or hide specific parts for a drawing view. You can also create scenes to show parts in different positions and to create drawing views of an assembly. There is no limit to the number of scenes in a file. To create a scene, issue the Create Scene command (**AMNEW** with the Scene option) icon from the Scenes toolbar as shown in Figure 8.62, right-click in the drawing area and from the pop-up menu, select Assembly Menu from the bottom of the pop-up menu (if it is not already the current menu) then select New Scene or select the Scene Tab from the browser, right-click in a blank area in the browser and choose New Scene from the pop-up menu.

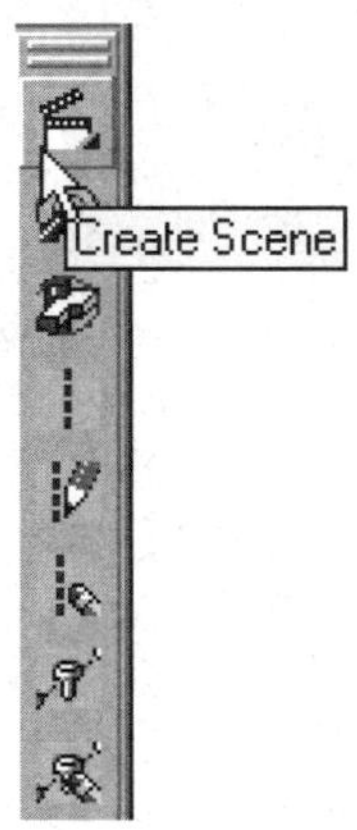

Figure 8.62

After issuing the command, type in a name for the scene and press ENTER. Then type in a value for the explosion factor and press ENTER. The explosion factor will be the distance between the parts. Any number is valid for the explosion factor, including zero. Press ENTER again to make this the active scene. This automatic movement only works for parts that have been constrained with the Mate or Insert assembly constraints. The Mate and Insert constraints have the normals facing toward each other. Think of them with regard to scenes as two magnets that are put together. As the magnets come together, they want to push away from one another. The part that appears first in the browser will be the part that stays stationary; the other part will move away from it. If the parts do not automatically explode, there could be one of three reasons:

1. The parts were not constrained with assembly constraints.

2. There was no Mate or Insert constraint used.
3. There was more than one Mate or Insert constraint used for a single part. If multiple Mates or Inserts were used on a single part, there could be two directions that the part could move. The result is that the part stays stationary.

If a part does not automatically move, you can use the Add Tweaks command (**AMTWEAK**) to move the parts. Tweaks will be covered in the next section. To switch between multiple scenes in a file, issue the Create Scene command (**AMNEW**) and select the scene name from the drop-down list, or double-click on the scene name in the browser under the Scene tab. To tell which scene is active, look in the browser under the Scene tab (the current scene's name will be white and the inactive scenes gray), or look at the lower left corner of the AutoCAD screen, which will show the current scene name.

After you create a scene, the parts will be held in a block and cannot be edited while a scene is current. To edit a part, select on the Assembly tab in the browser. Parts will revert back to the their positions before the scene was created. In this assembly mode, parts and assembly constraints can be added, edited or deleted. After editing the parts, make a scene active and all the changes will be reflected in that scene.

DELETING SCENES

To delete a scene issue the Mechanical Desktop Delete command **(AMDELETE** command with the Scene option) as shown in Figure 6.63 and type in the name of the scene to delete and press ENTER or right click on the scene name in the browser and select Delete from the pop up menu.

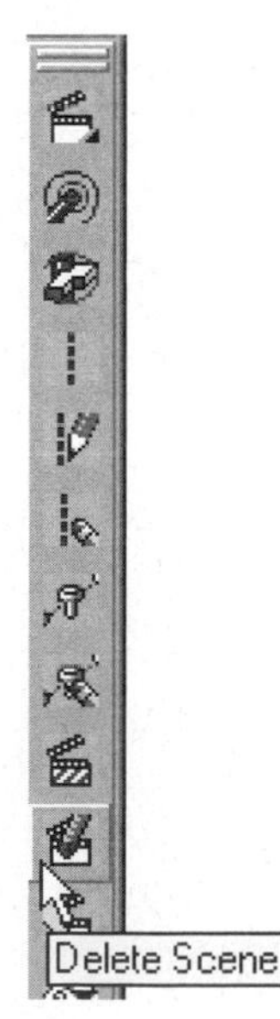

Figure 8.63

COPY SCENES

Once a scene has been created you can copy it and give it another name. All the characteristics (explosion factor, tweaks, trails and part visibility) of the original scene will be copied to the new scene. Issue the Copy Scene command **(AMCOPYSCENE)** from the Scenes toolbar, as shown in Figure 8.64, or right-click on a scene name in the browser and select Copy from the pop-up menu. At the command line, you will be prompted for a new name or to press ENTER to accept the default name. You will next be prompted to accept the original explosion factor or type in a new value. Press ENTER to accept the original explosion factor or type in a new value and press ENTER. The next prompt will ne to activate the scene. Press ENTER to make the new scene active or type "N" and press ENTER to not make it active.

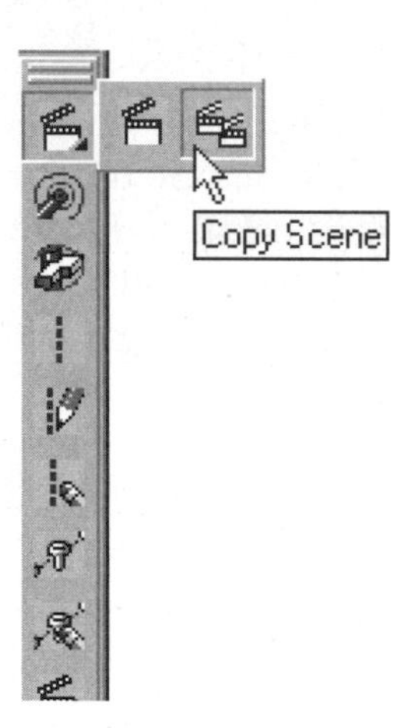

Figure 8.64

LOCK SCENES

After you create a scene with explosion factors, tweaks and trails, it is possible to return to the parts and suppress features. If you suppress a feature on a part and another part is constrained to the part using that feature, then the other part will lose its position. When the scene is updated, the part will return to its assembled position. To avoid this, a scene may be locked and its parts will stay in their current position. To lock a scene, make it the active scene and then issue the Lock Scene Position command **(AMLOCKSCENE)** by right-clicking on the scene name in the browser and selecting Lock Position. In the browser, locked scenes are represented by an icon that looks like a closed lock icon, as shown in Figure 8.65. Once a scene is locked, it cannot have an explosion factor or tweaks added or deleted. To make modifications to a locked scene, right-click on the scene's name and toggle the lock off by selecting Lock Position. Scenes can also be locked through the Desktop Options dialog box that was covered earlier in this section.

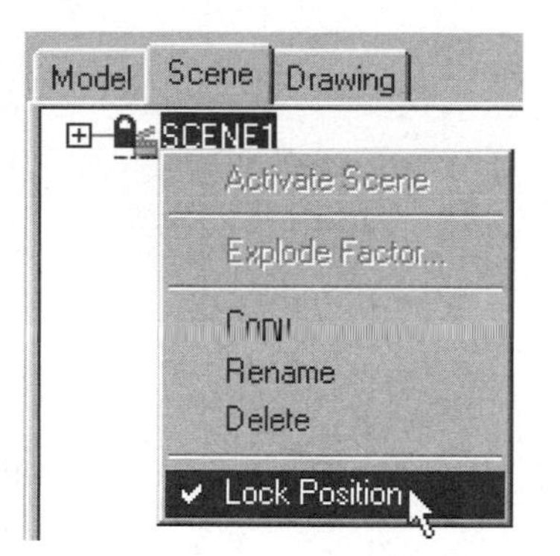

Figure 8.65

TWEAKING PARTS IN A SCENE

After creating a scene, not all the parts may be positioned where you want them. For example, in an exploded view you may need to reposition a part for clarity. Parts in a scene can be repositioned using the Add Tweak command. Once a scene is created, issue the Add Tweak command **(AMTWEAK)** from the Scenes toolbar, as shown in Figure 8.66, or right-click on the part name in the browser under the Scene Tab and choose Add Tweak from the pop-up menu. Then select the part to tweak and a dialog box will appear, offering three choices: Move, Rotate, Transform.

Move: Moves a part in the direction of an edge of a part that you select. After selecting Move, you will be prompted to:

```
Select reference geometry:
```

Select an edge. The part will then be moved in the direction and distance of the selected edge.

Rotate: Rotates a part around an edge of a part that you select. After selecting Rotate, you will be prompted to:

```
Select reference geometry:
```

Select an edge and then type in a value for the angle of rotation. The part will be rotated around the selected edge.

Transform: Moves or rotates a part about the *XYZ* axis of the current UCS or you can select two points to determine the direction. Type in a value and an arrow will appear to verify if you want to go in the positive or negative direction.

The base part in an assembly or a locked scene cannot be tweaked. Once a tweak has been created, it cannot be edited, it can only be deleted. To delete a tweak, issue the Delete Tweaks command **(AMDELTWEAKS)** from the Scenes toolbar, as shown in Figure 8.67, and select a part. If a part has multiple tweaks, they will all be deleted.

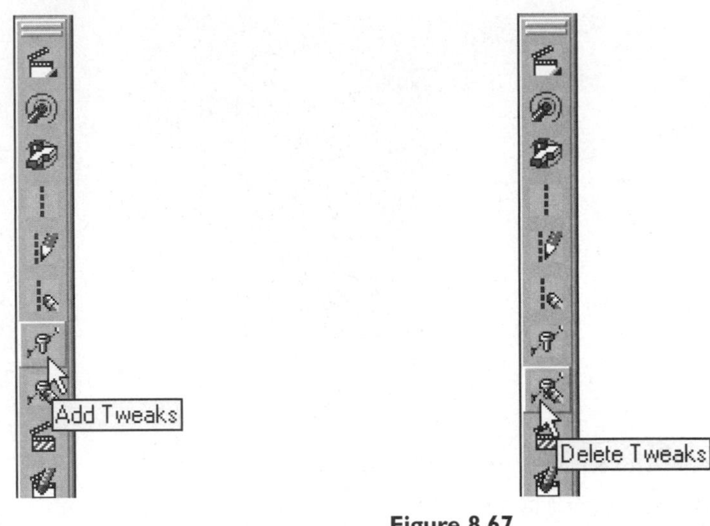

Figure 8.66

Figure 8.67

CREATING TRAILS

To add a line that shows how a part has moved in a scene, issue the Create Trail command **(AMTRAIL)** from the Scenes toolbar, as shown in Figure 8.68, or right-click on the part name in the browser under the Scene Tab and choose Add Trail from the pop-up menu. You will be prompted to:

```
Select reference point on part/subassembly:
```

Select a part. Where you select the geometry is where the trail will start. When an arc or circular edge is selected, the trail will go to the center of the arc or circular edge. If a line is selected, the trail will start at the closest endpoint. After the geometry has been selected, a Trail Offsets dialog box will appear, as shown in Figure 8.69. The dialog box contains two sections: Offset at Current and Assembled Position. The Current position refers to the part's place in the scene. The Assembled position refers to the part's position with a zero explosion factor. The Over Shoot and Under Shoot refer to the distance that a trail should go beyond or stop before the part's position. When creating trails, it may be easier to accept the zero defaults and then edit them if they do not look correct. The trails will take on the properties of the AM_TR layer. You can change the color and linetype of that layer as needed. This layer change would be a good change to make in your template file.

EDITING A TRAIL

After a trail has been created you can edit the trail by issuing the Edit Trail command **(AMEDITTRAIL)** from the Scenes toolbar, as shown in Figure 8.70, or right-click on the part name in the browser under the Scene Tab and select Edit Trail. You will be prompted to:

Create Trail

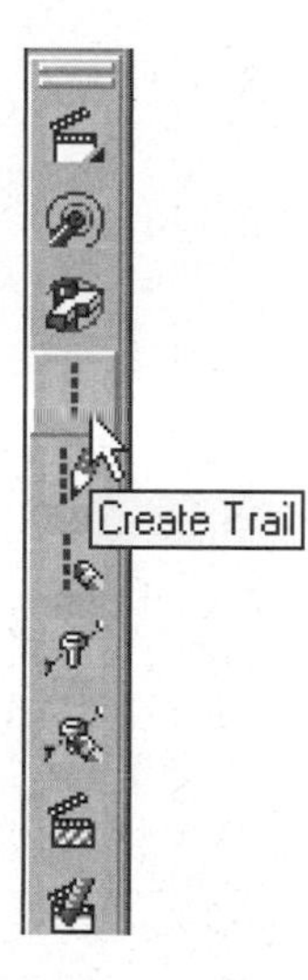

Figure 8.68

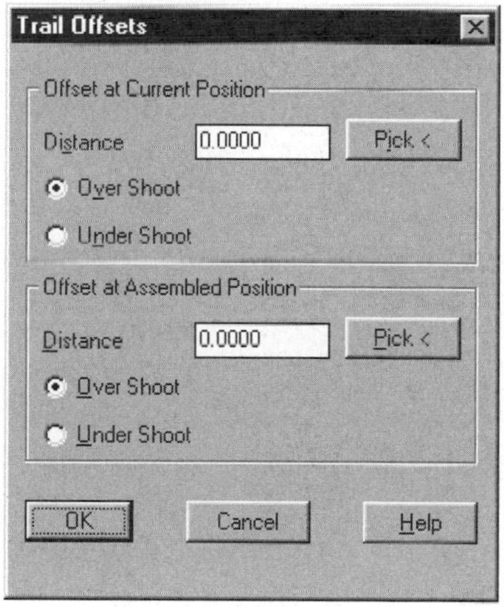

Trail Offsets
Offset at Current Position
Distance
0.0000
Pick <
Over Shoot
Under Shoot
Offset at Assembled Position
Distance
0.0000
Pick <
Over Shoot
Under Shoot
OK
Cancel
Help

Figure 8.69

```
Select trail to edit:
```

Select the trail to edit and the same Trail Offsets dialog box that was used to create the trail will reappear. Type in different values until the trail looks correct.

Figure 8.70

DELETING A TRAIL

To delete a trail, issue the Delete Trail command **(AMDELTRAIL)** from the Scenes toolbar or right-click on the part name in the browser under the Scene Tab and select Delete Trail. You will be prompted to:

```
Select trail to delete:
```

Select the trail and it will be deleted.

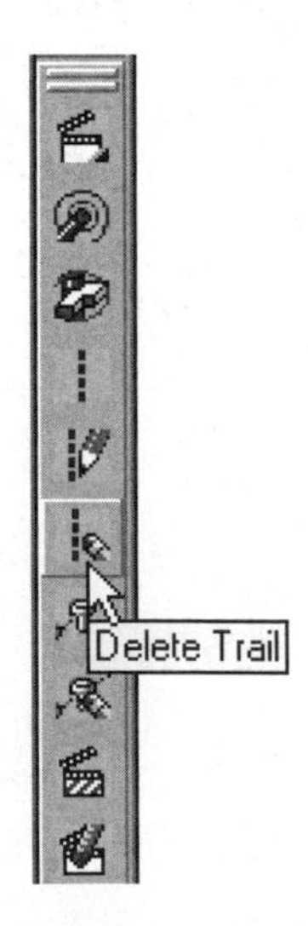

Figure 8.71

TUTORIAL 8.13—CREATING SCENES, TWEAKS AND TRAILS

1. Open the file \Md4book\Chapter8\TU8-13.dwg.
2. Issue the Create Scene command (**AMNEW** with the Scene option).
3. Type in a name "Exploded" and press ENTER. Type in a value of "2" for the explosion factor and press ENTER. Press ENTER to make this the active scene. When complete your screen should resemble Figure 8.72.

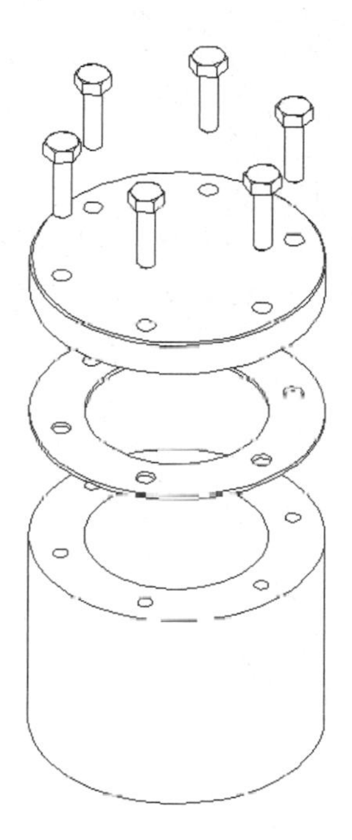

Figure 8.72

4. Repeat the Create Scene command.
5. Type in a name "Assembled" and press ENTER. Type in a value of "0" for the explosion factor and press ENTER. Press ENTER to make this the active scene. When complete, the parts will be back in an assembled position.
6. Make the Exploded scene current and change the explosion factor to "3".
7. Issue the Add Tweaks command **(AMTWEAK)**, select the rightmost bolt and select Transform from the dialog box.
8. Press in M and ENTER to move the part.
9. Press in X and ENTER to move the part along the *X* axis.
10. Type in "2" for the distance, press ENTER and press ENTER again to accept the default direction.

11. Press ENTER to move the part again.
12. Press Z and ENTER to move the part along the Z axis.
13. Press ENTER to accept the default distance of "2" and press ENTER again to accept the default direction.
14. Press X and press ENTER to exit the command.
15. Issue the Create Trail command **(AMTRAIL)** and select the bottom of the bolt that you just tweaked. Accept the default values in the dialog box.
16. Issue the Edit Trail command **(AMEDITTRAIL)** and edit the trail so that the Under Shoot at the assembled position is "1".
17. Change the linetype of the AM_TR layer to center and change the color to red. When complete, your screen should resemble Figure 8.73.

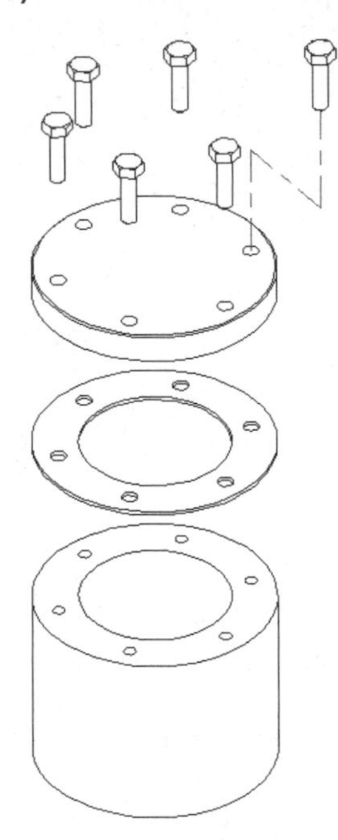

Figure 8.73

18. Issue the Copy Scene command **(AMCOPYSCENE)**.
19. Copy the Exploded scene and give it a new name Exploded2. Press ENTER to accept all the defaults and to make it active.
20. Practice tweaking a few of the other bolts
21. Then add trail to all the parts.
22. Save the file.

EXERCISE

For the following exercise, follow the instructions.

When constraining the handle to the screw, align the center of the handle to the work axis of the screw. For the offset between the screw and the handle, use the Mate constraint and offset it to a bottom point on the hole of the screw, use an offset distance of "1.3".

Exercise 8.1—Vice Assembly

1. Open the file \Md4book\Chapter8\ViceAssy.dwg.
2. Assemble the parts as shown in Figure 8.74.
3. Create a scene with an explosion factor of "3".
4. Add tweaks and trails to all the parts, as shown in Figure 8.75.

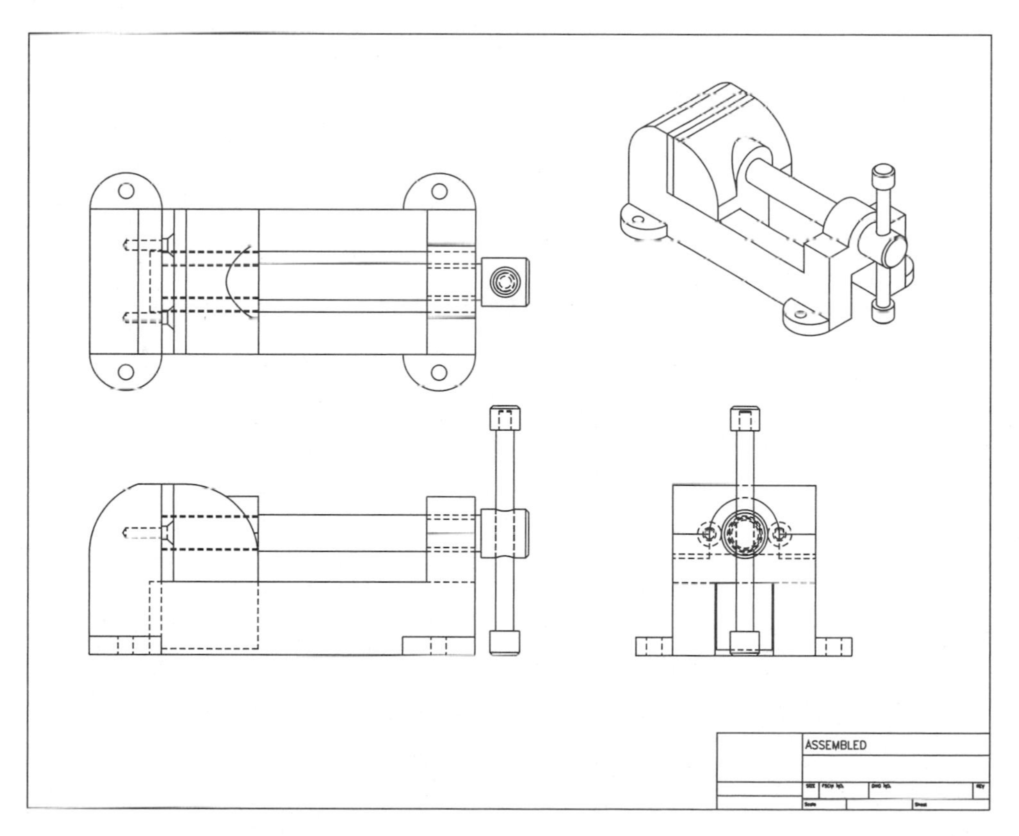

Figure 8.74

EXERCISE

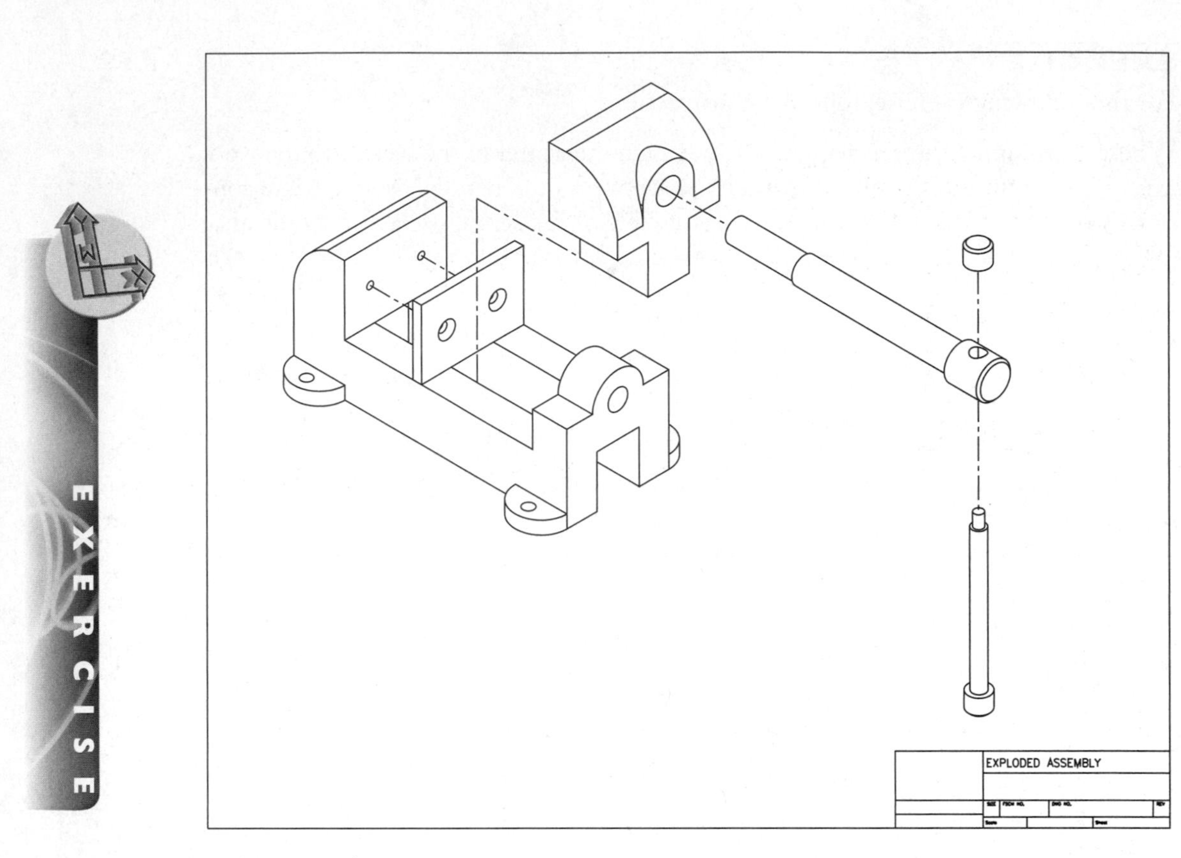

Figure 8.75

REVIEW QUESTIONS

1. A bottom up assembly has all the parts in the same file. T or F?
2. When you create an assembly, all the files must either be entirely inside or entirely referenced to the assembly file. T or F?
3. Edit in place is a technique that is used to edit local parts only. T or F?
4. Parts that are dragged in from another file are automatically referenced into the file. T or F?
5. Explain the difference between a Mate and a Flush constraint.
6. What does degrees of freedom refer to?
7. An assembly does not have to be fully constrained. T or F?
8. To change an assembly constraint to a different type (for example, Mate to Flush), the original constraint does not need to be deleted first. T or F?
9. The solid that is created from the Check Interference command is parametric. T or F?
10. Parts that are constrained with the Mate or Insert constraint will automatically be exploded when a scene is created that has an explosion factor greater than zero. T or F?

CHAPTER 9

Advanced Drawing Creation and Annotations

In this chapter you will learn how to create a part using tolerance modeling. Next you will create drawing views from parts that are in the same file. Each part's drawing views will be created in their own layout and all the drawing views will be in the same file. You will create a part using tolerance modeling. Then you will create drawing views based upon scenes and then you will add balloons and create a bill of materials to complete the drawing.

AFTER COMPLETING THIS CHAPTER, YOU WILL BE ABLE TO:

- Create a part using tolerance modeling.
- Create drawing views from multiple parts in the same file.
- Create drawing views based upon scenes.
- Add balloons to drawing views.
- Create a bill of materials.
- Export 2D drawing views.

TOLERANCE MODELING

In this section you will learn how to create tolerance dimensions and apply them to a part. Tolerance modeling can be used to show maximum and minimum dimensioning conditions of a part and then interference can be checked. There are three main steps to create a tolerance model: create tolerance dimensions, create drawing views and then set the tolerance condition for the part. The tolerance dimensions are also parametric dimensions. Tolerance dimensions can be used for the first profile or the profile that will be used to create a feature.

The first step is to create the tolerance dimensions. In Chapter 6 you learned about Power Dimensioning and Power Edit after selecting the geometry to dimension or edit with the Power Dimensioning and Power Edit command. A dialog box that resembles Figure 9.1 will appear. In the Power Dimensioning dialog box make the

Tolerances tab active and check the Enable box. Type in a value for both the Upper and Lower Deviation. If you want to apply a negative value to a deviation you must use the "-" symbol before the values as shown in the Lower deviation in Figure 9.1. Type in a value for the Expression or use the Copy from button to get a value for the Expression and then select OK. The part will use the value in the expression for the dimension. The tolerance dimension will be applied in the third step. Finish dimensioning the profile and create the part or feature from the profile.

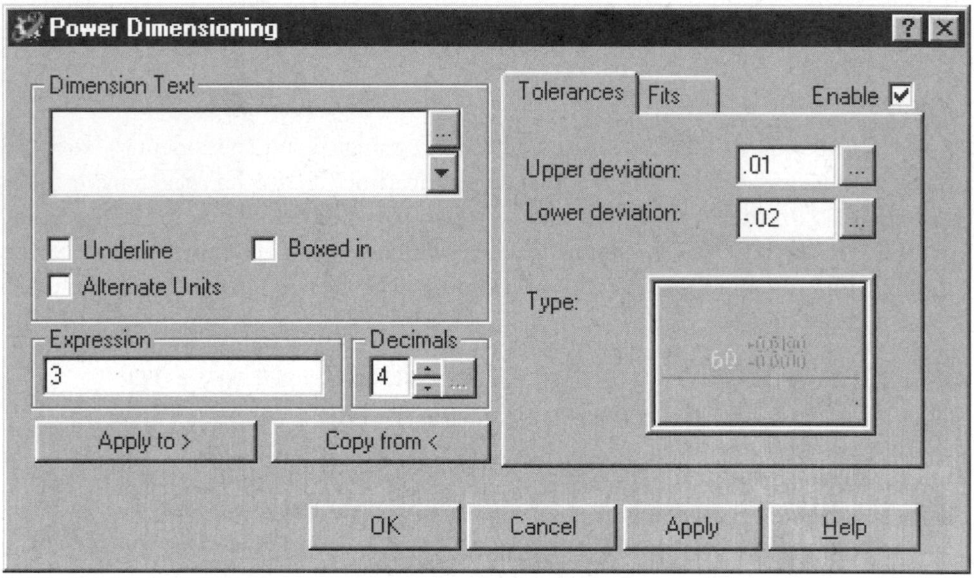

Figure 9.1

The second step is to create drawing views of the parts that need to be tolerance modeled. The tolerance dimensions need to be visible on the screen. Each part that needs to have a tolerance condition applied to it must have a drawing view created that shows the tolerance dimensions. Each part's drawing views can be created in a different layout. Layouts will be covered later in this chapter.

The third step is to apply a tolerance condition the part. You must be in Drawing mode. Issue the Tolerance Condition command **(AMTOLCONDITION)** as shown in Figure 9.2 from the Drawing Layout toolbar. The Transformation of Model dialog box will appear as shown in Figure 9.3. There are three areas that are explained next.

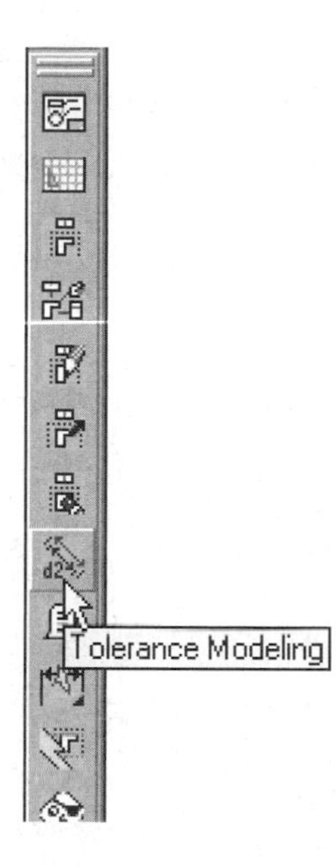

Figure 9.2

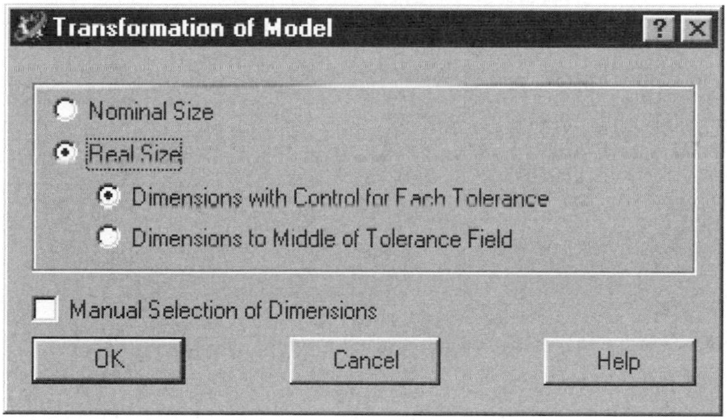

Transformation of Model

Nominal Size
Real Size
Dimensions with Control for Each Tolerance
Dimensions to Middle of Tolerance Field
Manual Selection of Dimensions
OK Cancel Help

Figure 9.3

Nominal Size: Regenerates the part, using the value that was input in the Expression area of the Power Dimensioning dialog box.

Real Size: Regenerates the part, using the real values of the dimensions using one of the following options.

> **Dimensions with Control for Each Tolerance:** When selected, the Manual Control of Dimension dialog box will be opened as shown in Figure 9.4. Select the condition or manually type in value to be applied to the highlighted dimension.

Dimensions to Middle of Tolerance Field: The part will be regenerated using the middle of the Upper and Lower deviation values.

Manual Selection of Dimensions: Select the tolerance dimension that you want to apply a condition to and the Manual Control of Dimension dialog box will be opened as shown in Figure 9.4. Select the condition or manually type in value to be applied to the highlighted dimension.

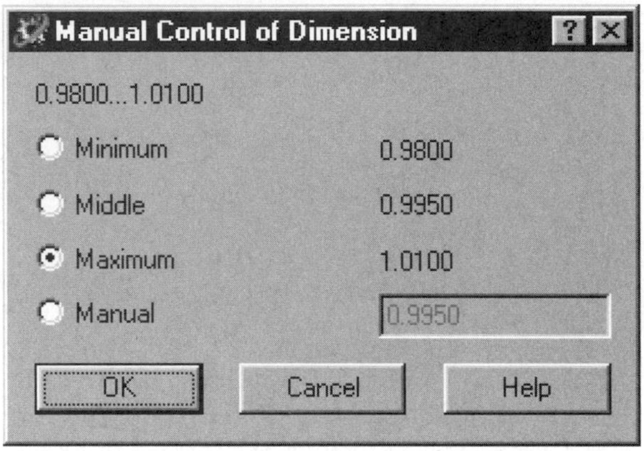

Figure 9.4

The Manual Control of Dimension dialog box, as shown in Figure 9.4, has the following options.

Minimum: Uses the value of the Lower deviation.

Middle: Uses the average of the Lower and Upper deviation.

Maximum: Uses the value of the Upper deviation.

Manual: Type in a value to be applied.

After selecting a tolerance condition, the part will be updated to reflect this new value. To change the tolerance condition, issue the Tolerance Condition command **(AMTOLCONDITION)** and select Nominal and select OK which will reset the part back to its original state. Then reissue the Tolerance Condition command and set the tolerance condition as needed.

TUTORIAL 9.1—TOLERANCE MODELING

1. Start a new file from scratch.
2. Set the precision to two decimal places for the standard dimension style.
3. Draw a square and profile it.

4. Issue the Power Dimensioning command **(AMPOWERDIM),** press ENTER and select the top horizontal line. In the Power Dimensioning dialog box type in a value of "1" for the Expression and set the Decimals to "2". Turn on tolerancing by selecting in the Enable area of the dialog box.
5. Type in a value of ".01" for the Upper deviation and "-.02" for the Lower deviation. Select OK to complete this dimension.
6. While still in the Power Dimensioning command press ENTER and select the left vertical line. In the Power Dimensioning dialog box type in a value of "1" for the Expression. Then select OK to accept the default tolerance values and complete the dimension.
7. Press ENTER twice to exit the Power Dimensioning command.
8. Extrude the Profile ".25" in the default direction.
9. Change to an isometric view using the **8** key.
10. Create a top, side and an isometric view as shown in Figure 9.5.

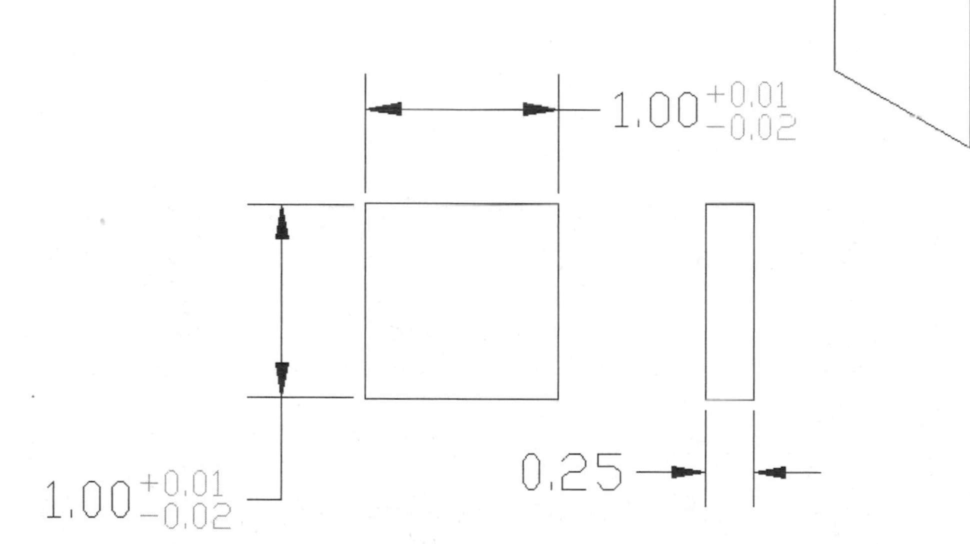

Figure 9.5

11. Issue the Tolerance Condition command **(AMTOLCONDITION).** If not already selected, make Real Size active as well as the option Dimensions with Control for Each Tolerance.

TUTORIAL

12. The 1.00 horizontal dimension should be highlighted. In the Manual Control of Dimension dialog box set the tolerance to Minimum, then select OK.
13. The 1.00 vertical dimension should be highlighted. In the Manual Control of Dimension dialog box set the tolerance to Maximum then select OK. When complete, your screen should resemble Figure 9.6.

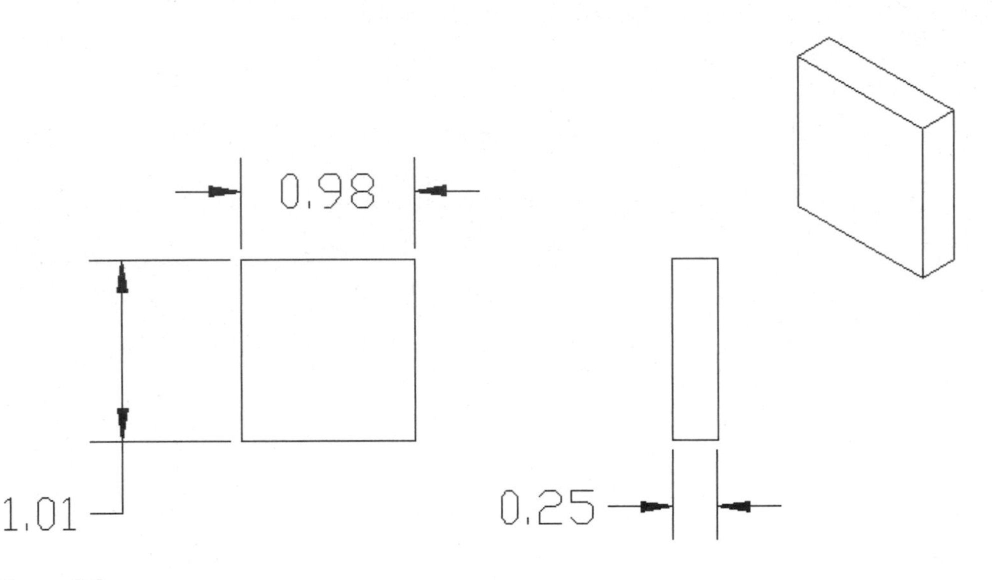

Figure 9.6

14. Make the Model mode active.
15. Use the AutoCAD DISTANCE command to query the part. The dimensions should match Figure 9-6.
16. Return to Drawing Mode.
17. Issue the Tolerance Condition command **(AMTOLCONDITION)**. Select Nominal Size and then select OK. The part and drawing views will now be back to a 1" square.
18. Save the file as \Md4book\Chapter9\TU9-1.dwg.

CREATING DRAWING VIEWS IN LAYOUTS

In Chapter 4 you created drawing views of single part and in the last couple of chapters you created multiple parts in the same file. In this section you will learn how to create drawing views from local parts and create them in separate layouts in the same file. To create drawing views from a part make the part active. Create a layout that the drawing views will be created in (to create or copy a layout right click on a layout name and select either New layout or Move or Copy then follow the prompts). Then issue the New View command and make the following settings in the Create Drawing View dialog box.

View Type = Base.

Data Set = Active Part.

Layout = Select the Layout name that the views will be placed.

Make other changes as needed and then select OK. Then create the drawing views as needed. To edit the views on a specific layout, make the layout active by double clicking on its tab on the bottom of the screen. Then edit the views as needed. There is no limit to the number of layouts that can be in Mechanical Desktop.

TUTORIAL

TUTORIAL 9.2— DRAWING VIEWS IN LAYOUTS

1. Open the file \Md4book\Chapter9\TU9-2.dwg.
2. Rename Layout1 by right clicking on the Layout1 tab on the bottom of the AutoCAD screen. Select Rename from the pop-up menu and type in "Base Plate" in the dialog box and then select OK.
3. Create a few drawing views of the active part (Base Plate blue part) in the layout named Base Plate.
4. Create a new layout by right clicking on the "Base Plate" tab on the bottom of the AutoCAD screen. A layout name of Layout1 will appear.
5. Rename Layout1 by right clicking on the Layout1 tab on the bottom of the AutoCAD screen. Select Rename from the pop-up menu and type in "Top Plate" in the dialog box and then select OK.
6. Make Model mode active.
7. Make the part "Top Plate" the active part.
8. Create a few drawing views of the active part (Top Plate green part) in the layout named Base Plate.
9. Make Model mode active.
10. Delete the fillets in both part.
11. Make each of the layouts active to see each layout's drawing views get updated.
12. Save the file.

CREATING DRAWING VIEWS FROM SCENES

In Chapter 8 you created assemblies and scenes from the assemblies. When you need to create drawing views of an assembly it is recommended that you create the drawing views from a scene instead of the Select option. Parts will automatically be added to each scene and then to the drawing views that used the scene. With the select option, new parts will not automatically appear in drawing in the drawing views. If you want the new parts to appear in the drawing views with the existing parts, the base drawing views will need to edited using Edit Selection. Set option of the Edit View command **(AMEDITVIEW)**. If you have multiple scenes in a file, drawing views from each scene can be placed in the same or separate layouts. If they are created in the same layout use the Base as the View Type for the first view of the scene. To create drawing views based from a scene issue the New Drawing View command **(AMDWGVIEW)** and in the Create Drawing View dialog box set the Data Set to the scene name that will appear in the drop down list. Then create drawing views as needed.

Note: There is no limit to the number of base views that can be created in the same layout.

TUTORIAL 9.3—CREATING DRAWING VIEWS FROM A SCENE

In this file there are two scenes (Assembled and Exploded) that have already been created and two layouts named Assembled and Exploded. The exploded scene has been tweaked and trails have been added.

1. Open the file \Md4book\Chapter9\TU9-3.dwg.
2. Issue the New Drawing View command **(AMDWGVIEW)**. Select Base as the type, for the Data Set select the scene name ASSEMBLED, and change the layout name to Assemble if it is not current, change the scale to ".5" and then select OK.
3. Select the top circular edge of the top cover for the plane and press ENTER. Orient the UCS by pressing *X* in reference to the world *X*.
4. Select a point in the upper left corner of the drawing and press ENTER to accept the position.
5. Create the front orthogonal view and an isometric view from this front view; create the isometric view ".75" scale relative to the parent. When complete, your screen should resemble Figure 9.7
6. Issue the New Drawing View command **(AMDWGVIEW)**. Select Base as the type, for the Data Set select the scene name EXPLODED, and change the layout name to Exploded, change the scale to ".375", uncheck Display Hidden Lines from the Hidden Lines tab and then select OK.
7. For the drawing view orientation use the current view as the plane, press V and press ENTER twice to accept the plane and *XY* orienation.
8. Locate the exploded view in the middle of the drawing border as shown in Figure 9.8.
9. Save the file.

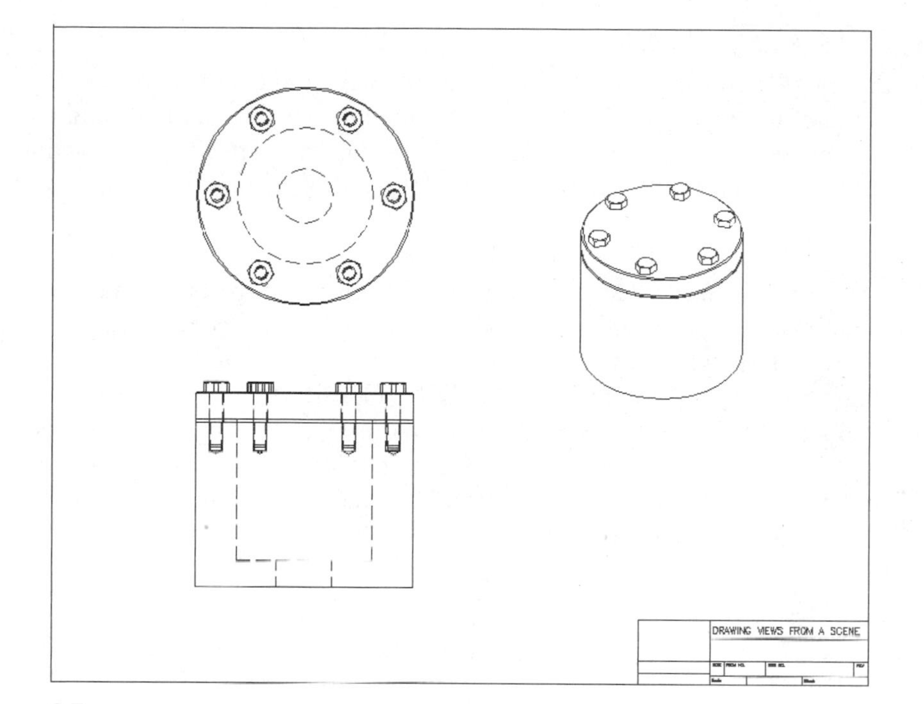

Figure 9.7

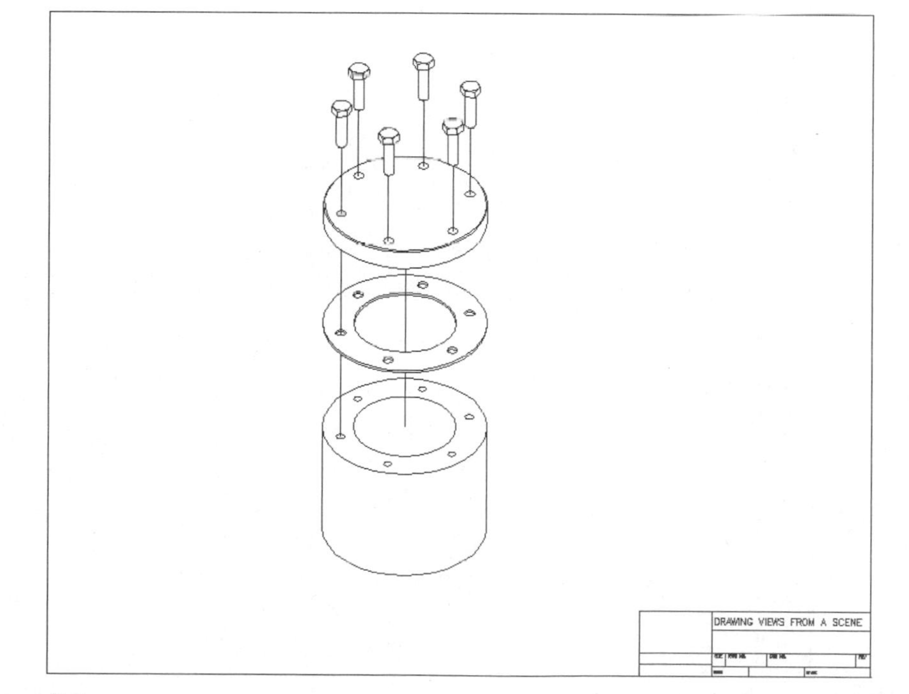

Figure 9.8

BILL OF MATERIAL

The final step in creating a drawing is usually to balloon the drawing views and generate a bill of material (parts list). There are numerous ways, standards and company requirements for ballooning drawing views and creating parts lists. This section will highlight the functionality of the options of the bill of materials commands in Mechanical Desktop. Practice different techniques and develop your own bill of material standards. The bill of material that Mechanical Desktop creates is associative to the parts in the file. As the quantity of parts change or parts are added or deleted will automatically be updated in the parts list in the layout. Before creating a balloon or a bill of material you need to establish a standard type. You can choose from the following standards; ANSI, BSI, CSN, DIN, GB, ISO or JIS or you can create your own standard based on one of these standards by right clicking on the standard name that you want to copy and choose Copy standard. To choose a standard, issue the Symbol Standard command **(AMSYMSTD)** from the Drawing Layout toolbar, as shown in Figure 9.9, and select a standard to use. A Symbol Standards dialog box will

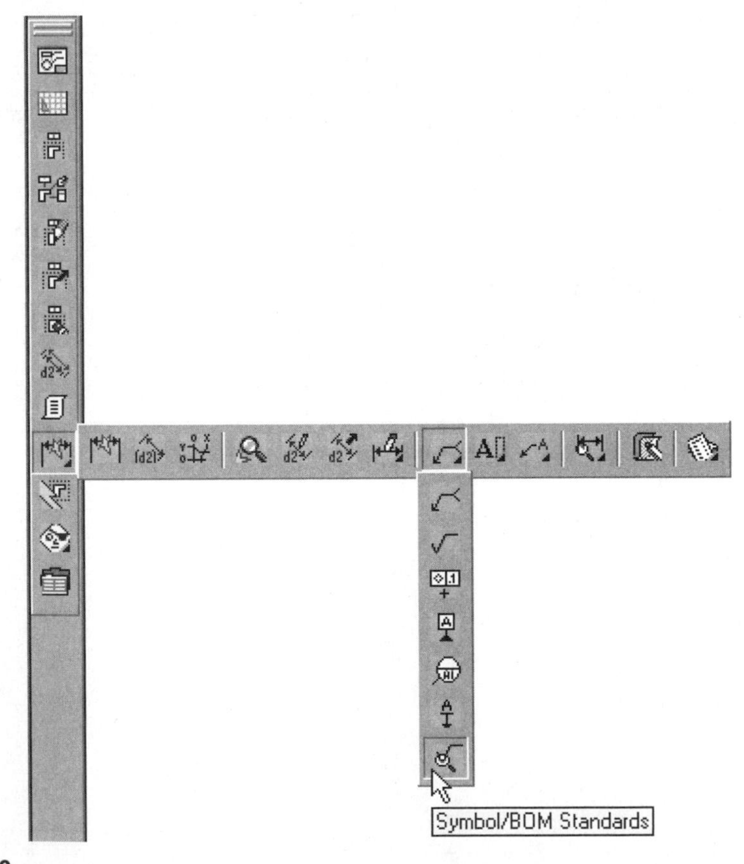

Figure 9.9

appear, as shown in Figure 9.10. Select the standard that you want applied to the bill of material. Figure 9.10 is expanded, showing the bill of material and balloon section. You can edit these sections by double clicking on the section you want to edit. Figure 9.11 shows the Properties for ANSI dialog box that will appear when the Parts List area is selected. You can edit these sections here or from the Bill of Material command **(AMBOM)** with the Standard Properties option, which will be discussed later in the chapter. When you open legacy files, or files created in previous versions of Mechanical Desktop, the Symbol Standard command **(AMSYMSTD)** will be invoked automatically to prompt you to specify the standard to be used for the bill of material in the drawing.

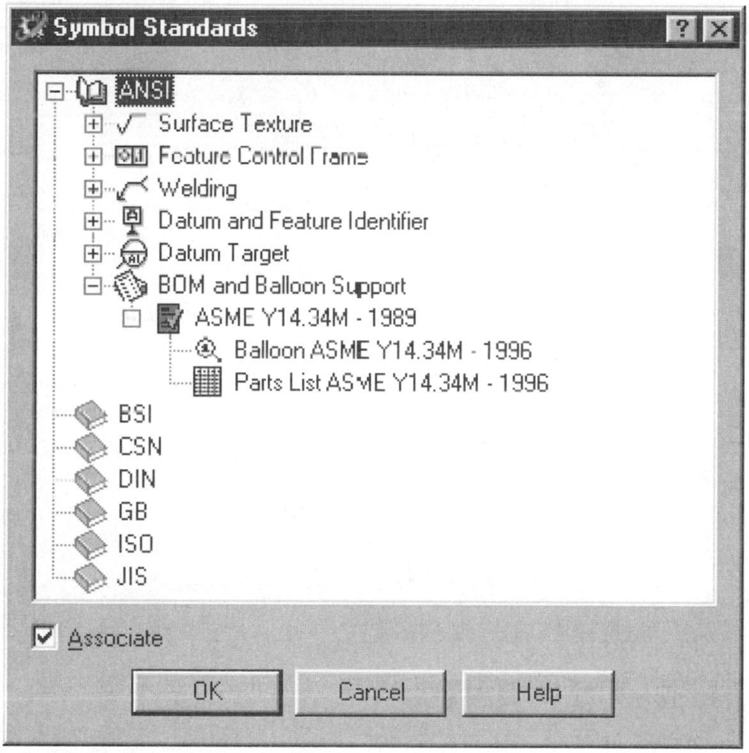

Figure 9.10

BILL OF MATERIAL ICONS

After you select a standard, the next step is to create a BOM database by issuing the Bill of Material **(AMBOM)** command—it is the top icon shown in Figure 9.12 from the Drawing Layout toolbar. Figure 9.12 shows the six bill of material commands for creating and editing bills of material. A description of each of the icons follows.

Figure 9.11

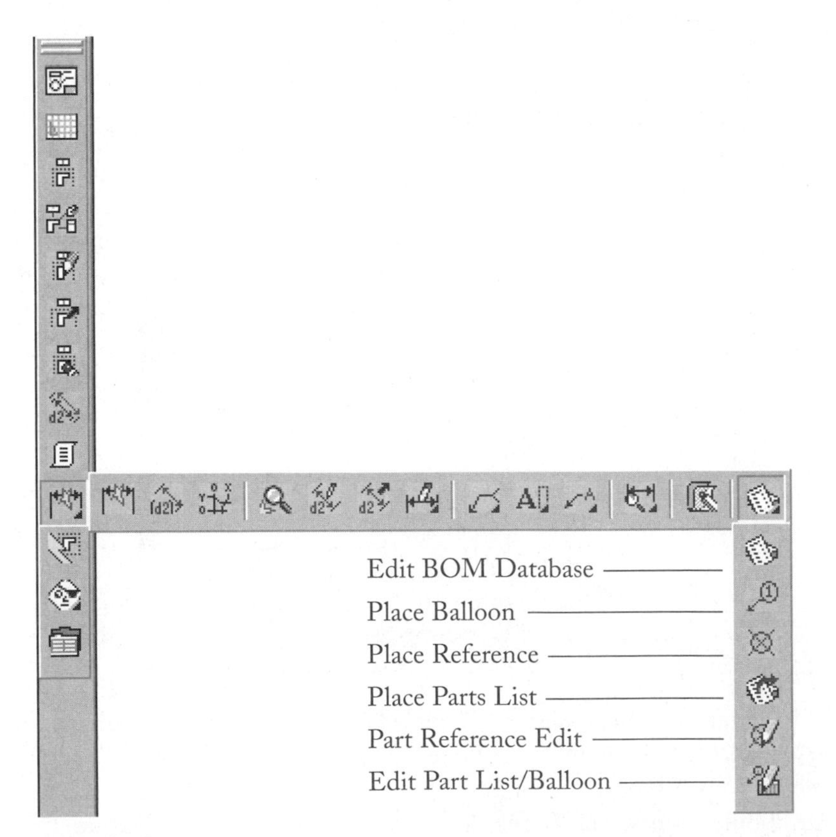

Figure 9.12

Edit BOM Database: (create and edit): This will issue the Bill of Material command **(AMBOM)**, which can perform eighteen functions, each will be described in the next section.

Place Balloon: A balloon or balloons will be created in the drawing. A drawing view must exist for this command to be valid and the Bill of Material command **(AMBOM)** command must have been issued first. The item number that appears in the balloon reflects the order in which the parts were created or placed in the file. There are five options for placing balloons: Auto, Collect, Manual, One and Renumber. After issuing an option a part reference a small circle with a "X" in it will appear at the selected location. This symbol holds the attributes for the bill of material and it will be referred to as a BOM symbol within this text.

> **Auto:** Allows you to use any AutoCAD selection technique to select all the BOM symbols that you want ballooned. Then press ENTER, and the leaders will attach to the BOM symbol and the balloons will orient themselves. To change the horizontal or vertical orientation of the balloons, press H or V and ENTER. Once the balloons are oriented correctly, press ENTER to accept this position. Once the balloons are in place, you can reposition them by moving, stretching or grip editing them.
>
> **Collect:** Balloons multiple parts using one leader. A new leader can be created or you can use a balloon from which the other balloons will come. Select a BOM symbol or symbols on the part(s) to balloon, press ENTER and then select a balloon to have the balloon come from or press N and ENTER to create a new leader. Then, depending on the direction of the leader, select an *X* or *Y* direction for the balloons to go. For example, if the leader is going up and to the right, the other balloons could only go up and to the right.
>
> **Manual:** Creates a balloon on a part or process that is not represented by a part or reference part. Select a point on the drawing for the balloon to start and manually fill in the information for the bill of material. The balloon will be assigned the first available number that is not used in the parts list.
>
> **One:** Create a balloon for an existing part. You will be prompted to select an object, select the BOM symbol of the part to balloon and then select the start and ending position of the balloon. This works in much the way as the Create Balloon command **(AMBALLOON)** worked in Mechanical Desktop 2.0.
>
> **Renumber:** You will first be prompted to "Enter starting item number <1>:" Type a number that the selected balloon will be. Then type a number that the balloons will increment. Then select the balloon to be renumbered.

Place Reference: This command creates attributes and associates them to an existing block or places them on a blank area on the screen. After the command is issued, a Part Ref Attributes dialog box will appear. Type in the information to be used in the bill of material; the same information in the column headings of the parts list will be asked here and that information will appear in the parts list. This can be useful for showing items in an assembly that are not parts. Representing paint or the painting process would be a good example for creating a part reference. Part references can be placed in the assembly or in the drawing mode. If the part reference is placed in the assembly, it will not appear when the view or the parts option is used for creating the parts list. After placing a part reference a small circle with a "X" in it will appear at the selected location. This symbol holds the attributes for the bill of material and it will be referred to as a BOM symbol within this text.

Place Part List: First you will be prompted to give the parts list a name. Type in a name or press ENTER to accept the default name. The parts list will be placed in the drawing. You will have the following choices.

> **All:** All parts in the file will appear in the parts list.
>
> **Parts:** You will select the parts to be shown in the parts list.
>
> **Range:** You will be prompted for row numbers to start and end the parts list.
>
> **Sheet:** All parts in the layout and all visible parts will appear in the parts list.
>
> **View:** Select a drawing view, and only the parts that are visible in the selected view will appear in the parts list. Referenced parts will not show up in the parts list.

Part Reference Edit: Select a part reference, and the information that was created for the part reference will appear in the same dialog box it was created in. Select any row and edit it.

Edit Part Lists/Balloon: Select a parts list and you can edit the properties, sort order and values of the list in the same dialog box they were created in. Changes made in the dialog box will automatically be reflected in the parts list.

After issuing the Bill of Material **(AMBOM)** the BOM dialog box will appear. Figure 9.13 shows an example of what the BOM dialog box will resemble with parts in the file. The entries that appear in the BOM dialog box represent the parts that exist in the file. Along the top of the BOM dialog box are eighteen icons, a few of them are also available from the toolbar as shown in Figure 9.12. These commands can be accessed from either location.

BOM DIALOG BOX ICONS

Figure 9.14 shows a description of the eighteen icons in the BOM dialog box. Each of the eighteen icons are explained in the following section.

Print: Prints the bill of material to the selected printer.

Add Parts: Any reference parts that do not appear in the database can be added with this icon.

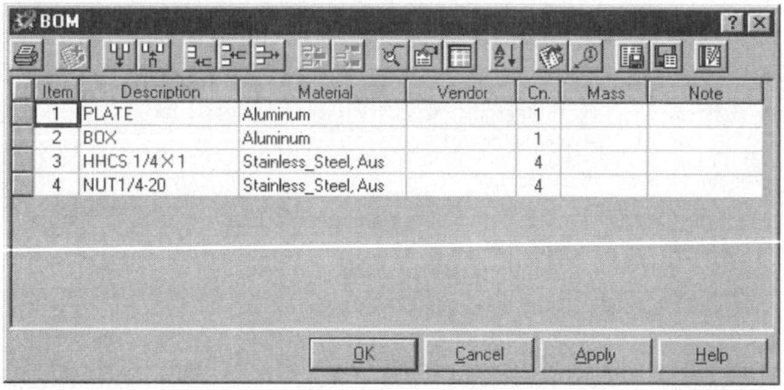

BOM

Item	Description	Material	Vendor	Cn.	Mass	Note
1	PLATE	Aluminum		1		
2	BOX	Aluminum		1		
3	HHCS 1/4X1	Stainless_Steel, Aus		4		
4	NUT1/4-20	Stainless_Steel, Aus		4		

OK Cancel Apply Help

Figure 9.13

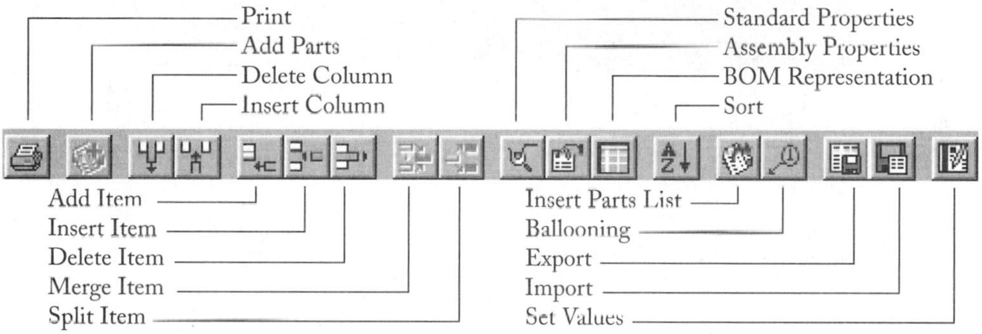

Print
Add Parts
Delete Column
Insert Column
Standard Properties
Assembly Properties
BOM Representation
Sort
Add Item
Insert Item
Delete Item
Merge Item
Split Item
Insert Parts List
Ballooning
Export
Import
Set Values

Figure 9.14

Delete Columns: Select a column in the BOM dialog box, issue the command, and the column will be deleted.

Insert Columns: Select a column in the BOM dialog box, issue the command, and a column will added before the selected column.

Add Item: A row will be added at the end of the database list.

Insert Item: Select a row in the BOM dialog box, issue the command, and a row will added above the selected row.

Delete Item: Select a row in the BOM dialog box, issue the command, and the row will be deleted.

Merge Items: Merges two or more rows. Merging is possible only if all data are the same.

Split Item: Splits a row. You can select part references, which have to be used to create a new row.

Merge Items: Two or more rows that contain the same data will be merged.

Split Items: A row that contains a part reference will be split into a new row.

Standard Properties: Allows you to control how the parts list and balloons will appear. If selected, a Standard Properties dialog box will appear, as shown in Figure 9.15. Along the top of the dialog box are four headings: Column, Caption, Width and Equivalents. These headings are used to create a parts list. To edit one of the areas, double-click on the name and type in a new value. As you select a specific cell, the information in the bottom half of the dialog box will change to show the settings for that cell.

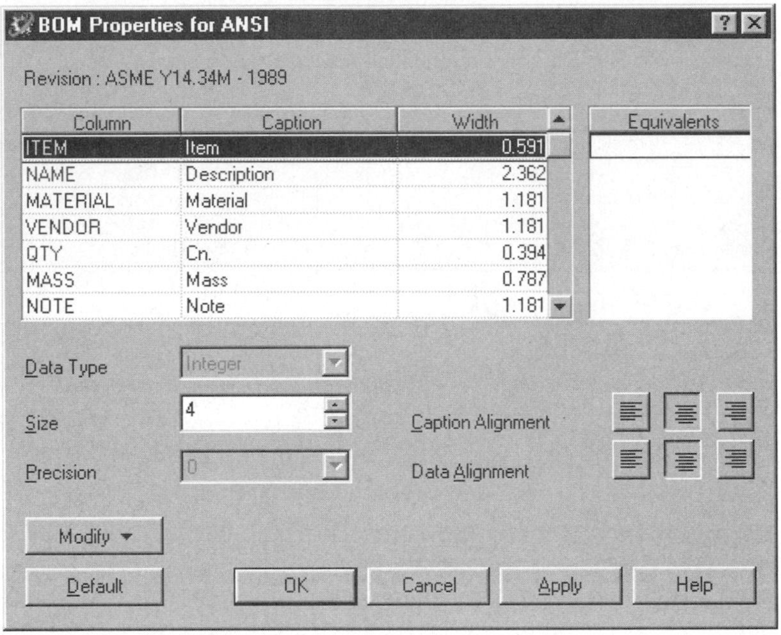

Figure 9.15

Column: The attribute name for the row that is used in the parts list. This attribute name is similar to an attribute in a block.

Caption: The name for the column in the parts list.

Width: The width of the column in the parts list, specified in units.

Equivalents: This section will have an attribute name that can also be used for the attribute name for the column.

Data Type: Text, integer and real are valid options.

Size: Sets the maximum character size of the data or the maximum amount of characters to be used in the highlighted column.

Precision: Choose the number of decimal places to be used. Mass uses the precision setting.

Caption Alignment: Choose left, center or right side alignment for the selected caption.

Data Alignment: Choose left, center or right side alignment for the selected cell.

Modify: From the drop-down list under Modify there are two choices: Balloon and Part List.

When the Balloon option is selected the Balloon Properties dialog box will appear as shown in Figure 9.16. The Balloon Properties dialog box is broken into these sections.

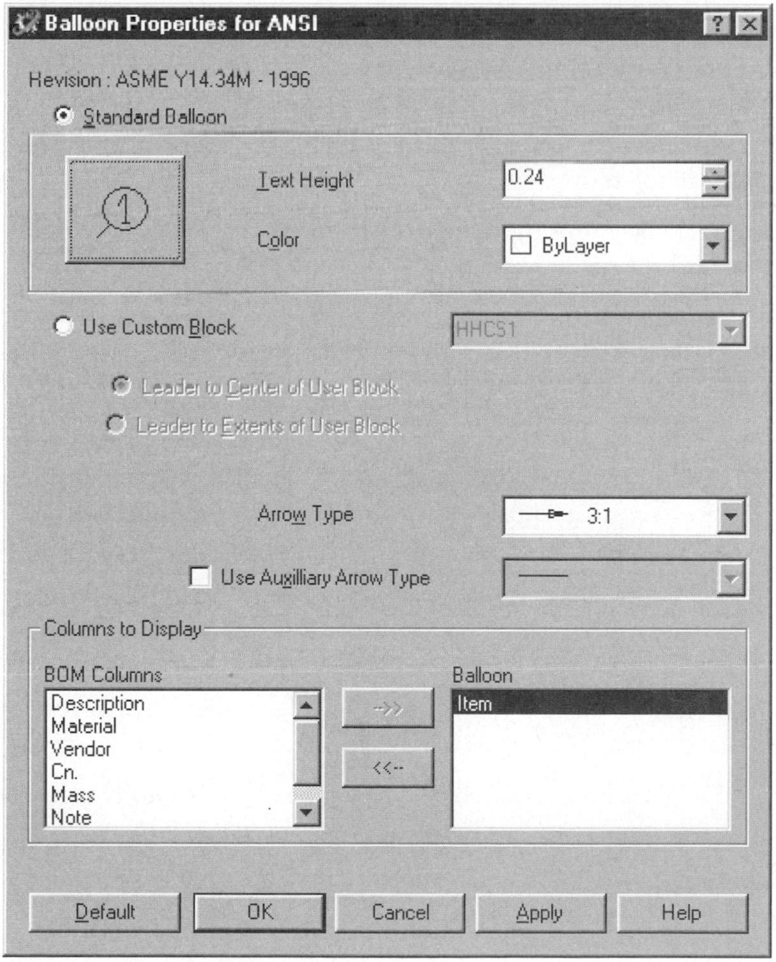

Figure 9.16

Figure 9.17

BALLOON PROPERTIES DIALOG BOX

Balloon: When the balloon icon in the upper left corner of the dialog box is selected, a pop-up menu will appear, as shown in Figure 9.17. Select from one of the five styles for the balloon to use.

Text Height: The height of the text in the balloon. Type in or select with the up or down arrow until the size is correct.

Color: Select the color to be used for the balloon bubble and text. The arrow color will be taken from the current layer.

Use Custom Block: When selected, you can select a block name in the file to use as the balloon.

Leader to Center of User Block: When selected, the leader will extend to the center of the user defined balloon (block).

Leader to Extents of User Block: When selected, the leader will extend to the outside of the user defined balloon (block).

Arrow Type: From the drop-down list, select an arrow style to use for balloon.

Use Auxiliary Arrow Type: From the drop-down list, select an arrow style to use for an alternate arrow type. This is invoked by picking off of the part edge.

Columns to Display: The bottom of the dialog box is split into two sections: BOM Columns and Balloon.

BOM Columns: The column choices for what could appear in the balloon.

Balloon: The column information that will appear in the balloon.

->>: To add a column attribute to the balloon, select it and then select the button.

<<-: To remove the column attribute from the balloon, select it and then select the button.

Default: If selected, this button resets all the settings in the dialog box back to the original settings.

Note. The diameter of the balloons is calculated from the text size. If you want smaller or larger balloons in your drawing, set a smaller or larger text height.

When the Parts List option is selected from the Modify button a Properties dialog box will appear similar to the one shown in Figure 9.18.

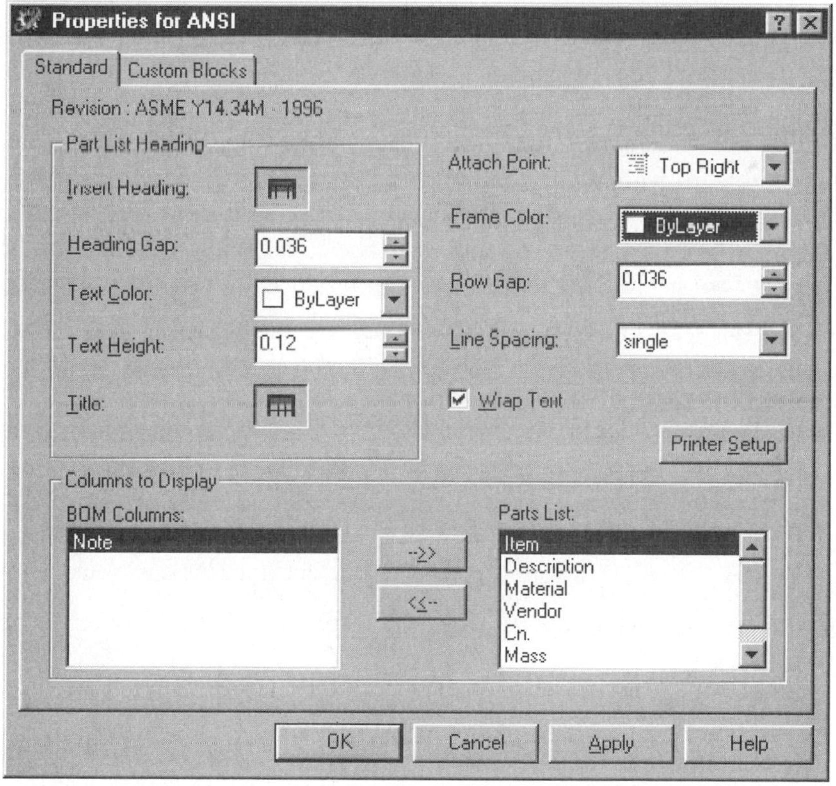

Figure 9.18

PARTS LIST DIALOG BOX

Inserting Heading: Check this box to include the heading in the parts list. When it is checked, you choose from an icon below this box to place the heading on top or on the bottom of the parts list. If the header is turned off, the title will also be turned off.

Heading Gap: Specifies the gap size between the heading text and the frame.

Text Color: Specifies the color of the text for the title.

Text Height: Specifies the height of the text for the heading and title.

Title: Type in a title to be used in the parts list. If you do not want a title, uncheck this box. The title can be turned off and the header can still be on.

Attach Point: From the drop down list select the point to attach the parts list.

Frame Color: Controls the color of the outside lines of the parts list.

Row Gap: The distance between the text in the parts list and the lines that define the rows.

Line Spacing: Defines how many lines of text each row should take up: one, two or three lines (single, double or triple).

Wrap Text: Checks if the text in the parts list should use word wrap.

Printer Setup: This will bring up the Print Setup dialog box. Change the settings to define what the printed parts list will look like when printed alone. These settings have no effect on how the parts list will appear in the Mechanical Desktop file. When a parts list is placed in the Mechanical Desktop file, it will print the way it appears.

Columns to Display: The bottom of the dialog box is split into two sections: BOM Columns and Parts List.

BOM Columns: The column choices for what could appear in the parts list.

Parts List: The column information that will appear in the parts list.

->>: To add a column to the parts list, select it and then select the button.

<<-: To remove the column from the parts list, select it and then select the button.

Default: If selected, this button resets all the settings in the dialog box to the original settings.

Custom Block tab: Under this tab you can specify a block that should be used for the parts list.

Assembly Properties: Modifies the Description, Material, Vendor, Mass, Note, and BOM name.

BOM Representation: When selected, a BOM Representation dialog box will appear, this is where you can control the appearance of external BOM tables.

Sort: Sorts the columns in ascending or descending order. Three modifiers can be set to alter the way the parts are listed.

Insert Parts List: First you will be prompted to give the parts list a name. Type in a name or press ENTER to accept the default name. The parts list will be placed in the drawing. You will have the following choices.

> **All:** All parts in the file will appear in the parts list.
>
> **Parts:** You will select the parts to be shown in the parts list.
>
> **Range:** You will be prompted for row numbers to start and end the parts list.
>
> **Sheet:** All parts in the layout and all visible parts will appear in the parts list.

Ballooning: A balloon or balloons will be created in the drawing. The same options are available here as through the Place Balloon command icon, discussed earlier in "Bill of Material Icons".

Export: Export data to the following formats: Microsoft Access (*.mdb), Microsoft Excel 97 (*.xls), Microsoft Excel 95 (*.xls), dBase 5 (*.dbf), dBase IV (*.dbf), dBase III (*.dbf), Microsoft FoxPro 2.5 (*.dbf), Microsoft FoxPro 2.6 (*.dbf), Visual Fox Pro 3.0 (*.dbf), Format text separated with tabulator (*.txt), Format text separated with semicolon (*.csv) and HTML Export (*.html).

Import: Import data to the following formats: Microsoft Access (*.mdb), Microsoft Excel 97 (*.xls), Microsoft Excel 95 (*.xls), dBase 5 (*.dbf), dBase IV (*.dbf), dBase III (*.dbf), Microsoft FoxPro 2.5 (*.dbf), Microsoft FoxPro 2.6 (*.dbf), Visual Fox Pro 3.0 (*.dbf), Format text separated with tabulator (*.txt) and Format text separated with semicolon (*.csv).

Set Values: Values of each column can be controlled here. For example, a parts item number can be changed here.

After setting up the parts list and a balloon style, balloon the drawing as needed and create a parts list. However, a drawing does not need to be ballooned for a parts list to be created. As parts are added or deleted to the assembly, the result will automatically be updated in the parts list but if a new part is added it will need to be ballooned.

Note: Changing a column's name will affect what appears in the parts list.

The text height and style for the text in the parts list are set through the Symbol Standard command (*AMSYMSTD*).

When opening a Mechanical Desktop file that was created in a previous release and that contains a bill of material, you will be prompted to select a standard to use.

Item numbers in the parts list and balloons are generated in the order the parts were created or instanced.

The part names that appear in the parts list are the names used as the definition names in the catalog.

The material assigned to the part for mass property information will also be used for the parts list.

If you make changes in the way a parts list should appear, erase the original parts list and recreate it.

The names used for the part names are being used from the part definition, not the browser name.

Make the changes to the bill of material setup and save them in a template file.

TUTORIAL 9.4—BILL OF MATERIAL

For this tutorial, a scene and a drawing view have been created and materials from mass properties have also been applied to the parts.

1. Open the file \Md4book\Chapter9\TU9-4.dwg.
2. Issue the Bill of Material command (AMBOM) and place a parts list, give it a name of Tutorial 9.4, use the "All" for the type of Parts list. Place the parts list in the upper right corner of the drawing. To complete the command, select OK in the BOM dialog box. When complete, your screen should resemble Figure 9.19.

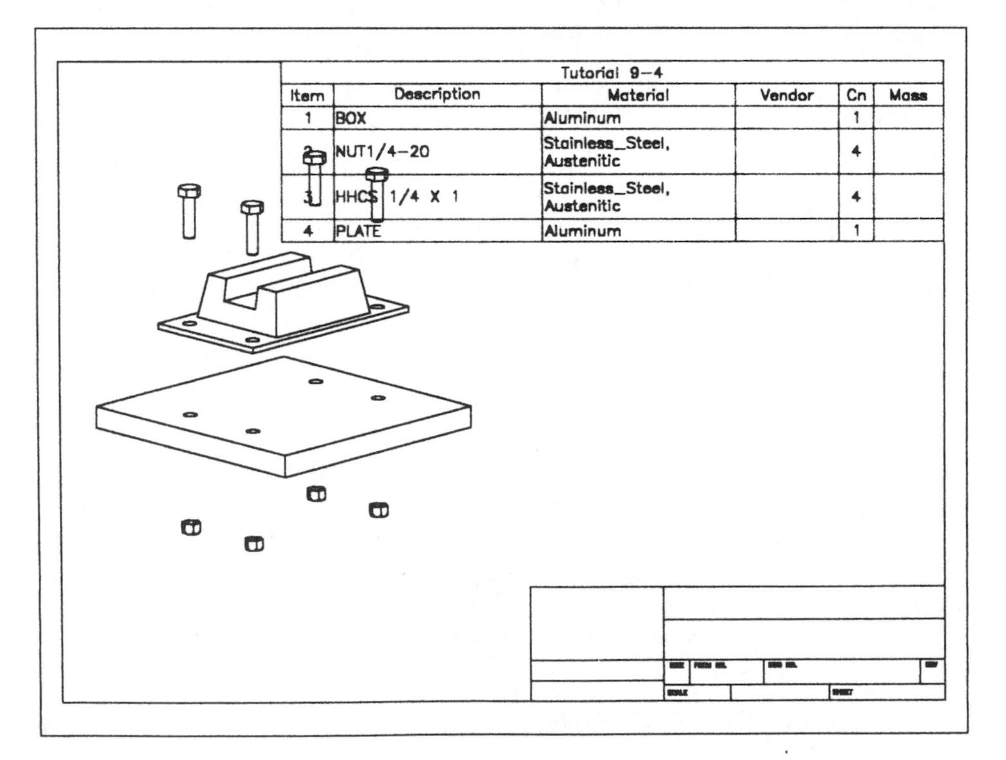

Figure 9.19

3. Issue the Bill of Material command **(AMBOM)** and delete the column Vendor. Select the column Vendor and then select the Delete Column icon, select Yes to delete the column. Select the Apply button and the Vendor column should be deleted from the parts list.
4. Select the Standard Properties icon in the BOM dialog box and change the width of the Item column to ".5", the Name column to "2" and the Material column to "1.5". To change the values width double click on the column width and type in a new value then press ENTER.
5. Select the Modify button in the Standard Properties dialog box and choose Parts List from the drop-down list.
6. Select Mass from the Parts List and select the <<- button to remove it from the list. Select OK twice to return to the BOM dialog box.
7. Select the column Cn. and select the Insert Column icon.
8. For the information for the new column change the column name to "Purchased", Caption to "Purch." and change the Width to ".75". Keep the defaults for Data type, Size and Justification. When complete, the BOM Properties dialog box should resemble Figure 9.20.

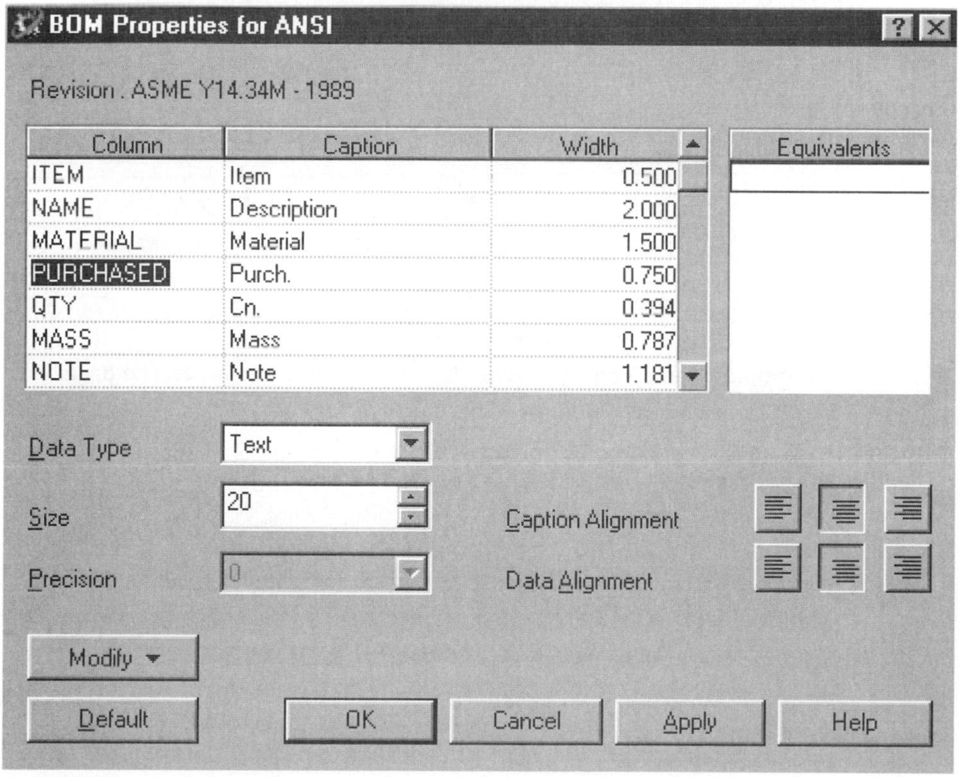

BOM Properties for ANSI

Revision . ASME Y14.34M - 1989

Column	Caption	Width
ITEM	Item	0.500
NAME	Description	2.000
MATERIAL	Material	1.500
PURCHASED	Purch.	0.750
QTY	Cn.	0.394
MASS	Mass	0.787
NOTE	Note	1.181

Equivalents

Data Type: Text

Size: 20

Precision: 0

Caption Alignment

Data Alignment

Modify

Default | OK | Cancel | Apply | Help

Figure 9.20

9. Select OK to return to the BOM dialog box.
10. Type "YES" in the Column Purch. For the rows HHCS 1/4X1 and NUT _-20. When complete, the BOM dialog box should resemble Figure 9.21.
11. Select the Standard Properties icon and in the BOM Properties dialog box select Modify and from the drop-down list select Parts List.
12. In the Properties dialog box select Cn. from the Parts List area, then select Purch. from the BOM Columns area. Then select the ->> button to insert the Purch before the Cn. column.

BOM

Item	Description	Material	Purch.	Cn.	Mass	Note
1	BOX	Aluminum		1		
2	NUT1/4-20	Stainless_Steel, A		4		
3	HHCS 1/4X1	Stainless_Steel, A	YES	4		
4	PLATE	Aluminum	YES	1		

OK Cancel Apply Help

Figure 9.21

13. To complete the command, select the OK button three times. The parts list should automatically be updated to reflect the changes.
14. Place a Part Reference by issuing the Part Reference command **(AMPARTREF)** and select a point in the middle of the plate. Fill in the Part Ref Attributes dialog box so that Description = Paint, Material = Blue Enamel, Purch. = Yes and the Reference Quantity = 2. When complete, the dialog box should look like Figure 9.22. Select OK, and the parts list should automatically be updated to reflect the change.
15. Issue the Create Balloon command **(AMBALLOON)** and type A and ENTERto use the Auto option. Select the BOM symbol for the rightmost bolt, nut and the plate and box. Then press ENTER to complete the selection process. Do not select the BOM symbol that represents the paint (its symbol is blue, while the others are cyan).

Part Ref Attributes

Description	PAINT
Material	BLUE ENAMEL
Purch.	YES
Mass	
Note	

Reference Quantity: 2

Sources OK Cancel Help

Figure 9.22

16. Move the mouse around and watch how the leaders change location as the mouse moves. Select a point in the right blank area above the title box. When complete, your drawing should resemble Figure 9.23.

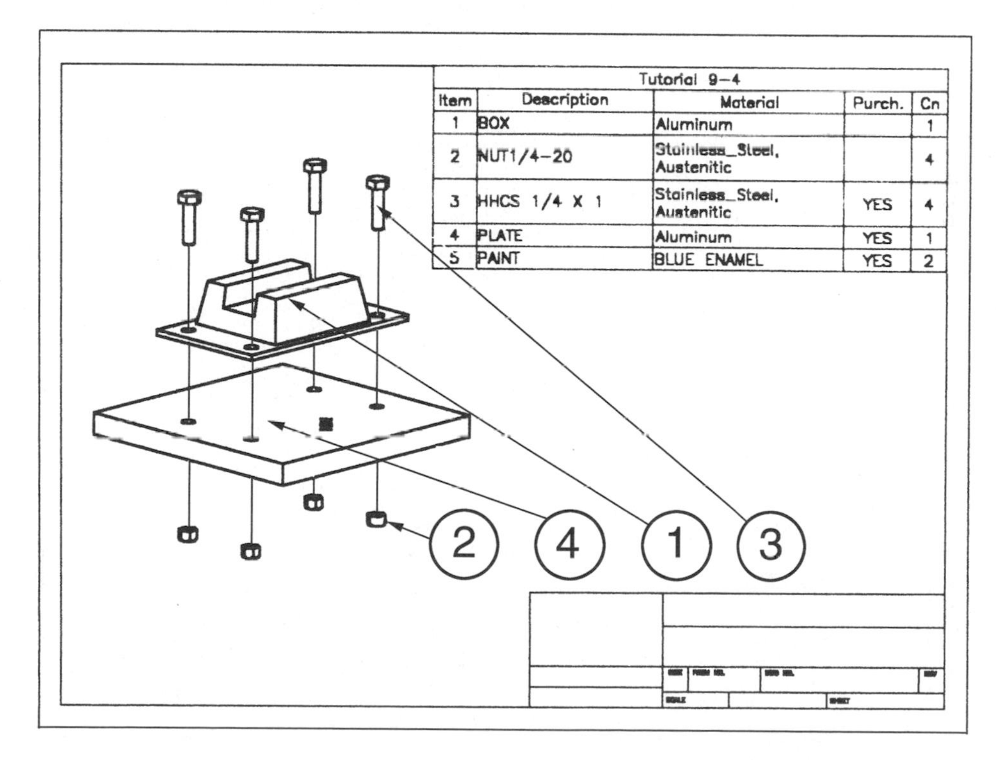

Tutorial 9-4				
Item	Description	Material	Purch.	Cn
1	BOX	Aluminum		1
2	NUT1/4-20	Stainless_Steel, Austenitic		4
3	HHCS 1/4 X 1	Stainless_Steel, Austenitic	YES	4
4	PLATE	Aluminum	YES	1
5	PAINT	BLUE ENAMEL	YES	2

Figure 9.23

17. Issue the Bill of Material command **(AMBOM)**.
18. Select the Standard Properties icon and in the BOM Properties dialog box select Modify and from the drop-down list select Balloon.
19. Change the Text height to ".18" and Apply the changes. Move the dialog box to see the change. Change the Text height to ".12". Then complete the command by selecting OK three times.
20. Issue the Create Balloon command and press C and ENTER to Collect the BOM symbol to add to a balloon. Select the BOM symbol for the paint (the blue symbol in the middle of the plate) and press ENTER.
21. Select balloon item 1, and then select a point to the right of the balloon and press ENTER.
22. Repeat the Create Balloon command and press C and ENTER to Collect the BOM symbol to add to a balloon. Select the BOM symbol for the paint (the blue symbol in the middle of the plate) and press ENTER.
23. Select balloon item 4, and then select a point to the right of the balloon and press ENTER.
24. Grip edit the starting location of the leaders so they point to the outside of all the parts. Then grip edit the balloon until your screen resembles Figure 9.24.

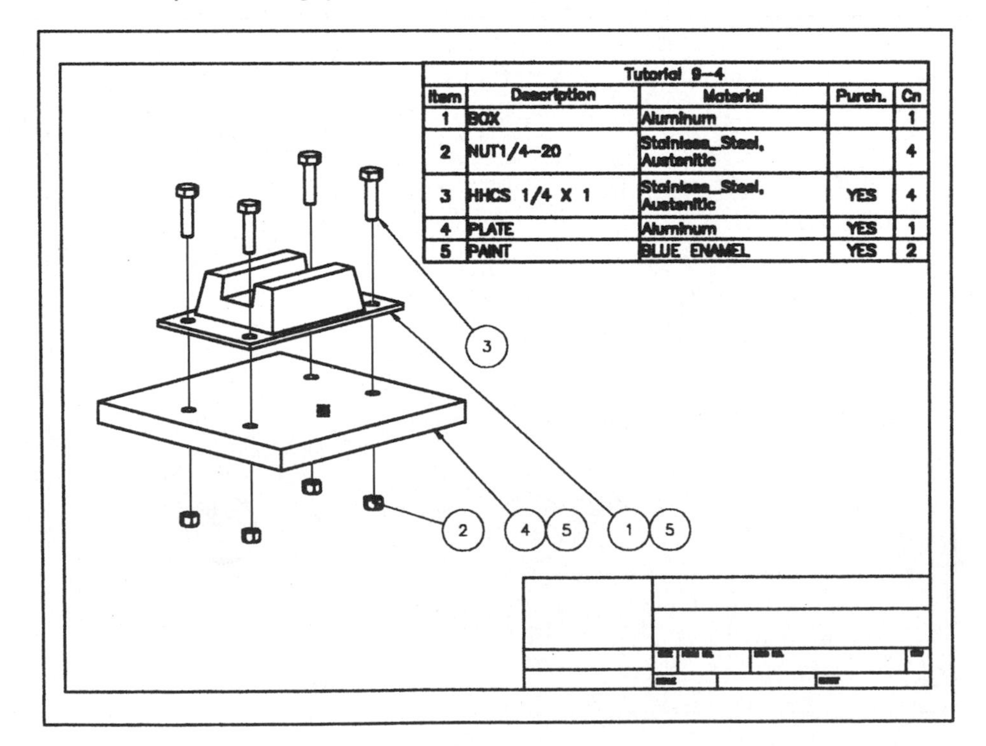

Figure 9.24

25. Experiment with different balloon types, adding and deleting columns and adding more reference parts.
26. When finished, save the file.

CREATING AN AUTOCAD 2D DRAWING FROM MECHANICAL DESKTOP DRAWING VIEWS

After creating drawing views you may need to share Mechanical Desktop drawings with someone who only has AutoCAD. If they open a Mechanical Desktop file they will only be able to view the file. If they need to modify the drawing views you can use the Drawing View Out command **(AMVIEWOUT)** to create an AutoCAD 2D file from Mechanical Desktop. Drawing views make the layout current that the views are in that you want to export. You can export selected or all the views as well as all the objects that are in paperspace. A new file will be created that will contain all 2D objects. This 2D file will have no relationship back to the mechanical Desktop file. To create a 2D AutoCAD drawing from Mechanical Desktop drawing views, issue the AMVIEWGOUT command from the Drawing pull-down menu as shown in Figure 9.25. After issuing the command give the new file a name and location. Then you will be prompted to select the drawing views to export and ENTER to export all the drawing views.

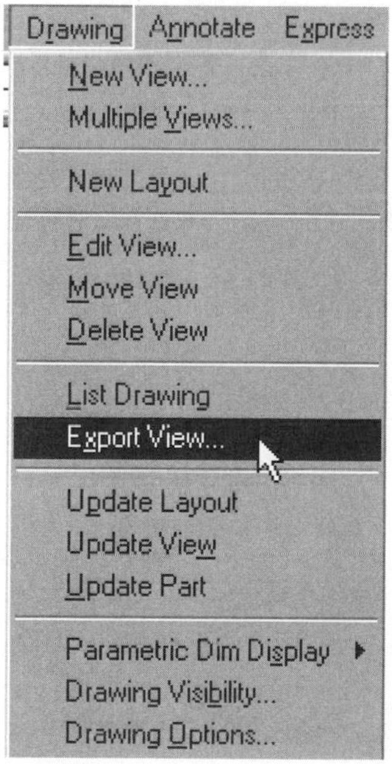

Drawing Annotate Express
New View...
Multiple Views...
New Layout
Edit View...
Move View
Delete View
List Drawing
Export View...
Update Layout
Update View
Update Part
Parametric Dim Display ▸
Drawing Visibility...
Drawing Options...

Figure 9.25

TUTORIAL 9.5—2D DRAWING VIEWS FROM MECHANICAL DESKTOP VIEWS

1. Open the file \Md4book\Chapter9\TU9-4.dwg.
2. If Layout1 is not active make it active.
3. Issue the Drawing View Out command **(AMVIEWOUT)**.
4. To export all the drawing views press ENTER.
5. Type in the file name and location of \Md4book\Chapter9\2DDRAWING.dwg.
6. Open the file \Md4book\Chapter9\2DDRAWING.dwg.
7. List a few objects to verify that they are AutoCAD objects.
8. Close all open files.

EXERCISES

For the following exercise, follow the instructions.

Exercise 9.1—Drawing Views with Balloons and a Bill of Materials

Open the file \Md4book\Chapter9\Guide.dwg. Create drawing views with a scale of 1/2 from the scenes Assembled and Exploded. Then add balloons and a bill of materials table, as shown in Figure 9.26. Then create two layouts named Assembled and Exploded. Create drawing views of the scene named Assembled in the layout named Assembled. Create drawing views of the scene named Exploded in the layout named Exploded. Create 2D drawing views of all the drawing views for the layout named Exploded.

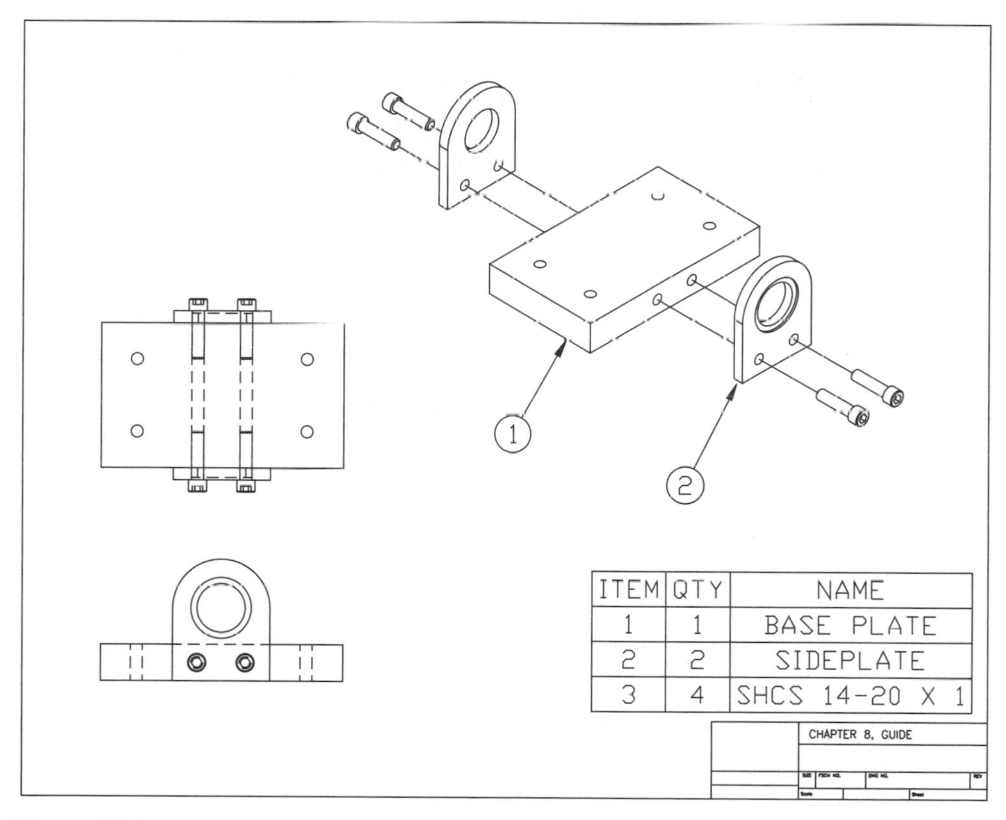

Figure 9.26

REVIEW QUESTIONS

1. A tolerance dimension can be placed with Mechanical Desktop New Dimension command **(AMDIM)**. T or F?
2. A tolerance condition is only applied to the drawing views, not the part itself. T or F?
3. A Mechanical Desktop file can have a maximum of two layouts. T or F?
4. There can only be one Base view per layout. T or F?
5. Why is it not recommended to use the Select option for the Data set when creating drawing views?
6. The properties of the default parts list can not be modified. T or F?
7. Balloons need to be placed before a parts list can be generated. T or F?
8. A bill of materials entry is fully associative to the parts. T of F?
9. Name three file formats that can be exported from a parts list.
10. Exported AutoCAD 2D drawing views of a Mechanical Desktop drawing views are fully associative back to the Mechanical Desktop file. T or F?

CHAPTER 10

Practice Exercises

In this chapter you will be guided through three exercises, intended to demonstrate different modeling and assembling techniques, not design principles. Through the exercises, you will apply the lessons that were learned in the previous chapters. Follow the steps to create the parts and assemble the parts. While working on the exercises, use any of the viewing techniques to better visualize the parts. When the exercises are complete, go back and experiment with other techniques in addition to creating drawing views.

EXERCISE 10.1—HAIR DRYER

In this exercise you will create the body of a hair dryer.

Figure 10.1 *Completed Hair Dryer*

1. Start a new file based on the template \Md4book\Chapter10\MD4Book.dwt.
2. Sketch, profile and dimension the geometry as shown in Figure 10.2.

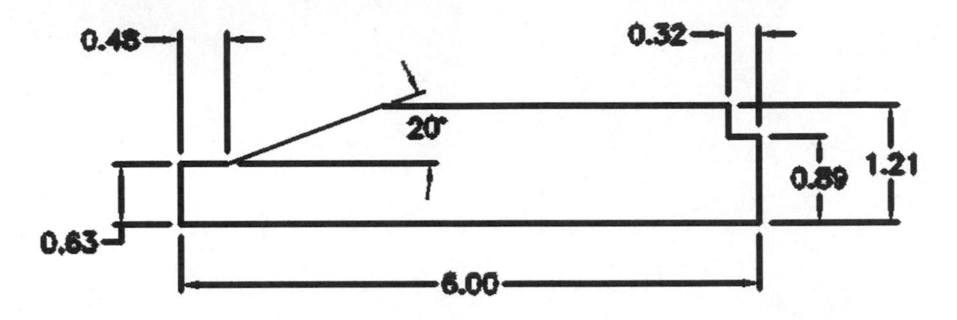

Figure 10.2

3. Change to an isometric view using the **8** key.
4. Revolve the profile 180° about the bottom horizontal line and flip the direction down. When complete, your screen should resemble Figure 10.3, lines are hidden for clarity.

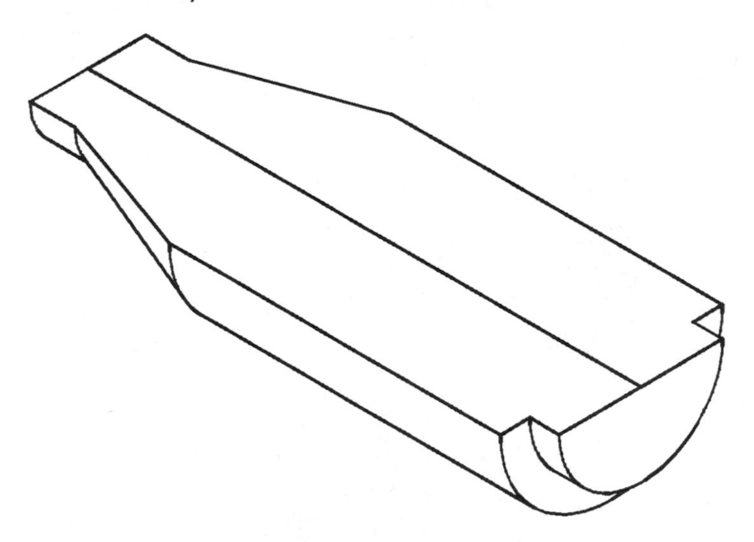

Figure 10.3

5. Sketch, profile and dimension the rectangle (the handle) as shown in Figure 10.4, lines are hidden for clarity. Also add a "YConstraint" to the top horizontal line of the rectangle and the revolution.

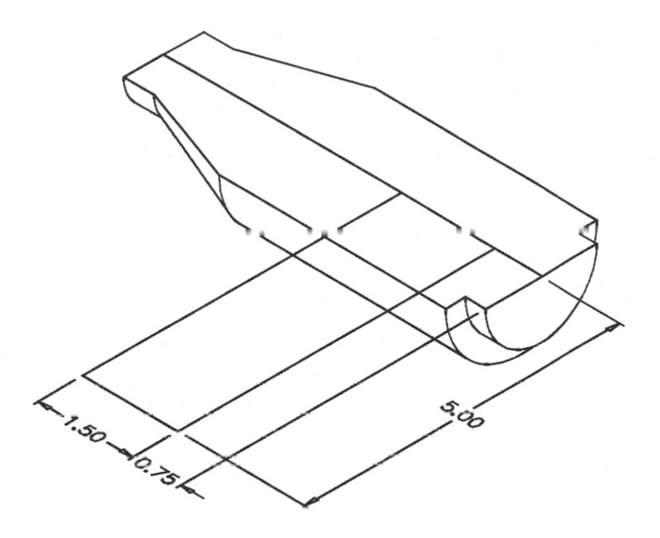

Figure 10.4

6. Extrude and join the rectangle ".375" with a draft angle of "-5" and flip the direction down. When complete, your screen should resemble Figure 10.5, shown with lines hidden.

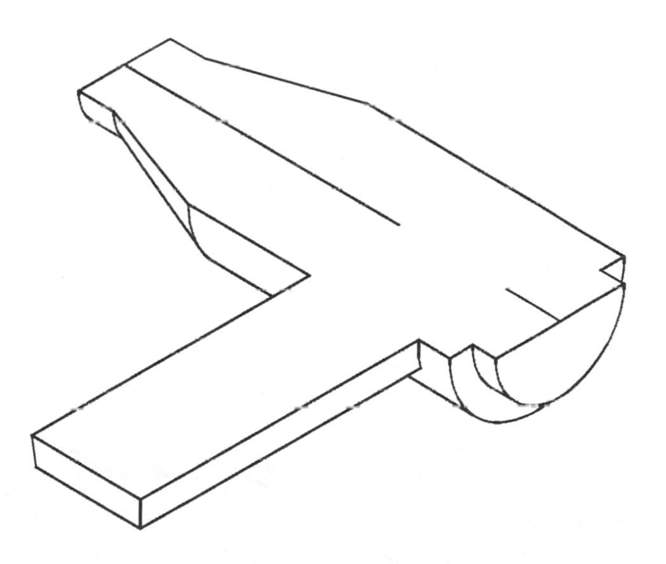

Figure 10.5

7. Add ".125" constant fillets to the five highlighted edges, as shown in Figure 10.6.

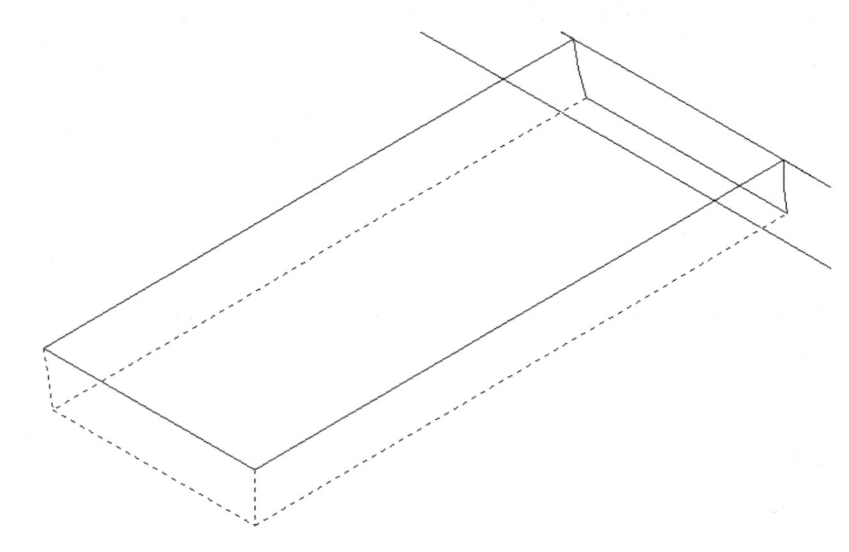

Figure 10.6

8. Shell the part ".03" and exclude the top face and the front arc that defines the front of the dryer. When complete, your screen should resemble Figure 10.7.

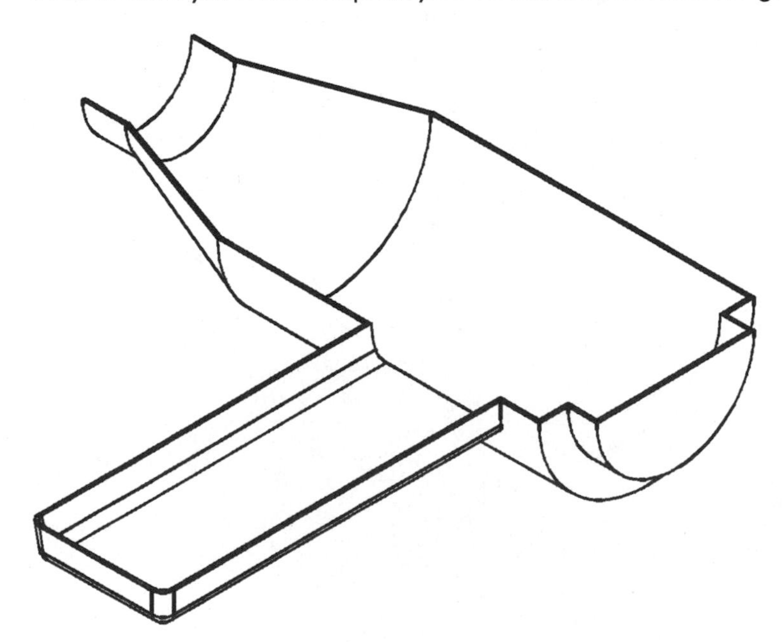

Figure 10.7

9. To create the top half of the dryer based on the bottom half, mirror the bottom half about the top plane of the bottom half. After selecting the plane to mirror about press ENTER twice to create a new part and except the default name. When complete, your screen should resemble Figure 10.8.

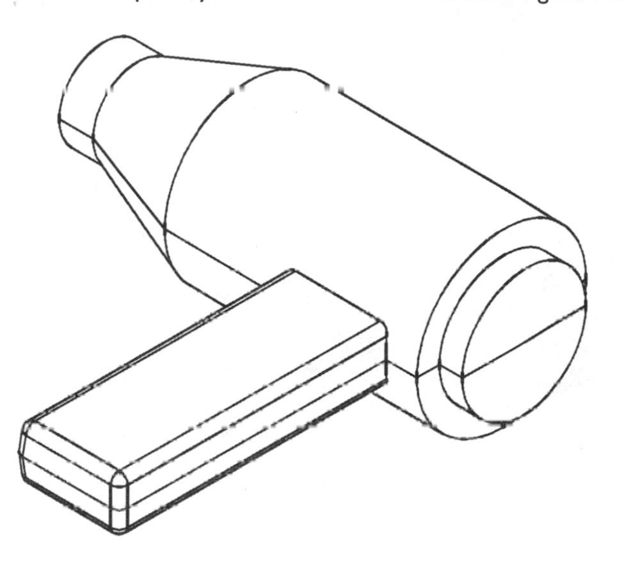

Figure 10.8

10. There are now two individual parts in the file. Combine both parts with the Join option using the AMCOMBINE command. Select the bottom half to join to the top half, because the top half is the active part. When complete, you should have a single part. You can verify this in the browser by expanding the combine feature in the browser to see Part1_1, as shown in Figure 10.9.

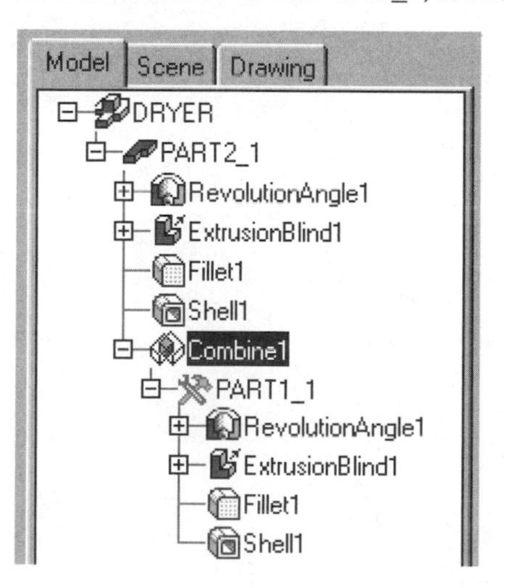

Figure 10.9

11. Add a ".2" fillet to the two highlighted edges shown in Figure 10.10. This will create one fillet going all the way around the dryer where the handle meets.

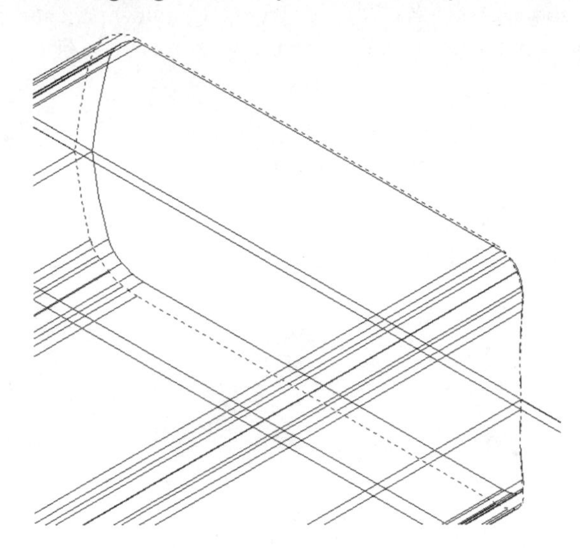

Figure 10.10

12. Make the back circular edge of the dryer the active sketch plane and orient the *X*, *Y* and *Z* as shown in Figure 10.11, shown with lines hidden for clarity.

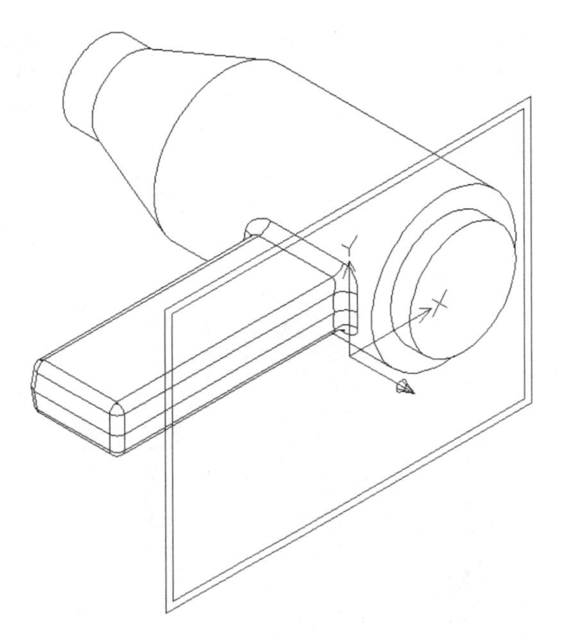

Figure 10.11

13. Sketch, profile and dimension the slot, as shown in Figure 10.12. Add an XValue constraint to an arc on the slot and the outside circular edge that defines the back of the dryer.

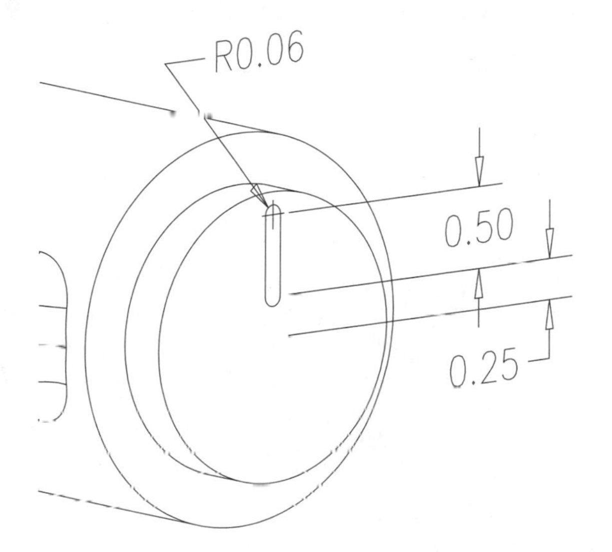

Figure 10.12

14. Extrude the slot with no draft angle, use the Cut and the Blind option. Use a value of .03 for the distance. Result as shown in fig 10.13.

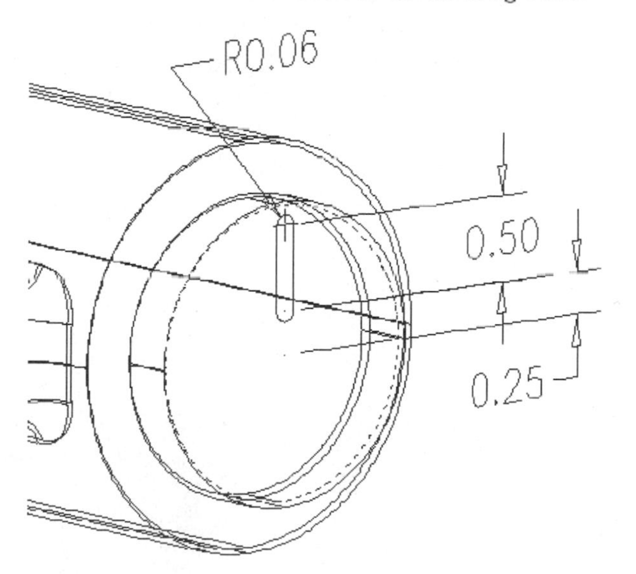

Figure 10.13

15. Array the slot with the polar option array with "8" instances and Full circle, Rotate as copied and then array the slot around the back circular edge. When complete, your screen should resemble Figure 10.14, shown with lines hidden and the work axis visibility turned off.

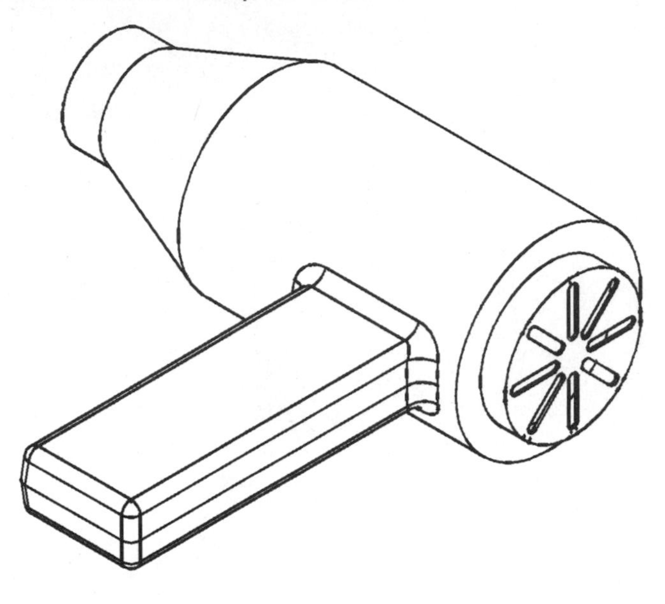

Figure 10.14

16. Turn off the visibility of the work axis.
17. Make the top plane of the handle (highlighted in Figure 10.15) the active sketch plane and accept the default orientation of the *X*, *Y* and *Z* axis. When complete, your screen should resemble Figure 10.15 shown with lines hidden.

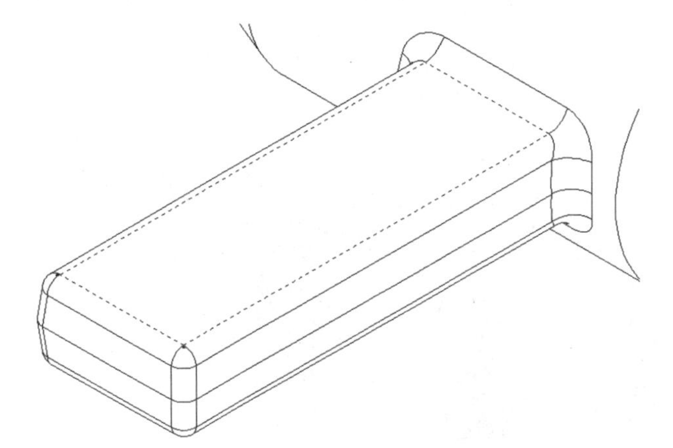

Figure 10.15

18. Sketch, profile and dimension the slot, as shown in Figure 10.16.

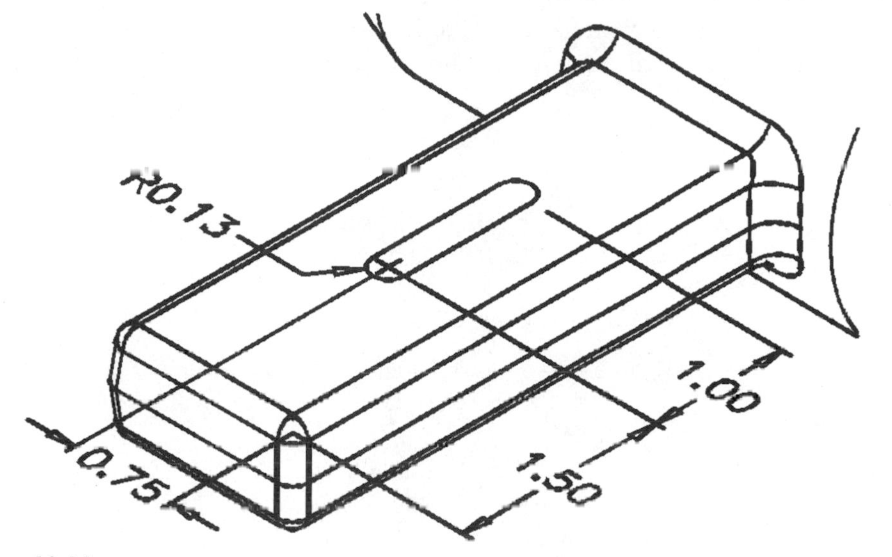

Figure 10.16

19. Extrude the slot a distance of .06 into the handle with the Cut option.
20. When complete, your screen should resemble Figure 10.17, shown with lines hidden.
21. Save the file as \Md4book\Chapter10\Dryer.dwg.

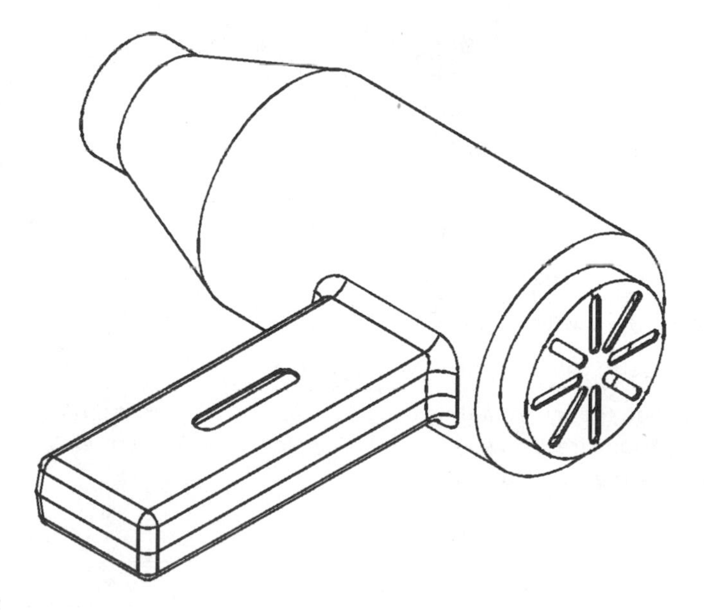

Figure 10.17

EXERCISE 10.2—CAR STAND

In this exercise, you will create a car stand that consists of three parts in a top down assembly (all the parts will exist in the same file).

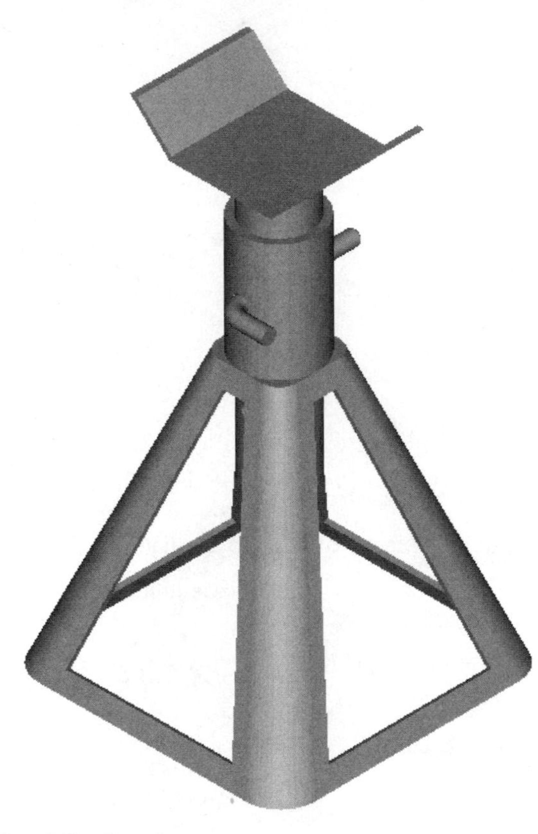

Figure 10.18 *Completed Car Stand*

1. Start a new file based on the template \Md4book\Chapter10\MD4Book.dwt.
2. Create a new part and give it a name of BASE STAND.
3. Sketch, profile and dimension a "2" square.
4. Change to an isometric view using the **8** key.
5. Extrude the square a distance of "7", with a "20°" draft angle and flip the extrusion direction so that it is going down into the screen. When complete, your screen should resemble Figure 10.19.

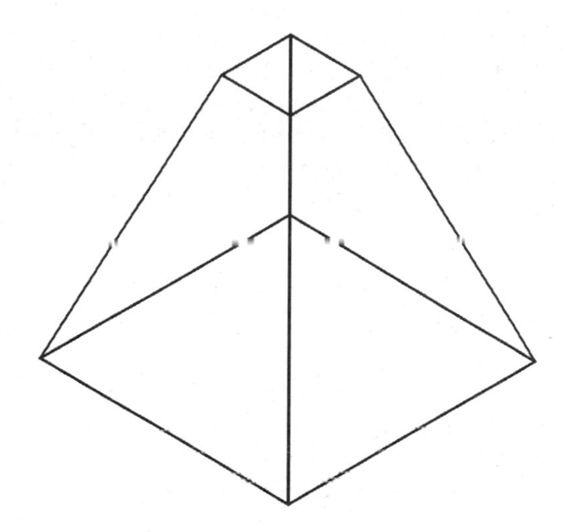

Figure 10.19

6. Create a linear fillet on each of the angled edges with a radius of "1" at the bottom and ".5" at the top of the part. When complete, your screen should resemble Figure 10.20.

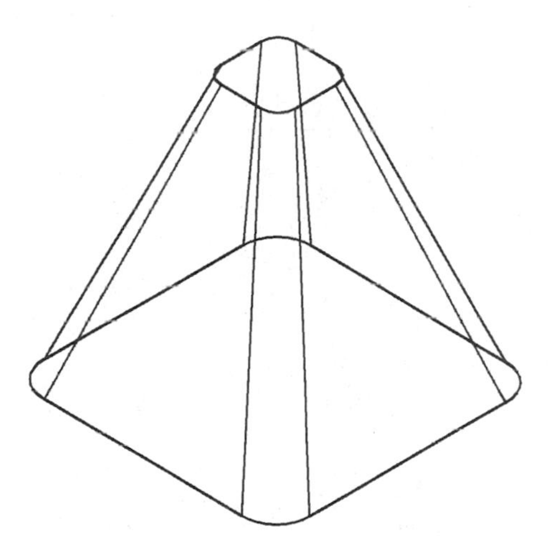

Figure 10.20

7. Shell out the part with a thickness of ".125" and exclude the bottom of the extrusion.

8. Shade and rotate the part to verify that the bottom of the part is open and the top is closed. When finished rotating the part, change to an isometric view using the **8** key.

9. Draw a circle on the top of the extrusion, profile it and add three tangent constraints to the circle and the top of the extrusion. Figure 7.21 shows the completed results with constraints shown. It does not matter what three edges the circle is constrained to since the edges are square. Note, since the top of the extrusion is still the active sketch plane, you do not need to make it the active sketch plane again.

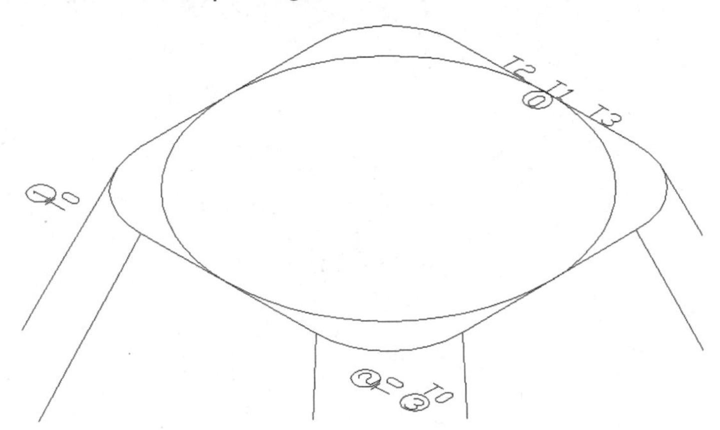

Figure 10.21

10. Extrude and join the circle a distance of "3" with no draft angle and accept the default direction of the extrusion out of the part.

11. Create a "1.625" diameter drilled through hole concentric to the cylinder. When complete, your screen should resemble Figure 10.22, shown with lines hidden.

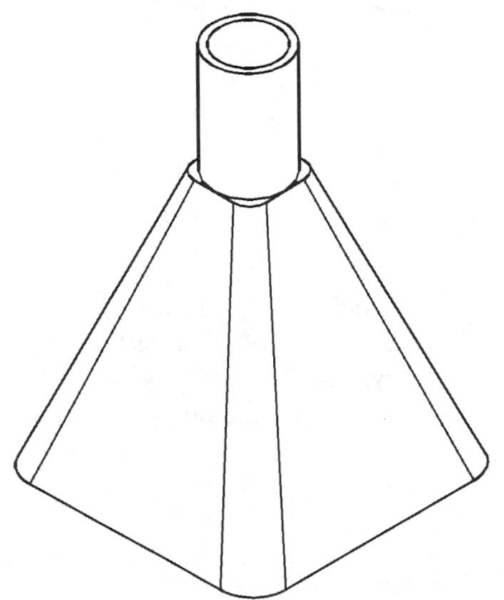

Figure 10.22

12. Create a work axis through the center of the cylinder.
13. Create a work plane using the On Edge and Planar Parallel options, check Create Sketch Plane if it is not already checked. Then create the work plane by selecting the work axis and then type in "ZX". Orient the *X*, *Y* and *Z* axis as shown in Figure 10.23.

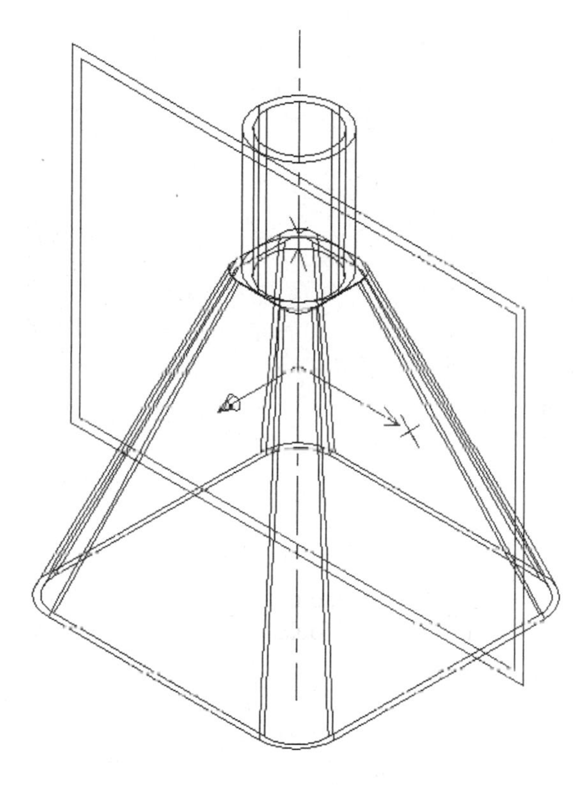

Figure 10.23

14. View the active sketch plane by using the **9** key.
15. Turn off the visibility of the work plane.
16. Draw a circle near the middle of the cylinder, profile it and dimension it as shown in Figure 10.24. Add an XValue constraint to the circle and a circular edge that defines the cylinder. This will fully constrain the part. A work point could also have been created on the outside of the cylinder to place this feature. When the holder is created, you will place a hole using the work point method.

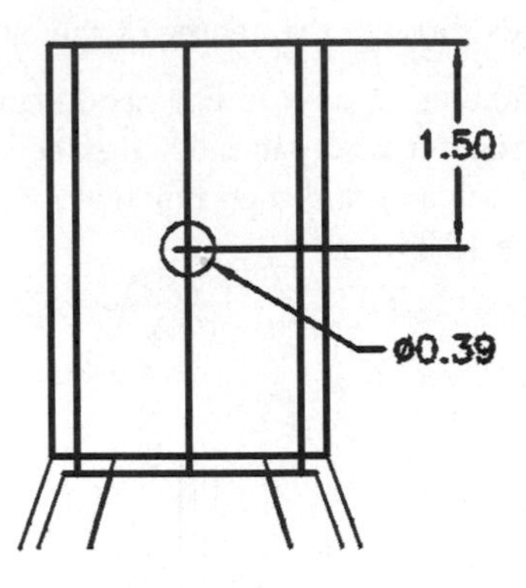

Figure 10.24

17. Change to an isometric view using the **8** key.
18. Extrude the circle-using, Cut as the operation and Mid-Through as the termination Mid Plane with no draft angle.
19. Make the front left face of the extrusion the active sketch plane as highlighted in Figure 10.25. Accept the default orientation of the *X*, *Y* and *Z* axis.

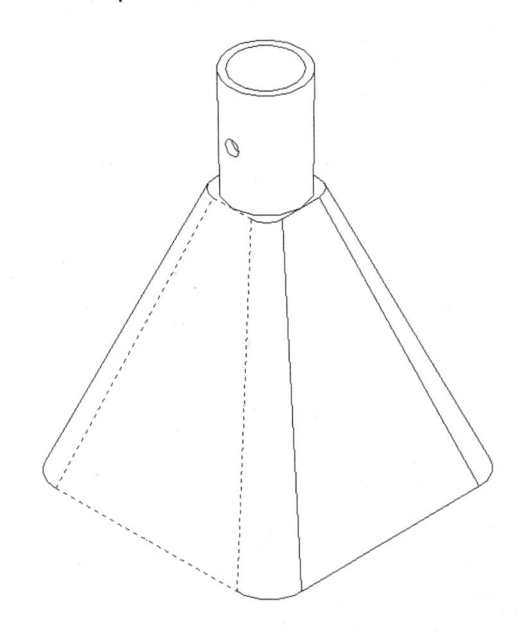

Figure 10.25

20. View the active sketch plane by using the **9** key.
21. Sketch, profile and dimension the geometry as shown in Figure 10.26.

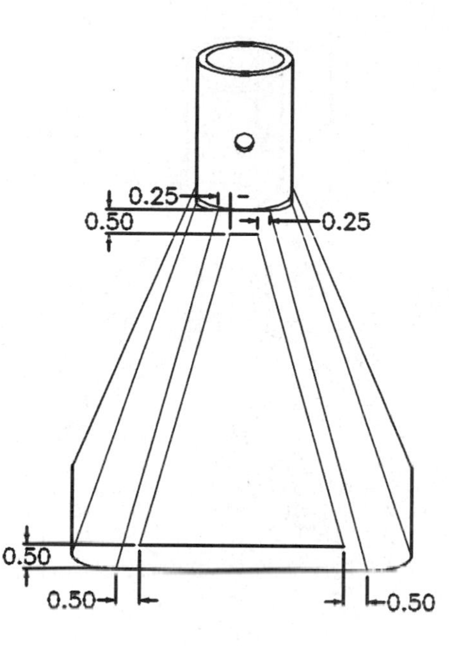

Figure 10.26

22. Change to an isometric view using the **8** key.
23. Extrude the profile ".25" to cut material from the side of the first extrusion.
24. Make the bottom of the base extrusion the active sketch plane and rotate the *Z* axis so it is pointing up and then accept the default orientation of the *X* and *Y* axis.
25. Array the cutout around the work axis with Polar Array, 4 instances, Full circle and Rotate as copied. Use the work axis to center the array.
26. Turn off the visibility of the work axis. When complete, your screen should resemble Figure 10.27, shown with lines hidden.

EXERCISE

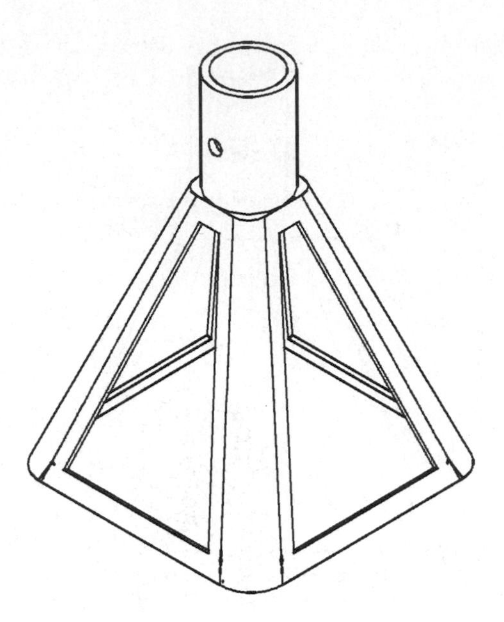

Figure 10.27

27. Save the file as \Md4book\Chapter10\Car Stand.dwg.
28. In the file Car Stand, create a new part named PIN.
29. Sketch the geometry shown in Figure 10.28.
30. Issue the 2D Path command (AM2DPATH) and select the right end of the horizontal line (near the work point shown in Figure 10.28) as the start point, type NO for the prompt:
    `Create a profile plane perpendicular to the path.`
31. Then dimension the path as shown in Figure 10.28.

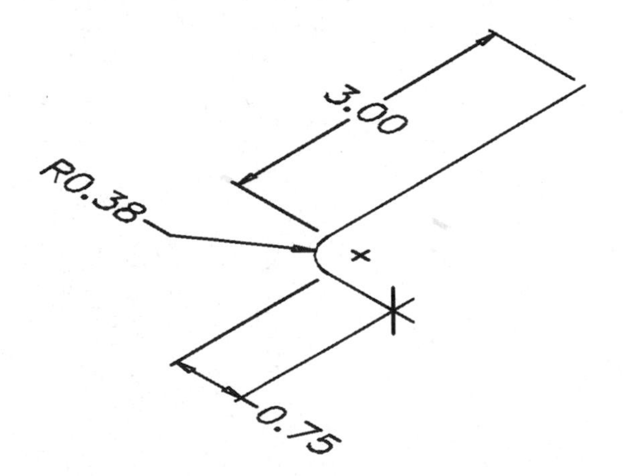

Figure 10.28

32. Create a work plane with the Normal to Start option and make it the Current Sketch Plane. Rotate the *X*, *Y* and *Z* axis until the *X* is pointing into the screen.
33. Turn off the visibility of the work plane.
34. Draw a circle near the middle of the work point, profile it.
35. Delete the Fix constraint that is on the circle. Add a ".375" diameter dimension, as shown in Figure 10.29. Then add a concentric constraint to the circle and the work point. This will fully constrain the part.

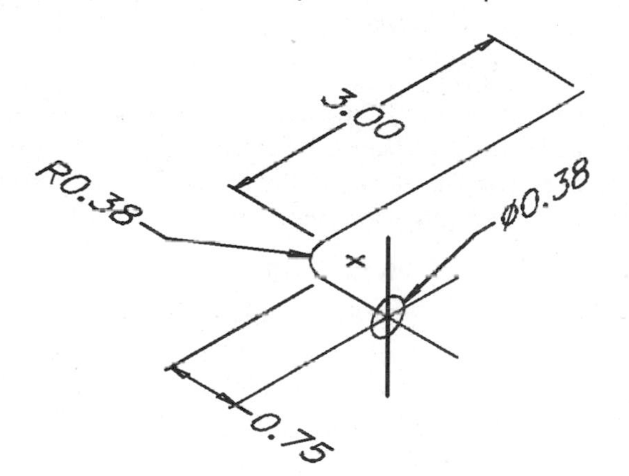

Figure 10.29

36. Sweep the circle along the path and accept the defaults in the dialog box.
37. Add a "0.0625" chamfer with the Equal Distance option to the circular edge at the end of the 3" long sweep.
38. When complete, your drawing should resemble Figure 10.30.

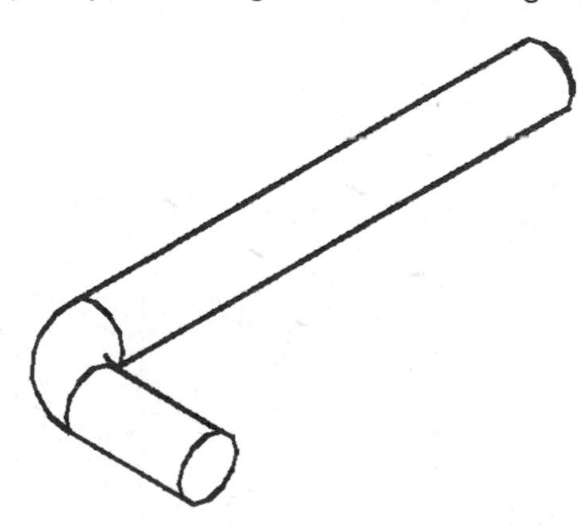

Figure 10.30

39. Save the file.
40. Issue the New Sketch Plane command **(AMSKPLN)** and type in *XY* to make the world coordinate system the active sketch plane.
41. In the file Car Stand, create a new part named HOLDER.
42. Draw a circle, profile, and dimension it with "1.625" diameter dimension.
43. Extrude the circle a distance of "7" and flip the extrusion direction so that it is going down into the screen.
44. Create a "1.375" concentric drilled hole through the center of the cylinder.
45. Place a work axis through the cylinder.
46. Create a work plane using On Edge/Axis-Planar Parallel and Create Sketch Plane should be checked. Select the work axis for the edge and type "ZX" for the plane to be parallel to. Rotate the *X*, *Y* and *Z* axis once until the *X* is pointing toward the command line.
47. Turn off the visibility of the work plane.
48. Draw, profile and dimension the geometry as shown in Figure 10.31. Add an Equal constraint between the top two horizontal lines and a YValue constraint between the bottom horizontal line of the profile and the top circular edge of the cylinder.

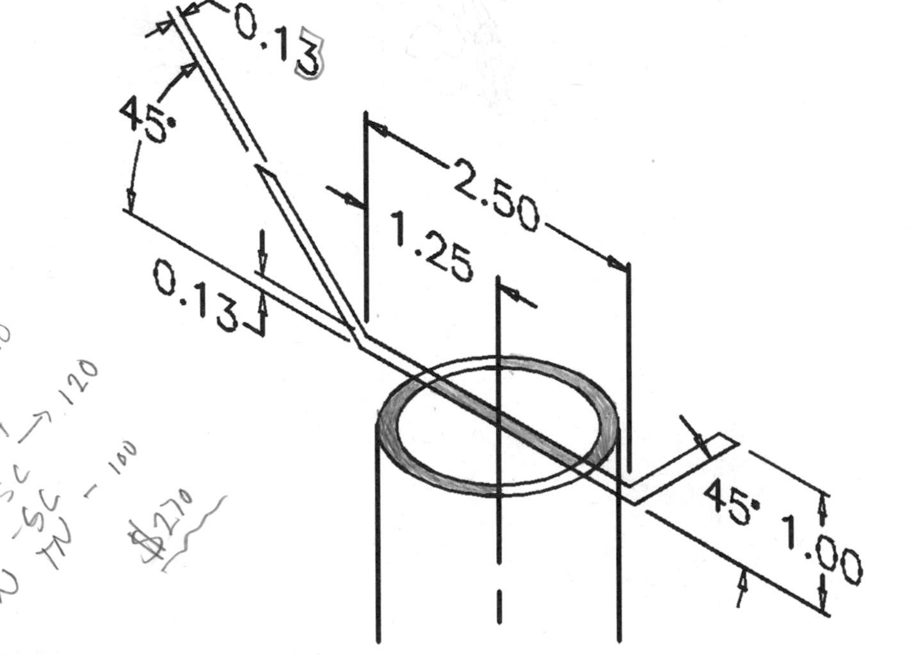

Figure 10.31

49. Extrude the profile "3" using Mid Plane as the termination and join it to the holder.
50. Turn the visibility of work plane back on.
51. Create a work plane that is Tangent-Planar Parallel and make the work plane the active sketch plane. Select near the southwest quadrant of the cylinder and select the existing work plane to be parallel to. Flip the direction of the offset of the work plane and then ENTER to accept the default orientation of the *X*, *Y* and *Z* axis.
52. Turn off all work planes and work axes. Then zoom into the top of the holder.
53. Create a work point and dimension it as shown in Figure 10.32.

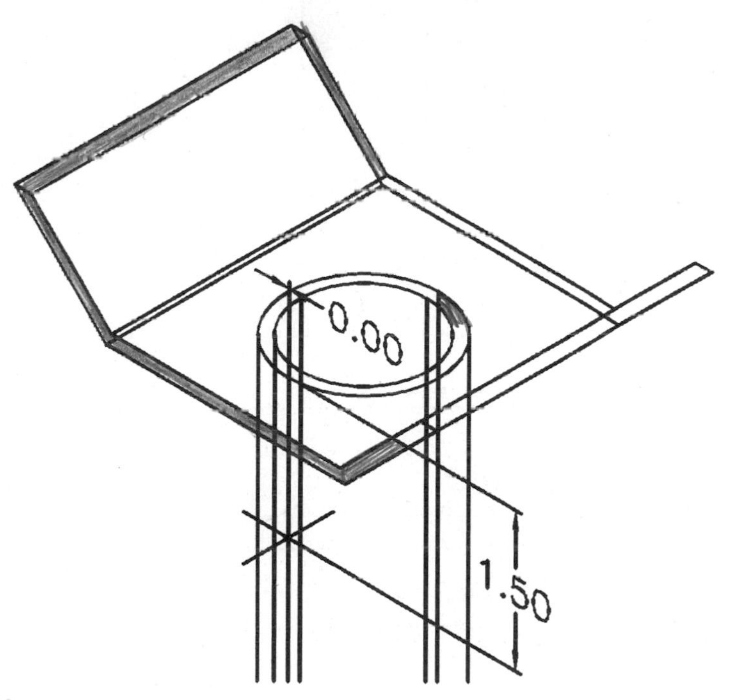

Figure 10.32

54. Create a ".38" diameter drill hole through the cylinder using the On Point option for the Placement.

55. Array the drilled hole with "4" Rows and a Spacing of "-1.25". When complete, your screen should resemble Figure 10.33.
56. Save the file.

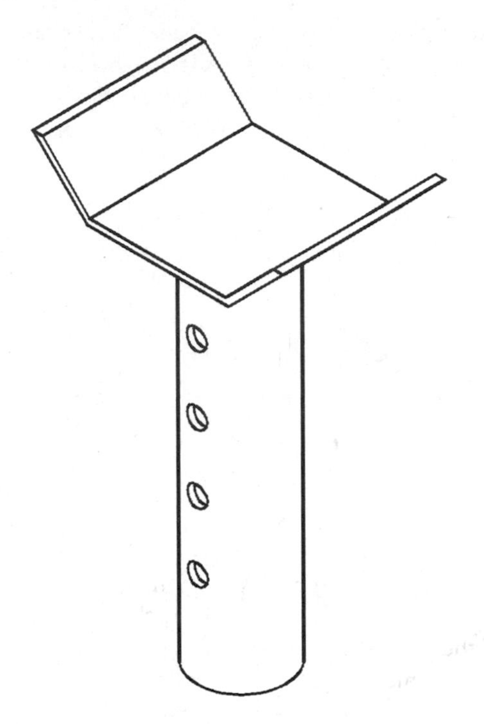

Figure 10.33

57. Before assembling the parts, reorder the HOLDER using the browser so it comes before the PIN.
58. To make it easier to select the parts geometry, move the HOLDER near the BASE STAND with the AutoCAD MOVE command. Then zoom in close to the BASE STAND and HOLDER.
59. Use the Mate constraint to align the center of the Base Stand and the Holder. Figure 10.34 shows the two centerlines of the parts. Press ENTER to accept the default offset distance of zero.

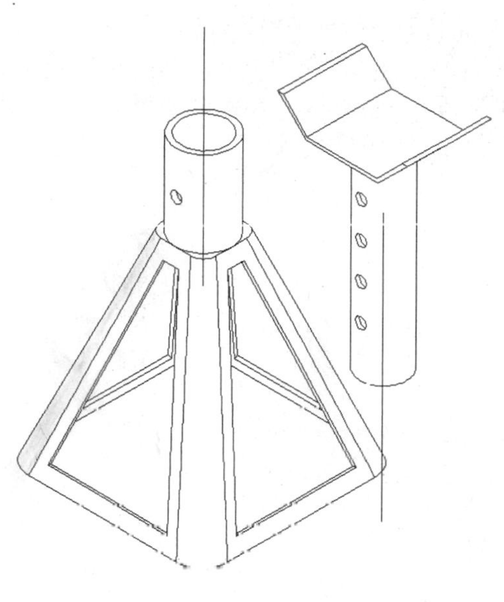

Figure 10.34

60. Repeat the Mate constraint and align the center of the hole in the BASE STAND and the hole that is second from the top in the Holder. Figure 10.35 shows the two centerlines of the parts (the centerline of the holder may appear off the part). Press ENTER to accept the default offset distance of zero. This will remove all the degrees of freedom.

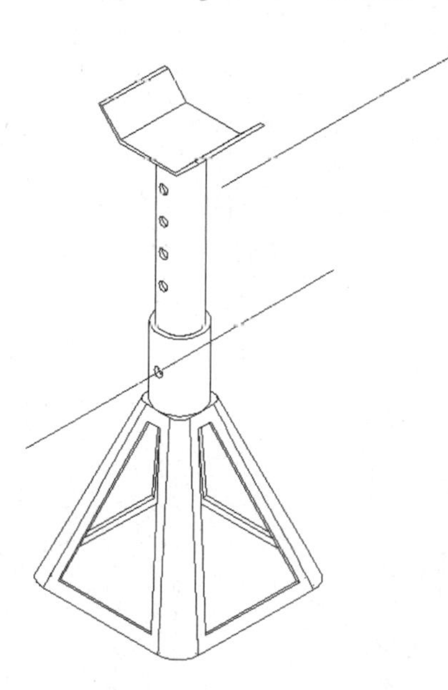

Figure 10.35

61. Move the PIN near to the BASE STAND.
62. Use the Mate constraint to align the center of the BASE STAND and PIN. Figure 10.36 shows the two centerlines of the parts. Press ENTER to accept the default offset distance of zero.

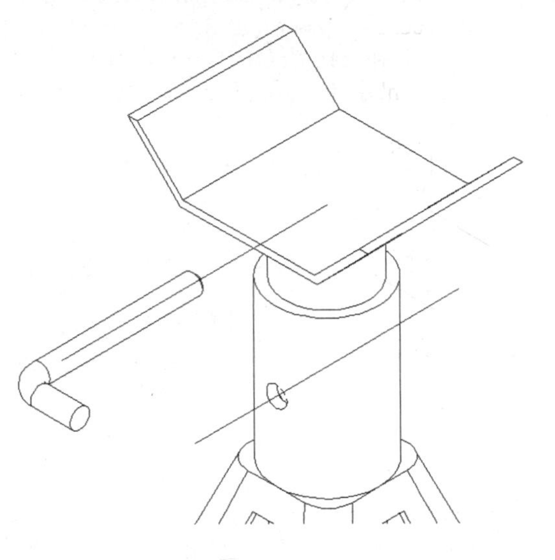

Figure 10.36

63. Repeat the Mate constraint and select the inside quadrant of the pin, using the quadrant object snap, as shown in Figure 10.37. Cycle through the options until the arrow points inward (this represents the tangent face at the inside

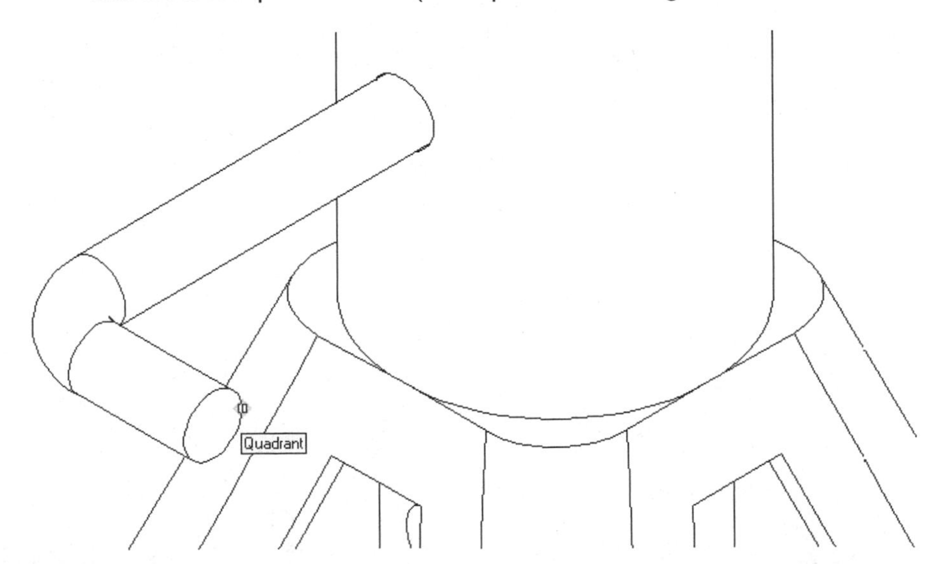

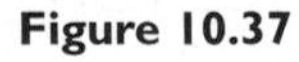

Figure 10.37

quadrant). Press ENTER to accept this option. Select the front quadrant of the bottom of the cylinder of the Base Stand, using the quadrant object snap as shown in Figure 10.38. Cycle through the constraints until the arrow points outward and press ENTER to accept this option. At this point, your screen should resemble Figure 10.39. Type in an offset value of ".1875" and ENTER. When complete, your screen should resemble Figure 10.40. If the PIN is on the opposite side of the cylinder, update the assembly (use the AMUPDATE command with the Assembly option) to switch the PIN to the correct side.

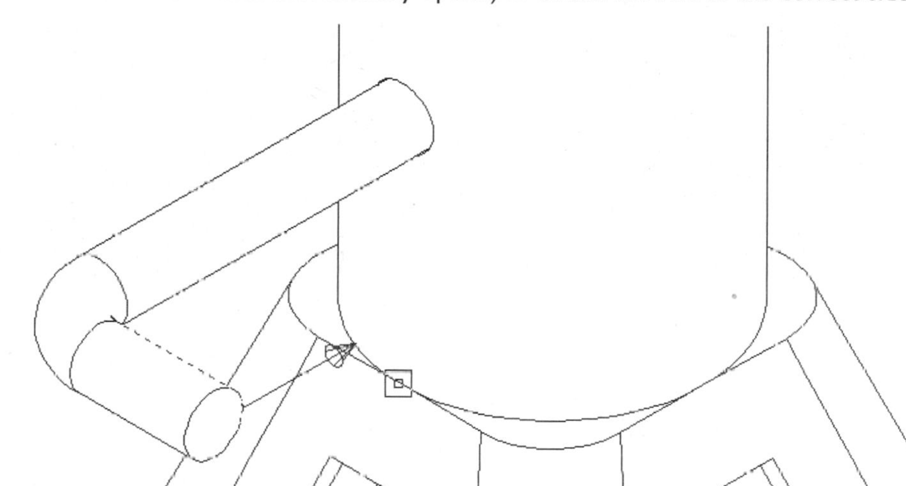

Figure 10.38

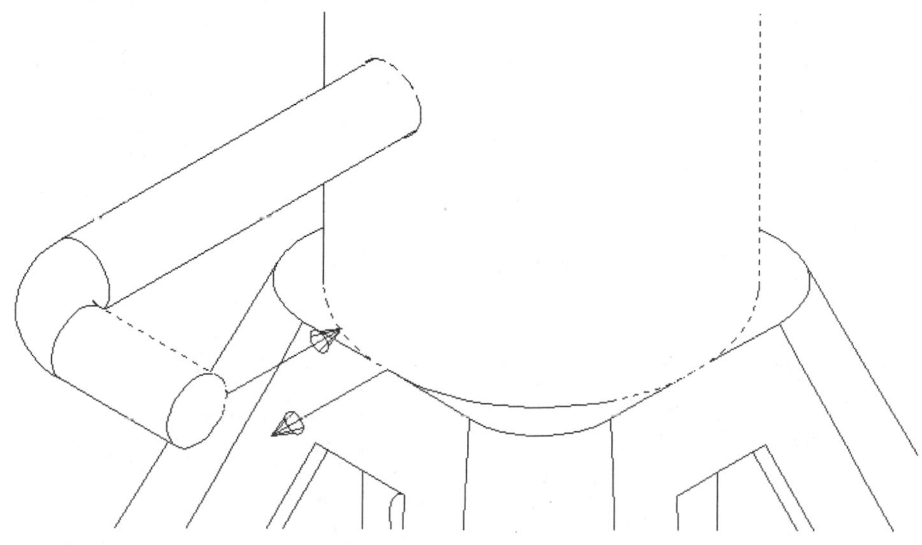

Figure 10.39

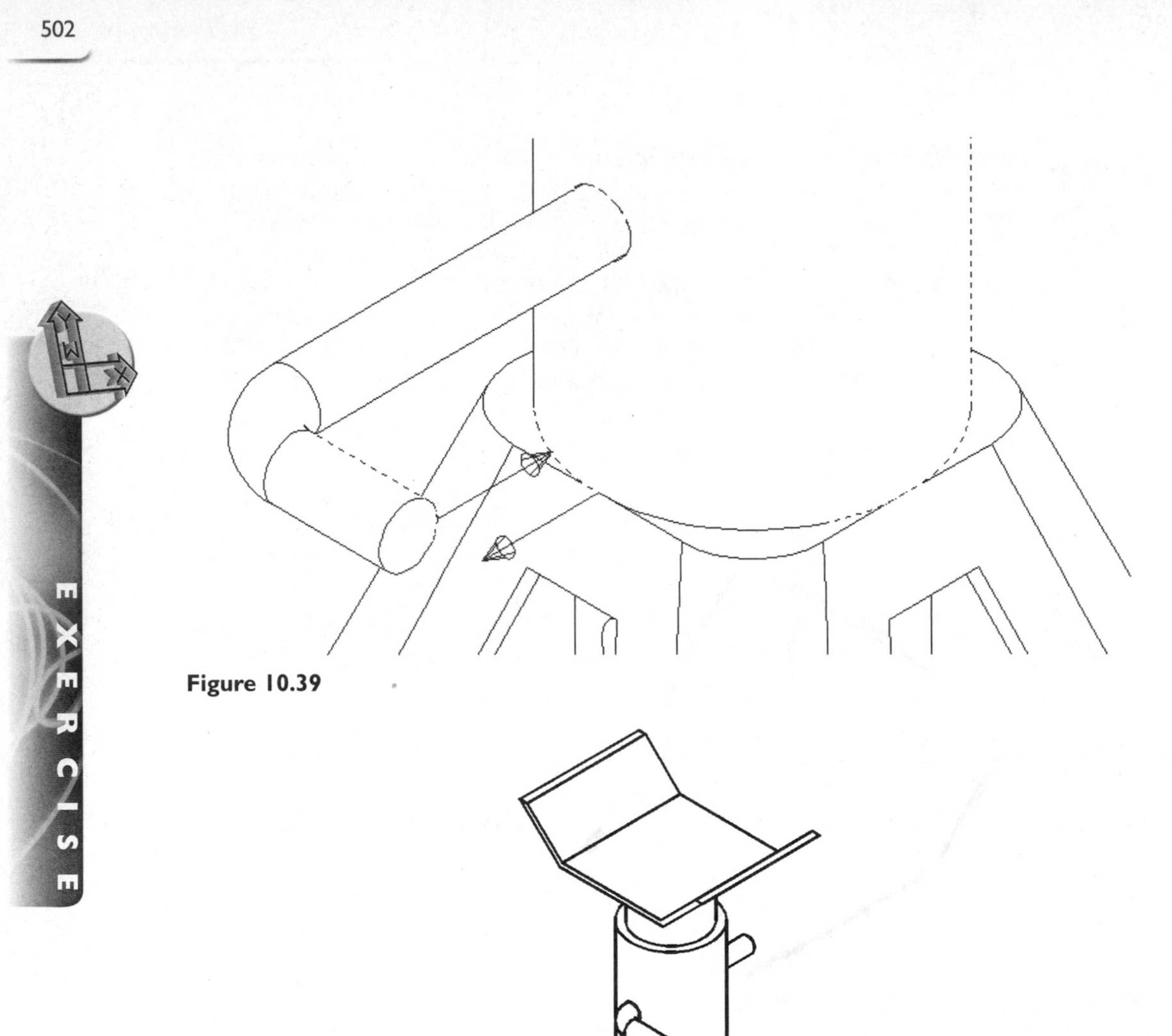

Figure 10.39

Figure 10.40

EXERCISE 10.3—WOOD PLANE

In this exercise, you will create a wood plane that consists of five parts in a bottom up assembly (each part is in its own file). You will create four of the parts in their own files. The fifth part is a flat head screw, which can be found on your CD at: \Md4book\Chapter10\10-32x1.dwg.

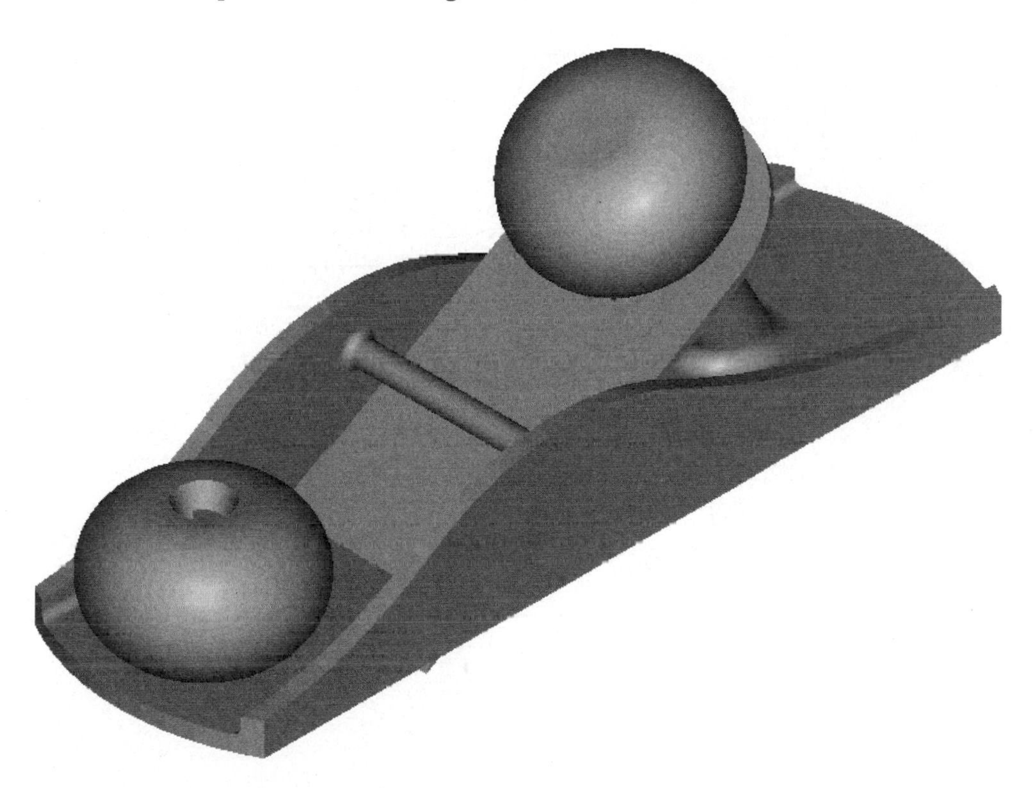

Figure 10.41 *Completed Wood Plane*

1. Start a New Part File (single part mode) for the base of the wood plane using the template \Md4book\Chapter10\MD4Book.dwt.
2. Sketch, profile and dimension the geometry as shown in Figure 10.42. The two construction lines should have tangent constraints applied to them and the arc they touch and the construction lines should be projected to the top of the .25 long vertical lines.

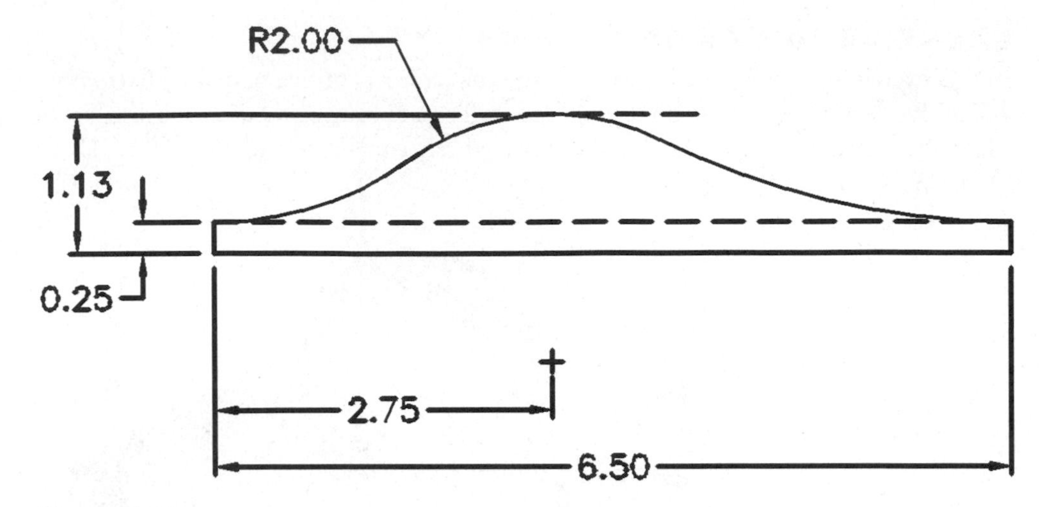

Figure 10.42

3. Change to an isometric view using the **8** key.
4. Extrude the profile "1.875" in the default direction.
5. To turn the part so that it is laying on its bottom use the ROTATE3D command and rotate the part 90° about the *X* axis.
6. Make the right side vertical face the active sketch plane and orient the *X*, *Y* and *Z* axis as shown in Figure 10.43.

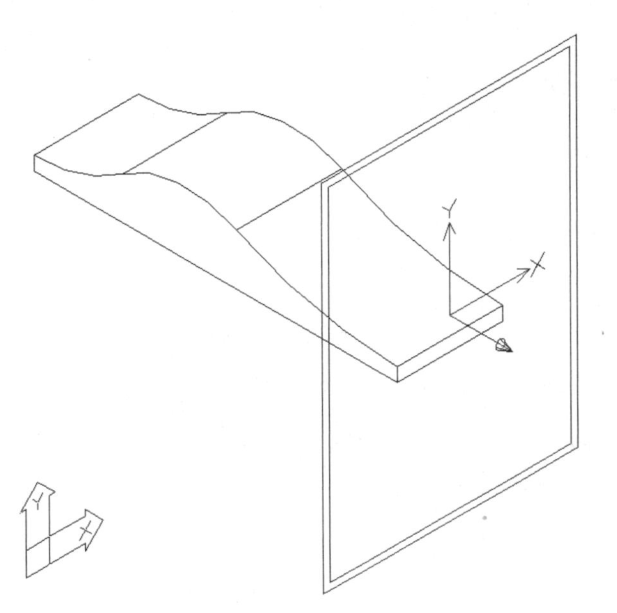

Figure 10.43

7. Sketch, profile and dimension the rectangle as shown in Figure 10.44.

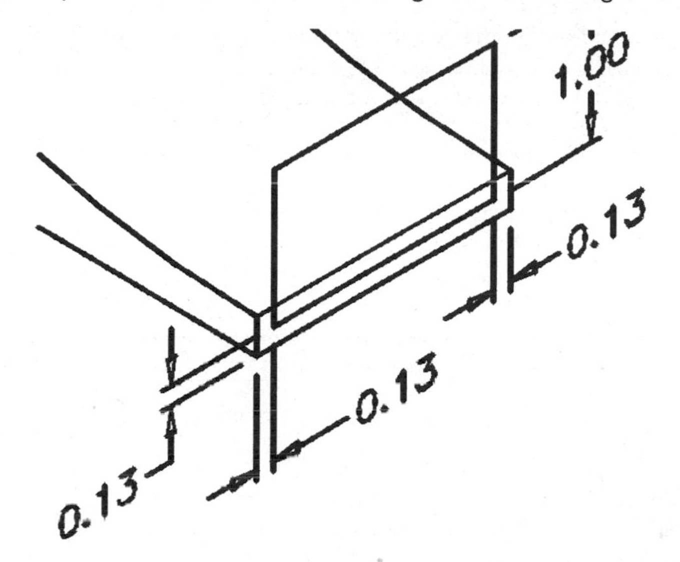

Figure 10.44

8. Extrude the rectangle through the part, cutting away the inside.
9. Make the bottom plane the active sketch plane and orient the *X*, *Y* and *Z* axis as shown in Figure 10.45. You will need to flip the *Z* axis and rotate the *X* and Y axis.

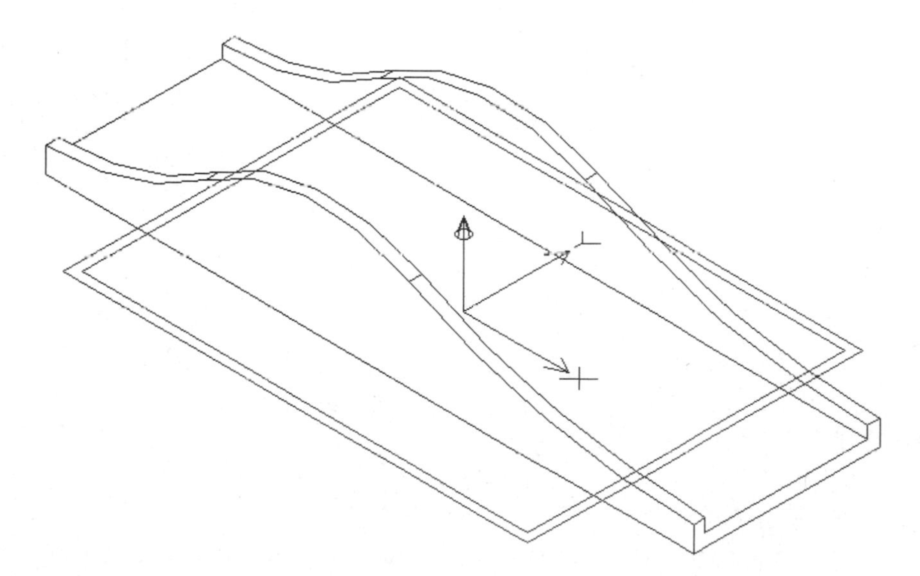

Figure 10.45

10. Change to sketch view (**9**) and press ENTER.
11. Sketch and profile the geometry shown with a width in Figure 10.46.

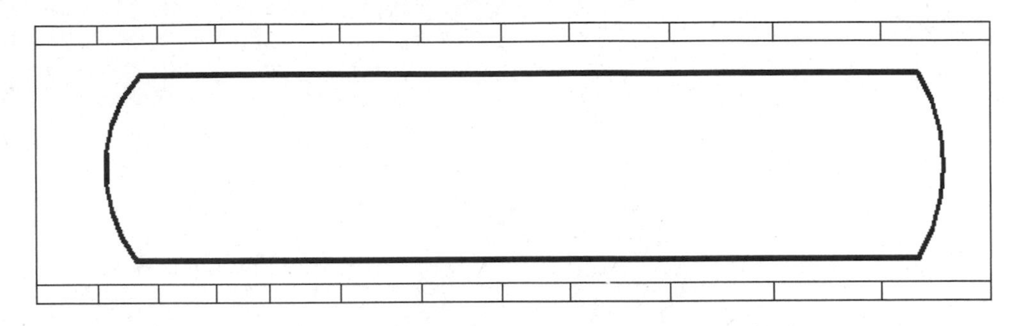

Figure 10.46

12. Add an XValue constraint to the left end of the top horizontal line and to the left side of the bottom horizontal line, add an equal constraint between the two horizontal lines, add a "6" dimension to one of the lines and add two tangent constraints to the arcs and outside vertical edges of the part. Add two collinear constraints to the lines and outside horizontal edges of the part and add a radius constraint to both arcs to fully constrain the profile. Figure 10.47 shows the fully constrained profile with constraints visible.

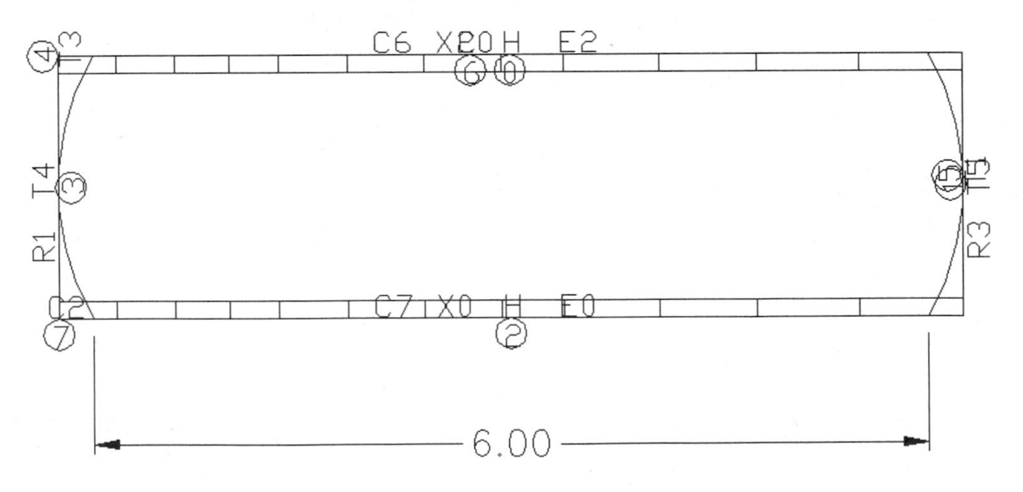

Figure 10.47

13. Change to an isometric view using the **8** key.
14. Extrude the profile through the part using Intersect as the operation. When complete, your screen should resemble Figure 10.48.

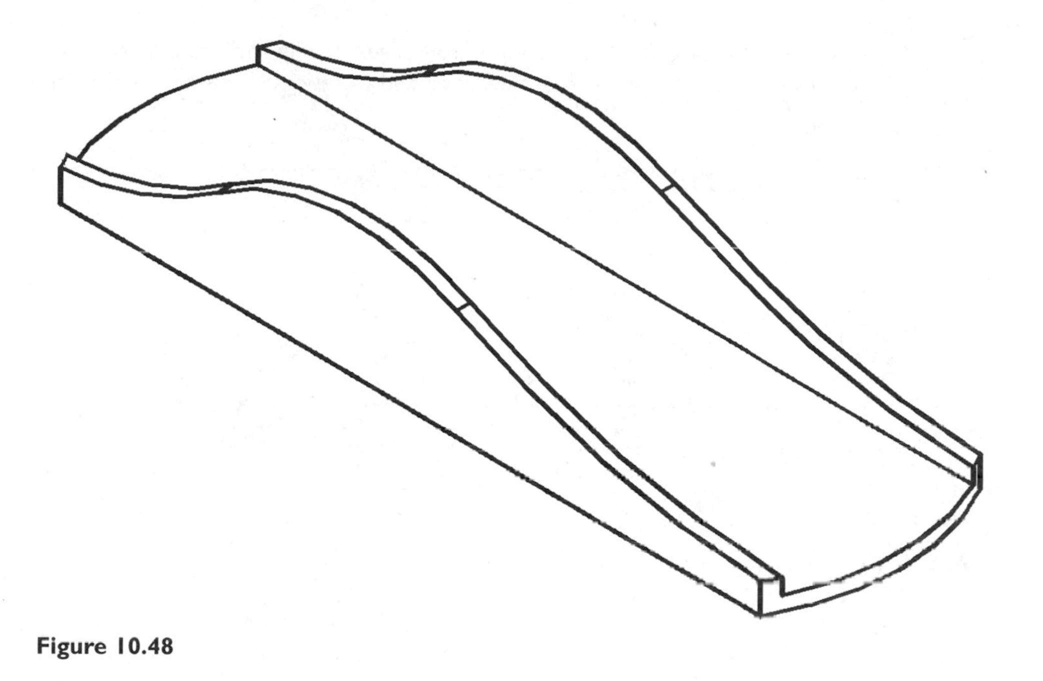

Figure 10.48

15. Next make the inside plane the active sketch plane as highlighted in Figure 10.49 and accept the default orientation of the *X*, *Y* and *Z* axes.

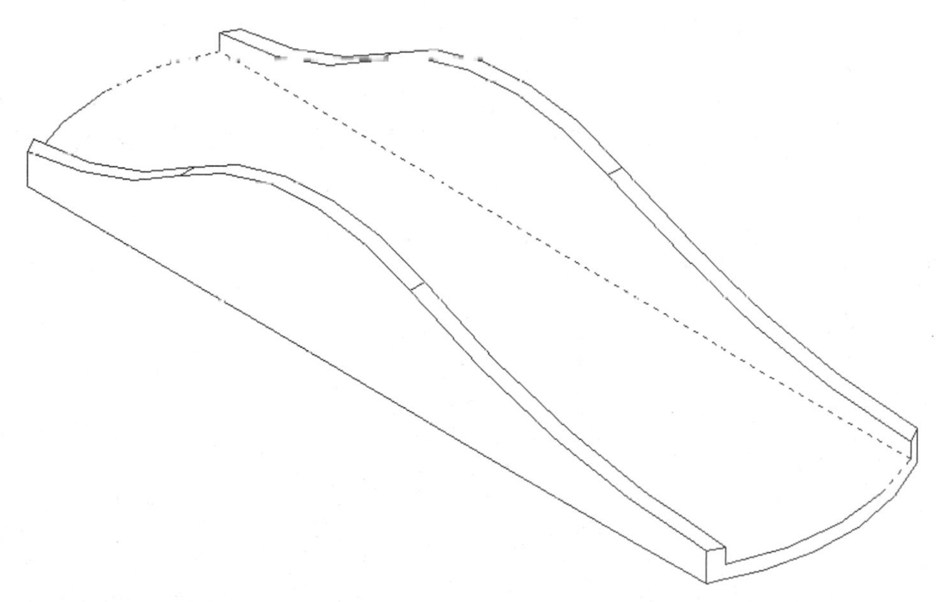

Figure 10.49

16. Draw two vertical lines close to the inside edges, profile them and select the two inside edges to close the profile. Then add the two dimensions as shown in Figure 10.50.

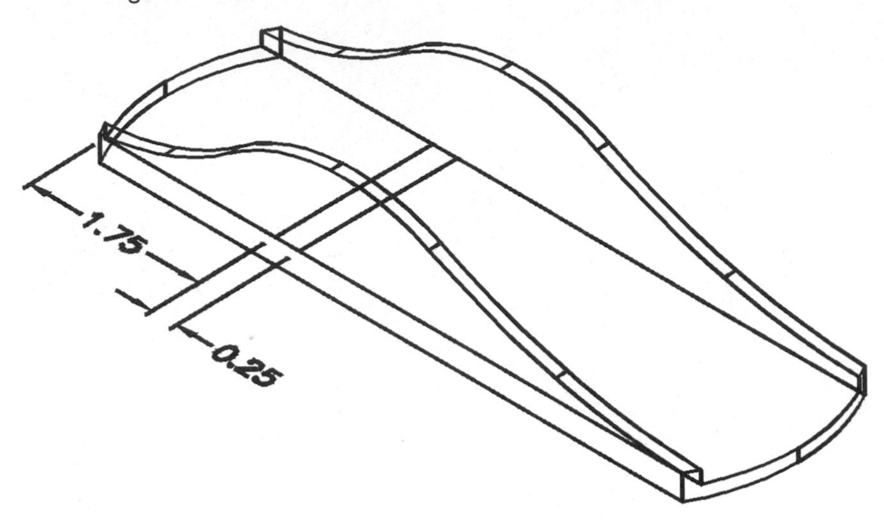

Figure 10.50

17. Extrude the profile with the Cut and Through options and remove the material through the bottom of the part.
18. Create a chamfer with the Distance x Angle option, setting the distance to ".125" and the angle to "60". Select the edge that is highlighted in Figure 10.51 and then ENTER. For the prompt:

    `Apply angle value to highlighted face.`

    Cycle until the inside vertical face is highlighted then press ENTER to accept this face.

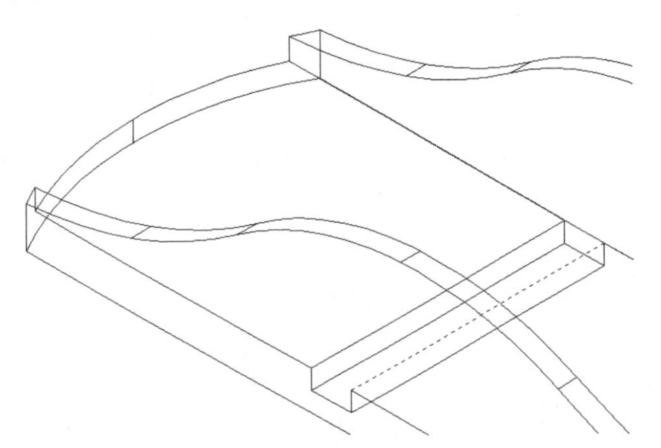

Figure 10.51

19. Change to a southwest isometric view by typing **88** and press ENTER.
20. Make the chamfered face the current sketch plane and orient the *X*, *Y* and *Z* axis as shown in Figure 10.52.

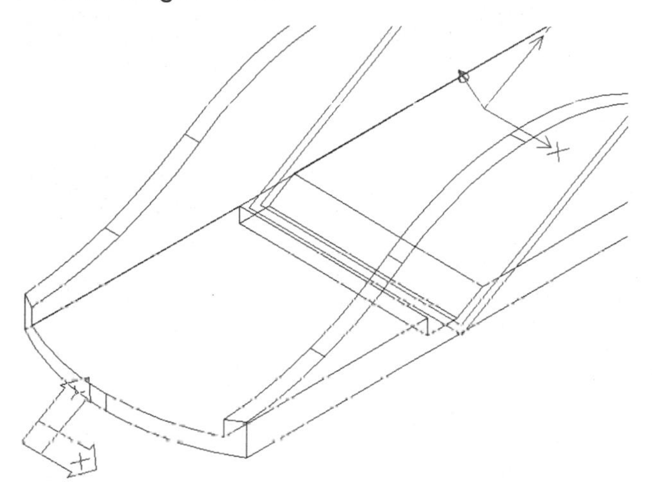

Figure 10.52

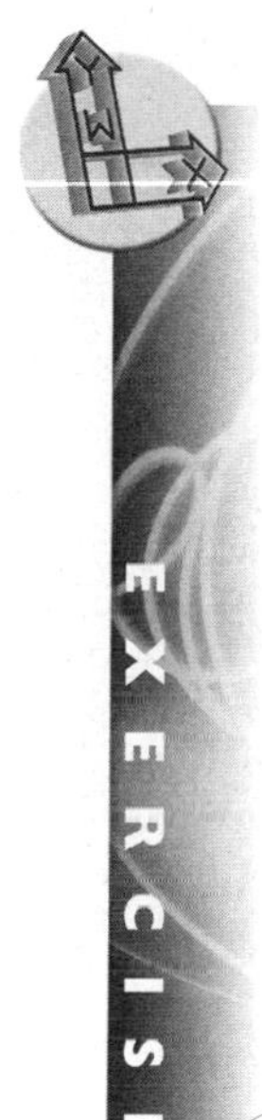

21. Change to the plan view (**9**) and press ENTER.
22. Draw a circle, profile it and dimension it as shown in Figure 10.53. To center the circle in the middle of the part add an XValue constraint between the circle and one of the bottom arcs in the middle of the base of the plane.

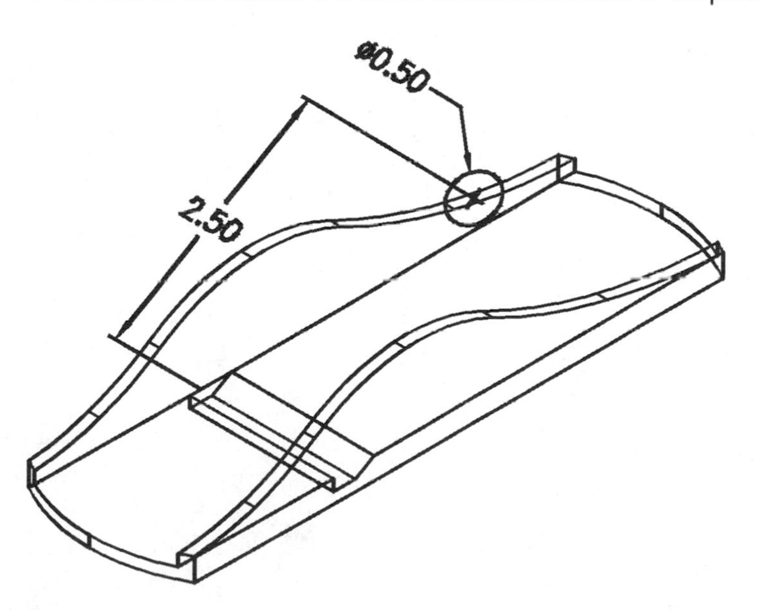

Figure 10.53

23. Extrude the circle using the Join Operation and To-Face/Plane termination and select the top plane. When complete, your screen should resemble Figure 10.54 shown with lines hidden.

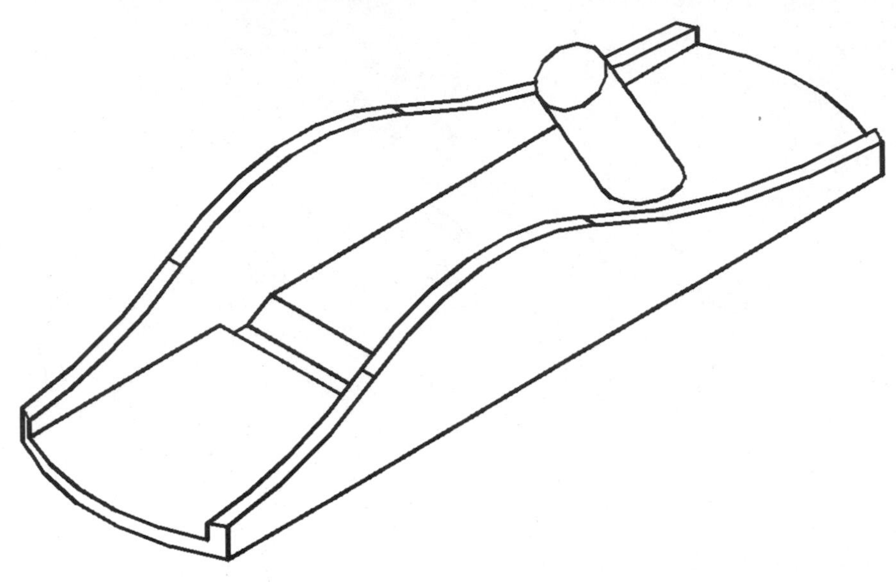

Figure 10.54

24. Make the right outside vertical face the active sketch plane, as highlighted in Figure 10.55, and ENTER to accept the default orientation of the *X*, *Y* and *Z* axes.

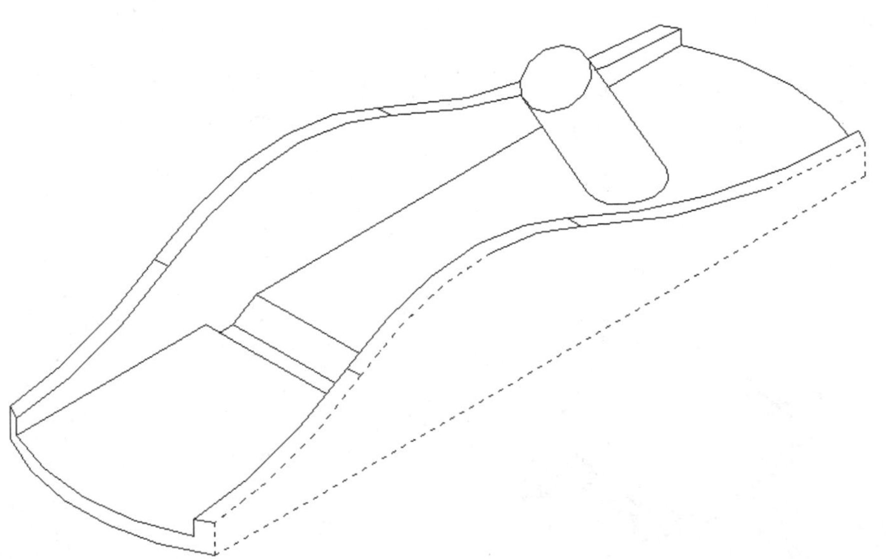

Figure 10.55

25. Draw a circle, profile it and dimension it as shown in Figure 10.56. Add an XValue constraint between the circle and one of the top arcs on the side of the plane.

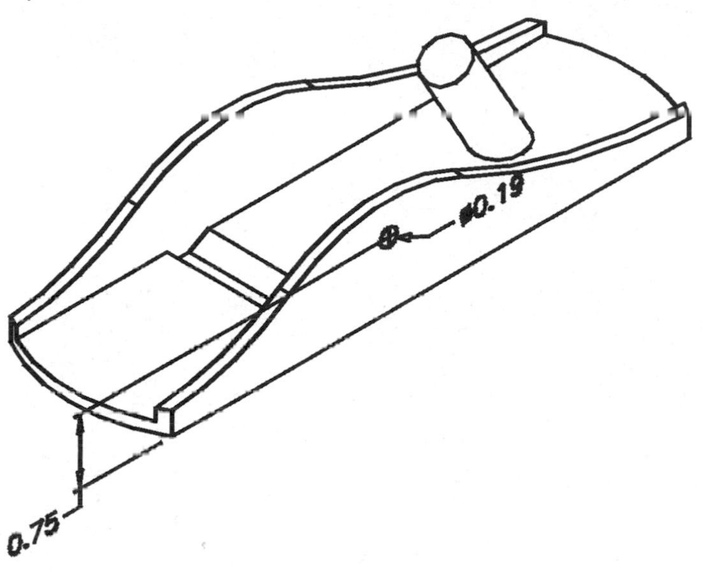

Figure 10.56

26. Extrude the circle using the Join Operation and To-Face/Plane as the Termination and select the opposite outside plane as the termination plane. When complete, your screen should resemble Figure 10.57, shown with lines hidden.

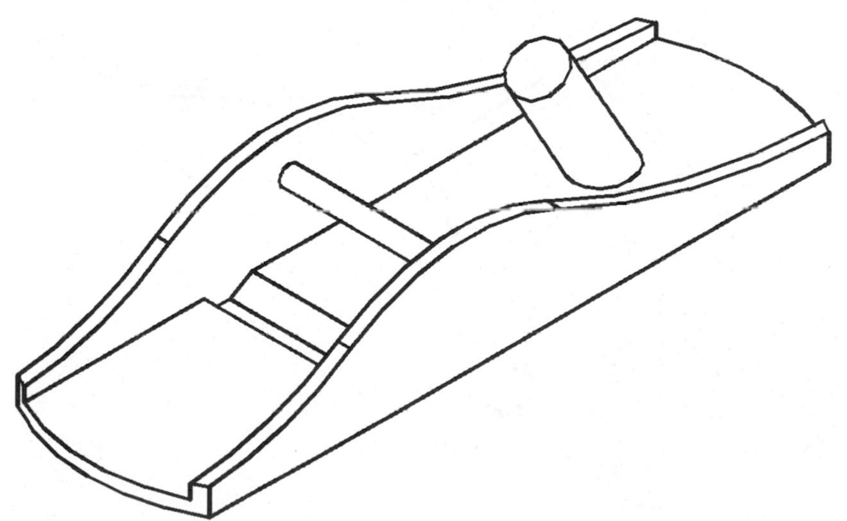

Figure 10.57

27. Make the inside face the active sketch plane, as highlighted in Figure 10.58 and ENTER to accept the default orientation of the *X*, *Y* and *Z* axes.

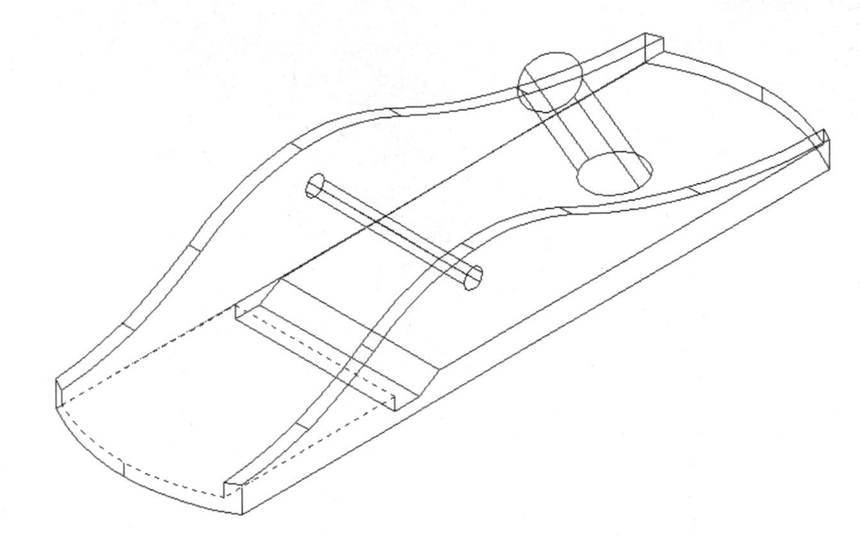

Figure 10.58

28. Place a work point and dimension it as shown in Figure 10.59.

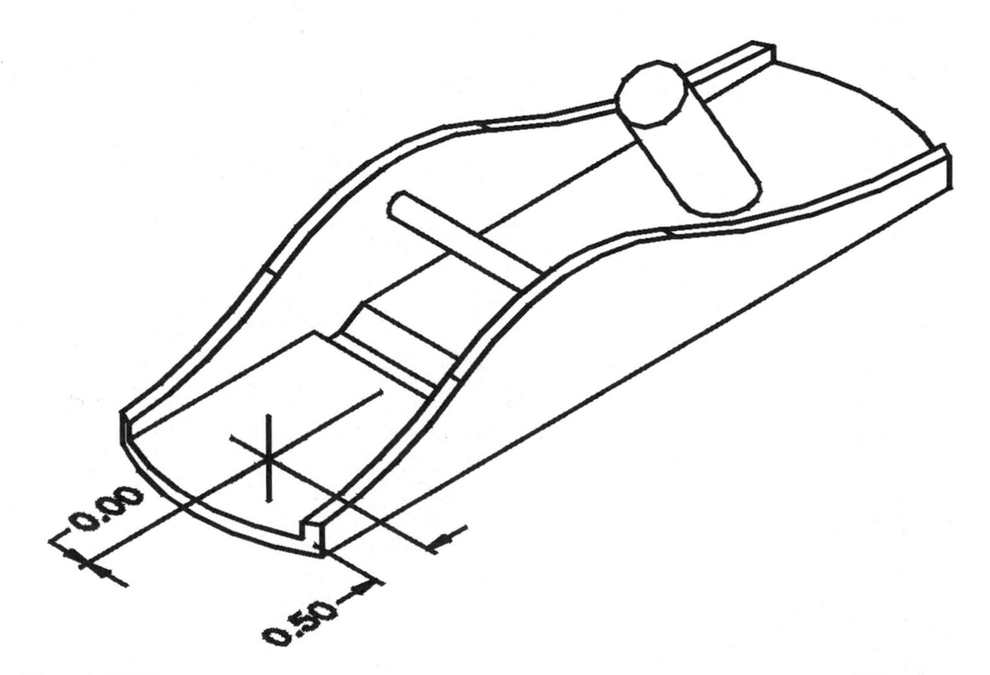

Figure 10.59

29. Create a #10-32 through tapped hole on the work point (.159 diameter for the drilled through hole, .19 major diameter for the tap and make the thread full depth).
30. Create a #10-32 x 1 blind tapped hole concentric to the angle extrusion (.159 diameter for the drilled through hole and .19 major diameter for the tap and make the thread full depth).
31. Add six ".06" constant fillets to the inside edges of the plane and the circular cross bar. You may need to rotate the part to clearly see all six edges.
32. Add a ".1875" fixed width fillet to the bottom of the angled circular extrusion. When complete, your screen should resemble Figure 10.60, shown with lines hidden.

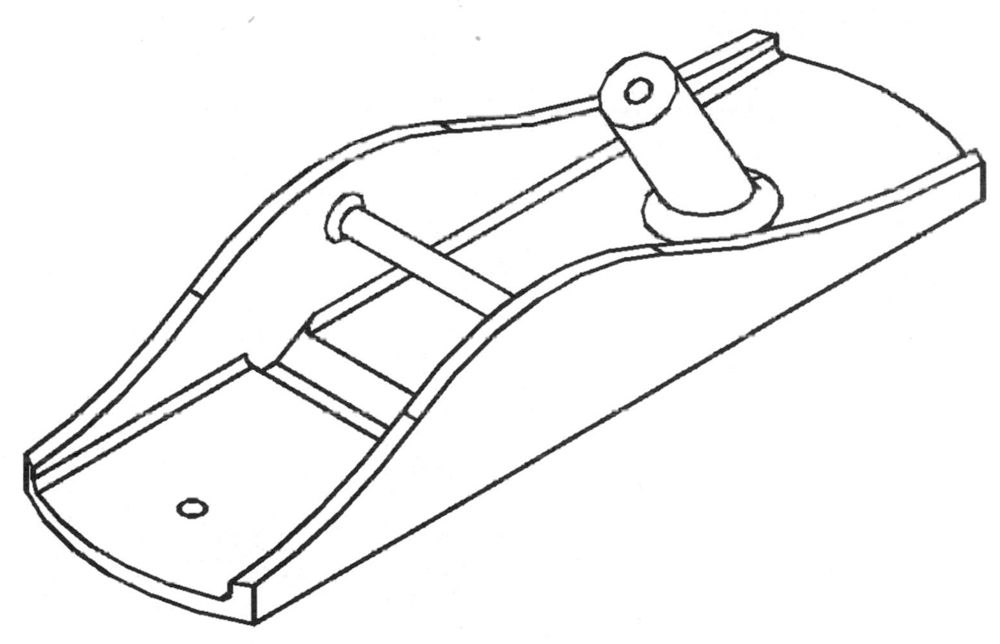

Figure 10.60

33. Save the file as \Md4book\Chapter10\Plane Base.dwg.
34. Start a New Part File (single part mode) for the blade of the wood plane using the template \Md4book\Chapter10\MD4Book.dwt.

35. Sketch, profile and dimension the geometry as shown in Figure 10.61.

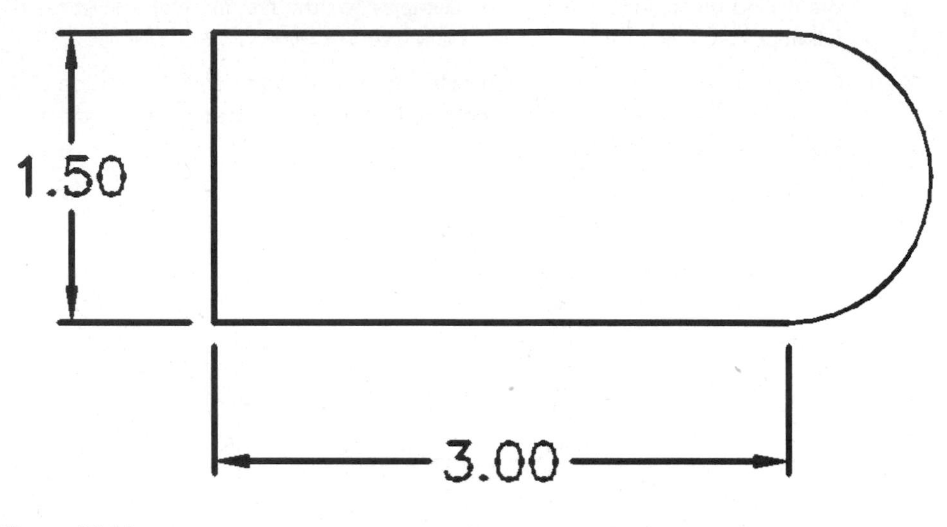

Figure 10.61

36. Change to an isometric view using the **8** key.
37. Extrude the profile ".0625" in the default direction.
38. Sketch, profile and dimension the slot as shown in Figure 10.62. Add a concentric constraint between the arc on the front of the slot and the arc on the blade.

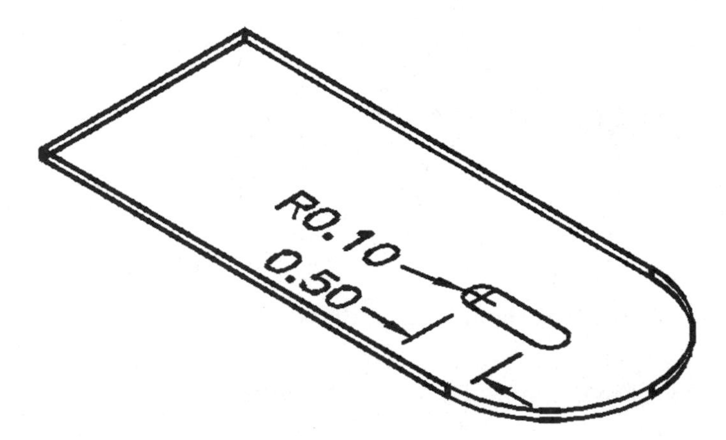

Figure 10.62

39. Extrude the profile with the Cut operation and Through as the termination.

40. Create a chamfer with the Distance x Angle option, setting the distance to ".0625" and the angle to "60". Select the edge that is highlighted in Figure 10.63. Press ENTER to accept the back face at the prompt:

    ```
    Apply angle value to highlighted face.
    ```

    When complete, your screen should resemble Figure 10.64.

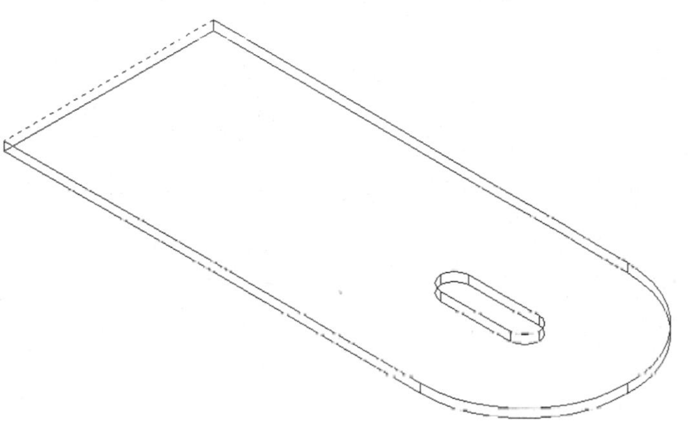

Figure 10.63

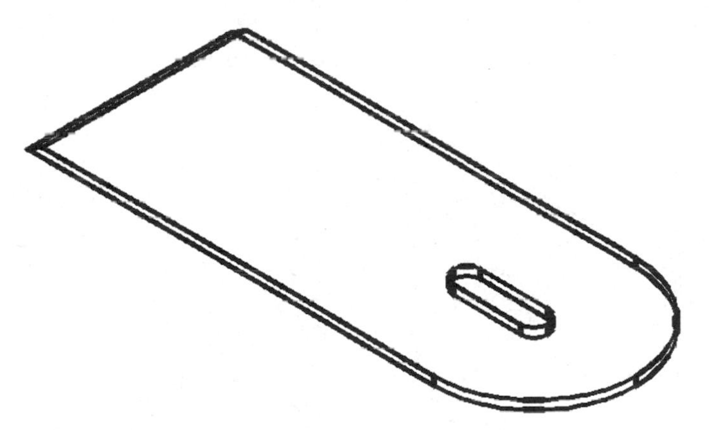

Figure 10.64

41. Save the file as \Md4book\Chapter10\Blade.dwg.
42. Start a New Part File (single part mode) for the front handle of the wood plane.

EXERCISE

43. Sketch, profile and dimension the geometry as shown in Figure 10.65.

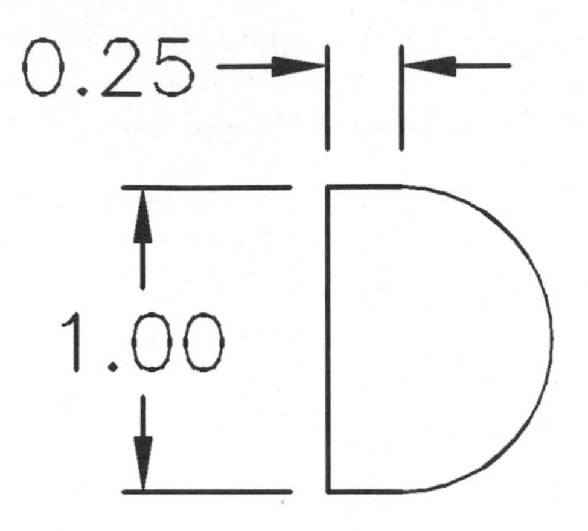

Figure 10.65

44. Change to an isometric view using the **8** key.
45. Revolve the profile 360° around the vertical line.
46. Create a through counter sink hole for a #10-32 flat head screw that is concentric to the top of the handle (the flat on the farside), (drill size Dia = ".2", C'Dia = ".4" and C'Angle = 45). When complete, your screen should resemble Figure 10.66.

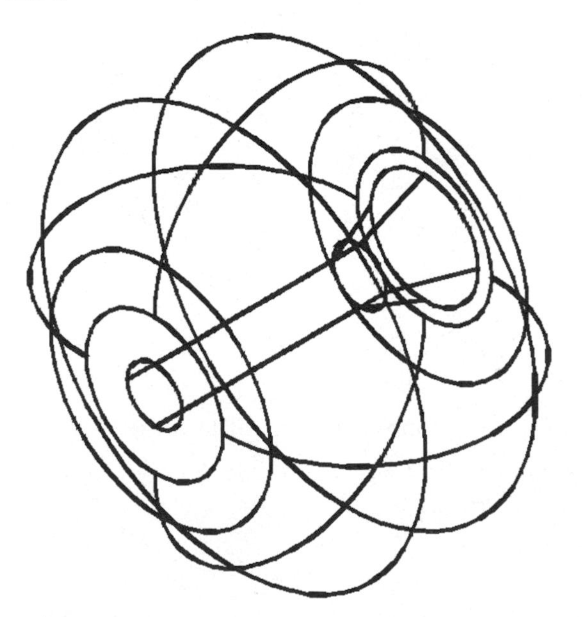

Figure 10.66

47. Save the file as \Md4book\Chapter10\Front Handle.dwg.
48. Using Windows Explorer, copy the file
    \Md4book\Chapter10\Front Handle.dwg.
    to a new file named
    \Md4book\Chapter10\Blade Handle.dwg.
    or
    If the file \Md4book\Chapter10\Front Handle.dwg. is current issue the AutoCAD Saveas command and type in a new name and path
    \Md4book\Chapter10\Blade Handle.dwg.
49. If the file Blade Handle.dwg is not current open the file
    \Md4book\Chapter10\Blade Handle.dwg.
50. Delete the countersink hole in the handle.
51. Make the front flat the active sketch plane, accept the default orientation of the *X*, *Y* and *Z* axes.
52. Draw a circle, profile it and dimension it with a ".19" diameter dimension and apply a concentric constraint between the circle and the circular edge.
53. Extrude the profile ".5" in the default direction joining it to the revolution.
54. Add a ".03" equal distance chamfer to the front circular edge of the cylinder. When complete, your drawing should resemble Figure 10.67 shown with lines hidden.

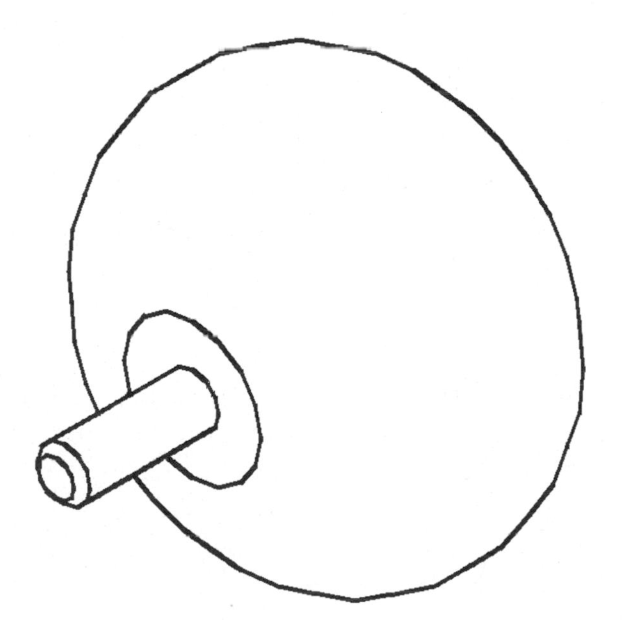

Figure 10.67

55. Save the file.
56. Close all open files.
57. Start a New File that will become the wood plane assembly.
58. Change to an isometric view using the **8** key.
59. From the Assembly Catalog dialog box, add the directory \Md4book\Chapter10\.
60. Attach one instance of each of the files; Plane Base, Blade, Front Handle, Blade Handle and 10-32x1. When complete, your screen should resemble Figure 10.68.

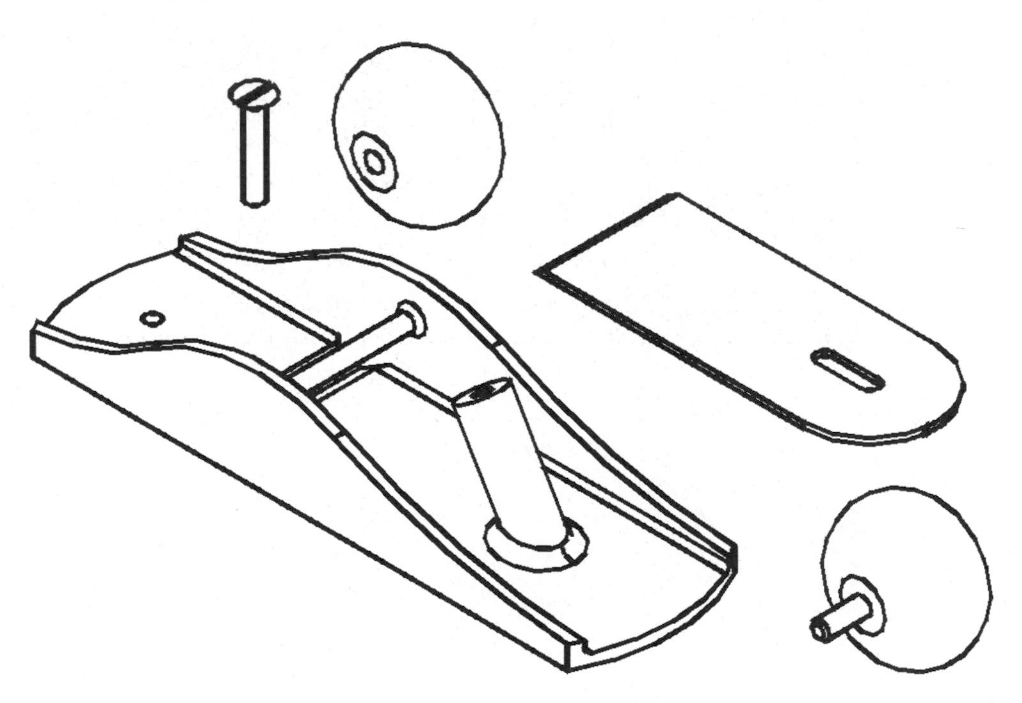

Figure 10.68

61. Use the Insert assembly constraint to align the bottom of the front handle and the front hole. Figure 10.69 shows the two centerlines and highlighted faces of the parts. Press ENTER to accept the default offset distance of zero.

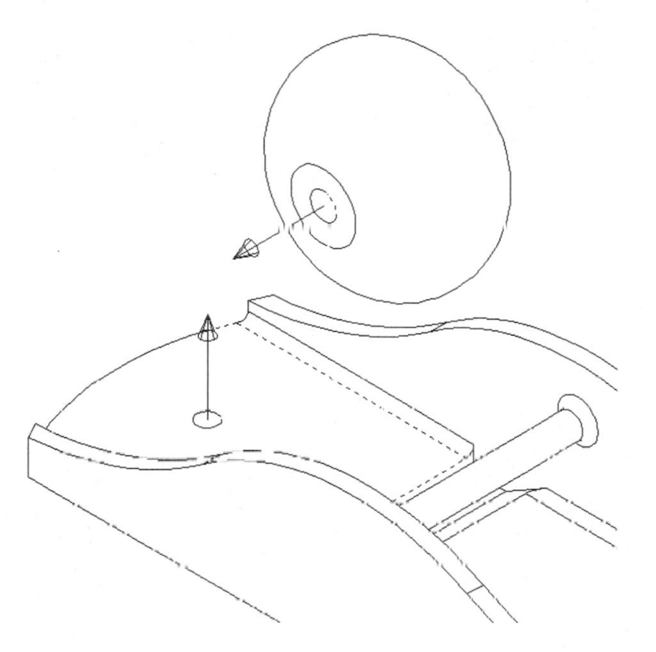

Figure 10.69

62. Use the Mate constraint to align the center of the screw and the front handle. Figure 10.70 shows the two centerlines of the parts. Press ENTER to accept the default offset distance of zero.

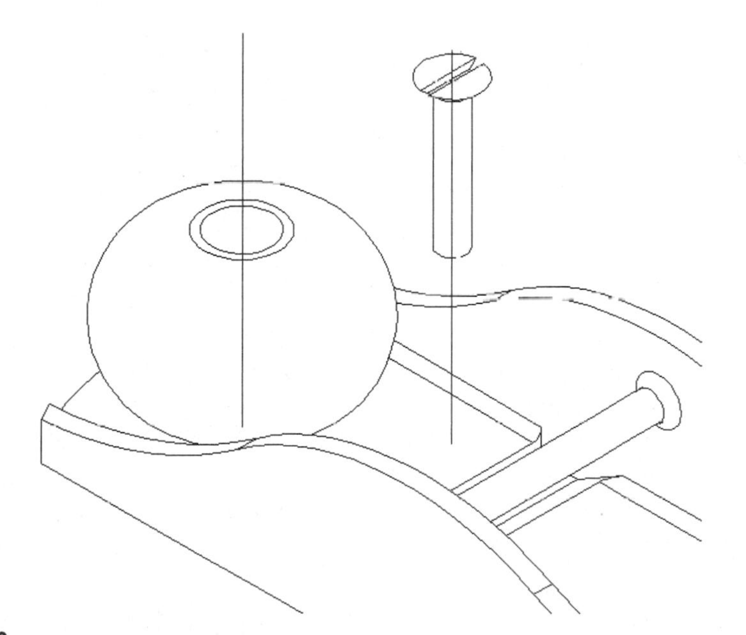

Figure 10.70

EXERCISE

63. To better select the geometry, move the parts so that they are not touching one another and change the AutoCAD variable ISOLINES to 2. To allow the change to take effect, regen the file.
64. Issue the Mate assembly constraint to align the center points of the top of the threads of the screw and the top of the drill diameter of the hole. Figure 10.71 shows the two center points of the parts. Press ENTER to accept the default offset distance of zero.

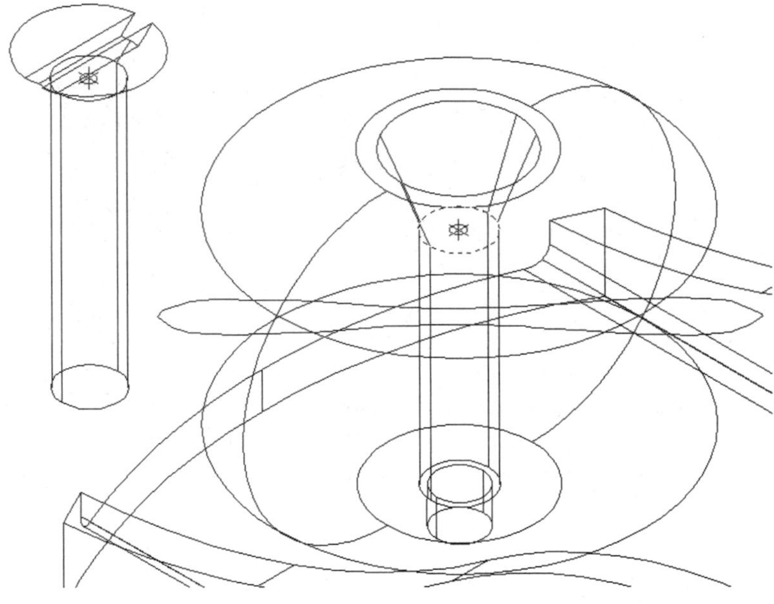

Figure 10.71

65. Change to a southwest isometric view by typing **88** and press ENTER.
66. Use the Mate constraint to mate the bottom of the blade and the top of the chamfer. Figure 10.72 shows the two mating faces of the parts. Press ENTER to accept the default offset distance of zero.
67. To position the blade from side to side, press ENTER to repeat the Mate constraint. Select the outside edge of the blade and the inside edge of the plane as shown in Figure 10.73. In Figure 10.73 the parts are moved away from one another and the lines are hidden for clarity. Type in an offset distance of ".0625".

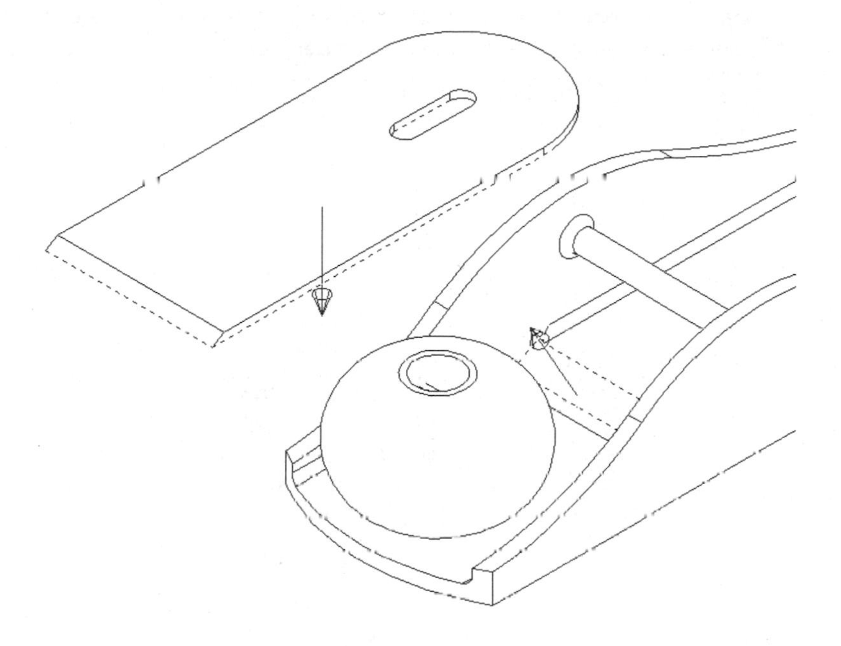

Figure 10.72

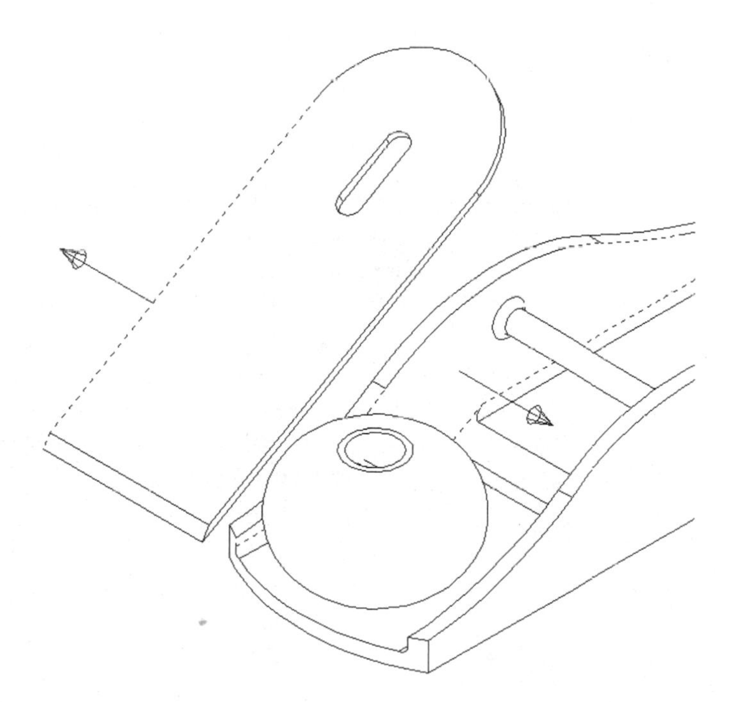

Figure 10.73

68. To position the slot in the blade to the tapped hole, repeat the Mate constraint. Select one of the top arcs of the slot in the blade and the tapped hole, as shown in Figure 10.74, to align the centerlines. Press ENTER to accept the default distance of zero.

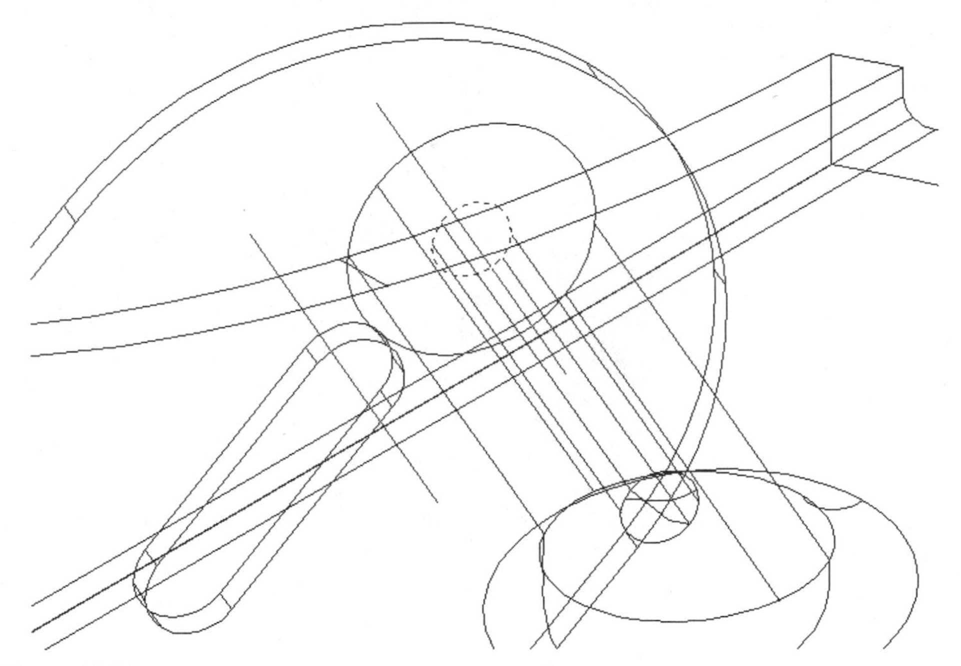

Figure 10.74

69. The last step is to position the blade's handle. We will not use the Insert constraint because the planes and the centerlines are on different parts. Instead, we will mate the two faces and centerlines in two steps. Repeat the Mate constraint and select the bottom of the handle and the top of the blade as shown in Figure 10.75. Press ENTER to accept the default offset distance of zero.

70. To better select the geometry, move the parts so that they are not touching one another. Use the Mate constraint to align the two centerlines of the handle and the tapped hole. Select the center of the handle and the tapped hole as shown in Figure 10.76. Press ENTER to accept the default offset distance of zero. When complete, your screen should resemble Figure 10.77.

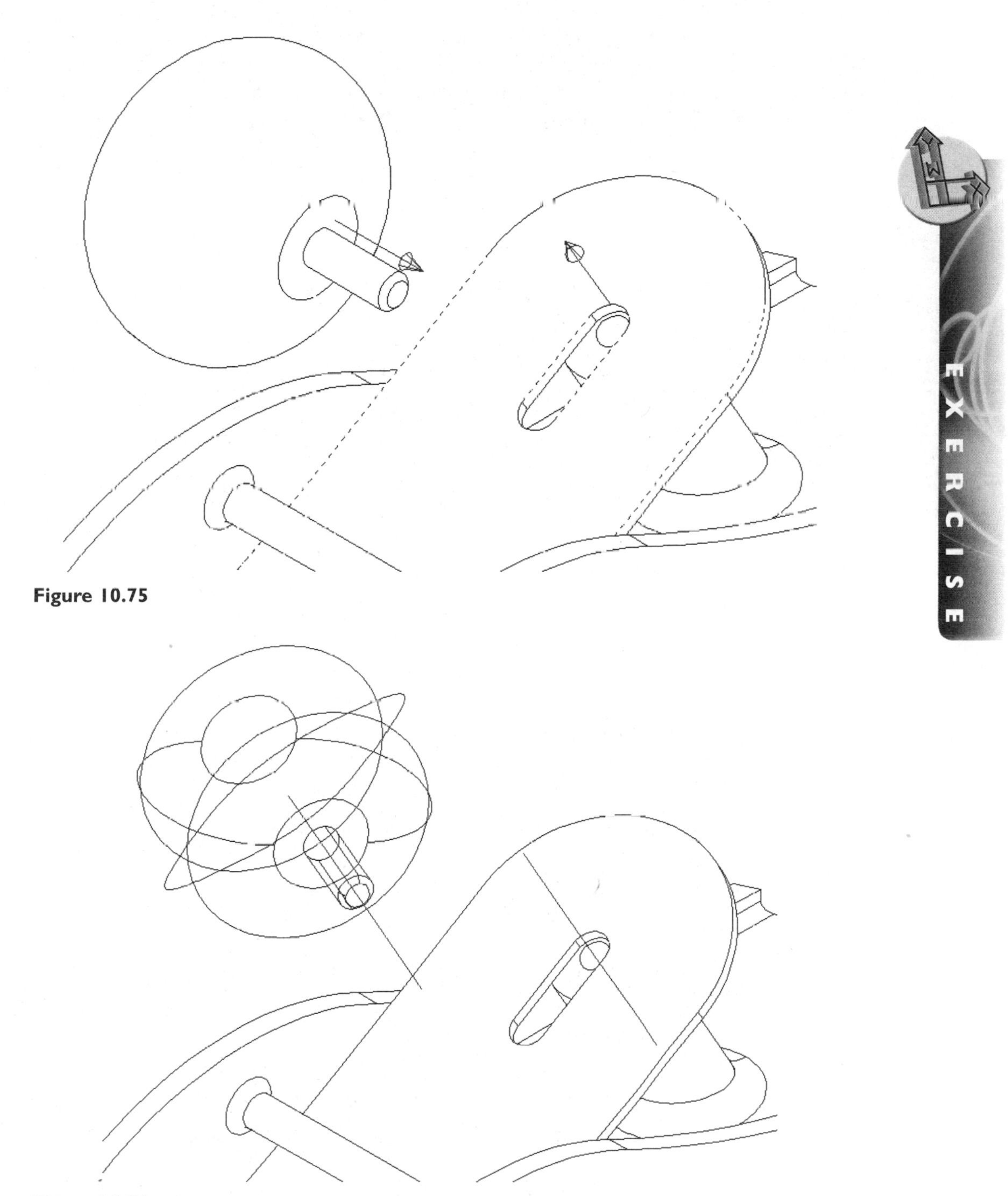

Figure 10.75

Figure 10.76

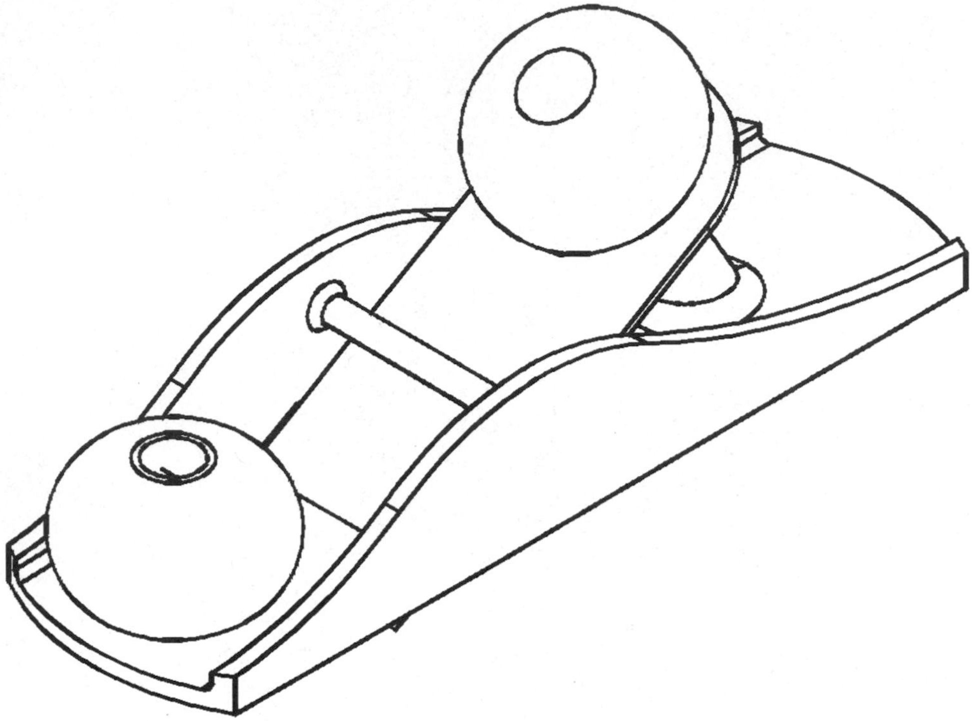

Figure 10.77

71. Rotate the assembly and you will see that the bolt extends beyond the bottom of the base. To edit the front handle's length activate the part Front Handle. Edit the feature and change the "1" dimension to "1.0625" at this point, your screen should resemble Figure 10.78.
72. Update the part with the AMUPDATE command.
73. Reissue the AMUPDATE command and Commit the changes back to the original part.
74. Update the assembly.
75. Rotate the assembly to make sure that the bolt no longer extends beyond the base.
76. Save the file as \Md4book\Chapter10\.PlaneAssembly.dwg.

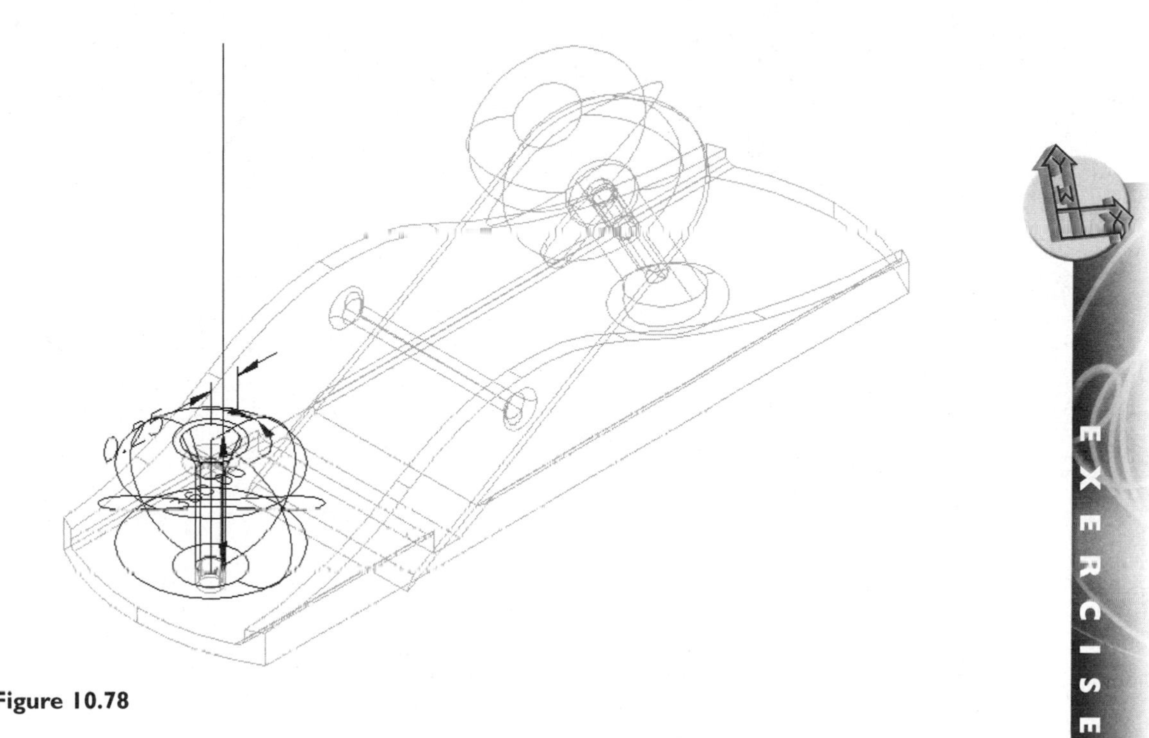

Figure 10.78

INDEX

License Agreement for Autodesk Press an imprint of Thomson Learning™

Educational Software/Data

You the customer, and Autodesk Press incur certain benefits, rights, and obligations to each other when you open this package and use the software/data it contains. BE SURE YOU READ THE LICENSE AGREEMENT CAREFULLY, SINCE BY USING THE SOFTWARE/DATA YOU INDICATE YOU HAVE READ, UNDERSTOOD, AND ACCEPTED THE TERMS OF THIS AGREEMENT.

Your rights:

1. You enjoy a non-exclusive license to use the enclosed software/data on a single microcomputer that is not part of a network or multi-machine system in consideration for payment of the required license fee, (which may be included in the purchase price of an accompanying print component), or receipt of this software/data, and your acceptance of the terms and conditions of this agreement.
2. You own the media on which the software/data is recorded, but you acknowledge that you do not own the software/data recorded on them. You also acknowledge that the software/data is furnished "as is," and contains copyrighted and/or proprietary and confidential information of Autodesk Press or its licensors.
3. If you do not accept the terms of this license agreement you may return the media within 30 days. However, you may not use the software during this period.

There are limitations on your rights:

1. You may not copy or print the software/data for any reason whatsoever, except to install it on a hard drive on a single microcomputer and to make one archival copy, unless copying or printing is expressly permitted in writing or statements recorded on the diskette(s).
2. You may not revise, translate, convert, disassemble or otherwise reverse engineer the software/data except that you may add to or rearrange any data recorded on the media as part of the normal use of the software/data.
3. You may not sell, license, lease, rent, loan, or otherwise distribute or network the software/data except that you may give the software/data to a student or and instructor for use at school or, temporarily at home.

Should you fail to abide by the Copyright Law of the United States as it applies to this software/data your license to use it will become invalid. You agree to erase or otherwise destroy the software/data immediately after receiving note of Autodesk Press' termination of this agreement for violation of its provisions.

Autodesk Press gives you a LIMITED WARRANTY covering the enclosed software/data. The LIMITED WARRANTY can be found in this product and/or the instructor's manual that accompanies it.

This license is the entire agreement between you and Autodesk Press interpreted and enforced under New York law.

Limited Warranty

Autodesk Press warrants to the original licensee/purchaser of this copy of microcomputer software/data and the media on which it is recorded that the media will be free from defects in material and workmanship for ninety (90) days from the date of original purchase. All implied warranties are limited in duration to this ninety (90) day period. THEREAFTER, ANY IMPLIED WARRANTIES, INCLUDING IMPLIED WARRANTIES OF MERCHANTABILITY AND FITNESS FOR A PARTICULAR PURPOSE ARE EXCLUDED. THIS WARRANTY IS IN LIEU OF ALL OTHER WARRANTIES, WHETHER ORAL OR WRITTEN, EXPRESSED OR IMPLIED.

If you believe the media is defective, please return it during the ninety day period to the address shown below. A defective diskette will be replaced without charge provided that it has not been subjected to misuse or damage.

This warranty does not extend to the software or information recorded on the media. The software and information are provided "AS IS." Any statements made about the utility of the software or information are not to be considered as express or implied warranties. Delmar will not be liable for incidental or consequential damages of any kind incurred by you, the consumer, or any other user.

Some states do not allow the exclusion or limitation of incidental or consequential damages, or limitations on the duration of implied warranties, so the above limitation or exclusion may not apply to you. This warranty gives you specific legal rights, and you may also have other rights which vary from state to state. Address all correspondence to:

Autodesk Press
3 Columbia Circle
P. O. Box 15015
Albany, NY 12212-5015